COVID in the Islands: A comparative perspective on the Caribbean and the Pacific

Yonique Campbell • John Connell
Editors

COVID in the Islands: A comparative perspective on the Caribbean and the Pacific

palgrave
macmillan

Editors
Yonique Campbell
University of the West Indies
Kingston, Jamaica

John Connell
School of Geosciences
University of Sydney
Sydney, NSW, Australia

ISBN 978-981-16-5284-4 ISBN 978-981-16-5285-1 (eBook)
https://doi.org/10.1007/978-981-16-5285-1

This Palgrave Macmillan imprint is published by the registered company Springer Nature Singapore Pte Ltd.
The registered company address is: 152 Beach Road, #21-01/04 Gateway East, Singapore 189721, Singapore

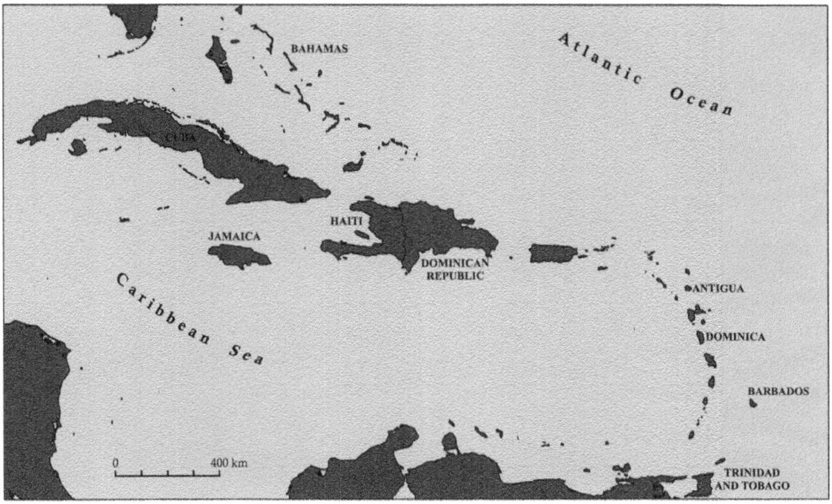
BAHAMAS
Atlantic Ocean
CUBA
HAITI
JAMAICA
DOMINICAN REPUBLIC
ANTIGUA
Caribbean Sea
DOMINICA
BARBADOS
0
400 km
TRINIDAD AND TOBAGO

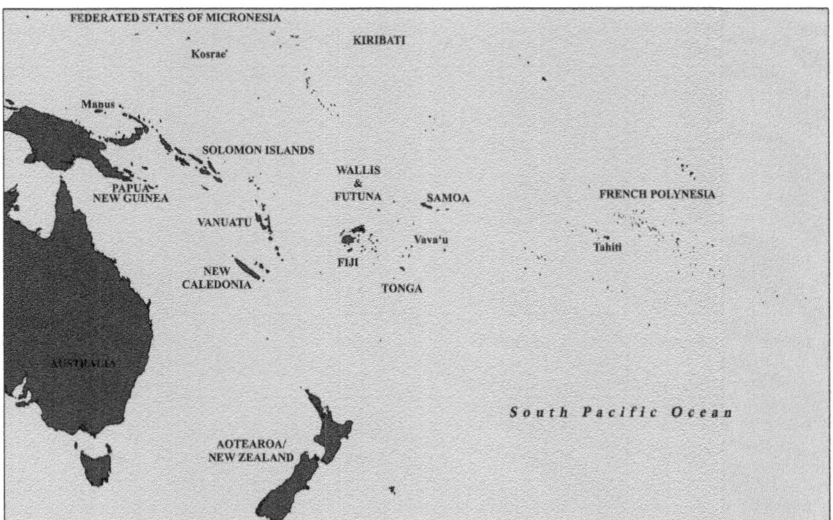
FEDERATED STATES OF MICRONESIA
KIRIBATI
Kosrae
Manus
SOLOMON ISLANDS
WALLIS & FUTUNA
PAPUA NEW GUINEA
SAMOA
FRENCH POLYNESIA
VANUATU
Vava'u
Tahiti
FIJI
NEW CALEDONIA
TONGA
AUSTRALIA
South Pacific Ocean
AOTEAROA/ NEW ZEALAND

Preface

This book considers the complex impacts of the COVID-19 pandemic in two island regions and within several islands and island states. Islands have unique but also quite different characteristics that vary across oceans and across scale and time. Apart from the parallel global challenges required to respond to COVID-19, most islands and small island states have been particularly affected by COVID-19 because of less diversified economies and a considerable dependence on tourism – prompting urgent challenges for governance, public health, economic management, and development. Island responses have already been directed to managing an unexpected but crucial self-reliance with repercussions across all economic sectors. Intriguing outcomes have taken different forms in different countries and islands. Even early on island states were working towards what was sometimes optimistically seen as a 'new normal'.

We compare and contrast the situation in the two main island regions dominated by Small Island Developing States (SIDS) – the Caribbean and the Pacific. While almost all the chapters are about SIDS, one concerns French Polynesia, formally an overseas collectivity of France, and another New Caledonia, but for simplicity and convenience we have nonetheless referred to SIDS throughout. We would like to have included islands in other realms but covering the Pacific and the Caribbean however fitfully and unevenly was challenge enough. We believe the book will nonetheless have resonance in Indian Ocean islands and elsewhere.

Meanwhile we the editors and most of the authors have been quickly learning new concepts and theories – and, we think, some basic epidemiology – in this complex interdisciplinary context. We are grateful to so many people who dropped other commitments to steer this book through in a very short time, some themselves working close to the frontlines, and all under particular exigencies. Many produced chapters despite considerable personal and practical difficulties. One potential author contracted COVID-19 and withdrew from the book. Another chapter was completed as Cyclone Yasa destroyed nearby islands. We hope that the results capture the first year of the new plague and that the book will be timely and of some practical value.

We thank Josh Pitt for his support and encouragement. The book is dedicated to the many friends at home and in other islands who experienced the full horrors of COVID-19 that we were so fortunate to escape ourselves.

Kingston, Jamaica
Sydney, NSW, Australia
March 2021

Yonique Campbell
John Connell

Contents

Notes on Contributors

Angela Apa is the Director of Kup Women for Peace, formed as a women's organisation about 20 years ago to bring an end to tribal fights in the Kup area of Simbu province, Papua New Guinea. The organisation provides services for women and families across the province and supports communities in addressing any form of violence.

Delvene Boso is the WorldFish Country Director for the Solomon Islands Program. Her research experience and interests lie in coastal fisheries management, particularly around scaling up of effective community co-management and the policy environment that supports it. She maintains partnerships with national institutional stakeholders and coordinates country-based project development efforts.

Yonique Campbell has a DPhil from the University of Oxford and is a Senior Lecturer at the University of the West Indies, Mona Campus, specialising in Public Policy. She serves as a Senior Policy Advisor to the Minister of Health & Wellness (Government of Jamaica). She is the author of *Citizenship on the Margins: State Power, Security & Precariousness in 21st- Century Jamaica* (Palgrave Macmillan).

Christine Clarke has a PhD in Economics from Rice University, Houston, and lectures in Economics at University of the West Indies, Mona, where she is the Chair of the Graduate Committee with responsibility for all postgraduate degrees. Her research areas include public finance, economic development, economics of sport and resilience. She has published two books on Jamaica's relationship with the International Monetary Fund and is a Director on the Board of the Bank of Jamaica.

John Connell is a Professor of Human Geography in the School of Geosciences, University of Sydney. His main interests are in migration and development in Pacific islands. He has written several books including *Islands at Risk* (2013). His most recent book is *The Ends of Empire. The Last Colonies Revisited* (with Robert Aldrich; Palgrave Macmillan 2020).

Olivia Dun is a Research Fellow in the School of Geography, University of Melbourne and the School of Earth and Environmental Sciences, The University of Queensland. She is a human geographer with a background in environmental science, migration studies and international development.

Peter D. Dwyer is an honorary Senior Fellow in Geography, University of Melbourne. Recent publications, jointly with Monica Minnegal, include 'We Are Fire Clan': Groups, Names and Identity in Papua New Guinea' (*Oceania*, 88, 2018) and 'COVID-19 and Facebook in Papua New Guinea: Fly River Forum' (*Asia and the Pacific Policy Studies*, 7, 2020).

Hampus Eriksson is a Senior Scientist at WorldFish and an Associate Professor at the Australian Centre for Ocean Resources and Security (ANCORS) at the University of Wollongong, Australia. His research focuses on testing and evaluating innovations that enhance fish-based livelihoods and sustainability in small-scale fisheries and food systems.

Carol Farbotko is an adjunct fellow in Geography at the University of the Sunshine Coast, Queensland. She is a cultural geographer with research interests in climate mobilities and the politics of climate risk, centred on Tuvalu.

David Fazzino is an Associate Professor and Chair in the Department of Anthropology at Bloomsburg University of Pennsylvania. His primary research interests are in food systems, historic preservation, and development.

Lystra Fletcher-Paul holds a Bachelor of Science degree from The University of the West Indies, St Augustine Campus, Trinidad and Tobago and a PhD in Biometrics and Plant Science from the University of British Columbia, Vancouver. She has over thirty years of experience working in the agricultural sector, mainly across Latin America and the Caribbean and is presently a lecturer in Biometrics at UWI, St Augustine.

Eriko Fufurefa is Director of the Kafe Urban Settlers' Women's Association, based in Goroka, Eastern Highlands, Papua New Guinea. Eriko founded the organisation with women in her community to provide local services, address violence towards women and children and lobby on human rights issues at provincial and national level.

Chelcia Gomese is a senior research analyst for WorldFish based in Solomon Islands. She has built her professional career through work on approaches to gender and small-scale fisheries development. Her primary interest is in the relationship between Traditional Ecological Knowledge and Resource Management and Solomon Islands culture.

Ainka Granderson is a climate adaptation specialist whose expertise includes community-based and ecosystem-based adaptation and resilience. She holds a Ph.D. in Geography from the University of Melbourne, Australia. She is Senior Technical Officer and Resilience Lead at the Caribbean Natural Resources Institute (CANARI) in Trinidad and Tobago and previously worked in government and academia.

Anthony Harriott is Professor of Political Sociology and Director of the Institute of Criminal Justice and Security, University of the West Indies, Mona, Jamaica. His many publications include *Organized Crime and Politics in Jamaica: Breaking the Nexus* and *Police and Crime Control in Jamaica: Problems of Reforming Ex-Colonial Constabularies.*

Charlotte Heinzlef is an Assistant Professor in the University of French Polynesia and a member of the UMR 241 EIO (Pacific Island Ecosystems). Her research focuses on climate change, risk, implementation of risk management strategies, resilience and adaptation to climate change.

Olivier Hoffer is a Lecturer at the University of New Caledonia, Noumea, specialising on urban development issues, and an associate researcher with the UMR-GRED research team (IRD – French National Research Institute for Sustainable Development).

Nick Howlett is a strategic communications specialist and is currently the Manager of the Communications and Digital Assets Program of the Vanuatu Tourism Office. During the pandemic, Nick has spearheaded tourism innovations for Vanuatu's urban and rural enterprises. Nick holds an MA in Communications from Griffith University.

Wendy-Ann Isaac is a Senior Lecturer in Crop Science in the Department of Food Production, Faculty of Food and Agriculture. She received her BSc. MSC., and PhD. Degrees from the University of the West Indies, St Augustine Campus, Trinidad and Tobago and a MAppl. Sci. in Agronomy from Lincoln University, New Zealand. Her research and teaching revolves around sustainable vegetable production, integrated weed management and innovative farming technologies.

Tisha Jack is an experienced Caribbean event planner and consultant based in Trinidad and Tobago. She holds an MSc (Distinction) in International Event Management from Surrey University (UK), a BSc in Hotel Management from the University of the West Indies and Certificate in Virtual Event and Meeting Management from the Event Leadership Institute. She is the Lead Event Manager with her brand TCJ Events that works with individuals and corporate clients to create and execute impactful events.

Fotuosamoa Jody Jackson-Becerra is from Savai'i, Samoa, but resident in Auckland and involved in connecting communities through indigenous storytelling, and on the importance of culture and celebrating traditional knowledge in disaster risk reduction. Her previous research interests involved seasonal employment and the impact on those left behind in Samoa.

Alecia Johns completed her DPhil at Oxford University in 2015, where she examined the conceptual foundations of political candidacy as a human right and the attending duties of the state which flow from that characterisation. She has published a number of articles on human rights law and currently practises law in the British Virgin Islands.

Leslie-Ann Jordan is a Senior Hospitality and Tourism Lecturer at the University of the West Indies (UWI), Department of Management Studies, St. Augustine, Trinidad. She holds a Ph.D. in Tourism Policy and Planning from the University of Otago, New Zealand. She also holds a Professional Certificate in Event Management from the George Washington University and the Arthur Lok Jack Graduate School of Business, UWI. Her publications include *Sports Event Management: The Caribbean Experience* which was published by Ashgate, London in 2010. Her research interests include tourism development in small island developing states with special reference to the Anglophone Caribbean.

Jackie Kauli is Senior Research Fellow in the Creative Industries Faculty at Queensland University of Technology, Brisbane. Her research focuses on applied theatre and community development that address human rights and gender inequalities.

Taukiei Kitara is the President of the Tuvalu community in Brisbane and a council member for the Pacific Islands Council for Queensland, and has represented Tuvalu civil society at several international climate change conferences. He is currently studying for a Masters in Global Development at Griffith University.

R. Clive Landis is a Professor of Cardiovascular Research, Pro Vice Chancellor of The University of the West Indies (UWI) and Principal of the Barbados UWI Cave Hill Campus. He is also former Director of the George Alleyne Chronic Disease Research Centre (GA-CDRC) and Chairman of The UWI COVID-19 Task Force. Dr. Landis earned a Bachelor of Science in Biochemistry from Birmingham University, UK, a Master of Science in Microbiology aand a PhD from Loyola University, Chicago. He completed postdoctoral training at Cancer Research UK before joining Imperial College London as The British Heart Foundation Lecturer in Cardiovascular Medicine in 1996, before he relocated to Barbados in 2003.

Jacqueline Lau is a post-doctoral research fellow at the Centre of Excellence for Coral Reef Studies, James Cook University, Townsville, Australia, and at WorldFish. She studies ecosystem services, access and environmental justice in coastal communities in Papua New Guinea, and is increasingly focused on climate change resilience in small-scale fisheries in the Pacific.

Loïc Le Dé is a Senior Lecturer at Auckland University of Technology where he leads the Disaster Risk Management and Development programme. He has a strong interest in developing participatory approaches and tools for disaster risk reduction. Loïc's work focuses on the Pacific, including Samoa, Tonga, Vanuatu and New Caledonia.

Nicole Leotaud has more than 25 years of experience working in environmental governance and management across the Caribbean. A conservation biologist by training, she is currently Executive Director of the

Caribbean Natural Resources Institute (CANARI) in Trinidad and Tobago and has also held positions in other non-profit organisations as well as the government, private sector and academia.

John Marazita III is director of the Environmental Mobility Research Unit and a doctoral candidate at the University of Geneva. He is a cultural geographer interested in climate migration as a durable solution to disaster displacement in small island states. He recently published, with IDMC, *Displacement in Paradise: Hurricane Dorian slams the Bahamas* (2020).

Tracy Marshall is a doctoral candidate in the Department of Geography, at The University of the West Indies, St. Augustine Campus, Trinidad and Tobago. Her research is focused on Disaster Risk Reduction (DRR) in Small Island Developing States (SIDS) by better understanding the reason for inaction by households in reducing disaster risk. She has Master's degree in Risk, Crisis and Disaster Management from the University of Leicester, United Kingdom.

Michelle McLeod is at the University of the West Indies (UWI), Mona Campus, Jamaica. Her tourism industry experience spans over 28 years. Her research interests include knowledge networks, destination and policy networks, and service productivity. Dr. McLeod's two co-edited books are *'Knowledge Networks and Tourism'* and *'Tourism Management in Warm-water Island Destinations'*.

Sarah Mecartney is currently working as a strategy adviser with the Pacific Community (SPC) in New Caledonia. She is part of a growing regional professional cohort practicing foresight and Pacific ways of knowing and learning to navigate COVID-19 and geopolitical disruptions. Sarah holds an MSc in Geography from the University of Sydney.

Ashley Meredith is the National Cultural Anthropologist for the Federated States of Micronesia at the National Archives of Culture and Historic Preservation. She carries out capacity building for conducting ethnographic surveys with the country's five historic preservation offices, and conducts research on traditional routes, wayfinding, inter-island connections, and food systems.

Monica Minnegal is an Associate Professor in Anthropology at the University of Melbourne. Her research focusses on how people renegotiate relationships when caught up in globalising phenomena. Her most recent book, jointly with Peter Dwyer, is '*Navigating the Future: an ethnography of change in Papua New Guinea*' (ANU Press, 2017).

Apisalome Movono, originally from Fiji, is a Senior Lecturer at the Institute of Development Studies at Massey University, which he joined in 2020 after around ten years in various positions at the University of the South Pacific. His research draws on postmodernist techniques and seeks to improve understandings of resilience, sustainable livelihoods, climate change and tourism development amongst Pacific Island communities.

Anna Naupa is a Pacific policy and development specialist based, in Vanuatu, and currently works in the field of human development and skills. She has been active in COVID-19 preparedness and response private sector training programs and has written of the interface between formal and informal private sector in Vanuatu. Anna holds an MA in Geography from the University of Hawai'i, Manoa and an MPA from the Harvard Kennedy School.

Kenisha V. Nelson is a lecturer in the Faculty of Education and Liberal Studies at the University of Technology, Jamaica. She is also a research assistant at the None-in-Three Research Centre Jamaica where she conducts research on gender-based violence. She has a Ph.D. in Psychology from Cardiff University, Wales. Her research interests include topics on occupational health, stress and responses to mental health related problems.

Liz Pechan is the co-owner of The Havannah, a boutique resort in Vanuatu and an active private sector representative to Vanuatu's COVID-19 preparedness, response and recovery task teams. Liz holds a BA in Geography from the University of Sydney. She wrote of early experience of the pandemic impact in ANU's *Devpolicy Blog* Series.

Umba Peter is a Male Advocator with the Kafe Urban Settlers' Women's Association in Papua New Guinea. He works with men and youth, and often integrates sports in youth advocacy programs. He is an experienced facilitator in using drama and participatory learning strategies for advocacy.

Kirstie Petrou is a human geographer and a research associate at the Climate and Sustainability Policy Research Group at Flinders University. Her research interests include urbanisation, migration and development in the Pacific. Kirstie's recent book, *If Everyone Returned, the Island Would Sink* (2020), focuses on migration in Vanuatu.

Gregoire Randin is a recent MA graduate in International Development from the University of Sydney. His MA thesis consisted of an ethnography investigating 'the moral politics of environmental stress in rural Fiji' and he was an unanticipated resident in Fiji from the beginning of the COVID-19 outbreak.

Anouk Ride is a scientist for WorldFish based in Solomon Islands and a Research Associate with the Australian National University, Canberra. For the past decade, she has conducted research and multimedia productions on aid and development in the Asia-Pacific region. Her interests focus around participatory research methods and research that enables local people to identify their own solutions to conflict, gender inequality and social exclusion.

Trevor Riley is chief executive officer of the Shipping Association of Jamaica (SAJ), and its subsidiaries and related companies. He has an MBA from Kingston University, UK, and an LLB (Hons) degree from the University of Technology, Jamaica. He is a Member of the Chartered Institute of Marketing (MCIM) and holds a certificate in Negotiation Techniques from the Harvard Programme on Negotiation.

Arlette Saint Ville is a lecturer in the Geography Department, at The University of the West Indies, St. Augustine Campus, Trinidad and Tobago. Her research centres around food security, and explores the interplay between food insecurity, natural resource governance and social capital. She completed a PhD in Natural Resource Sciences at McGill University.

Regina Scheyvens is Professor of Development Studies at Massey University, where she combines a passion for teaching about international development with research on tourism, empowerment and sustainable development. Her work mainly relates to small island states in the South Pacific. Her two most notable books are *Tourism for Development: Empowering Communities* (Pearson, 2002), and *Tourism and Poverty* (Routledge, 2011).

Damien Serre is a Professor in the University of French Polynesia and director of the UMR 241 EIO (Pacific Island Ecosystems). He is leading research in the area of resilience to risks and climate change with a special focus on resilience assessment and critical infrastructure management. He is editor of the *Journal of Water and Climate Change*.

Diana E. Spencer holds an MBA in International Business and a BA in Production and Operations Management. She is a Lecturer at the Caribbean Maritime University in Jamaica and is currently undertaking a PhD at the University of the West Indies.

Andrew J. Spencer has a PhD from Bournemouth University. He is currently Senior Lecturer and Deputy Executive Director of the Mona School of Business and Management at the University of the West Indies, Jamaica, and was most recently the Executive Director of the Tourism Produce Development Company (TPDCo).

Sarah Sutcliffe is a PhD candidate at the Centre of Excellence for Coral Reef Studies, James Cook University, Townsville, Australia, in partnership with WorldFish. She studies food and nutrition security in small-scale fishing communities, primarily in the Pacific, with a focus on food system adaptations to social and environmental change.

Verena Thomas is Associate Professor in the Creative Industries Faculty at Queensland University of Technology, Brisbane. Her research focuses on communication and social change, and examining and applying participatory approaches in research and advocacy.

Ijah Thompson graduated from the University of the West Indies, with a M.B.B.S and is currently pursuing a MPhil/Ph.D. in Social Policy, building on his recent training in Organizational Behaviour at the same institution. He has been heavily involved in general medicine and workplace wellness consulting and his research interests include stress and burnout at work, workplace wellness practices, remote working, behaviourally guided workplace policy, and appreciative cultures and gratitude at work.

Christopher Tufton The Honorable Christopher Tufton, Member of Parliament for West Central St. Catherine since 2016, is Jamaica's Minister of Health and Wellness. Prior to this he was Minister of Industry, Investment and Commerce and Minister of Agriculture and Fisheries. As Minister of Agriculture he spearheaded the expansion of greenhouse farming technology and the creation of agro-processing facilities, and redefined the sector's strategic importance to the Jamaican economy, for which he was named by the influential *Jamaica Gleaner* as Man of the Year for 2010. After temporarily leaving Government in December

2011, he served as the Co-Executive Director of the Caribbean Policy Research Institute, a University of the West Indies Think Tank. A graduate of the University of the West Indies and Georgia State University he completed his Doctorate in Business Administration at the Manchester Business School in the UK in 2002. He is a business strategist and has lectured and conducted business consultancy in Jamaica, the United Kingdom and the USA, in several areas including International Marketing, Business Strategy and Entrepreneurship.

Collin Tukuitonga is the inaugural Associate Dean (Pacific) and Associate Professor of Public Health in the Faculty of Medical and Health Sciences, University of Auckland, New Zealand. He is a Public Health Medicine Specialist with comprehensive experience in the Pacific region, New Zealand and internationally, and was previously Director General of the Pacific Community, the largest scientific and technical organization in the Pacific region.

Peni Turagabaleti is currently undertaking research on the socio-economic implications of COVID-19 in Suva (Fiji) at the University of the South Pacific, after having finished a short-term contract with the Fiji Development Bank (FDB). He has also worked with the Secretariat of the Pacific Community (SPC) on forest rehabilitation.

Jillian Tutuo is a nutritionist with extensive experience in public health research and nutrition-related policy intervention for the prevention of obesity and non-communicable diseases in the Pacific. She is currently a Research Fellow with WorldFish, an International NGO working to promote food and nutrition security within the Pacific food system to address the double burden of malnutrition in Solomon Islands and Vanuatu.

Leody Cruzat Vainikolo has two degrees in Forestry from the University of the Philippines at Los Banos, Philippines. She is currently the Officer-in-Charge and Deputy Director for the Ministry of Agriculture, Food and Forests (MAFF) Vava'u Branch, Kingdom of Tonga, with the primary responsibility of promoting food and nutrition security.

List of Figures

List of Tables

CHAPTER 1

Introduction: COVID-19 and Small Island States

Yonique Campbell and John Connell

> Historically, pandemics have forced humans to break with the past and imagine their world anew. This one is no different. It is a portal, a gateway between one world and the next. We can choose to walk through it, dragging the carcasses of our prejudice and hatred, our avarice, our data banks and dead ideas, our dead rivers and smoky skies behind us. Or we can walk through lightly, with little luggage, ready to imagine another world. And ready to fight for it. (Arundhati Roy, 3 April 2020)

Suddenly and catastrophically, early in 2020, the island world experienced an unprecedented shock. In different places and at rather different times the COVID-19 pandemic was crossing the horizon. We focus here on the earliest phase of the pandemic and its impact on Small Island Developing States (SIDS) in the Pacific and Caribbean regions. In a few

Y. Campbell (✉)
University of the West Indies, Kingston, Jamaica
e-mail: yonique.campbell02@uwimona.edu.jm

J. Connell
School of Geosciences, University of Sydney, Sydney, NSW, Australia
e-mail: john.connell@sydney.edu.au

Y. Campbell, J. Connell (eds.), *COVID in the Islands: A comparative perspective on the Caribbean and the Pacific*,
https://doi.org/10.1007/978-981-16-5285-1_1

SIDS deaths mounted. Almost everywhere economies ground to an abrupt halt and social lives dramatically altered, but not necessarily irrevocably. Only remote islands and island states were spared the pandemic, and some of them only initially. Nonetheless, without exception, the SIDS have survived rather better than metropolitan states.

For a pandemic that erupted in the interior of China late in 2019, the new coronavirus proved a dramatic shock for SIDS and other island territories. In many respects it represents the epitome of globalisation – rather more so than the Spanish Flu of 1918 that extinguished around 50 million lives. Even in the very few SIDS that escaped the virus, at least at the end of 2020, it caused major problems. No SIDS had experience of such contemporary epidemics as SARS and Ebola. This was uncharted territory. None were equipped to deal with a pandemic. To try to escape the virus, borders had quickly to be sealed: a universal but sometimes temporary or selective response in the SIDS. Islands retreated into themselves, and seas made strong borders.

Often hazard-prone, SIDS are particularly vulnerable to external shocks because of the small size and limited diversity of their economies, heavy reliance on tourism and the service sector and limited resources and capacity to respond to crises. COVID-19 amplified these problems, posing a clear and present danger to the livelihoods of citizens, island social and economic well-being, and public health systems. Several Pacific SIDS experienced damaging cyclones, and other hazards, at much the same time: a tyranny of overlapping disasters (Webb, 2020). These impacts created multiple problems for island political, economic and social systems at every scale.

On most islands changes cut across many interconnected spheres, sometimes in quite new ways, presenting novel problems that less frequently brought novel solutions. Many SIDS quite quickly closed their borders to tourists and others. By the end of March all Pacific SIDS were sealed off: an international social distancing. In some small isolated territories like Tokelau it took almost a year before overseas Tokelauans were allowed home. Consequently however some Pacific SIDS, like the Solomon Islands, Vanuatu, Tonga and Samoa, managed to entirely escape COVID-19 in the first phase. Few could escape completely: the Solomon Islands experienced its first case in October and Vanuatu in November after some citizens were repatriated from the Philippines and the United States. By then at least they had better-informed and organised health care systems in place, and the virus was contained. Where borders stayed open policies could be draconian and dramatic; PNG fitted all foreign workers coming into the country with electronic ankle bracelets during their quarantine period. Most Caribbean SIDs closed their borders towards the end of March, and nationals from some countries including Trinidad and Jamaica were effectively locked out. Most Caribbean SIDs also closed their borders towards the end of March, and subsequently employed control entry protocols, but none could completely escape the virus.

Largely in response to various facets of globalisation, the pandemic spread reasonably predictably into and through the islands. Fortunately island SIDS had a little more time than most to get organised (Murphy et al., 2020). The virus invariably first appeared in capital cities, that were also air and sea transport hubs, and waystations for tourists. In some archipelagic SIDS such as New Caledonia (Chap. 7) it never reached outlying islands or it arrived belatedly as in French Polynesia (Chap. 6). SIDS closely linked to metropolitan states were some of the first to be affected; the earliest cases in Guam and the Northern Marianas – the first islands in the Pacific to record COVID-19 – were the outcome of the United States military and naval presence. Unlike many other SIDS, Guam was unable to close its borders and control immigration since it had to abide by United States law (King, 2020), the same problem that faced other politically dependent territories. Eventually Guam recorded the highest mortality rate of any of the SIDS (Table 1.1). When, much later in the year, the Marshall Islands recorded its first cases they were all associated with the American military base on Kwajalein.

Larger SIDS with more wide-ranging global connections tended to be quickly affected – and, rather later, to have the greater number of cases and deaths. Airline connectivity was a key factor (Filho et al., 2020). This was evident for both the Caribbean and the Pacific. Residents returning from the US and the UK contributed to early increases in COVID-19 in Jamaica and Bahamas. In the Caribbean, Puerto Rico and the Dominican Republic, both closely linked to the United States, had the highest incidence, whereas more isolated SIDS such as Cuba and Haiti fared better (although numbers in several of these states are unreliable). Bahamas, so close to Florida, like other dominant tourist destinations, had particularly high rates. The French island territories of Guadeloupe and St. Martin had similarly high rates, and the latter's island partner, Dutch Sint Maarten – followed by Aruba – had the highest rate of all in the Caribbean. Smaller Caribbean SIDS – Dominica, St Kitts, Grenada and St Vincent had no deaths. In the Pacific the outcome was quite different; early in 2021, six

Table 1.1 Small Island Developing States (SIDS) in the Caribbean and Pacific: Coronavirus cases (December 2020)

	Population	*COVID-19 cases*	*COVID deaths*	*Deaths per million*
American Samoa	56,800	3	0	–
Anguilla	15,000	15	0	–
Antigua and Barbuda	98,000	184	6	61
Aruba	107,000	6350	52	480
Bahamas	393,000	8021	175	445
Barbados	287,000	1036	7	24
British Virgin Islands	30,000	114	1	33
Cayman Islands	66,000	374	2	30
Cook Islands	15,300	0	0	–
Cuba	11,300,000	17,096	162	14
Curaçao	164,000	4505	19	115
Dominica	72,000	109	0	–
Dominican Republic	10,850,000	188,969	2432	224
Federated States of Micronesia	105,500	1	0	–
Fiji	895,000	53	2	2
French Polynesia	278,900	17,635	126	451
Grenada	112,000	139	0	–
Guadeloupe	400,000	8886	154	385
Guam	176,600	7474	125	707
Haiti	11,400,000	10,365	238	21
Jamaica	2,961,000	13,992	323	109
Kiribati	118,700	0	0	–
Marshall Islands	54,600	4	0	–
Martinique	375,000	6263	44	117
Montserrat	5000	13	1	200
Nauru	11,700	0	0	–
New Caledonia	273,000	41	0	–
Niue	1600	0	0	–
Northern Marianas	56,600	128	2	35
Palau	17,900	0	0	–
Papua New Guinea	8,935,000	834	9	1
Puerto Rico	2,860,000	146,970	1692	59
Samoa	198,600	2	0	–
Sint Maarten	43,000	1641	27	628
St. Kitts and Nevis	53,000	34	0	–
St. Lucia	183,000	576	6	33
St. Vincent and the Grenadines	111,000	388	0	–

(*continued*)

Table 1.1 (continued)

	Population	*COVID-19 cases*	*COVID deaths*	*Deaths per million*
Solomon Islands	712,000	17	0	–
Tokelau	1500	0	0	–
Tonga	99,800	0	0	–
Trinidad and Tobago	1,400,000	7343	130	93
Turks and Caicos Islands	39,000	1062	6	154
Tuvalu	10,600	0	0	–
U.S. Virgin Islands	104,000	2180	24	231
Vanuatu	294,700	1	0	–

Source: Based on Worldometer

Note: Non-Island SIDS and smaller islands are excluded. Population Data are from the Pacific Community for Pacific states and from Worldometer for Caribbean states. All totals are rounded. Deaths are calculated per million although most SIDS have fewer than a million people. *None* of these data are necessarily accurate because of doubts over population totals, the number of cases, whether deaths can be attributed to COVID-19, the coverage of various testing regimes and poor health infrastructures

states – Tuvalu, Kiribati, Tonga, Samoa, Nauru and Palau were yet to have a single case, true also of various dependent territories, including the Cook Islands and Niue. Only Guam and French Polynesia stood out as having significant death rates.

Collectively the SIDS, and other island states, fared relatively well. Proportionately they fared better than several western European states, suggesting that the tropics were not necessarily synonymous with disease (nor with ineffective management). Small countries (those with fewer than 10 million people) generally proved better able to handle the crisis. Like islands they could more easily secure external borders, while larger states – with several large cities – had fewer hard borders. Six of the top-ten performing countries in 2020 were island states, exemplified by New Zealand and Australia, with obvious comparative advantages in managing border closures (Leng & Lemahieu, 2021). Some of these differences must be attributed to political decisions, social and public health policies and practices, but especially the extent and comprehensiveness of border closures; otherwise the difference between similar SIDS, notably New Caledonia and French Polynesia, would probably have been trivial. The French connection proved problematic in the Caribbean where both Guadeloupe and Martinique had significant case numbers and high mortality rates, but the COVID mortality rate in Martinique (deaths as a percentage of cases) was half that of Guadeloupe. Without more detailed analysis the reasons for such variations remain opaque. At this stage more subtle explanations for variations in the incidence of COVID-19 are impossible to disentangle and measure with small populations, but are likely to involve the extent of urbanisation (and nucleated villages), access to clean water, compliance with regulations and a host of other factors, not least management capability. In the Pacific SIDS at least, the absence of care centres (that have been sites of spread elsewhere) may have been beneficial. Care centres are also not particularly widespread in the Caribbean. Confidence and trust in government – and thus some degree of social cohesion – and relatively even income distribution have been seen as significant (Elgar et al., 2020), and may also prove to have been so in at least some SIDS.

COVID-9 was also differentiated by gender. Globally women have been more likely to experience morbidity and mortality from COVID-19 (Connor et al., 2020; Dang & Nguyen, 2021). In Jamaica for example, where more patients were from urban places, up to January 2021 there was a greater number of female patients, 8309 compared to 6834 men. Deaths were however lower among women (154) than men (183). For French Polynesia, late in 2020, there were higher confirmed cases (51%) and deaths (57%) among women but greater hospitalisation among men (59%) (Chap. 6). Data are inadequate to examine whether the incidence of cases, and mortality rates, were generally greater amongst women and/or an older cohort, or how such incidences reflected patterns of work, socialisation or social distancing.

The spread and extent of COVID-19 demonstrated emphatically that there were advantages in being an island state, and that more remote islands and territories fared particularly well in terms of lives: a rare benefit from being small and isolated. By contrast, tiny land-locked states such as Andorra and Luxembourg fared much worse and, early on, San Marino had the worst COVID mortality rate in the world. At the end of 2020 just two non-island states, North Korea and Tajikistan, had reported no COVID-19 cases, but this was widely regarded as infamous 'fake news'. But, however distant from infection islands and island states might initially have been, they were certainly not spared the impacts of the virus.

Public Health

The global pandemic had fundamental health impacts for small islands, and revealed that many were poorly prepared to cope not only with its ramifications. Investment in public health has been weak; in Papua New Guinea (PNG), where tuberculosis and malaria are widespread, and outbreaks of polio have recently occurred, only 2.5% of the budget is spent on health. Other SIDS were almost as parsimonious. WHO statistics show that average health spending in high income countries is as much as 70 times higher than that of low-income countries. Although SIDS generally have lower expenditures on health care, there are occasional deviations. Bermuda had one of the highest levels of per capita expenditure on health, surpassing that in the US and Canada. Most Pacific and Caribbean SIDs, even the middle-income states, spent less than 6% of their budget on healthcare. Among those with low expenditures were Grenada (4.4%) St. Vincent and the Grenadines (4.5%), and St. Lucia (4.4%) (WHO, 2019). Even before the COVID-19 pandemic, health systems of many SIDS were stretched, underfinanced, and, in some cases such as PNG, on the verge of collapse.

Many SIDS, and particularly their outer islands, have poor public health systems, and high levels of chronic non-communicable diseases (NCDs) that contribute to co-morbidity. The Caribbean has the highest NCD mortality rate in the Americas, which means that many people, including the steadily aging older population (Eldemire-Shearer et al., 2011; Samuels & Unwin, 2007), are susceptible to being severely ill from COVID-19. A rapid assessment survey by WHO in May 2020 found that 75% of countries reported interruptions to NCD services as a result of COVID-19 (The Lancet, 2020). In PNG people living with HIV/AIDS had reduced access to hospitals as transport costs rose (UNDP, 2020). NCDs have had a disproportional effect on those in lower income brackets, who generally also have difficulties accessing quality healthcare and proper nutrition (Chap. 3). Infectious diseases are also more common among lower income groups in some Caribbean and Pacific SIDs, posing a double burden. Socio-economic status and health outcomes are inextricably linked. Despite the obvious need to tackle NCDs and infectious diseases, social inequalities and limited capacity in public healthcare systems are a major problem in some island states.

The effects of having poor health systems were also reflected in limited Intensive Care Unit facilities and poor access to PPE, and even to basic masks. The pandemic revealed that such acronyms as PPE (Personal Protective Equipment) were quite unfamiliar, largely because their use had never hitherto been significant or even necessary. Most SIDs were, therefore, confronted with questions of how to best procure PPE equipment and how to deploy limited financial, technical and human resources to expand testing capacity, engage private labs and organise workers around core functions such as contact tracing and information sharing. These proved particularly challenging in the Caribbean as some SIDs, such as the Bahamas (Chap. 12), started to reopen their borders and case numbers increased.

Even where COVID-19 was absent, planning for its possible arrival placed new pressures on unprepared staff. In the Solomon Islands, nurses went on strike to demand that their allowances be paid, only to be criticised by the government for holding the nation to ransom at a critical time. In some Pacific SIDS retired nurses were requested to return to work. In PNG – and no doubt elsewhere – nurses were themselves particularly fearful of catching the disease, and worried about transmitting it into the community. As one observed: 'Our relatives stopped visiting us as they were apprehensive too. I felt discriminated against' (quoted in Magick, 2020: 10). The social pressures on health workers were enormous. Simultaneously in some SIDS, as in PNG, people became more fearful of hospitals where infection was located and failed to attend for other health problems (Chap. 25), had less income to pay for health care and transport costs and were even fearful of fellow villagers who had ventured into town (Chap. 22). Such concerns were universal across the SIDS especially in the early days.

However in the Caribbean, where the University of the West Indies (UWI) had taken the lead in developing a regional network of research and testing services, necessitated by previous experiences with HIV and Zika, SIDS were better able to cope, especially in the initial stages (Chap. 4). Influenza pandemic plans developed in 2006 also proved useful for some Caribbean SIDs (Chap. 2). Other regional organizations such as the Caribbean Disaster Emergency Management Agency (CDEMA) shared information in a timely fashion, estimated damages and needs, and coordinated on border closures. These efforts solved some problems, but testing capacity became a challenge once the number of cases started to increase, borders reopened and new variants began to emerge. Citizens in several Caribbean countries complained about delays in receiving test results.

National governments still confronted seemingly insurmountable challenges. Indeed the tension between lives and livelihoods, between health and economics, was constantly debated – as case numbers changed – and borders stayed closed or were reopened. That tension and that debate are well reflected in the account of Jamaica's Minister of Health and Wellness, Christopher Tufton (Chap. 2). They were no less challenging or critical elsewhere. At different scales and for different issues that debate was never ending.

The 2019 measles epidemic in Samoa, where 60 people died, indicated how poorly placed some small states were to cope with an epidemic (Craig et al., 2020). The private hospital systems in some Caribbean islands even showed reluctance, largely based on lack of preparedness and capacity, at a basic level, to have any interaction with COVID-19 patients. This made referrals a necessity, but limited health facilities and human resource shortages also made referrals challenging. Pacific SIDS especially spend a disproportionate percentage of health finance on medical referrals and evacuations, within and outside the country, that also involve high indirect costs – to patients, their families and society more broadly. Absent transport connections after border closures enhanced this problem for some Pacific SIDs. Although, limited health facilities, human resource shortages and scarce specialisms have meant that referrals to metropolitan states have been frequent, especially for dependent territories such as the Cook Islands, where one account was entitled 'Feel the pain, get on the plane' (Marsters, 2012). The closure of borders and the absence of planes meant that was no longer possible, putting additional pressure on island health care facilities, readily evident in Kiribati, where this was compounded by local hospitals running out of supplies. Many returned to traditional naturopathy (Chap. 10). Niue, with one flight per fortnight to New Zealand, was experiencing similar problems of gaining access to referral opportunities.

Even in the larger SIDS, specialist and hospital-based care is usually limited to urban areas where they are reasonably resourced and staffed. Ordinarily people had to travel long distances, and at some cost, to receive that care, but in the pandemic that conflicted with lower incomes, poorer transport connections, fear of infection and social distancing. The dramatic closure of borders as a first response, and rapid global spread of the virus, caused social anxiety about the readiness of island health care systems to respond to any, however minor, increases in COVID-19 cases. Fragmented news of high levels of mortality in China and Europe brought fear and epistemological reflection. In Fiji one church leader humorously reminded people that God would look after them but still wanted them to wash their hands (Maclellan, 2020a). In PNG, as in most island states, the resort to prayer was a similarly immediate response, but combined with concerns about the effectiveness of western science (Chap. 24). Fear of visiting hospitals, to avoid possible infection resulted in people avoiding assessment or care for other conditions. The avoidance of care by HIV patients, 30% in some instances, became an issue in Jamaica, partly driven by fear of getting the disease. In Fiji, UNAIDS, Fiji Positive Network and the Rainbow Pride Foundation established a social media campaign to raise awareness about stigmatisation, discrimination and the importance of ensuring people's right to privacy. While these strategies might have helped address the fears of those infected, they did not solve the problem of the diversion of resources to fight COVID-19, and the resultant worsening of other problems. Late in 2020 it was becoming evident that the inevitable focus on COVID-19 could prevent efforts to control and eradicate malaria (and also TB) in several Pacific SIDS.

Unlike larger states, because the COVID-19 virus was absent from many SIDS, it was also absent from the densely populated informal settlements where many people lived (and often worked) in large extended families, at high densities in already somewhat insanitary circumstances, where social distancing was impossible and the spread could have been very rapid. Where such circumstances did occur, as in PNG and Jamaica (Chap. 27), the escape was fortuitous. Equally fortunately, most SIDs had few aged care homes, as older people were cared for by extended families, so avoided some 'superspreader' contexts. However where migrants from the SIDS were concentrated in densely populated urban areas overseas, and co-morbidity existed, as in Auckland, New Zealand, the possibility of infection was significantly increased (Chap. 3). Such circumstances were also apparent elsewhere, notably in Hawaii, but there, New Zealand and the United States mainland (Berta et al., 2020) islanders themselves were taking the lead in resolving these problems.

Economies in Crisis

Despite several SIDS having largely escaped the virus at least until late in 2020, COVID-19 devastated the critical island tourism industry (and much else besides). Border closures brought an abrupt end to land-based tourism, cruise ships (derided as 'floating petri dishes) departed, and resorts, hotels and hire car businesses closed their doors. In the Pacific the continued shut-down of the sector brought down even the largest resorts. Beyond tourism, the impact in terms of growth prospects, debt sustainability, commodities trade, cross border financial flows, was reflected in limited export capacity as well as long-standing infrastructural and labour market challenges. Even before the onset of COVID-19 many governments faced revenue challenges. Macroeconomic issues were widespread. By mid-2020 most economic forecasts were expecting GDP growth to be negative for virtually all Pacific countries for the year, in some cases with double-digit negative GDP growth, and much the same was true for the equally open economies of the Caribbean (Chap. 5). Those countries who had most benefited from tourism experienced the greater economic downturns. Thus in the Pacific, Fiji experienced one of the greatest downturns, whereas Tonga, with few tourists, one of the least (Howes & Surandiran, 2020b). Indeed Tonga was able to expand its agricultural output in some areas (Chap. 11) benefit from remittances and achieve some economic stability.

With the devastation of tourism, and facing not just a public health but an economic crisis, every one of the SIDS was constantly forced to balance and rebalance lives against livelihoods, graphically depicted for Jamaica (Chap. 2). Policies that found the right mix between economic recovery and public health were seen as the ideal, but that mix was always indeterminate. Because the viability of many SIDS depended so heavily on international tourists, and that could not easily be influenced, and the waves and spikes of COVID-19 were unknowable, public health policies at least initially took precedence over economic recovery. It proved a delicate and crucial balancing act.

But economic revival was essential. Many SIDS were preoccupied with questions of debt sustainability and erosion of gains made under agreements with the IMF and economic reform programmes. Jamaica, for example, had only recently reduced its high debt-to-GDP ratio, through a successful 7-year reform programme, which involved agreements with the IMF, a debt exchange programme and strong economic management through a civilian oversight body – the Economic Programme Oversight Committee. That resulted in a positive economic outlook. However, having implemented measures to combat the spread of COVID-19, between July–September 2020, the Jamaican economy contracted by 11.3% compared to the previous quarter. That was broadly the situation elsewhere particularly where the tourism industry, that had initially tried to retain key staff, began to shed its remaining workers.

Problems of austerity brought different responses but few SIDS had the capacity to provide or administer wage relief or income support. Governments lacked the social welfare systems or large financial safety nets of larger economies; at the same time they were obliged to spend more on social safety nets in the first six months of the crisis, even more than on health, support for businesses and food security (Howes & Surandiran, 2020a). Many SIDS had versions of Vanuatu's Economic Stimulus Package that provided support for salaried workers in businesses, gave grants to small and medium enterprises, and developed an Employment Stabilisation Payment Scheme to support wage and salaried workers. Fiji and Solomon Islands granted tourism businesses five-year tax holidays. In Cayman Islands the government's medium-to-long-term tourism plan aimed to provide incentive packages for businesses, and new ways of helping citizens retrain to meet the demands of a 'new normal' in the workplace. The Barbados Government, through its Employment and Sustainable Transformation Plan, invested in tourism firms, with a view to enabling them to re-engage all their workers on 80% of their normal salary for up to two years.

Few SIDS provided direct cash payments to people to ward off the early economic shocks. Where this occurred, as in Timor-Leste, it proved a particularly effective way of providing social assistance because of low administrative costs, and the multiplier effect on local markets and jobs (Wilson et al., 2020). Jamaica's stimulus package had various components, including assistance to people who lost their jobs, small businesses, and to the tourism sector. Cash payments proved difficult to administer because many claimants did not use the formal banking system and money had to be transferred by other means. Substantial economic packages of this kind could not easily be sustained over time in poorer economies. Only occasionally did they extend effectively to the relatively poor, as in Antigua and Barbuda where responses to the economic crisis included policies to reduce the costs of electricity to the public and the cost of fuel to fisherfolk, at least for a limited time, alongside expansion of safety net programmes. Most governmental interventions and 'compassionate grants' focused on the formal sector, with strict eligibility requirements, largely ignoring the poor and vulnerable in the informal sector – especially semi-subsistence farmers and fishers – who didn't meet criteria. Social protection policies and practices thus favoured the formal sector, especially the public services, so emphasising inequality. In practice that was little different from the exclusion of the 'gig economy' from assistance in much wealthier countries.

Most SIDs also reduced or suspended taxes, revised their budgets as well as their economic and surplus targets and sought external financial assistance. As an early response to the pandemic, many SIDs including Dominica, Grenada, St. Lucia, Jamaica and the Bahamas all made requests to the IMF for emergency financial assistance under the Rapid Credit Facility to help address the balance of payment and other challenges posed by COVID-19. Most economic polices focused on short-term objectives; vital band-aids in uncertain times. Even before the full onslaught of COVID-19, governments made desperate attempt towards helping businesses and households adapt to the economic downturn conditions. Few had the money and revenue capacity to provide a significant boost to the economy at the same time. And most, like Fiji, had to immediately boost funding for health services; more money ($40 million) was directed to health care rather than to support the disintegrating tourism sector and just $5 million went to small- and medium-sized enterprises (Gounder, 2020). Health came first.

As early as April 2020, in reviewing the steps taken by the Fijian government to revive the economy Neelesh Gounder (2020) suggested: 'This could be an opportunity for us as a country to strategically reposition our economy for a sustainable future. This would require reliable, honest, transparent and accountable political leadership'. In practice, in Fiji as elsewhere, strategic repositioning was absent. While that could be attributed, as some of those who responded to Gounder's article suggested, to the lack of 'reliable, honest, transparent and accountable leadership' – certainly lacking in many SIDS – along with informed leadership. In practice the impact of COVID-19 was forever changing, no-one could predict when (and where) the pandemic might be over, and the re-opening of existing economic activities dominated planning. With unemployment rates and poverty increasing, politicians, business owners and employees sought the quickest possible return to 'normal' rather than contemplate a different vision of the future, that would require restructuring.

Tourism

In numerous SIDS, in both the Caribbean and Pacific, including Vanuatu, Palau, the Bahamas, Barbados, St. Lucia, Antigua and Barbuda and Jamaica, 30–60% of GDP comes from tourism. Tourism had increasingly dominated island economies in this century. Despite leakages, labour-intensive tourism generated considerable employment, often the bulk of all formal employment, through an interlinked and considerable informal sector. As tourist numbers slumped, related activities – restaurants, handicraft markets, bars and car hire -also collapsed, along with linkages to local fisheries and agricultural production. Revenues and taxation from the industry were significantly reduced. In Vanuatu, as incomes fell, share taxis simply dropped fares just to survive. In 2017, hotels and resorts in Fiji's main tourism areas had spent US$39 million on fresh produce from local suppliers (ACIAR, 2020). That time and those linkages were over.

Some resorts and other businesses provided direct assistance to employees who had lost jobs and to nearby villages who supplied produce. In Fiji five Marriott International Fiji Hotels set up "Solia Lesu" (Fijian for 'Give Back') to provide 100 meals per day to workers, their communities, first responders, healthcare workers and quarantine centres. Such benevolent schemes, while useful, could hardly be relied on for long term relief. Most petered out through lack of resources, and, as in Samoa, workers gradually returned to their home villages to re-engage in more 'traditional' locally-based economic activities (Chap. 15; Connell, 2021; Connell & Taulealo, 2021). Support for workers, within or outside tourism, was sometimes scant and, in Barbados, accusations of a lack of generosity toward 'poor black people' to ease the burden of the crisis were made against the wealthy capitalist class. One of the Barbados Prime Minister's special advisers, Trevor Prescod, complained that 'certain institutions in this country that extract the wealth from the first day of slavery, still own the lands, still own the businesses, still [take] the money out of the resources of this country, and put it into their expansionist programme abroad, and they don't even want to give back a cent' (Clarke, 2020). Similar accusations surfaced elsewhere, as in the Bahamas where cruise ships disappeared without trace and international resorts were reluctant to restart (Chap. 12), despite a sense that a crisis demanded universal commitment to its resolution: not least a frequently repeated 'we're all in this together'.

The tension between lives and livelihoods was probably greatest in the tourist-dependent Bahamas. As the Minister of Tourism and Aviation stated in September:

> You don't have to be a genius to figure out that 50 percent of our [tax] receipts come from tourism, so we have to get it up and running. This country cannot run out of tourist receipts. We have a country to run. There are expenses to pay for, and we cannot borrow indefinitely. We have to get the business of the Bahamas, which is predominantly tourism, back up and run-

ning so that the Government of the Bahamas can fund what it needs to do. Nobody will lend you money if you don't have a business. We have to get real quickly. (quoted in Hartnell, 2020)

Stark choices had to be made and, here, health issues were marginalised, sadly with a spike in case numbers as tourism began to open up (Chap. 12). In Jamaica where a similar opening-up was planned local people objected strongly (Chap. 14). Opening up in French Polynesia for similar economic reasons to those in the Bahamas, and following a French rather than a more cautious Pacific model, brought a serious spike in cases (Chap. 6). In New Caledonia, debates over opening up were embroiled in intense political divisions between those who sought independence and those who opposed it, resulting in borders being more effectively sealed. Here the indigenous population were well aware of the many lost lives from previous epidemics following European contact (Chap. 7). Similar perspectives existed in Kosrae (Federated States of Micronesia) (Chap. 9), and in Samoa where households still remembered lost victims of the Spanish flu (Aiono-Le Tagaloa, 2020). Everywhere such memories encouraged and enabled border closures without local resistance.

Where tourism had completely ended, as in Fiji and Vanuatu, workers had little alternative but to return to their home villages and regenerate agriculture and fishing; for some a less ordered and commercial world was welcome, but for others paying bills was now vastly more difficult and adjusting to a differently structured existence was not always easy (Chaps. 13 and 15). Such an alternative was no longer possible in the Bahamas and several other Caribbean SIDS. In the Pacific too skilled workers, such as chefs, were often international migrants whose future was uncertain.

In several Caribbean SIDS, as in the Bahamas (Chap. 12), elaborate schemes were devised to try and resuscitate even fragments of the tourism industry. Grenada sought to designate particular beaches for tourists and exclude the rest of the public from them (Telesford, 2020). Bubbles (airbridges or airline corridors) and corridors of various kinds were contemplated, one covering the north coast of Jamaica where hotels offered quarantine facilities (Chap. 14), and some, as in Fiji, involving elite tourists in private planes travelling to private islands (Connell, 2021). As virus case numbers grew, in the United States especially, and international airlines left the Pacific, it was largely wasted effort.

By May 2020, in Fiji, barely three months after border closure, half of all tourism businesses had closed down or were 'hibernating' (with no ongoing activity), and 35% were functioning but with a reduced workforce. It was anticipated that 29% of all the tourism businesses would have become bankrupt within the following six months (Wainiqolo, 2020). The firms that were most likely to close down were small and medium-sized locally-owned activities without international linkages. Those with international linkages were hardly spared but were usually better placed to survive large losses. The cruise ship industry, most significant in the Bahamas, was severely impacted. Before the crisis, Bahamas had the highest level of direct cruise sector expenditures, US$405.75 million, providing 9004 jobs and accounting for US$ 155.71 million in total wages (González, 2020). Jamaica had the second highest level of direct expenditures with US$244.53 million. All that became instant economic history.

Since the two main sources of tourism in Pacific SIDS – Australia and New Zealand – were largely free of the disease (with almost all cases coming from overseas), island states hoped and anticipated that a travel 'bubble' opening between New Zealand and Australia, would eventually extend to the Pacific and enable tourism to resume. Fiji was particularly enthusiastic about a 'bula [hello] bubble' to get tourism from Australia and New Zealand restarted. Other Pacific SIDS were more circumspect, fearing the import of COVID. So too were Australia and New Zealand and the idea remained a dream.

Similar enthusiasm was evident in the Caribbean, chastened however by increasing case numbers in the 'market' countries, notably the United States and the United Kingdom, but also mainland Europe. Antigua and Barbuda and St. Lucia were among the first in the Caribbean to reopen their borders to international travellers. Cautiously, these were followed by Jamaica, St. Barts, Guadeloupe, St. Vincent and the Bahamas. Others came later and some excluded travellers from the US. But tourists too were cautious, and quarantining on holiday held no attractions, hence numbers were few. Some SIDs such as Barbados developed specific strategies to woo longer-term 'tourists/residents'. There were costs. In the Bahamas, for example, the reopening of the tourism sector too soon increased the number of imported cases, putting even more pressure on the health care system (Chap. 12). That had proved dramatically so in French Polynesia; when borders were opened to US and French tourists, COVID-19 subsequently spiked, and borders had to be quickly closed again, but the damage had already been done and case numbers and deaths increased through local transmission (Chap. 6). Returning nationals also posed problems, with the Solomon Islands and Vanuatu acquiring their only cases that way.

All SIDS sought to revive tourism, or at least maintain hotels and restaurants in place by encouraging an unfamiliar domestic tourism. That had some success in larger states – such as New Caledonia and Fiji – where there were affluent local residents (some frustrated at their inability to travel further). In New Caledonia that was so successful that some rural hotels were more profitable than ever before (Chap. 7). Most SIDS, such as Vanuatu and Samoa, were simply too small to enable facilities to remain open for the few possible local tourists, although they hosted special events and slashed prices (Connell, 2021). It was more successful in Fiji, centred on a 'Love our Locals Fiji' campaign, because of the larger concentration of well-paid workers, especially with international agencies. Caribbean SIDs also increased their focus on promoting domestic tourism but, unlike OECD countries where domestic tourism accounted for 70% of the tourism market, this proved little different from the Pacific and the gaps left by the devastation of international tourism could not be filled. With the rare exception of Vanuatu (Chap. 13), where the absolute collapse of the industry with little chance of early revival forced time for contemplation, and despite a degree of lateral thinking aimed at boosting domestic tourism, neither governments nor the industry contemplated new directions.

Beyond Tourism

Business closures extended far beyond tourism, following the drop in wage incomes and consequent consumer spending and a loss of confidence. In some SIDS, such as Vanuatu, small businesses were still making employment cuts in September 2020, having initially held on to staff in the hope of an improved economic climate. In most SIDS, the majority of laid off workers came from the accommodation and food services sectors, but other sectors were also hit. Poor households were more impacted by these dramatic changes. In mid-2020 it was estimated that the population of Pacific SIDS living in extreme poverty could increase by as much as 40% in the short-term, with most of these being in PNG, Solomon Islands and Vanuatu. Before the virus, a quarter of the population of PNG and Solomon Islands already lived in poverty as did 14% of those in Vanuatu (Hoy, 2020). By contrast, where financial services were significant, as in Bermuda and the Cayman Islands, the advantages of 'long-distance' digital economies were apparent, as business was unhindered.

Construction activity declined alongside small service businesses such as hairdressers and clothes stores. Supporting services for the events industry similarly faded away (Chap. 18). Many small stores in urban and rural areas closed or lost most of their trade because of reduced activity in the local economy (ACIAR, 2020). That produced a crisis of under-employment in the informal sector as markets collapsed, market prices fell, demand for services fell and informal sector jobs disappeared. That was particularly problematic in PNG where large urban food markets were closed, supply chains disrupted, nutrition worsened and rural incomes consequently reduced. Small businesses had fewer resources to fall back on, informal services has no markets, and both found it harder to call on government support.

The entertainment sector also suffered from COVID-19 restrictions, with severe consequences for employment. Jamaica, Trinidad and Barbados all had to cancel major entertainment events, which included Trinidad and Tobago Carnival, Crop Over (Barbados), Jamaica Carnival, Reggae Sumfest and Rebel Salute (Jamaica). Many in the entertainment and events industry, due to cancellation of events, were laid off and forced to find alternative ways of surviving (Chap. 18). The pandemic emphasised the need to recognise the events industry with more seriousness, yet there was little evidence that the pandemic was changing the way governments thought about the role and socio-economic value of entertainment and events.

Some SIDS actively sought to develop new employment opportunities, and even to attract workers into the countries. In Trinidad and Tobago, the events component of the tourism industry switched to virtual and digital mode requiring more input from skilled IT workers (Chap. 18). Barbados instigated a scheme offering twelve-month visas to people willing to reside there and work remotely. Between July and December 2020, 3300 people had applied for a 'digital nomad' visa to live and work in Barbados for up to a year. As the founder of a video-marketing agency, who had moved with his family from New York, remarked: 'I'd never been to the island but it has managed the coronavirus well and we can live a very normal life. I can run my business from here and enjoy surfing, scuba diving and golf every day' (quoted in Rowlinson, 2020). That boosted the high-end property market. Much smaller Montserrat later sought the same model, a 'Remote Workers Stamp', but was likely to be disadvantaged by less adequate physical and electronic communications. Here at least were relatively rare examples of industries and SIDS engaged in innovative and creative restructuring.

Remote working – on a small-scale in SIDS – reflected a general expansion in the use of digital technology. With exceptions, economic activities (such as tourism) were less likely to require digital inputs or enable working from home. However use of social media boomed, and proved valuable both for disseminating information on COVID-19 (more positively than in some developed countries) as had become invaluable for 'conventional' disasters (Khosla & Pillay, 2020; Mow et al., 2017; Dwyer & Minnegal, 2020) and for keeping in contact with relatives overseas so enabling 'online agency' (Enari & Faleolo, 2020). That was particularly crucial for seasonal Pacific guestworkers stranded in Australia and New Zealand (Chap. 17). Remittances were almost exclusively sent through digital channels, rather than physically carried: digital literacy was now of real practical value, and offered hope for future economic development.

Maritime transport was scaled down but did not simply end. In the Pacific one outcome was the rise in prices of imported goods (including food), reduced opportunities to export commodities and decreased availability of all imported goods in outer islands. Both import and export supply chains faced challenges, and disrupted supply chains made exporting more difficult (where storage of agricultural products was problematic) and increased the price of imports (Chap. 23). Somewhat fortuitously some islands were able to take advantage of government transport services that were seeking business (Chap. 11). Caribbean countries have relied heavily on international trade and imports and are major transhipment points. While ports remained open, to facilitate cargo, workers were affected by COVID-19 protocols and supply and logistic chains disrupted. Educational supplies, food, essential services and employment were all affected, as they were in the Pacific where services dwindled. Few economic activities in SIDS had the employment structure or capacity to enable working from home, while those who remained at work had to engage in social distancing, often in difficult circumstances.

Airline transport virtually came to a halt, with only a few repatriation flights to some Pacific SIDS, although the Caribbean retained more of its air transport industry. While few SIDS had their own airlines, those that did were badly affected economically. By May, Fiji Airways, the largest airline based in the Pacific, had laid off all its cabin crew, expatriate executives and pilots: more than half the workforce. The airline industry in the Caribbean suffered similar problems. Following travel restrictions, between March and April, there was a 94% decline in flights in Caribbean airports. Belize and Antigua and Barbuda reported a 97% decline while

Saint Kitts reported a 90% decline. In Bahamas, flights to the national airport plummeted from 329, 520 in August 2019 to 7900 a year later. Sharp declines in services were especially challenging for small regional carriers that were already unprofitable in previous years and their situation worsened with the pandemic. LIAT Airlines in Antigua and Barbuda was liquidated in June 2020. For a small island economy dependent on tourism such losses were likely to be disadvantageous for eventual opening up.

Employment

The simultaneous crash of work, wages and commerce brought an immediate rise in unemployment. Many who remained in formal employment experienced reduced hours and reduced wages. In Barbados about 30% of workers, across all income groups, reported working fewer hours during the pandemic (Garavito et al., 2020). Numbers of unemployed grew rapidly. In Fiji between February and June 2020, 115,000 people – around a third of the formal workforce – became unemployed or had their hours sharply reduced (ACIAR, 2020). There too the Fiji Trade Union Congress reported that almost half of all workers were earning as much as 30% less that pre-pandemic levels resulting in a weekly household income less than half that of the national poverty line (Radio New Zealand, 31 August 2020). Even holding on to jobs offered little security.

However, employment losses in all SIDS were invariably of most significance in the informal sector. Because of the numbers involved (in small-scale marketing, transport, food sales, domestic work, vending and other informal commercial activities) and because few had any effective alternative source of income, or formal status in urban areas, and were not covered by social insurance, many were thrust into poverty. Women were especially disadvantaged by losses of jobs in tourism, small business and the informal sector. Conscious of this, some resorts sought to retrench people as equitably as was feasible (Chap. 13). In PNG women in female-headed households and older people were most likely to have lost work (World Bank, 2020a). Youth too disproportionately lost employment, critical where most SIDS had a demographic youth bulge; so-called 'Generation COVID' faced dire employment prospects (World Bank, 2020d). Furthermore, losses were most significant in the Melanesian states where poverty was already significant. Some states did target unemployed youth in development programmes; Fiji boosted its tree-planting initiative through establishing 150 tree nurseries alongside a 'Cash for

Tree Planting' programme specifically aimed at providing opportunities for newly unemployed people to plant trees in their communities.

As the pandemic continued, employment prospects worsened and activities that had been hanging on began to close their doors. While in Barbados 55% of middle-income workers and 66% of high-income workers were told by employers that they would be rehired (Garavito et al., 2020) when that might occur was always uncertain and even those percentages and possibilities were bleak. Unemployment insurance claims also increased but that possibility was unavailable in most SIDS.

Despite the decline of markets, not every part of the informal sector disappeared. One sex worker indicated in Fiji that, despite losing international clients and prime working hours, and the activity being illegal, competition was increasing:

> Times are hard, this is a reality and people will do anything to survive. People are resorting to sex work in times like these. The socio-economic impact of COVID-19 forced sex workers and others who had no other choice to go out on the streets just to earn enough to pay bills and put food on the table. We are already facing financial difficulties with people getting laid off, so one could only imagine what our women are going through, so the only option for quick cash is sex work. (quoted in *Fiji Sun*, 2 November 2020)

But, as that account went on to indicate, circumstances in the informal sector were desperate and workers – almost exclusively women – in this industry were often bullied and defrauded.

The loss of employment by women especially, in hotels, restaurants and markets, further emphasised the wider gendered impact of COVID-19. The percentage of women in the informal sector, where labour is precarious, was high, and globally, women are over-represented in the informal and 'high risk' sector (Chen, 2011), which includes domestic work, manufacturing, tourism, agriculture and health. In most SIDS they were also paid less than men. Adding to gendered challenges and precarity in the informal sector, COVID-19 responses such as stay-at-home orders, and the closure of schools increased the pressure of unpaid care work. The burden of unemployment also tended to fall on younger workers, adding to already high youth unemployment levels.

Remittances

Further complicating the employment picture was what was initially feared to be a parallel fall of remittances. It was initially widely assumed that remittances to SIDS – so important a component of national and household incomes in most – would fall after COVID-19 as islander migrants were among the first to be displaced from employment in metropolitan destinations. Remittance flows to Pacific island states were initially predicted to fall by 20% in 2020 (World Bank, 2020c). That was expected to be compounded by the loss of guestwork for seasonal workers in Australia and New Zealand, for example from Vanuatu and Samoa, as borders closed.

Flows of remittances are invariably boosted during crises in homelands, but in this case the crisis was at both ends of the migration chain. An initial decline followed employment problems of long-term migrants from Tonga and Samoa, disadvantaged in destination labour markets, such as New Zealand and Australia, where unemployment had risen. However, stimulus packages in New Zealand and Australia weakened an initial downturn, and what had been a sustained rapid growth of remittance in recent years in some of the major Pacific recipients – Fiji, Tonga and Samoa – simply continued and grew – after a brief pause early in the year (Howes & Surandiran, 2020c). Similar trends occurred in the Caribbean SIDS (Chap. 5). In Jamaica remittances also initially fell but then increased by 30% between April and September 2020 (Bank of Jamaica, 2020). It is unlikely that many in the informal sector and those who are over 65 with no pension or coverage under the National Insurance Scheme would find it possible to survive the effects of measures to limit the spread of the virus without this increase. Remittances helped out in Kosrae (Chap. 9) but were most valuable in Tonga and Samoa, two SIDS particularly dependent on remittances, where social obligations are unusually strong even over generations, and at cost to the livelihoods of the sender, as amongst Samoan migrants in New Zealand (Chap. 16). Guestworkers largely sustained remittance flows despite their own lower earnings, as they reduced their own consumption (World Bank, 2020c). All that was facilitated by institutional changes; in Australia and New Zealand the removal of the ANZ Bank's fee for electronic transfers led to an immediate increase in remittances.

In Pacific SIDS a growing component of household incomes in several states came from the remittances and savings of migrant workers engaged in agricultural guestwork in New Zealand and Australia. While their presence was essential to the sustainability of these food systems and more broadly for the resilience of their host communities, their marginalisation limited their ability to protect themselves during the pandemic. Migrant workers experienced adverse impacts such as unemployment, short working hours, isolation, poor quality of living, social discrimination and mental pressures while their dependents at home faced financial problems due to the limited or reduced cash flow from these working relatives and the need to take on activities that would have been done by returned workers. Nor could they cross Australian state borders to access available work. With limited access to social protection, stranded workers were assisted by employers, churches and NGOs but many were fearful of living and working where a virus, absent in their home countries, existed (Chap. 17). At the same time potential workers were unable to travel to New Zealand or Australia to fill vacancies. That changed later in 2020 when both countries made special arrangements for charter flights to get workers from Pacific SIDS, with New Zealand recruiting 2000 seasonal agricultural workers from three countries – Fiji, Samoa and Vanuatu – and slightly smaller numbers going to Australia. Such was the pressure from potential employers in both countries that these migrant workers were singled out for priority migration, even sometimes before those of stranded citizens overseas.

At Home: Back to Basics?

The collapse of urban labour markets brought urban-rural migration of the unemployed – especially from informal settlements, and resorts, to rural 'home' villages. Honiara, the capital of Solomon Islands, was reported to have lost a third of its population within a few months, and probably contracted by 20% (World Bank, 2020b). Tuvalu and Solomon Islands were perhaps the only countries to deliberately direct and encourage people to return to home islands for their own safety from COVID and because resources were scarce at the centre (Chap. 8). Extended families proved to be the most basic safety net of all. Throughout the SIDS people sought the security of home. Rural population expansion, sometimes as in Tuvalu (Chap. 8) after decades of decline, brought some tensions through competition for scarce rural resources, and disputes over land and marine rights, and access to fisheries. Arguments over overfishing affected the lives of rural people experiencing their own concerns as market prices collapsed with much greater competition, circumstances also true of many coastal communities (Bennett et al., 2020).

Some who returned experienced a sense of pleasure and pride in returning to familiar old ways, at least for a break (Chap. 15; Leweniqila & Vunibola, 2020), while others simply increased pressure on scarce resources. Many who returned to Solomon Islands coastal villages no longer had the fishing skills to survive there (Chap. 21). Return to rural areas brought increased tension over 'idle' and disputed land (ACIAR, 2020). Agricultural production expanded but not quickly enough to cater for too many new arrivals. Greater pressure on scarce resources even contributed to theft of food plants (Chap. 21). Choice of fast-growing species was crucial, as in Tonga (Chap. 11). There and elsewhere governments permitted the extension of fishing seasons, encouraged the resumption and renewal of agriculture and provided training, seed distribution and cheap equipment. Antigua and Barbuda, where 80% of food was imported, distributed seedlings in an attempt to stimulate self-reliance. The Tuvalu government held workshops to teach youth indigenous food production methods, such as taro planting and sap collection from coconut palms, and even food storage (Milko, 2020; Chap. 8). Pride came from reviving local agriculture, but it was almost too late in Kosrae where local knowledge was in steep decline (Chap. 9). A return to subsistence occurred and bartering revived in Samoa and Fiji (below). Foods usually sold to hotels in, for example, Solomon Islands were sold on Facebook (ACIAR, 2020). Food plants however gained greater precedence, as in those parts of Vanuatu where cash crops had begun to dominate agricultural systems, but the unexpected demand meant that seeds and planting materials were often in short supply, partly because of disrupted imports of seeds and fertilisers (ACIAR, 2020; Maclellan, 2020b). Backyard agriculture boomed in and around urban areas, well evident in Fiji (Chap. 20), and local media eagerly portrayed groups of workers, some from organizations and agencies such as the Pacific Community and the Pacific Conference for Churches in Suva, working together on new agricultural plots. Collective self-help was the order of the day in many places, including highlands PNG (Chap. 25), but there was also intense competition, notably in a profusion of roadside market stalls.

Pressure on fisheries also increased, with size limits being ignored and intrusions into marine protected areas, as fish were more immediately available than newly planted garden foods. This, too, happened in Caribbean SIDs such as Belize, where survival and food security concerns have led to overfishing. At the same time, with declines in fish and seafood production some fishers in Caribbean SIDS were unable to sell their products. In some more remote islands in Fiji, access to fuel for outboard motors became difficult (and costly), so that traditional boatbuilding of wind-powered canoes was experiencing a minor resurgence. Theft of kava, sandalwood trees and food plants, especially when grown around urban areas in Fiji and Vanuatu (and probably elsewhere) occurred, as too in rural Solomon Islands (Chap. 21). Initially crime decreased in places like Jamaica, but increased shortly after as desperation ensued (Chap. 27). In rural Jamaica too, theft of livestock, crops and farming equipment increased. Pressure to find alternative means of surviving carried new risks for some of those who relied on scarce land resources, now complicated by the arrival of the 'newly minted' in search of rural sustainability and safety-nets.

Revival of local agricultural systems was almost inevitable, and in rural areas especially diets became more likely to be composed of local foods – with some early indications that the shift in the structure of nutrition had benefited health (Chap. 22). That was less evident in urban areas where access to land was limited and incomes had fallen (Farrell et al., 2020) and in some Caribbean SIDS where schools closed and children who had previously benefited could no longer access feeding programs (Byron et al., 2021). In the light of climate change, Caribbean SIDs emphasised the importance of ensuring sustainable local agricultural systems to offset threats to food systems (Rhiney et al., 2018; Robinson & Wren, 2020). This took on greater importance given the increases in food prices and food shortages that followed the pandemic.

In many rural areas, where access to local foods was adequate, life continued much as before. In remote islands, in the Solomon Islands (World Bank, 2020b) and in Vava'u, Tonga (Chap. 11), there was less change in the structure and content of food consumption, despite a decline in imported processed foods. But that was not always so; in the tiny island of Ahus (PNG) people had less access to food from local markets, so resumed processing and eating sago – regarded as a 'famine food' – limited their meal sizes and ate fewer meals (Chaps. 21 and 22). A third of households across the Caribbean reported coping strategies of eating less preferred foods, skipping meals or reducing food intake, partly because food insecurity followed lack of market access (Chap. 23). That pattern was particularly common in urban areas where food insecurity increased, with people eating fewer meals per day and consuming cheaper and less nutritious food (ACIAR, 2020; UNDP, 2020; World Bank, 2020a). As one Fijian resident said: 'We are now cooking less and consuming smaller amounts of food' (quoted in Dean, 2020: 101). That pattern was widespread, particularly in Melanesia, and even more evident in the Solomon Islands where two thirds of households experienced food insecurity, some going without food for a day or more (Wairiu et al., 2020). Less adequate foods emphasised the problem of NCDs.

Social Lives, Social Worlds

The impact of COVID-19 was as serious and as complex for social changes and, more intriguingly, as it had done for agriculture, brought some revival of past traditions. Fresh food prices went down in smaller SIDS but in PNG they went up – as transport declined and large markets were closed – hence with already reduced purchasing power this resulted in less nutritious diets (ACIAR, 2020). In Fiji and Samoa particularly there was revitalization of more traditional exchange relationships and a revival of bartering, alongside an increase in roadside market stalls. In Fiji, beyond more common exchanges of fish and taro, pigs could be swapped for roofing iron, baked goods for children's toys, food for tutoring of children unable to attend to school because of COVID-19 restrictions and, famously, 'two piglets for a kayak' (Tora, 2020). As many of 165,000 people joined a Facebook group, 'Barter for Better Fiji', while smaller groups catered for specific towns, with a similar pattern emerging in Samoa (Darmadi, 2020; Boodoosingh, 2020). That was necessary in at least one remote atoll in PNG when petrol costs rose and it became impossible to access ATMs (Chap. 22). A similar revival of bartering occurred in the Solomon Islands, as 'Trade Bilong Iumi' (Chap. 21), and also in Bali (Laula & Paddock, 2020). In parts of PNG some people were returning to the use of shell money, as 'real' money became scarce (Chap. 24). As one Solomon Islander pointed out

> People from the Weather Coast needed some taro. So they went fishing and then went on Facebook and said 'We've got some tuna and we need some bags of taro or cassava, and they actually exchanged the goods. This is what we have been practising back in the olden days. That's how our ancestors have survived. (quoted in Maclellan, 2020b)

Such changes and adaptations, sometimes strengthening local organisations, as in PNG (Chap. 25), enabled a degree of flexibility and resilience.

Conversely however, some traditional ties and obligations weakened. Sharing did not always work, when there was less food available, and some became greedy (Chap. 22). Theft of foods and plants was not uncommon. Border closure, cancelling a special international treaty, ended cultural ties between PNG people and Australians through the Torres Strait, separating families, ending trade and exchange and preventing PNG residents accessing fishing areas and health care across the border (Faa, 2020). Much the same was true for Shortland Islanders in western Solomon Islands, with kin in Bougainville, PNG, who conventionally traded and purchased goods across the border. Everywhere such restrictions pushed up costs.

By contrast with greater local cooperation in some places, in several SIDS, notably Fiji, people began stockpiling essential items while disregarding safety guideline set by government. Panic buying and toilet roll hoarding appeared universal. Some shopkeepers in Fiji cheated consumers through price gouging on face masks, hand sanitisers and food. Developing laws and guidelines that both allowed civil and political rights, and enabled safety from COVID-19, was difficult (Chap. 26). Guidelines were to be broken. Curfews were often flagrantly ignored; so too regulations on the size of gatherings (Dean, 2020; Chap. 27). In Grenada many people took to the streets in defiance of social distancing after the annual Carnival was cancelled claiming they 'needed to release the stresses of lockdown' (Telesford, 2020: 3). In Jamaica (Chap. 27) many people claimed exceptionalism: laws over assembly and social distancing did not apply to them.

In the larger archipelagic SIDS, mobility was restricted as lockdowns occurred in various places to try and limit the spread of COVID-19, usually beyond the capital cities. PNG introduced restrictions on domestic air travel, ended cross-province travel, and closed food markets in some urban centres to discourage local mobility. Some provinces, such as Bougainville, effectively closed themselves off to the rest of the country. In Milne Bay province people were not allowed to travel between islands. After COVID-19 was detected early in November in Port Vila, the capital of Vanuatu, the main island of Efate was closed off from the rest of the country. Lives were inevitably more local.

In a sense the islands were returning to an older order of more small-scale, localised – more cooperative – lives and more 'traditional' lifestyles with fewer benefits from imported goods. Strangers, potentially disease-ridden, were dangerous. Social lives were less gregarious; in Port Vila, kava bars closed and the price of kava fell by 30% over a few months. People remained at home and no longer drunk socially. Much the same happened in the Caribbean where bars and restaurants were ordered closed at the start of the pandemic and thereafter had to observe curfew protocols. The same was true for the entertainment sector. The family became the main site of almost everything, from entertainment to survival and food.

New epistemological questions emerged. Owing to the multiple information channels that challenged facts and scientific evidence, people sought answers beyond the scientific community for explaining the emergence of the pandemic, certainly at the beginning when little was known and frightening consequences seemed inevitable. Some looked beyond science to religion and religious practices for answers. The pandemic was interpreted and (mis) understood by different groups in different ways not just because of fear, and the rapid spread of misinformation, but because people's worldviews have always been shaped by socio-political and historical forces beyond science and modernity. Western science had failed to solve the COVID problem. In PNG some turned towards religion, others recognised sorcery (Chaps. 24 and 25). In Kingston, Jamaica, congregational church adherents knew that a higher power would save them (Chap. 27). In some of the most threatened areas, such as the overcrowded informal settlements of PNG's capital, Port Moresby, most residents felt it was a hoax and continued much as usual (Ezebilo, 2020). People viewed the existence of the virus, and causes and solutions, in different ways, and religion and cultural practices resulted in different responses, from beliefs to practices. Some straddled the line between science and religion, relying on God for protection in the church but taking necessary precautions elsewhere.

Domestic violence (DV) increased, as too did social tensions, especially in urban areas with men losing employment and women losing both employment and market access. Quarantine and stay at home orders, with women experiencing greater home and caring roles, and less income, were particularly challenging for women who suffered partner violence, with home the most likely place for a woman to be attacked or killed (Agüero, 2021). Against the tide, DV declined somewhat in parts of PNG, where it had previously been high; although men stayed at home they no longer had the income to spend on beer and drunkenness declined (Chaps. 22 and 25). NGOs generally reported significant rises in DV, with the Fiji

Ministry of Women recording a tenfold increase between March and April 2020. Almost half the women reported a correlation between COVID-19 and violence linked directly to restrictions on movement and strain on families (Radio New Zealand, 24 September 2020). That occurred as many women were experiencing a greater economic and emotional load, and extra home and caring roles, undoing previous gains for women in access to employment. Tonga recorded a 54% increase in DV cases during lockdown. In Samoa, where a 150% increase from the previous years had occurred in helpline calls, the President of the Samoa Victim Support Group, Silliniu Lina Chang, said: 'Families were fighting over the rationing of food; neighbours were stealing from each other's plantations; domestic violence was rife as couples were forced to stay in close proximity to each other during lockdown, and with the daily struggle for food and other necessities, stress and anger were the usual triggers of violence' (quoted in UN Women, 2020). Such problems occurred where access to care was more limited and resources for survivors were under greater strain.

That was no different for the Caribbean where gender-based violence was also prevalent. National surveys conducted between 2016 and 2019 showed that 46% of women in Guyana, Jamaica, Grenada and Trinidad and Tobago experienced at least one form of violence (UN Women, 2020). Women's groups and activists in the Caribbean warned their respective governments about the need to address the issue of DV during COVID-19. One activist, Bonita Montague, in Guyana reported that COVID-19 measures had 'increased the time of enclosed contact with abusers and the abused, and spoke of a case involving a woman whose husband, due to the lockdown, is no longer employed, and now spends his days drinking, to the point where he becomes violently abusive' (Guyana Chronicle, 2020). Local legal systems were forced to rethink issues of freedom of movement and equity.

Mental health was considered to have worsened. That prompted Shamima Ali, the Director of the Fiji Women's Crisis Centre, to argue that although Fijians might be resilient 'we must stop romanticizing the Fijian way of life, the traditional mechanisms' since not everyone had land and rural homes to return to (quoted in Doherty, 2020; Dean, 2020). In PNG, where the incidence of DV was already very high, the loss of cash income (as trading opportunities declined) heightened exposure to different forms of violence, exacerbated by the diversion of already thinly-spread police resources to other COVID-19 crisis response tasks (ACIAR, 2020). That typified a widespread pattern where more remote islands and places were particularly disadvantaged by less adequate transport provision, a reduction in already scarce government services and limited IT access. A survey conducted by the Pan American Health Organization (PAHO) and WHO revealed that countries in the Americas were falling short in implementing mental health services during the pandemic. PAHO, owing to concerns about mental health in the region, called for deinstitutionalisation of psychiatric care and a focus on strategies to allow for treatment of mental health conditions within communities (PAHO, 2020).

In most SIDS, some schools closed down for long periods, but online services were not robust enough to deliver online education, especially to remote locales (to which some people had returned), and, obviously, where students were without internet connections. Closure of schools also posed burdens on parents often ill-equipped to manage home-schooling. More than half of all households in PNG reduced the number of children attending school (World Bank, 2020a). Although computer and internet access is higher in the Caribbean, only about 50% of households have a computer for home use, while internet coverage and reliability exacerbate the challenges of online teaching and learning (Byron et al., 2021). From Fiji there was evidence that some girls did not later return to secondary schools but were pulled into domestic 'duties'. Likewise, when schools reopened in Solomon Islands many children did not return since parents could no longer afford school fees, bus fares and uniforms (Wairiu et al., 2020). That pointed to rural-urban and emerging class divisions, and long-term repercussions for human capital, taking SIDS somewhat further away from equity.

Public Policy: Assistance from Afar

These shocks, challenges and responses, on an unprecedented scale, raised critical questions for public policy and policy-makers, even beyond the obvious requirements for immediate economic policies to meet the sharp rise in unemployment and poverty. Most island governments took the same precautionary policies as elsewhere: closing borders, then schools, sporting facilities, restaurants, markets and beaches, encouraging mask wearing and social distancing, travel restrictions, and developing financial strategies to support the unemployed and small business, albeit with more limited finance. Most SIDs, in their response to COVID-19, followed standard WHO protocols. Mask-wearing was one area that revealed the extent to which national governments relied on the WHO for guidance and scientific direction. Many countries waited, despite evidence from China that mask-wearing for the general public was a useful prevention measure, until the WHO endorsed it. Testing and contact tracing policies were also introduced although not easy to implement.

The political environment has had some impact on policy and implementation success. Political risk calculations are different across SIDs, and policy responses which involve trade-offs between health and economic well-being and between individual liberties and collective responsibility are inherently political and politicised. All the SIDS focused on here have democratic systems of government- but their political systems and culture vary- and they have all had to contend with public opinion, opposition parties and interest groups. Opposition parties often blame governments for not coming up with appropriate solutions and the public have varied interpretations of the virus itself and state responses, whether related to curfews, mask wearing, lockdown and other protocols or vaccines. In Jamaica, there was a measure of cautiousness at making earnest attempts to vaccinate a hesitant population. A proposed policy by the Prime Minister of Antigua and Barbuda, Gaston Browne, to introduce a mandatory vaccination programme was met with strong resistance from the opposition. In St. Vincent and the Grenadines, Ralph Gonsalves, who has consolidated his power – having been in power for 20 years- was able to implement a policy that required a choice between COVID-19 vaccination for central government employees or bi-weekly testing. Calculations of political and economic risks in SIDs were based on different configurations of political power.

While independent states had the ability to close their borders as they saw fit (and most acted very quickly), French and American Caribbean and Pacific territories had more complex relationships (including transport) with metropolitan states and more limited ability to make unilateral decisions. Thus French Polynesia and New Caledonia were caught up in complex debates with Paris about border closures. Much as in the relationship between Greenland and Denmark, and between the Netherlands and the Dutch Caribbean islands, that cast a different light on the relationships between islands, diseases and geopolitics, with metropolitan states tending to exercise more pressure on their dependencies for economic reform (Grydehoj et al., 2020; Rojer, 2021). It ultimately contributed to high COVID-19 rates in Guam and French Polynesia. Other overseas territories such as the British Virgin Islands found themselves at odds with the UK over allegations of corruption in the distribution of COVID-19 funds for families affected by the pandemic. Tokelau gratefully deferred scheduled discussions over independence from New Zealand until the crisis was over. In New Caledonia the pandemic was caught up in arguments about independence, as the territory moved to a referendum on its status, and thus about the nature and extent of support and disruption from Paris (Chap. 7). At different scales and levels the pandemic was political.

COVID-19 also led to reflections about the meaning and purpose of regionalism, size, geopolitics and the need to draw on collective resources within the regions. The Caribbean Public Health Agency, the Caribbean Disaster Emergency Management Agency and the Caribbean Community Implementation Agency for Crime and Security (IMPACS), all regional inter-governmental agencies in CARICOM, became important means by which member states could share information, manage borders and coordinate international assistance and inter-regional efforts. Some regional organisations that had been struggling to establish their purpose and authority suddenly found themselves in the centre of the response to the pandemic.

Multilateralism and external aid were also important policy responses in the Caribbean although challenged by America's inward looking policies, its decision at the beginning of the pandemic to prevent the export of PPE, its withdrawal from the WHO, and its unwillingness to participate in COVAX (the plan for global access to vaccines). Even physical aid delivery had proved difficult where social distancing and quarantining is involved, especially in the virus free Pacific states. Despite metropolitan countries undergoing their own economic problems, SIDs have received aid from the World Bank, the United Nations and other international organizations. The World Bank provided aid towards food security, improving testing capacity, relief to the poor and small businesses and access to medical equipment, in most Caribbean and Pacific SIDS. That was hugely beneficial to these countries, but perhaps as a prelude to long-term health care provision and its funding.

Bilateral relations were important, notably in securing PPE, loans and access to vaccines. China responded quickly in the Pacific and the Caribbean region, partly to counter its poor image because of the origin of the virus and partly to increase its presence in the region. It provided grants for medical supplies to several Pacific island SIDS, quickly dubbed 'mask diplomacy'. Acrimony between China and the US also exists in the Caribbean. China is an important player in the region, much to the chagrin of the US, providing low-interest loans for major infrastructure projects. In keeping with its longstanding relationship with Jamaica, Cuba also sent nurses to assist governments in its initial response to COVID-19. Australia and New Zealand responded to the needs of Pacific SIDS. Both initially focused on rapid financial support, directed to two key outcomes: to support health preparedness by providing critical equipment into existing national health systems, such as oxygen machines, ventilators and PPE, and, secondly, to build economic preparedness through direct budgetary support and with a range of partners, including Pacific governments, and the Pacific Islands Forum Secretariat, to ensure that supply chains for food, fuel, and medical supplies remained operational around the region (Kings, 2020; Australia, 2020). Both countries sought to tackle the shorter-term COVID-19 impacts as well as longer-term strategic issues, including health system strengthening, and social protection, social inclusion, and the range of underlying issues that lead to fragility and vulnerability within countries and particularly for the most disadvantaged communities (Kings, 2020). That intended redirection towards social protection – and health system strengthening, necessary in its own right – might not have happened without COVID-19. The pandemic quickly demonstrated the limited capacity of several SIDS with insufficient funding for recovery and development, and limited personnel, to manage the distribution of relief during crises.

The UK announced an aid decline in November 2020, most likely to be in Pacific SIDS where the UK had belatedly returned in the 2010s. The UK confirmed that it would procure vaccines on behalf of its Overseas Territories. Criticisms from the UK public threatened to derail aid to larger, independent countries in the Caribbean. France and the Netherlands provided support to their Caribbean territories. At the same time that international aid donors were devising ways to meet island needs, in the midst of their own financial crises, many international volunteer services, such as the Peace Corps and JICA (Japan), were quickly closing down their operations and repatriating their volunteers, resulting in a reduction in technical human resources and capacity building. In smaller SIDS like Samoa such NGOs are core providers of essential services (Tierney & Boodoosingh, 2020). They remained cautious about when their activities might resume. Despite such disappointments and setbacks, most SIDS found international support necessary and useful.

The Diversity of SIDS

The economies of SIDs have collapsed but mortality rates have been lower than in metropolitan states. In most, if not all, SIDS, COVID-19 has had a far greater economic impact than health impact, insofar as these can be compared. Islands have fared better than many larger states because external and internal borders could more easily be closed (and monitored), and people were relatively compliant. Conceivably in smaller states trust may be greater where governments have had to place more trust in science. Indeed, in terms of mortality, most SIDS fared better than developed countries with greater state capacity, scientific expertise and more educated populations. They also fared better than those states, such as Brazil, the United States and Russia, where populism was virulent and political agendas marginalised health crises.

Significantly, especially in Pacific SIDS, governments were relatively cautious about opening up borders and economies – and, more significantly, because of their economic structures, were without pressures from large corporations to do so. There was some pressure and disagreement but the consensus was that lives took precedence over livelihoods. The ability of many places to return to subsistence may have been a factor. The outcome was that the independent SIDS fared rather better than those territories, like French Polynesia and Guam, where borders could not be easily closed, international transport connections were greater and metropolitan states influenced local decision-making. Nonetheless international tourism was largely consigned to history, unemployment soared – especially of women and youth and from the informal sector – and the displaced returned to more subsistence-oriented lives in rural areas. Without remittances from overseas kin many households would have foundered. There were still vast differences between how some SIDS, such as those spared the virus, withdrew from the world – if not by choice – and how others sought to rejoin it as fast as possible.

Linking all of them was the idea that the pandemic imposed on islands an opportunity to rethink the future, to 'build back better', and to see it, as Arundhati Roy had done, as a portal to a more optimistic future. Every chapter that follows indicates some of the ways in which this was happening at village and international levels, and how that might mean entirely different things in different places. As the pandemic remains in place it is unsurprising that the following chapters raise as many questions as answers since the greatest challenge facing most SIDS is uncertainty. Nonetheless they offer a reflection on global issues at a local scale, and how local people have responded to one of the greatest challenges of our times: certainly the greatest of the twenty-first century. While the chapters are retrospective all offer at least hints of important changes and even different futures as small states, islands and households coped with and responded to the first year of the COVID-19 pandemic.

Acknowledgements We thank Sophie Webber for her helpful comments on an earlier version of this Introduction.

References

ACIAR. (2020). COVID-19 and Food Systems in the Indo-Pacific: An Assessment of Vulnerabilities, Impacts and Opportunities for Action. *Canberra: ACIAR Technical Report No, 96*, 94–126.

Agüero, J. (2021). COVID-19 and the Rise of Intimate Partner Violence. *World Development, 137*, 105217.

Aiono-Le Tagaloa, T. (2020, November 27). Devastated by Diseases in the Past, Samoa Is on High Alert After *Recent* Coronavirus Scares. *The Conversation.*

Australia. (2020). *Partnerships for Recovery: Australia's COVID-19 Development Response.* Department of Foreign Affairs and Trade.

Bank of Jamaica. (2020). *Remittances Bulletin.* http://www.boj.org.jm/uploads/news/remittance_bulletin-_october_2020.pdf

Bennett, N., Finkbeiner, E., Ban, C., Belhabib, D., Jupiter, S., et al. (2020). The COVID-19 Pandemic, Small-Scale Fisheries and Coastal Fishing Communities. *Coastal Management, 48*(4), 336–347.

Berta, O., Berman, E., & Latior, A. (2020). COVID-19 and the Marshallese. *Oceania, 90*(1), 53–59.

Boodoosingh, R. (2020, June 30). *Bartering in Samoa During COVID-19.* ANU DevPolicyBlog.

Byron, J., Martinez, J., Montoute, A., & Niles, K. (2021). Impacts of COVID-19 in the Commonwealth Caribbean. *The Round Table, 110*(1), 99–119.

Chen, M. (2011). Women and Informality: A Global Picture, the Global Movement. *Martha SAIS Review, 21*(1), 71–82.

Clarke, S. (2020, November 18). Wealthy Businesses not Giving Back, *Nation News.* https://www.nationnews.com/2020/11/18/wealthy-businesses-not-giving-back/

Connell, J. (2021). COVID-19 and Tourism in Pacific SIDS: Unwelcome Lessons from Fiji, Vanuatu and Samoa? *The Round Table, 110*(1), 149–158.

Connell, J., & Taulealo, T. (2021). Island Tourism and COVID-19 in Vanuatu and Samoa: An Unfolding Crisis. *Small States and Territories, 4*(1), 105–124.

Connor, J., Madhvan, S., Mokashi, M., Amanuel, H., et al. (2020). Health Risks and Outcomes That Disproportionately Affect Women During the COVID-19 Pandemic: A Review. *Social Science and Medicine, 266*, 113364.

Craig, A., Heywood, A., & Worth, H. (2020). Measles Epidemic in Samoa and Other Pacific Islands. *Lancet Infectious Diseases, 20*(3), 273–275.

Dang, H., & Nguyen, C. (2021). Gender Inequality During the COVID-19 Pandemic: Income, Expenditure, Savings and Job Loss. *World Development, 140*, 105296.

Darmadi, G. (2020, June 1). Fijians Turn to Bartering System as Coronavirus Shutdowns Cause Mass Unemployment, *ABC News*.

Dean, M. (2020). COVID-19 and Fiji: A Case Study. *Oceania, 90*(1), 96–106.

Doherty, B. (2020, June 29). Quarter of Businesses in Pacific Fear they will not Survive Covid-19 Pandemic, *Guardian*.

Dwyer, P., & Minnegal, M. (2020). COVID-19 and Facebook in Papua New Guinea: Fly River Forum. *Asia Pacific Policy Studies, 7*, 233–246.

Eldemire-Shearer, D., James, K., Morris, C., et al. (2011). Chronic Disease and Ageing in the Caribbean: Opportunities Knock at the Door. *The West Indian Medical Journal, 60*(4), 471–477.

Elgar, F., Stefaniak, A., & Wohl, M. (2020). The Trouble with Trust: Time-Series Analysis of Social Capital, Income Inequality and COVID-19 Deaths in 84 Countries. *Social Science and Medicine, 263*, 113365.

Enari, D., & Faleolo, R. (2020). Pacifika Collective Well-Being During the COVID-19 Crisis: Samoans and Tongans in Brisbane. *Journal of Indigenous Social Development, 9*(3), 110–126.

Ezebilo, E. (2020). COVID-19 Pandemic as Perceived by Residents of Informal-Built Areas Segment of Port Moresby. *Spotlight, 13*(17), 1–5.

Faa, M. (2020, December 7). Torres Strait and PNG Families Separated by Australia's COVID Border Closures, *ABC Radio*.

Farrell, P., Thow, A.-M., Wate, J., Nonga, N., et al. (2020). COVID-19 and Pacific Food System Resilience: Opportunities to Build a Robust Response. *Food Security, 12*, 783–791.

Filho, W., Lütz, J., Sattler, D., & Nunn, P. (2020). Coronavirus: COVID-19 Transmission in Pacific Small Island Developing States. *International Journal of Environmental Research and Public Health, 17*, 5409.

Garavito, M., Beuermann, D., Alvarez, L., & McCaskie, A. (2020, October). The Consequences of COVID-19 on Livelihoods in Barbados: Results of a Telephone Survey. *Inter-American Development Bank*. https://publications.iadb.org/publications/english/document/The-Consequences-of-COVID-19-on-Livelihoods-in-Barbados-Results-of-a-Telephone-Survey.pdf

González Moncada, V. (2020). Impact of COVID-19 on Transport and Logistics Connectivity in the Caribbean. *ECLAC*. https://repositorio.cepal.org/bitstream/handle/11362/46507/1/S2000603_en.pdf

Gounder, N. (2020, April 17). *Fiji's Economic Freefall, and the Government's Response*. ANU DevPolicy Blog.
Grydehøj, A., Kelman, I., & Su, P. (2020). Island Geographies of Separation and Cohesion: The Coronavirus (COVID-19) Pandemic and the Geopolitics of Kalaallit Nunaat (Greenland). *Tijdschrift voor Economische en Sociale Geografie, 111*(3), 288–301.
Guyana Chronicle. (2020, April 16). *COVID-19 'Lockdown' Triggers Spike in Domestic Violence*. https://guyanachronicle.com/2020/04/16/covid-19-lockdown-triggers-spike-in-domestic-violence/
Hartnell, N. (2020, September 8). Bahamas Must 'Get Real' Over Tourism Re-opening, *The Tribune*.
Howes, S., & Surandiran, S. (2020a, August 24). *What are Pacific Governments Spending on in Response to COVID-19*. ANU DevPolicy Blog.
Howes, S., & Surandiran, S. (2020b, August 28). *Constraints on the Pacific Response to COVID-19*. DevPolicy Blog.
Howes, S., & Surandiran, S. (2020c, November 16). *Pacific Remittances: Holding up Despite COVID-19*. ANU DevPolicy Blog.
Hoy, C. (2020, June 15). *Poverty and the Pandemic in the Pacific*. ANU DevPolicyBlog.
Khosla, V., & Pillay, P. (2020). COVID-19 in the South Pacific: Science Communication, Facebook and 'Coconut Wireless'. *Journal of Science Communication, 19*(5), 1–22.
King, R. (2020). *Guam*, COVID-19 Island Insight Series No. 7. University of Strathclyde Centre for Environmental Law and Governance, University of Prince Edward Island Institute of Island Studies and Island Innovation.
Kings, J. (2020, May 8). *Pivoting New Zealand's Aid Programme to Respond to COVID-19*. ANU DevPolicy Blog.
Laula, N., & Paddock, R. (2020, July 20). With Tourists Gone, Bali Workers Return to Farms and Fishing, *New York Times*.
Leng, A., & Lemahieu, H. (2021, February 1). Looking for the Keys to Covid "Success", *The Interpreter*.
Leweniqila, I., & Vunibola, S. (2020). Food Security in COVID-19: Insights from Indigenous Fijian Communities. *Oceania, 90*(Suppl.1), 81–88.
Maclellan, N. (2020a, March 25). God will Protect us, but he Still Wants us to Wash Our Hands, *Inside Story*.
Maclellan, N. (2020b, June 11). Smart Harvest, *Inside Story*.
Magick, S. (2020). Honouring Our Health Frontliners. *Islands Business, 46*(12), 8–11.
Marsters, E. (2012). Feel the Pain, Get on the Plane. In *Cook Islanders' Experience of Seeking Health Across a Transnational Field*. University of Auckland.
Milko, V. (2020, December 29). Amid Pandemic, Pacific Islands Work to Offset Food Shortages, *The Diplomat*.

Mow, I., Shields, C., Sasa, H., & Fitu, L. (2017). Towards a People Centred Early Warning and Disaster Response System in Samoa: The use of ICT by Samoans During Disaster. *Electronic Journal of Information Systems in Developing Countries, 81*(1), 1–18.

Murphy, M., Jeyaseelan, S., Howitt, C., Greaves, N., et al. (2020). COVID-19 Containment in the Caribbean: The Experience of Small Island Developing States. *Research in Globalization, 2*, 100019.

PAHO. (2020). *Health Care Expenditure and Financing in Latin America and the Caribbean [Fact Sheet].* https://www.paho.org/hq/dmdocuments/2013/FactsheetHEFJan31.pdf

Rhiney, K., Eitzinger, A., Farrell, A., & Prager, S. (2018). Assessing the Implications of a 1.5°C Temperature Limit for the Jamaican Agriculture Sector. *Regional Environmental Change, 18*(8), 2313–2327.

Robinson, S.-A., & Wren, C. (2020). Geographies of Vulnerability: A Research Note on Human System Adaptations to Climate Change in the Caribbean. *Geografisk Tidsskrift-Danish Journal of Geography, 120*(1), 79–86.

Rojer, G. (2021). Rising from the Ashes of the Pandemic: The Case of the Dutch Caribbean. *Small States and Territories, 4*(1), 125–136.

Rowlinson, L. (2020, December 12). A Workplace in the Sun, *Financial Times*, 3.

Roy, A. (2020, April 4). The Pandemic Is a Portal, *Financial Times.*

Samuels, T., & Unwin, N. (2007). The 2007 Caribbean Community Port-of-Spain Declaration on Non-communicable Diseases: An Overview of a Multidisciplinary Evaluation. *Revista Panamericana de Salud Pública, 42*, e193.

Telesford, J. (2020). *Grenada*, COVID-19 Island Insight Series No. 3. University of Strathclyde Centre for Environmental Law and Governance, University of Prince Edward Island Institute of Island Studies and Island Innovation.

The Lancet. (2020). COVID-19, a New Lens for Non-communicable Disease. *The Lancet,* 396(1052). https://www.thelancet.com/journals/lancet/article/PIIS0140-6736(20)31856-0/fulltext

Tierney, A., & Boodoosingh, R. (2020). Challenges to NGOs' Ability to Bid for Funding Due to the Repatriation of Volunteers: The Case of Samoa. *World Development, 136*, 105113.

Tora, T. (2020, May 7). Two Piglets for a Kayak: Fiji Returns to Barter System as Covid-19 Hits Economy, *Guardian.*

UN Women. (2020). *Across the Pacific Crisis Centres Respond to COVID-19 Amid Natural Disasters.* UN.

UNDP. (2020). *Socio-economic Impact Assessment of COVID-19 on Papua New Guinea.* UNDP.

Wainiqolo, I. (2020, July 7–8). COVID-19 and the Fiji Economy: An Opportunity to Reorient Public Spending, *Pacific Economic Monitor.*

Wairiu, M., Iese, V., Walelenea, J., Teva, C., et al. (2020). *Assessing Nutrition and Socio-economic Impact of COVID-19 on Rural and Urban Communities in Solomon Islands*. USP Centre for Environment and Sustainable Development.

Webb, J. (2020, July 20–22). Samoa and Tonga: Recoveries Interrupted, Risks Realized, *Pacific Economic Monitor*.

WHO. (2019). *Global Spending on Health: A World in Transition*. https://www.who.int/health_financing/documents/health-expenditure-report-2019.pdf?ua=1

Wilson, B., Saeed, I., de Jesus, A., Leahy, C., & Tweedie, P. (2020, November 13). *Timor Leste's COVID-19 Cash Transfer: Good if You Got It*. ANU DevPolicyBlog.

World Bank. (2020a). *Papua New Guinea High Frequency Phone Survey on COVID-19*. World Bank.

World Bank. (2020b). *Solomon Islands High Frequency Phone Survey on COVID-19*. World Bank.

World Bank. (2020c). *Pacific Labor Mobility, Migration and Remittances in Times of COVID-19*. World Bank.

World Bank. (2020d). *Pacific Island Countries in the Era of COVID-19: Macroeconomic Impacts and Job Prospects*. World Bank.

CHAPTER 2

Public Health and Political Imperatives: Balancing Lives, Livelihoods, and Democracy in Jamaica's COVID-19 Experience

Christopher Tufton

Introduction

Jamaica registered its first COVID-19 case the same day that the Minister of Finance opened the 2020–2021 budget debate. That timing, while coincidental, in many ways set the stage for Jamaica's experience in responding to and coping with the COVID-19 pandemic. Since early January, we—I as Minister of Health and Wellness, with the team at the Ministry of Health and Wellness (MOHW), the state entity responsible for public health, including managing epidemics and pandemics—had been readying ourselves for COVID-19. It was clear that it was only a question of time before cases started to present themselves. But Patient Zero's positive test result being confirmed the same day as the budget was announced was an inauspicious coincidence.

C. Tufton (✉)
Ministry of Health and Wellness, Kingston, Jamaica

Y. Campbell, J. Connell (eds.), *COVID in the Islands: A comparative perspective on the Caribbean and the Pacific*,
https://doi.org/10.1007/978-981-16-5285-1_2

The announcement of the budget would, in normal times, signal a key phase in the political cycle, which in Jamaica, follows a fairly consistent pattern. Without being affected by external factors, the country begins gearing up for an election in the fourth year in office, after the budget is read. We—the Jamaica Labour Party (JLP) administration of Prime Minister Andrew Holness, who had been voted into office in February 2016—were on track to follow that pattern. The Finance Minister made his sectoral presentation on March 11, 2020, where he announced a budget that, without being imprudent or irresponsible, was good. It included a series of revenue measures designed to provide an economic stimulus to Jamaica's economy through a reduction of taxes totalling JA$18 billion for the 2020/21 fiscal year, and no new taxes. It was a budget that reflected Jamaica's progress in economic management over the past 8 years, having emerged from decades of stagnation, the only country in the world to lower its debt-to-GDP ratio without a bailout. And, it was a favourable position for the administration to go into an election with, an election that, according to the cycle pattern, was expected by May or June.

That first COVID-19 case signalled the advent of the pandemic, and with it, the budget was effectively undermined. Billions of dollars had to be reallocated towards the C19 response, and some tax cuts were reversed. The country's economic growth prospects, modest as they were, were shattered, as were the historically low unemployment figures. And there was no prospect of an election any time soon.

Lives

Patient Zero may have been Jamaica's first diagnosed COVID-19 case, and the date of that diagnosis became widely used as the beginning of the pandemic in Jamaica, but the test result did not mark the start of Jamaica's coronavirus journey. That diagnosis actually marked the transition to the second stage of the epidemic. Preparations had begun some 6 weeks before, in January, when what came to be called COVID-19 was still thought to be just a new virus wreaking havoc in China.

It first occurred to me that the world was on the brink of something serious at the beginning of the third week of January. I had seen videos of unofficial footage that people in Wuhan, China, where the virus first broke out, had taken with their phones. At that time, the Chinese authorities were still insisting that the virus was likely not as deadly or infectious as the SARS outbreak, and was still preventable and controllable. Yet here was video showing catastrophic scenes of hospitals overwhelmed with sick people. Then we saw Italy, which is a source country of many tourists who visit Jamaica. There was no question that this novel coronavirus was going to eventually reach Jamaica. From the moment that we, at the Ministry of Health and Wellness (MOHW), were in agreement that this virus was heading towards us, we had two goals: delay the virus's arrival in Jamaica as long as possible to put the necessary measures in place. It wasn't a matter of if, but when, and we needed to push that when back as far as we could.

Our first goal was to keep COVID-19 out of Jamaica for as long as possible, primarily by way of travel bans from the worst affected countries. We stopped cruise ships with passengers with flu-like symptoms from docking—a move that sent shock waves through the tourism-dependent Caribbean and the cruise shipping industry. We were widely criticized at home and abroad when we wouldn't let the Italians disembark the Costa Luminosa, the second cruise ship to pose an issue. That vessel then went to Cayman, where an ill passenger was medically evacuated; that person later tested positive for COVID-19 and died. As the ship continued to sail, an Italian passenger was the first to fall ill on board. Fourteen more Costa Luminosa passengers later died from COVID-19. By the time we diagnosed that inevitable first case—a Jamaican national living in the UK who had come to the island for a family funeral—the public health system was on its way to being equipped to deal with the projected influx of cases.

The next immediate goal was to contain the spread and save the ill from dying, and to flatten the curve as long as possible to continue to ready the hospitals and other aspects of treating large numbers of cases. Schools were shuttered and borders were closed. We issued a work from home order, though it was up to each employer to determine if a worker's in-person presence was necessary. We mandated social distancing, people were to stay six feet away from the nearest person. Frequent handwashing and hand sanitizing were encouraged. Taxis were ordered to carry fewer passengers, and buses to carry only seated passengers. Gatherings of over ten persons were banned, and we ordered churches and bars closed. We required people over 65 to stay indoors. We made masks compulsory in mid-April, when the science incontrovertibly showed that they were effective in preventing the virus' spread. We encouraged Jamaicans to "tan a yu yaad" (stay home), and there were curfews, mostly from late evening through to early morning. There was never a total lockdown, largely because of the economic repercussions that would have wrought, and the lack of capacity to enforce it.

In those first few months, the public health imperatives of the COVID-19 response were the government's priority. Virtually every containment measure that the MOHW recommended was adopted by Cabinet. No expense was spared. The health and stability of the economy, relative to previous years in Jamaica, allowed for these massive, unplanned expenditures. Six billion dollars was allocated for the Ministry of Health and Wellness for its response efforts, some 10% more than would have obtained without COVID-19. Within this there was J$775 million in financial support to the health authorities; infrastructure upgrades of more than J$89 million, including equipping hospitals with an additional 63 intensive care and high dependency units; $1.7 billion on medical equipment, such as personal protective gear and COVID-19 test kits; and more than $500 million to procure prescription drugs. Several facilities across the island were retrofitted to quarantine and/or isolate and care for persons who were suspected or confirmed to have COVID-19.

What money could not buy however, in that very short term, was greater capacity in the public health system, in terms of sustainably staffing all these rollouts, and bridging some of the pre-existing gaps in what was an under-resourced public health system.

Livelihoods

By December 2020, COVID-19 had caused 260 deaths, and infected over 10,000 people.[1] Jamaica boasted one of the lowest infection and death rates in the world. Every life lost was one too many. The disruption to livelihoods and the economy was as, if not more, devastating. The pandemic and the measures to contain it brought a 9% reduction in real GDP,[2] the greatest economic decline in Jamaica's history. Fifty-seven percent of Jamaican households saw a reduction in income between the onset of the coronavirus in March and September, and some 40,000 households, 5% of

[1] As at December 3, 2020. COVID-19 Clinical Management Summary for Thursday, December 3, 2020, Ministry of Health and Wellness, www.moh.gov.jm/covid-19-clinical-management-summary-for-thursday-december-3-2020/

[2] Maria Emilia Cucagna and Suzette Johnson, "Return to paradise: A poverty perspective on Jamaica's COVID-19 recovery response," World Bank Blogs, November 17, 2020, https://blogs.worldbank.org/latinamerica/return-paradise-poverty-perspective-jamaicas-covid-19-recovery-response

all households, sought government aid.[3] Whereas the Cabinet had been virtually unanimous in its agreement about increased budgetary allocations to the MOHW, community quarantines, curfews, and work from home orders, other priorities began to surface. The COVID-19 crisis did not happen in a vacuum: Jamaica's fragile economy was only just recovering from decades of stagnation, and poverty levels were still high. The pandemic and response measures were detrimental to employment, growth, and the economically vulnerable. The government had allocated J$10 billion to a Covid Allocation of Resources for Employees (CARE) programme, which gave financial relief in different forms to several different groups and sectors, including to the self-employed, small businesses, and to displaced tourism workers. Jamaica had never had a direct cash transfer like that in its history. But the economic fallout was dire. A survey done in June and July found that eight in ten households with children experienced a reduction in income, with households losing an average of 46% of their income.[4] There was a widespread realization that if economic needs were not moved up on the priority list, the country would face not only severe economic stress, but the social and psychological consequences of same. Which is what happened.

We were looking at an increase of unemployment and poverty levels, and a decrease in government tax revenues. Business had to be given more space in which to operate, to, if even partially, bring about some resumption of aggregate demand. For the private sector, from the largest players to the micro entrepreneur, the border closure, social gathering restrictions, shutdown of beaches and rivers, and curfews had paralyzed their businesses. Activities across tourism, restaurants and bars, and retail were all negatively affected. The private sector players themselves pressed their cases. The tourism sector wanted to re-open the borders, the entertainment sector to remove curfews and restrictions on gathering. A private sector lobby in any democracy has to be at least heard out, and their views

[3] According to the research findings released by STATIN, cited in Corey Robinson, "Staring Down A Recession - COVID Swallows Up Job Gains Made Under IMF Programmes," *Sunday Gleaner*, November 8, 2020, http://jamaica-gleaner.com/article/lead-stories/20201108/staring-down-recession-covid-swallows-job-gains-made-under-imf. In 2011 Jamaica had 748,326 households. "Number of Households by Parish: 2001 and 2011," 2011, STATIN, https://statinja.gov.jm/Census/PopCensus/Completed/NumberofHouseholdsbyParish.aspx

[4] Caribbean Policy Research Institute, "Socioeconomic Impact of Covid-19 on Jamaican Children," Kingston: CAPRI & UNICEF, 2020.

accommodated. But in a small country such as Jamaica, private sector interests can wield disproportionate amounts of power, based on their personal connections to high ranking government officials, their previous or prospective support of the party that forms the government, or the significance of their sector to the economy. An unfortunate perception took hold, that the tourism lobby was pressuring the government to reopen the borders to tourists, that the government was bending to their pressure, and that the tourists would bring COVID-19 to Jamaica. In fact, most of our incoming travelers were Jamaicans coming from the US. The few tourists were not the cause of the surge that began in August. Tourists being allowed in didn't have a direct effect on the surge, in that the source of infection was not with them, but in opening the beaches and attractions to them, we were forced to open them to all Jamaicans. Further, the Jamaicans coming in did not observe the quarantine rules, and we were unable to enforce them, even with the app, which turned out to be largely ineffective.

The COVID-19 challenge in Jamaica transitioned from a straightforward attempt to control and contain the virus, to a far more complex balancing of often competing interests. We had effectively suppressed the virus's spread within the island, with the border closures that were necessary to keep the virus out, and the restrictions of the movement of people that facilitated the virus's transmission chains. By May, some 8 weeks in, the disruption to people's economic survival became too much, and the lines between lives and livelihoods began to be more sharply drawn.

Politics in a Pandemic

2020 was to have been an election year in Jamaica. A Westminster parliamentary democracy, the head of government has the constitutional prerogative to call an election at any time, but the recent trend is for the election to occur with a few months past the fourth year in office. What this meant was that the political climate in which the pandemic occurred, and the context in which the pandemic response was situated, was competitive and acrimonious. Issues were unnecessarily politicized, constraining the government's response options, and, in some instances, thwarting the response's effectiveness.

The opposition People's National Party (PNP) was internally splintered, and struggling in the polls. Their weakened leaders, having not managed to create any platforms of their own on which to campaign, needed something to oppose, and the government's management of the pandemic, and pandemic-related issues became, not surprisingly, an irresistible target. We didn't expect the entire government response to go completely unchallenged, and the several doctors in the PNP would have had a relevant basis on which to evaluate the response measures. But the administration, particularly in the first several months of the pandemic, acted almost entirely based on the MOHW's technocrats' decisions and advice, which we had complete confidence in, though it was sometimes criticized as overly cautious and conservative. This rendered much of their criticism on issues such as testing, PPE supplies, and hospital readiness flat.

Where they did make inroads, and where those inroads did have public health consequences, was in the politicization of the border closure. That hypocritical posturing forced the re-opening of the borders to be expanded and accelerated far beyond the country's capacity, one of the critical turning points in losing the tight grip we had on the situation up to that point.

Jamaica closed its borders relatively early. First was the travel ban on people coming from China. This was no small matter: the Chinese-Jamaican relationship is very important to Jamaica. China is Jamaica's largest bilateral donor and lender. All of the country's major infrastructural projects, namely roads and hospitals, are financed by China, whether through grants or through long term, low-interest loans. The Chinese are extremely sensitive to anything that could be perceived as a slight on their nation's image. It was immediately a diplomatic crisis. The border closure was entirely a public health measure, to keep the virus out, and it was largely successful. Where other countries' infections rates were soaring, Jamaica was doing well flattening its curve.

Among the tens of thousands of Jamaicans who were overseas, whether travelling, studying, or temporarily resident in another country, there were several thousand cruise ship workers stranded at sea. They were now stranded and wanted to come home. Jamaica's borders were closed, and they were stuck on the ships, as were over 100,000 other cruise ship workers from countries all over the globe. The cruise companies too were anxious to get these workers off their ships.

Just as we can pinpoint Jamaica's COVID-19 trajectory to the first case diagnosed, April 2, 2020, could be said to have been the beginning of us reopening our borders before we were ready, a move that effectively took full control of the virus out of our hands. The plan was to take the Jamaicans at a pace that the public health system could manage, and we were preparing for that when the Marella Discovery II cruise ship docked at Port Royal, outside of Kingston, to refuel. There were 45 Jamaican crew members on board, who hoped to disembark. It was the first time in 21 days that those aboard were seeing land, and they were longing to get off the ship and go home. An exemption had to be sought for the ship to dock and people to disembark; a request to land the Jamaicans was sent by the ship's captain. The Jamaican health authorities proceeded to identify quarantine facilities for them, so as to present the full picture to the Prime Minister when the request was presented. While those arrangements were being made on the ground in Jamaica, the ship's captain, frustrated at having received no update, set sail. The ship went on to the Dominican Republic where 32 Dominican workers were given permission to disembark. It crossed the Atlantic, was denied docking by Portugal, and finally was allowed to set anchor in England, though at a considerable distance from shore.

It was an unfortunate turn of events. The distraught workers took to social media and the radio with their situation, and the opposition picked up on it and used it as an opportunity to politicize the crisis. The PNP attempted to portray the border closure as a deliberate effort to keep Jamaicans out of their own country. The plight of those Jamaicans was dire, and there was and never would have been an intention to deliberately not allow Jamaicans to come home. They misrepresented the sequence of events as the government having rejected the Jamaican cruise workers. The prime minister would almost certainly have allowed them to land. They were eventually repatriated, the first set of stranded Jamaicans to come home, under what we called "controlled re-entry." TUI the travel company who operated the ship, arranged a special charter flight and they landed in Kingston May 6, 2020. They were tested on arrival and placed in state quarantine for 2 weeks. Despite having been at sea for several weeks, six tested positive on arrival, and another six returned positive test results in follow-up testing. Our caution had not been unwarranted.

Then came word that a Royal Caribbean vessel, Adventure of the Seas, was heading towards Jamaica with 1027 Jamaicans that it intended to land. These were Jamaicans who had been working on Royal Caribbean cruise ships all over the world, and had all been consolidated on to one ship. Royal Caribbean was doing this with all of its workers that it wanted to return to their homes. The government had been engaged with the cruise line for several weeks, negotiating the Jamaicans' return. But we were not ready nor equipped to handle such large numbers of people. After the experience of Marella Discovery II returnees the MOHW had calculated it could manage 150–200 returnees every 2 weeks. No agreement had been reached, but the ship proceeded on its way. Some of the cruise workers contacted local media, a few engaged an attorney, and the opposition again seized the opportunity to turn it into a political issue, accusing the government of rejecting its own people. As the ship drew nearer, and the public outcry grew louder, we accepted that we had no choice but to make the necessary arrangements. Royal Caribbean had won, by essentially bullying their way in.

On May 19, 2020, the ship docked in Falmouth, and the 1027 Jamaicans disembarked, 200 at a time, over the course of a week. They were tested and then went into 2 weeks state quarantine. Despite several assurances from Royal Caribbean that they were all C-19-free, there were 28 positive test results from that group. As soon as that 1027 had been landed the Carnival Glory docked on May 26, 2020. Of the 258 Jamaicans who came off that ship, three tested positive for C-19. Just over 300 more cruise ship workers returned to Jamaica over the next month, from three more ships; of those, three tested positive. In the end, of the seven sets of repatriated cruise ship workers, only one set, the 202 who arrived on the Disney Fantasy on May 30, 2020, had no-one test positive for C-19.

From that point a full border reopening was inevitable. On June 1, 2020 commercial flights began to land, and the border was open including to foreign visitors, with testing at the airport, and quarantine requirements for all, regardless of the test result, and a tracking device installed on each person's phone. And still we managed to keep the virus under control, prevent its spread, and suffer few fatalities. It was unexpected that we would do as well as we did, but of course we welcomed it. Except for one workplace cluster outbreak of some 200+ people, Jamaica's rate of infection was far lower than projected, and the C-19 curve was flattened. In early July our numbers were enviable: of the 737 confirmed cases, only 14 or 1.9 percent had been critically ill, and 11 or 1.5 percent, moderately ill. One hundred and twenty days after the first COVID-19 case was confirmed, Jamaica had a mortality rate of 1.4%, compared to the international rate of between three and 4%. And we registered an 80% recovery rate, a hospitalisation rate of less than 1%, and a reproductive rate of less than one.

Up to the end of July, we had ten deaths, still one of the lowest rates for any country in the world, and fewer than 1000 total cases, most which were recovered. The government, having managed the crisis with unprecedented transparency and accountability, had earned the trust of the

Jamaican people. While we had tried to avoid being self-congratulatory, and endeavoured to not become complacent, we thought we had a good handle on the situation, and the mindset was we were in recovery phase. We thought we had come through the hard part, and were about to enter a "new normal" phase, where the emphasis would be on rebuilding the economy, heralded by the renewal of the mandate through the polls. On August 11, 2020, the election date was set for September 3, 2020. The day the election was called it was clear that there had been a marked uptick in cases, but, at that time, no-one anticipated the surge in numbers that, we later realized, was already in train.

Infections Surge, Restrictions Relax

The day after the election, on September 4, 2020, Jamaica was officially declared to be in community spread. The source and progress of the virus could no longer be identified on a case by case basis. The dramatic rise in numbers in the second week of August was attributed to the emancipation and independence day holidays, on August 1 and 6 respectively, and the festivities in the days between, for which the government had relaxed several restrictions on public gatherings, beaches and rivers, and curfews. It wasn't a spike, but a sustained exponential increase in positive diagnoses, and, unfortunately, deaths. The decision to go ahead with the election, while it was taken before it became clear that we were indeed heading toward community spread, would probably have been taken anyway. There was a great deal of criticism that we had held the election in a pandemic; some public health experts recommended against it, and chastised us for having gone ahead with it. But there was no telling what would happen with the virus, and the government needed a renewed mandate with the difficult times and the difficult decisions ahead.

We were still the government, after a landslide victory. We expected to win, but no one could have predicted the outcome: of the 63 seats we won 50. We had gone into the election with the house at 33–30, and among the 17 new seats that we won were several upsets of longstanding PNP members of parliament that no one ever thought could be beaten. The turnout was abysmal, 37%, compared to 48.37% in 2016, but the mandate was incontrovertible. I was sworn back in as Jamaica's Minister of Health and Wellness for a second term. In effect, the work had never stopped. I had campaigned, of course, but the usual break that comes when parliament is dissolved never happened.

Another surge was expected, given the campaigning, and the movement of people on election day. There was a widespread public sentiment, fueled by the opposition, that the government had created an illusion of empathy and transparency, which we abandoned after winning the election. But the pressure to remove restrictions, and further reopen the economy, was sustained. By this point we accepted that testing at the airport was not viable. In a sense, it never really was, but it was the only way we could see, at the time, to get the Jamaican people to trust that reopening the borders could happen safely. We could never have been adequately prepared to do it in a way that would have had a meaningful effect. Jamaica simply did not have the staff or equipment to test the thousands of people arriving every day or properly manage the pre-test requirements, the technology to adequately track people's movements, or the scope in the security services to manage breaches of quarantine orders. No amount of money could buy that kind of capacity in the span of time given.

By the third week of September the country counted some 70 deaths and over 5000 cases; we were reporting over a hundred new infections each day, some days over two hundred. We were now faced with a new challenge: how to convey to the Jamaican people that we had not lost control, that the situation we were in was fully anticipated, and that we were dealing with it, in terms of the public health response, according to best practice. Our hospitals were fully equipped to deal with the expected cases needing hospitalization. We had augmented the cadre of contact tracers by several hundred health care workers. We were following a testing protocol that, however, counterintuitive, was designed to identify the cases where they most likely were. We continued to increase hospital and testing capacity. We struggled with keeping the public's confidence. By October Jamaica was a mirror image of where we had been in April: major restrictions and few infections, versus widespread infections and fewer and fewer restrictions. But we did not follow the trajectory of the US and Europe whose surges were unabated, and where the death rate was far higher.

The pandemic's other effects were beginning to become clearer. The psychological and emotional repercussions on the elderly, who having been restricted to home since March, were lonely, some to the point of despondency. Research findings came in showing that children were psychologically dislocated amidst growing tensions in homes that were now under economic pressure. The COVID slide – the term coined to denote the learning loss that would have occurred when schools closed – was a concern, as it became clear that significant numbers of children had not accessed remote school at all, or had not been able to access it properly because of the lack of access to the internet, or to an appropriate device. There were suggestions that interpersonal violence was increasing, including domestic abuse, and abuse of children. Even if we had weathered the virus itself relatively well, we were not spared the side effects.

No Playbook

When the first COVID-19 vaccine was administered in England on December 8, 2020. Again there was a coincidence, similar to that of March 10, as that same day, Standard and Poor's Global Ratings gave Jamaica a 'B+' rating. In their analysis S&P stated their expectation that Jamaica would return to fiscal surpluses in the short-term due to the government's demonstrated commitment to prudent fiscal policy management. The report acknowledged the economic fallout from the pandemic, particularly the disruption in tourism and other productive sectors, but noted Jamaica's ability to fund the COVID-19 public health, economic, and social response as it did. Jamaica, said S&P, was expected to rebound in 2021, and fully recover in financial year 2022/2023. Jamaica's infection numbers and deaths had trended down since October, and while there were several dozen new cases reported each day, the death rate remained low.

By then, most businesses had reopened. A pilot of face-to-face learning in a handful of public schools had been successful; not a single case of coronavirus had occurred. Even though there had been breakouts in several elderly nursing homes and facilities with very vulnerable people, and in two prisons, there had been only one death among those cases. A night-time curfew and gathering limits were still in effect, and masks were mandatory. We were as close to "normal" as we had been since mid-March, and with a vaccine finally created, the expectation was that would continue. There was also an expectation that there might be another surge after Christmas, but we felt secure that nothing extraordinary or uncontrollable awaited us.

Politically it was a tale of two cities. With such a large majority, we were confident in our actions and decisions, and energized by the renewed sense of purpose of getting Jamaica back on track, "recovering stronger," as our campaign slogan had said. The PNP were in shambles, a situation that kept going from bad to worse, and then even worse. After their humiliating loss at the polls, the party leader Peter Phillips resigned, and a new party president, Mark Golding, was elected. The internal divisions continued to fester, and at the time this chapter was being finished, they were caught up in an embarrassing fiasco concerning a Senate seat.

As the time of year headed to that period where one has an opportunity to reflect on the year just gone, and with the prospect that we were indeed heading back to normalcy, the inevitable question of "what would we have done differently?" came to the fore. Could anyone come up with a solid answer that they could stand by? The medical experts, the technical people at the MOHW, the policy makers, the private sector stakeholders, all who were responsible for the decisions made in managing and coping with the C19 pandemic. Noone knew what was going to happen, or how best to address it. It was a brand new disease. There hadn't been a global pandemic in one hundred years, and that was sufficiently long ago that it may as well never have happened. Each decision that was taken was contextualized by known unknowns, and unknown unknowns. Even in retrospect there was no way to see what might have been the outcome if a different decision had been taken, because there were so many unknown and unexpected variables that constantly arose.

Going forward, however, there were some things I was certain of. One was the importance of a plan. Jamaica didn't have a COVID-19 plan, of course, but it did have a detailed pandemic influenza response plan that had been prepared in 2006 and updated in 2007. That plan was based on the principles of rapid implementation of global and national actions, guidance from the WHO, and dependence on early disease recognition, high-level political support, and transparent communication—all which we had followed. It was adapted for COVID-19, and gave some structure to what we did, especially in the very beginning, before COVID-19 got to Jamaica, and was crucial to our management of the pandemic thereafter. Jamaica would always have a plan going forward, whether for another pandemic, or other natural disaster or shock.

Because of COVID-19 public health finally got its due. When I was first appointed Minister of Health in 2016, I had little appreciation for the sector, its importance for national development, and the parlous state of Jamaica's public health care system. It was chronically under resourced, and perennially ranked lower than other sectors in terms of priorities for Jamaica's never-enough resources. It took a global pandemic that brought the country and the world to a standstill, for the sector to receive long-needed funds for some basic needs, and to be considered a priority area for government policy and programming. The fracturing of that old normal, while accompanied by massive societal and economic costs, was a boon to the public health system, and is one aspect of the old normal to which we should not return.

CHAPTER 3

COVID-19 in Pacific Islands People of Aotearoa/New Zealand: Communities Taking Control

Collin Tukuitonga

Introduction

The COVID-19 pandemic seriously disrupted the lives and livelihoods of people around the world, but Aotearoa/New Zealand was widely regarded as having adopted an effective elimination strategy for COVID-19 led by a popular Prime Minister and strongly supported by science-based advice. An important part of the New Zealand response included early and decisive lockdowns restricting the movement of people within the country and between New Zealand other countries. New Zealand also restricted the number of people coming into the country with strict border control measures and mandatory quarantine and testing of all arrivals. The Prime

C. Tukuitonga (✉)
School of Public Health, Faculty of Medical and Health Sciences,
University of Auckland, Auckland, New Zealand
e-mail: c.tukuitonga@auckland.ac.nz

Y. Campbell, J. Connell (eds.), *COVID in the Islands: A comparative perspective on the Caribbean and the Pacific*,
https://doi.org/10.1007/978-981-16-5285-1_3

Minister was widely commended for the strong, empathic leadership and clear communication throughout the pandemic.

While there were concerns about the negative economic consequences of the border restrictions, lockdowns and public health measures, most New Zealanders supported the strategy adopted by the Government. The 'go hard, go early' policy was widely regarded as a critical part of the New Zealand response. The strategy received worldwide attention as a leading example of the actions needed to protect its citizens from COVID-19, which remained a risk to New Zealand through its continuing spread in other parts of the world.

When WHO declared the COVID-19 pandemic on 11 March 2020, New Zealand had already made good headway with its pandemic response guided by the New Zealand's Influenza Pandemic Plan (NZIPP). The NZIPP has been in place for many years, developed in response to previous outbreaks of H1N1 Influenza in 2009 and Severe Acute Respiratory Syndrome (SARS) in 2003. The New Zealand Influenza Pandemic Plan: A framework for action (NZIPAP) set out the all-of-government measures to be taken to prepare for and respond to an influenza pandemic (New Zealand Ministry of Health, 2017). New Zealand was ready. Indigenous Maori and people from the Pacific Islands resident in New Zealand were regarded as at elevated risk of COVID-19 infection, hospitalisation and death, resulting from socio-economic disadvantage, crowded and damp housing combined with a higher prevalence of co-morbidities such as diabetes, heart disease and delays in access to health care being responsible for the elevated risk of viral outbreaks, including COVID-19. The second phase 'Auckland August Cluster' affected mainly Pacific and Maori people with three quarters of all cases reported in Pacific Islands people. This chapter discusses the COVID-19 pandemic in NZ with a focus on Pacific Islands people living in New Zealand.

Pacific Islands People of Aotearoa/New Zealand

Pacific Islands people have been in Aotearoa/New Zealand since the 1950s when they came from Samoa, Tonga, Fiji to work in the factories as part of a flourishing NZ economy based on manufacturing. People from the Cook Islands, Niue and Tokelau had always had free access into Aotearoa/New Zealand having automatic NZ citizenship. Migrants were initially motivated by economic factors, to earn higher wages and provide financial support to family who remained in the home country. Pacific migrant workers were initially primarily employed as production line workers on vehicle and domestic appliance assembly lines. Others occupied positions in the service sector, such as hospital kitchen hands and laundry workers, and the service sector gradually became more important (Spoonley and Bedford, 2012).

From small origins, Pacific Islands people, most of whom are Polynesians like indigenous Maori, made up 381,642 (8%) of the total 4,700,000 NZ population by 2018. Moreover it was among the fastest growing ethnic groups in Aotearoa/New Zealand, resulting in a growing influence in politics, arts, music and culture, and a slow but demonstrable increase in their participation in business, the private sector and public institutions in New Zealand. Several have become Members of Parliament. Most live in Auckland, the location of the largest international airport and port, and the largest city in the country; in 2018 the Pacific Islands population in Auckland was 243,978 (16%) of the total population of 1,571,718.

Of the 381,642 total Pacific Islands population in New Zealand, approximately 182,424 (48%) identified as Samoan, 82,053 (22%) identified as Tongan, 79,763 (21%) as Cook Islands Maori, 30,867 (8.0%) as Niuean and 19,722 (5.2%) as Fijian. Many individuals identified with more than one ethnic group leading to the total number exceeding 381,642 people.

The Pacific Islands population in New Zealand is very young and projected to increase rapidly for some time to come. The median age was 23.4 years compared with 41.4 years for the national total. More than two thirds of the Pacific Islands population in 2018 were born in New Zealand, indicative of high fertility rather than a declining rate of migration. Nonetheless, a third (34%) of the Pacific Islands population were born overseas compared with 17% of the European population and 77% of the Asian population. Pacific Islanders have steadily becoming increasingly ethnically diverse.

Life expectancy at birth for Pacific Islands women was 6 years less and 5 years less for men compared with Non-Maori, Non-Pacific women and men. Ethnic differences in life expectancies have narrowed very slowly between Pacific Islands people and other New Zealanders. Ethnic differences in health and social indicators reflect pervasive and preventable inequities which exist in the country despite decades of government social policies seeking to address the socioeconomic circumstances of Pacific Islands people and other groups in the country.

Pacific Islands people have long been disproportionately represented in the lower socio-economic groups in New Zealand. The median income for Pacific Islands people in 2018 was 23,400 NZD compared with 34,600 NZD among Europeans, 24,300 NZD among Maori and 28,500 NZD among Asian groups. Similarly, unemployment rates in 2018 were 7% among Pacific Islands adults compared with 8% among Maori, 3% among Europeans and 4% among Asians. COVID-19 resulted in significant job losses which predominantly affected the service and hospitality industries where most employees were Pacific Islands and Maori people. In the second quarter of 2020, unemployment rates were however 6% for Pacific Islanders, 7% for Maori, and 3% for Europeans. Increasingly social support agencies reported growing numbers of Pacific families experiencing severe economic hardship and seeking support from NGOs and government agencies. At the same time development and hardship could not be reduced to economic variables alone. Wealth and prosperity were seen by many Pacific islander households as defined and measured in how safe, happy, healthy and productive families were. Spirituality and involvement in church and community provided its own wealth and prosperity.

Socio-Economic Determinants of Health

Poor health and higher risks for communicable and non-communicable diseases (NCDs) arise from inequities across a wide range of socio-economic factors relating to housing, education, employment, income and social position. Pacific Islands people were more likely than other groups in New Zealand to live in larger households, have greater fluidity of household members, and experience higher rates of household crowding; around 40% of Pacific households are officially overcrowded (Statistics New Zealand, 2018). Home ownership by Pacific Islands people is the lowest of all the ethnic groups in New Zealand. While large overcrowded Pacific households may reflect cultural preference this is more likely to be a response to economic hardship and the high cost of housing in New Zealand.

The significant relationship between household crowding and the transmission of infectious diseases disproportionately affects Pacific Islands populations, for whom 17% of all hospitalisations can be attributed to crowded living arrangements. In the context of COVID-19, self-isolation in crowded households presented considerable challenges for Pacific Islands people, transmission risks were heightened due to a higher percentage of households living in multigenerational families, many of whom were also at high risk due to age and underlying health conditions or disability.

By the end of the twentieth century it was well evident that Pacific peoples adverse socio-economic conditions, increased health risks, problems in health services utilisation, and transitional demographic characteristics contributed to a poorer health status relative to the total New Zealand population (Tukuitonga & Finau, 1997). NCDs such as diabetes, heart disease and some cancers had become the leading cause of death, disease and disability among Pacific Islands people in New Zealand, reflecting multiple risk factors such as smoking, unhealthy diets and reduced levels of physical activity. Pacific Islands children and young people have the highest rates of obesity in New Zealand and more than two thirds of all Pacific Islands adults were obese (New Zealand Ministry of Health, 2020). Many Pacific Islands people with diabetes and high blood pressure remain undiagnosed and approximately half of those who are diagnosed with diabetes do not receive the recommended quality of care. As a result, diabetes complications were higher resulting in unsustainable demand for kidney dialysis and other services.

The severity of COVID-19 and the risk of death increases amongst those with one or more underlying health conditions such as obesity, diabetes and cardiovascular disease. Pacific Islands people have a high rate of underlying diseases. Compared to non-Māori non-Pacific peoples, Pacific Islands peoples have twice the rates of hospitalisation for cardiovascular disease and three times that for respiratory conditions. Diabetes was a particular concern, with 20% of Pacific peoples aged 20–79 estimated to have diabetes, compared to 10% of Māori, 8% of Asian and 6% of NZ Europeans (New Zealand Ministry of Pacific Peoples, 2016).

At the same time, the risk of community transmission of infectious diseases was high for Pacific Islands and Maori populations. Due to their high representation in jobs involving close contact with people (such as in supermarkets, in security, and in myriad health care environments) and in manufacturing roles, many Pacific Islands workers did not have work from home options during New Zealand's lockdowns. This increased exposure risk in the workplace was further compounded by the location of the outer urban suburbs of Mangere and Otara, more deprived than most Auckland suburbs, where the majority of residents are Pacific Islands people; 68% of residents of Mangere and 78% of those of Otara identify as Pacific islanders. Both suburbs are close to the airport national border and to the managed isolation and quarantine hotel facilities. They provide local examples of the importance of social determinants of health, where poor and localised health outcomes tended to reinforce racism, stereotypes and stigma.

High rates of unmet need for health care in Pacific Islands people were also attributed to access barriers related to cost, language and communication, access to translation services, health literacy, work commitments, transport and caregiver costs, discrimination, and racism (Ryan et al., 2019). Furthermore, under-representation of Pacific Islands peoples in medical, nursing and many other roles across the health workforce constrained the delivery of equitable, culturally safe, and appropriate health services for Pacific Islands people. These barriers to care had implications for Pacific Islands peoples' access to key public health messages (and their understanding of these messages) thereby impacting their engagement with health protection activities and the need for early presentation for testing for COVID-19, required to prevent spread in the community.

A review of the health sector's response to the 2019 measles outbreak in New Zealand noted the difficulty for Pacific Islands communities in accessing translated information and a "clear difference in access to health information for Pacific and Māori groups" compared to the rest of the population (Sonder & Ryan, 2020). Growing evidence of a digital divide and the 'digital exclusion' of Pacific Islands people (lower access to the internet at home, work or school than other groups and lower levels of digital literacy) suggest that access to digital technology is a barrier for Pacific Islands people's access to health information and remote alternatives to face-to-face care, as well as the ability to connect for online work and education. In addition to the greater risk of COVID-19 transmission and complications, the existing socioeconomic inequities experienced by Pacific Islands communities tended to intensify during and beyond the pandemic, with anecdotal evidence suggesting that Pacific Islands peoples were likely to be more deeply affected, and for longer, during periods of economic downturn and recession.

Pacific Islands people have had major problems accessing health care services in New Zealand, with considerable disparities between Pacific islanders and others and continued negative trends. Various studies found that Pacific people had a lower rate than others in the use of primary care services, such as GP visits and filing prescriptions, despite research revealing that 19% of Pacific peoples reported their health as being "not so good" or "poor" compared to 8% of Europeans (Wright & Hornblow,

2008). Radical transformation of the health system at the end of the century provided the opportunity to establish ethnic specific primary health services, hence Pacific communities were able to establish Pacific-specific health services which were potentially better suited to their needs (Tukuitonga, 1999). Further improvements occurred in this century but were slowed by the limited Pacific presence in the health workforce. The establishment of Pacific-owned primary health care clinics changed the health care landscape in Auckland and strong advocacy by Pacific Islands health professionals improved the availability of a better quality of health care for Pacific Islands people. Nonetheless, persistent, pervasive and preventable inequities continued to exist for many Pacific Islands people. Reducing ethnic inequities in health remains an important priority in Aotearoa/New Zealand.

COVID-19 in Aotearoa/New Zealand

The first case of COVID-19 was confirmed in New Zealand on 28 February 2020 in a person who returned to the country from Iran. Most of the cases of COVID-19 in the first wave were reported in New Zealanders returning from overseas travel (Fig. 3.1). As a result, the number of COVID-19 cases reported in Pacific Islands people was low. The total number of cases by November 2020 was 1973 consisting of 1305 imported cases, 664 locally acquired cases and four then under

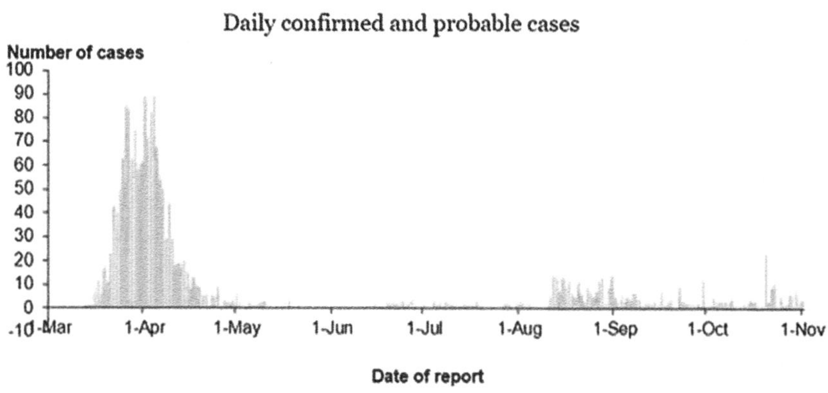

Fig. 3.1 New Zealand confirmed and probable COVID-19 cases over time, November 2020

investigation (Jefferies et al., 2020). There had been three distinct clusters since the outbreak was first reported in February with a number of 'incursions' with front-line health care workers and security personnel infected at their workplaces. While these incursions caused some disquiet in the community, they did not result in any significant community transmission.

A total of 1569 cases were reported in the first 'wave' of cases reported between February and August, of which 132 were reported in indigenous Maori and 76 in Pacific Islands people. There were 20 deaths. Most of these were returning travellers and the age groups most affected were 20–29 years followed by 50–59 years. Pacific Islands people made up 5% of all cases compared with their being 8% of the total New Zealand population. This under-representation came as a relief to New Zealand because Pacific Islands health and community leaders in New Zealand were concerned about the potential for the virus to spread quickly among Pacific Islands communities and indigenous Maori populations.

In response to the rise in the number of confirmed cases, NZ progressively implemented a number of public health measures as part of the national 'elimination' strategy. All borders and entry ports of New Zealand were closed to all non-residents on 19 March 2020, with returning citizens and residents being required to self-isolate. After 10 April, all New Zealanders returning from overseas were required to go into 2 weeks of managed isolation. A four-level alert level system was introduced on 21 March to manage the pandemic. The Alert Level was initially set at Level 2, but was subsequently raised to Level 3 on 23 March due to concerns about the potential spread of the virus in the community. On 25 March, the Alert Level was moved to Level 4, putting the country into a nationwide lockdown. Level 4 lockdown restricted people to their homes, all public venues were closed and all public gatherings were cancelled. All businesses were closed except for essential services such as supermarkets, pharmacies, medical clinics, petrol stations and lifeline utilities. All schools were closed. Lockdown included public health messages for the community such as social distancing, hand hygiene and coughing etiquette that were part of regular messages relayed to the public on radio, television, print media and social media outlets. The Director General of Health and the Prime Minister or Minister of Health made daily announcements on case numbers and policy responses. Despite adverse effects on businesses and the economy, most New Zealanders supported the Government strategy with a focus on public health and personal safety. Pacific Islands communities were provided additional information through dedicated radio and television messages in their first language. The Alert Level was moved back down to Level 3 on 27 April, partially lifting some lockdown restrictions, and down to Level 2 on 13 May, lifting the rest of the lockdown restrictions while maintaining physical distancing and mass gathering size limits. The country moved down to Level 1 on 8 June, removing all remaining restrictions except border controls.

New Zealand has been recognised globally for its elimination strategy based on science and evidence-based interventions. Regular and transparent information flow from the political and health leaders were a key factor in the effective strategy. The national response has also been widely supported by the citizens and the country has seen unprecedented social cohesion and widespread support work together as a 'team of five million'. The personal popularity of the Prime Minister has been an important part of the national response. Promotion of misinformation and conspiracy theories by fringe political parties linked to the October national elections did not adversely affect the national strategy. Indeed the Prime Minister and the Labour Government were praised for their response to the pandemic and had a particularly convincing win in the election. Similarly, ongoing concerns by the private sector about the negative impact on business and the economy and negative narratives by some media outlets and politicians did not affect the overall effectiveness of the elimination strategy.

The Auckland August Cluster

On 11 August, four cases of COVID-19 from an unknown source were unexpectedly reported in a Pacific Islands family in Auckland, the first from an unknown source in 102 days when no cases were reported. The 'index family' was reported to have been exposed to a worker at a cool storage facility near Auckland airport where the father of the family also worked. No direct link was established between the family and the workplace and no obvious source ever found. At noon the following day, the Auckland region moved up to alert level 3, while the rest of the country was moved to level 2. On 30 August, Auckland moved down to "Alert Level 2.5", a modified version of Alert Level 2 with limitation on public gatherings, funerals, and weddings. On 23 September, Auckland moved down to Alert Level 2, after the rest of New Zealand had moved to Alert Level 1.

The second wave consisted of a total of 404 cases were reported between 11 August until it was closed on 3 November (Fig. 3.1). There were 179 cases linked to the "Auckland August Cluster" with 118 cases (66%) reported among Pacific Islands people, 39 (22%) among Maori and the remaining 22 (12%) in other ethnic groups. A multiethnic evangelical church became the centre of focus of this cluster because most cases were reported from church members and household or workplace contacts. The cluster became a focus for the Pacific Islands communities being in South Auckland where the majority of the population lived.

Pacific Communities Respond

With encouragement and support from public health officials and Pacific health professionals, the Pacific Islands communities affected by the outbreak responded by working together to limit the spread of the virus. Pacific social support agencies and Pacific personalities responded in a coordinated way to support case identification and isolation, contact tracing and public health measures provided by the national health services. Pacific community activities were strongly supported by Pacific Members of Parliament with support from national leaders. The Pacific MPs were also able to secure dedicated additional resources to support activities by and for Pacific communities. A Pacific Expert Group was established to provide technical and policy advice to Pacific Ministers. All government agencies provided specific and additional support for Pacific communities.

Despite socio-economic disadvantages and difficulties accessing health services, the response to the Auckland August cluster by Pacific communities proved a triumph over adversity. In response to the threat of COVID-19 Pacific Islands families and communities, church and community leaders, health care providers and media came together to inform and support their communities to prevent the spread of COVID-19. Churches became focal points for the distribution of food parcels and social support and centres for COVID-19 testing. Church leaders became important and visible champions for public health measures such as hand washing, social distancing and limitations on mass gatherings. Churches rapidly transitioned to online services and congregations advised to remain at home. Pacific media were responsible for the dissemination of information on radio, TV, print and social media, including information translated into several Pacific languages.

Pacific Islands primary health care providers were able to respond promptly to the needs of their communities. For example, community COVID-19 testing centres, including mobile clinics, were set up in strategic locations throughout the community which allowed Pacific Islands people to get tested without the traditional geographical barriers that existed in accessing health care. As a result, testing rates for COVID-19 were highest among Pacific Islands people. The Auckland Regional Public Health Service established a 'Pacific Team' to support the national contact tracing efforts with staff able to speak one of the Pacific languages. This improved the efficiency and reliability of contact tracing activities.

During the pandemic, routine and non-urgent health care services were either postponed or delivered remotely. Pacific Islands primary health care providers worked together to provide ongoing health care through telephone and the internet. Pacific doctors and nurses also contributed to the dissemination of information translating official advice into Pacific languages. Pacific clinicians also provided much needed reassurance for Pacific communities who were concerned about family safety and wellbeing. Anecdotal evidence suggested heightened levels of anxiety and other mental disorders among Pacific families who were confined to their homes for extended periods. Police and justice information showed increased rates of Violence Against Women and Girls in Pacific households.

The August Auckland cluster affected Pacific Islands people more than other ethnic groups. The response from Pacific communities was unprecedented and highly effective in containing the spread of the virus further, whereas Pacific communities in Auckland are usually regarded as passive and dependent. The response to the COVID-19 cluster was an excellent illustration of empowered and informed people concerned about a shared health threat. Pacific communities quickly adopted such public health measures as social distancing, hand washing and face masks. Such response occurred in communities often poorly served by mainstream health and social service providers, and at a time when they were prominent victims of the downturn in the New Zealand economy, with significant losses of jobs.

The Pacific Islands community's response to the August cluster in Auckland showed the importance of involving affected communities and supporting ownership and active participation by them from the earliest stages of the pandemic. Disadvantaged communities were not only at high risk of infection, hospitalisation and deaths from COVID-19 but many were often regarded as passive recipients of government services without active involvement in the design and delivery of these services. An important enabler for Pacific Islands communities was the delivery of information in their first language (which was done in nine different Pacific languages) and the availability of information on a 24-hour basis especially via social media. Engaging with language and culture were crucial. Informed communities became empowered communities able to participate fully in the overall response. Provision of social services, food and financial support was critical in engaging and supporting community participation and enabling Pacific Islands people to help disseminate information within their respective communities. Social support was essential for many Pacific Islands families given their fragile financial position as a result of job losses caused by COVID-19. The Government also allocated financial resources to support activities with and for Pacific Islands people e.g. options for isolation outside the family home was provided to some families as a result of crowded housing and inability of some people to isolate effectively in the family home.

Perhaps unremarkably, Pacific islander communities elsewhere experienced similar problems, as in the United States. Amongst Islander communities and indigenous Hawaiians (in Honolulu) infection rates were influenced by a poor understanding of public health issues and regulations, inadequate access to welfare, poverty and the need to work (and no possibility of doing that from home). Islanders had similar underlying health problems and co-morbidities to those in New Zealand, could not afford to stop work (often in essential jobs such as meat plants), and were ineligible for Medicaid. They tended to live in overcrowded multigenerational households and had ten times the infection rates of all other ethnic groups (Derauf et al., 2010). In Utah, where there was a significant concentration of Pacific Islanders, they had an infection rate double that of the state as a whole, partly a function of their being regarded as an essential workforce in the service sector where exposure to the virus was relatively high, partly a function of stigma in admitting to having COVID-19 and partly from living in large multigenerational households (Vaughn et al., 2020). Language and cultural issues disadvantaged poor and relatively recent migrants, as they did amongst refugees in tower blocks in Melbourne, Australia, and similar ethnic minority groups in the United Kingdom (Steyn et al., 2020). At this stage there are no indications of whether there were different infection rates amongst Islanders of different national origins.

Looking to the Future

In many countries, like New Zealand and the United States, people of Pacific Islands descent and indigenous populations are at high risk of COVID-19 infection and death. Modelling work has shown that Pacific Islands people and indigenous Maori people were at increased risk of COVID-19 infection and hospitalisation for COVID-19 (Steyn et al., 2020). Increased risk of infection and death is probably due to a combination of socioeconomic conditions, related to incomes, diets and accommodation, and to the high prevalence of comorbidities such as obesity and diabetes. Overcrowded, damp and cold houses are associated with the spread of infectious diseases, including COVID-19. Public health measures such as social distancing and home isolation are difficult to implement in overcrowded situations.

However, Aotearoa/New Zealand was fortunate that its political leaders adopted strong leadership on COVID-19 pandemic elimination based on science with clear and decisive actions in response to changing risk levels. Prime Minister Jacinda Ardern was widely commended for decisive leadership and a focus on protecting public health despite some resistance from the business community, media and some political parties. The NZ elimination strategy was widely regarded as world leading. The outcome might have been different had the more vulnerable and affected Pacific islander communities not worked cooperatively and effectively to establish public health practices. Nonetheless infection rates were higher for Pacific peoples, as they had been for earlier epidemics (Steyn et al., 2020). Likewise, in Auckland, as in Honolulu, Salt Lake City (Utah) and elsewhere, language and culture were critically important and emphasised the need for health services to work with Islanders as partners and listen to members of affected communities.

Aotearoa/New Zealand continued to experience 'incursions' at the border with small numbers of cases reported in returnees in the managed quarantine facilities throughout 2020 but no further community outbreak was reported. Surveillance and vigilance remained crucial to ensure that cases were identified early and contacts traced promptly. Early and prompt response was vital in order to reduce the risk of community transmission, and possible spread to the Pacific islands, since Auckland is the main transit point for travellers to the islands and from overseas.

National success consequently had two incidental advantages. Firstly, it reduced concerns in several Pacific island states, most like Samoa and

Tonga without any COVID-19 cases. to which their citizens were intermittently being repatriated, often after seasonal employment or medical referrals to New Zealand. Secondly, the quick and effective response contributed to a better than expected economic recovery, so that although the economy contracted it experienced fewer problems than countries such as Japan and the United Kingdom, and unemployment rates were lower than initially predicted and anticipated.

COVID-19 still posed elevated risk to Pacific Islands people in New Zealand in light of their adverse socioeconomic circumstances and prevalence of comorbidities such as obesity and diabetes. However the national strategy and Pacific Islands communities response and experiences during 2020 helped better inform Pacific Islands people of the risks and the measures that needed to be taken to protect themselves and their families. It is even plausible that success at overcoming COVID-19 may encourage progress in other areas of public health. Anticipated introduction of the vaccination programme restored the confidence of health officials and the wider population that the pandemic could be contained and possibly eliminated. Given their risk profile, Pacific Islands and indigenous Maori people were among the priority groups for receiving the COVID-19 vaccination early. The early rollout of the COVID-19 vaccination programme by the NZ Government to several Pacific island states (including the Cook Islands, Niue and Tokelau) in early 2021 was expected to be an important intervention for Pacific Islands people, that would allow travel bubbles to be developed, and some restoration of the island economies.

COVID-19 has been the most important threat to global public health and economies in decades. All nations were affected in different ways with developed nations in Europe and the USA being most severely affected. The pandemic has re-defined how people live, work, play and communicate and will have stimulated new norms in business, learning and work where remote and electronic means reduce person to person contacts. The full impact of COVID-19 on how people interact with one another is yet to be fully appreciated. Much still depends on the timing, practice and outcome of vaccination. At least at the end of 2020 New Zealand was poised to be perhaps the first large country to ride out the pandemic storm.

References

Derauf, D., Miller, D., & Brown, T. (2010). *The Fierce Urgency of Now – Engaging Pacific Islander Communities in Hawaiʻi to Contain COVID-19*. East West Center.

Jefferies, S., French, N., Gilkison, C., Graham, G., et al. (2020). COVID-19 in New Zealand and the Impact of the National Response: A Descriptive Epidemiological Study. *The Lancet Public Health, 5*(11), e612–e623.

New Zealand Ministry for Pacific Peoples. (2016). https://www.mpp.govt.nz/assets/Uploads/Contemporary-Report-Web.pdf

New Zealand Ministry of Health. (2017). *New Zealand Influenza Pandemic Plan: A Framework for Action* (2nd ed.). Ministry of Health.

New Zealand Ministry of Health. (2020). Annual Data Explorer 2019/20: New Zealand Health Survey [Data File]. https://minhealthnz.shinyapps.io/nz-health-survey-2019-20-annual-data-explorer/

Ryan, D., Grey, C., & Mischewski, B. (2019). *Tofa Saili: A Review of Evidence About Health Equity for Pacific Peoples in New Zealand*. Pacific Perspectives Ltd..

Sonder, G., & Ryan, D. (2020). *Health Sector Response to the 2019 Measles Outbreaks*. Ministry of Health.

Spoonley, P., & Bedford, R. (2012). *Welcome to Our World? Immigration and the Reshaping of New Zealand*. Dunmore Publishing.

Statistics New Zealand. (2018). *Living in a Crowded House: Exploring the Ethnicity and Well-Being of People in Crowded Households*. Retrieved from http://www.stats.govt.nz/assets/Uploads/Reports/Living-in-a-crowded-house-exploring-the-ethnicity-and-well-being-of-people-in-crowded-households/living-in-a-crowded-house-exploring-the-ethnicity-and-well-being-of-people-in-crowded-households.pdf

Steyn, N., Binny, R., Hannah, K., Hendy, S., et al. (2020). Estimated Inequities in Covid-19 Infection Fatality Rates by Ethnicity for Aotearoa New Zealand. *New Zealand Medical Journal, 133*, 1521.

Tukuitonga, C. (1999). *Primary Healthcare for Pacific People in New Zealand*. Ministry of Health.

Tukuitonga, C., & Finau, S. (1997). The Health of Pacific Peoples in New Zealand up Until the 1990s. *Pacific Health Dialog, 4*(2), 59–67.

Vaughn, K., Fitisemanu, J., Hafoka, I., & Folau, K. (2020). Unmasking the Essential Realities of COVID-19: The Pasifika Community in the Salt Lake Valley. *Oceania, 90*(1), 60–67.

Wright, S., & Hornblow, A. (2008). Emerging Needs, Evolving Services: The Health of Pacific Peoples in New Zealand. *Kotuitui: New Zealand Journal of Social Sciences Online, 3*, 21–33.

CHAPTER 4

Coronavirus and CARICOM: The Benefit of a Regional University in a Coherent Pandemic Response

R. Clive Landis

Introduction

The COVID-19 pandemic erupted across the globe at a time of growing populism and nationalism, that has arguably exacerbated the pandemic (Leonhardt & Leatherby, 2020). The UN and the World Health Organization urgently called for international solidarity and cooperation in tackling the global COVID-19 health threat (UN News, 2020; Eurasia Group, 2020), but this was hard to achieve as the legal frameworks governing international trade in a pandemic buckled under populist sentiment serving overtly nationalistic interests (Williams et al., 2020; Wilson et al., 2020). The UN Secretary General remarked at the 31st Special Session of the General Assembly on December 3rd 2020 that when facts are rejected

R. C. Landis (✉)
University of the West Indies (UWI), Cave Hill, Barbados
e-mail: clive.landis@cavehill.uwi.edu

Y. Campbell, J. Connell (eds.), *COVID in the Islands: A comparative perspective on the Caribbean and the Pacific*,
https://doi.org/10.1007/978-981-16-5285-1_4

and guidance ignored "and when countries go in their own direction, the virus goes in every direction" (United Nations, 2020).

The largest geopolitical grouping of Caribbean countries is the Caribbean Community (CARICOM), which includes most of the English-speaking Caribbean along with Dutch-speaking Suriname and French-speaking Haiti in a 20-member bloc. CARICOM is one of the oldest surviving integration movements in the developing world with mature regional institutions dedicated to the economic, political, human and social development of its citizens (CARICOM, 2020). CARICOM's institutions are supported by security and disaster management agencies as well as a regional university.

This chapter describes the relative success of CARICOM countries in controlling the COVID-19 pandemic and offers possible explanations. One likely explanation is the strong sense of regionalism and solidarity that has enabled a systematic and scientific approach to be adopted across CARICOM. The practice and resilience gained from dealing with prior viral epidemics and annual weather-related disasters is another factor that may have conditioned Caribbean citizenry to participate in an all-of-society public health response. When the multilateral actions are considered as a whole they fit nicely onto the six pillars of the WHO's Health Systems Framework, suggesting that engagement of the Health Systems Framework in a pandemic setting may help deliver a more coherent response against this or other emerging virus pandemics.

CARICOM and UWI: Regional Partners in Perpetuity

The regional heart of The University of the West Indies (UWI) is based on two foundational documents: the founding charter of 1948 (the Royal Charter) and the "Grand Anse Declaration" of 1989. The *"Grand Anse Declaration and Work Programme for the Advancement of the Integration Movement"* was promulgated by the Caribbean Community (CARICOM) on July 31st 1989 at the 10th CARICOM Heads of Government Conference at Grande Anse, Grenada (CARICOM, 1989). The "Grand Anse Declaration" essentially sets out a perpetual covenant between the University and the region it serves. Annex 1 of the declaration enshrines a leading role for the University in developing the human resource potential of the Caribbean, with a corresponding commitment by the donor governments to support the institution indefinitely. The Annex concludes with the statement: "DECIDED that, in view of the major role which the University of the West Indies is being called upon to play, it should remain a regional institution indefinitely."

The Grand Anse Declaration therefore consolidated the regional character of the University by defining its role in the CARICOM integration movement while giving permanence to its relationship with member governments. UWI therefore belongs to a select group of just two truly regional universities in the world that have multiple campuses hosted by sovereign independent countries: the University of the West Indies founded in 1948 and the University of the South Pacific (USP) founded in 1968. Each has a federal funding structure, funded by twelve Pacific island states in the case of USP and seventeen Caribbean states and territories in the case of UWI (Fig. 4.1). It is notable that both UWI and USP share an explicit mission to be active partners in the social, economic and political development of their respective regions (Thaman, 2007; UWI, 2017).

A dissenting viewpoint is sometimes encountered that the administrative structure of the UWI is too complex or unwieldy due to the multiple layers of reporting by the many campuses, to their host country on the one hand and to the federated whole on the other. This dissenting narrative was notably absent during the unprecedented global pandemic of COVID-19: in fact, the pandemic of 2020 demonstrated as never before the value of a truly regional university capable of assisting member countries in mounting a coherent regional response at this time of health crisis. But before we can consider how UWI organized itself to interface with the CARICOM institutions in response to coronavirus in 2020 we must go back to 2016 when another virus was present: Zika.

Zika: Practice Makes Perfect

Zika was the first public health emergency that UWI had confronted with a Task Force of the kind it has deployed since 2015 in response to natural disasters (e.g. tropical storm Erika in 2015, hurricanes Irma and Maria in 2017 and hurricane Dorian in 2019). In weather-related disasters, the team might include structural engineers, water quality experts, psychological counsellors, and development economists for example (UWI, 2015). The array of expertise that UWI can mobilise is quite awesome because of the wingspan of the university encompassing a full range of

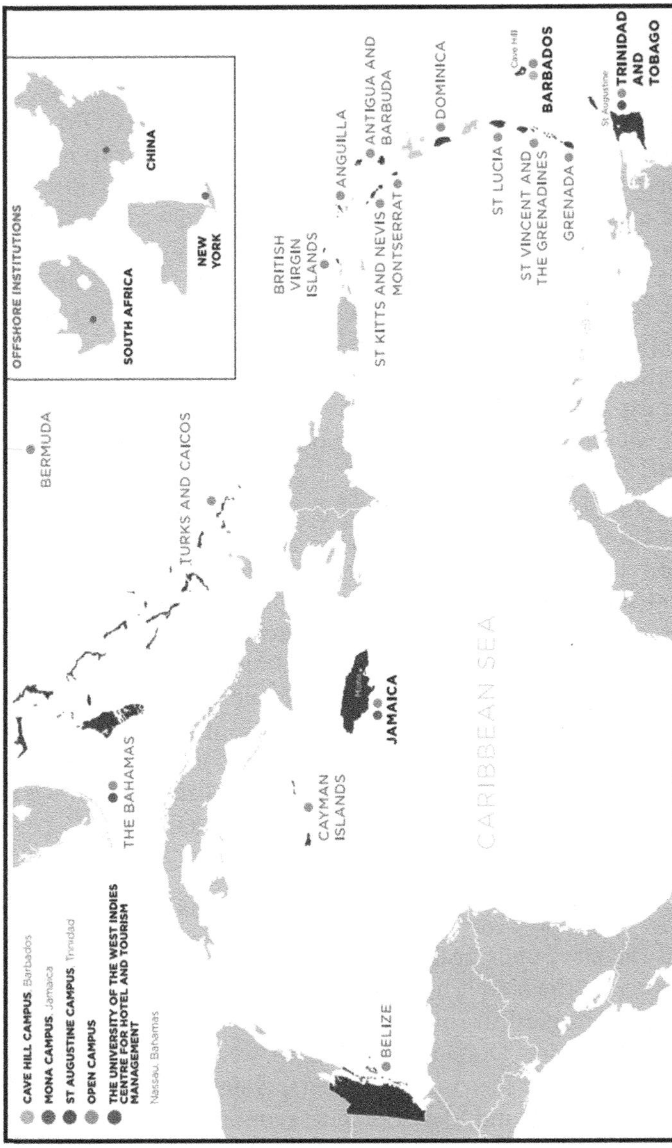

Fig. 4.1 The University of the West Indies

Faculties across the region. To face the Zika epidemic, the UWI Vice Chancellor announced formation of a Task Force within 1 week of the Health Emergency of International Concern (HIEC) declared on February 1st 2016. The Task Force established four founding principles that would later underpin an equivalent Task Force to COVID-19: first, the response must be immediate, drawing on existing talents and capacity within UWI (i.e. not waiting for resource mobilisation); second, UWI will respond as one regional institution, with equal representation on the Task Force drawn from each of the Campuses plus 'UWI Centre'; third, the response must be collaborative and regional, cooperating closely with the public health agencies that operate in the region, especially the CARICOM Health Desk and the Caribbean Public Health Agency (CARPHA), and fourth, the university will set out deliberately to be an authoritative source of accurate information on the pandemic.

Objective and trustworthy communication is key when confronting a viral pandemic. The work of the Zika Task Force was therefore underpinned by a website containing a rich store of information designed to serve policy makers, healthcare professionals and the public (www.uwi.edu/zika). Although Zika was not accompanied by quite the same scale of misinformation and conspiracy theories as COVID-19 would be, UWI nonetheless recognised that accurate information was essential in the public health response, to which end a new information vehicle, UWI*tv*, was piloted. The kick-off Zika Symposium organised by the Task Force in March 2016 under the theme *"One Health, One Environment"*, was live-streamed by a media team from the Cave Hill Campus that would go on to form the nucleus of UWI*tv*. The art of the live-stream, webinars and video vignettes, honed during the work of the Zika Task Force, has since become a byword for UWI*tv*'s media output during COVID-19.

The collaborative model pioneered with the Zika Task Force seeks to leverage the research expertise of the UWI as a public good during a health crisis. Membership in the Zika Task Force was designed to tackle the epidemic head-on, comprising experts in epidemiology, entomology, pediatric medicine, child health, neurology, critical care medicine, environmental resource management, communication and continuing education (Landis, 2019). Research out of the Task Force demonstrated how this pernicious neurotropic virus was targeting the most vulnerable human life in Caribbean communities: the unborn child. In Jamaica, the largest country in the English-speaking Caribbean for which reliable birth cohort data was available, research showed that 50 babies were born with microcephaly (very small head circumference) in the year 2016–17, well above the expected annual caseload of 1–2 per year (James-Powell et al., 2017; Samms-Vaughn et al., 2006).

The legacy of the UWI Zika Task Force has extended beyond the time and space of the Zika epidemic. International research collaborations have been especially strong in the field of child health through a "ZikAction" consortium (Christie, 2016) and through genomic collaborations investigating the mosquito virome. A collaboration between the UWI Cave Hill Campus and the Medical University of Vienna yielded a diverse array of viruses harboured within field-trapped mosquitoes in the Caribbean and Europe, including a hitherto unknown virus discovered in Barbados: MaV-BRB (Mosquito-associated Virus – Barbados) (Thannesberger et al., 2020). The UWI Mona Campus also contributed sequence analysis of Zika strains in Jamaica towards a global genomic database, to help monitor the evolution of viruses as they move from country to country (Metzky et al., 2017). A legacy of the Zika Task Force therefore has been to embed UWI into global biosecurity networks to foster early detection, prevention and monitoring of emerging virus outbreaks.

Laboratories as Lynchpin

The ability to test for viruses with the Polymerase Chain Reaction (PCR) test is *sine qua non* when executing an effective public health response to an emerging virus outbreak, as the COVID-19 pandemic has so ruthlessly demonstrated (Schneider, 2020). The quantitative PCR test, known as real-time (rt)- PCR, is a high complexity diagnostic test that must be conducted by a competent laboratory; hence public health laboratories have become the lynchpin of the pandemic response to the novel coronavirus. From the moment that the first COVID-19 case was confirmed in a CARICOM member country on March 10th 2020, the region has performed to world class standards in rt-PCR testing for Severe Acute Respiratory Syndrome Coronavirus-2 (SARS-CoV-2), not only in the fidelity of its testing but also through the rapid turnaround of results that is so critical for contact-tracing and containment.

What was so remarkable about the almost seamless laboratory response to the coronavirus pandemic, is that the Caribbean had never had PCR testing on anything like this scale or precision before. But the success was no fluke; it was earned through decades of painstaking work building laboratory capacity and quality systems piece by piece during the HIV pandemic. And at just the right moment, all the pieces of the puzzle clicked together and the Caribbean was able to meet the COVID-19 pandemic head-on with a federated laboratory network of some dozen national public health laboratories, trained and supplied with the WHO's approved PCR test kit for SARS-CoV-2 and co-ordinated through the CARPHA reference laboratory in Trinidad.

The era of the HIV pandemic saw a dramatic expansion of laboratory capacity in the Caribbean driven by a coalition of home-grown and international partner organisations. The main organisations that funded or supported laboratory expansion in the HIV era included National Health Ministries; regional bodies such as the CARICOM Council for Human and Social Development (COHSOD), CARICOM Regional Organisation for Standards and Quality (CROSQ), The Caribbean Public Health Agency (CARPHA), Organisation of Eastern Caribbean States (OECS), Pan American Health Organization (PAHO), Pan Caribbean Partnership Against HIV, and The UWI; global bodies such as UNAIDS, Global Fund to Fight AIDS, Malaria and Tuberculosis, Centers for Disease Control and Prevention (CDC), and the President's Emergency Plan for AIDS Relief (PEPFAR); development partners such as the European Development Fund and Caribbean Development Bank; professional societies such as Caribbean MedLabs Foundation (CMLF), Caribbean Cytometry & Analytical Society (CCAS), Caribbean HIV/AIDS Regional Training Network (CHART), and the Caribbean Association of Medical Technologists; and, finally, private companies, foundations and philanthropic donors.

The HIV epidemic has been recognised by global development agencies as a vehicle to drive laboratory capacity-building in developing countries, not just for HIV but for other disease agents, such as Mycobacterium tuberculosis, Neisseria gonorrhoea, C. trachomatis, Herpes simplex, Human Papilloma Virus, and Influenza (Abayomi & Landis, 2008; Nkengasong et al., 2010; Peter et al., 2010). The Caribbean was no exception and the CDC was especially active across the region seeking to leverage the HIV epidemic to strengthen molecular diagnostic capacity more generally (Alemnji et al., 2012, 2018). The CDC established a close working relationship with three home grown organizations – CCAS, CMLF and CHART – that over a span of 15 years hosted annual HIV training workshops rotating through the Caribbean. The week-long workshops were organized under the auspices of the Caribbean Cytometry &

Analytical Society (CCAS), a professional society that I founded in 2006 with the aim to improve access to antiretroviral treatment in the Caribbean by removing technological and social barriers to care (Landis, 2007). As a charitable organization (registered Barbados charity # 660) CCAS has raised in excess of US$ 0.75 M to host the Continuous Professional Education (CPE) accredited workshops. The workshops listed in Table 4.1, organized by a cadre of volunteers drawn from the UWI, the Barbados Ministry of Health and the private sector, trained over fourteen hundred professionals in HIV diagnostics and care between 2004 and 2018.

CCAS workshops served as an annual meeting point at which updates on laboratory capacity strengthening could be benchmarked. A

Table 4.1 Cumulative deaths per Million: Caribbean vs. other regions (at Dec. 10 2020)

Country/Region	*Deaths/Million*
United States	882.59*
South America	779.70
European Union	666.47
Belize	460.24
The Bahamas	414.50
Canada	347.89
Dominican Republic	217.37
Montserrat	200.04
Suriname	199.24
Guyana	195.79
Turks and Caicos Islands	154.97
Bermuda	144.52
Jamaica	91.18
Trinidad and Tobago	87.17
British Virgin Islands	33.07
Cayman Islands	31.25
Antigua and Barbuda	30.63
Barbados	24.36
Haiti	20.43
Cuba	12.01
Saint Lucia	10.89
Anguilla	0
Dominica	0
Grenada	0
St. Kitts & Nevis	0
St. Vincent & the Grenadines	0

retrospective analysis of regional data collected at CCAS workshops was able to demonstrate 10 year trends for steadily increasing laboratory capacity in the Caribbean (Landis et al., 2018). These trends are illustrated in the heatmap shown in Fig. 4.2. However, an island archipelago such as the Caribbean presents additional challenges distinctive for Small Island Developing States (SIDS), particularly the human resource constraint and transport and logistics challenges. Put simply, it is often not sustainable to offer high throughput or high complexity molecular tests on every island; the solution therefore rests in the creation of an inter-island laboratory referral network. The need to establish a regional

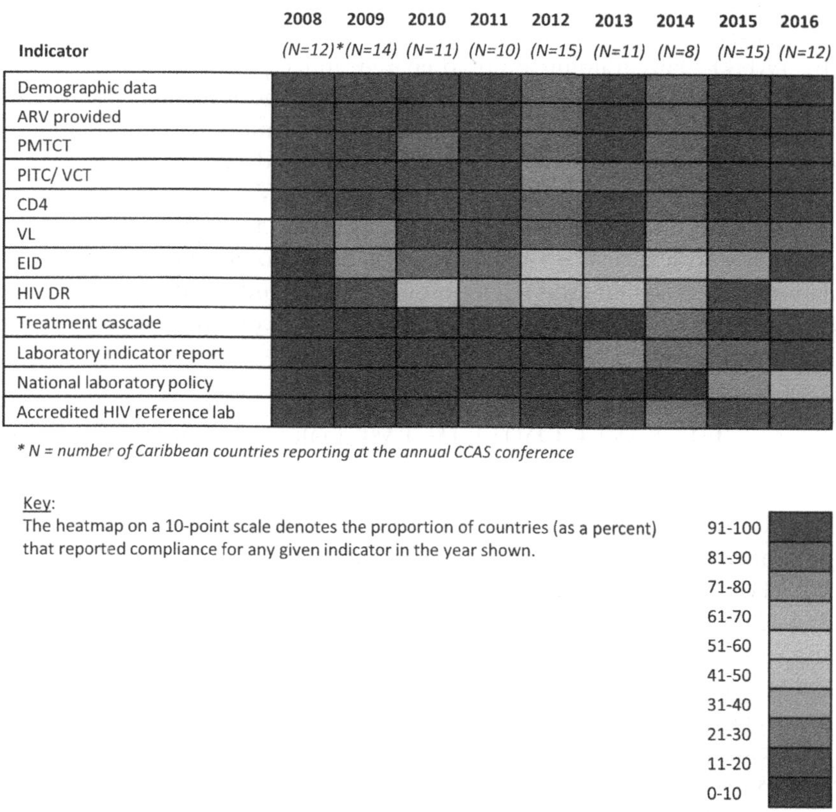

Fig. 4.2 Indicators of Caribbean laboratory and monitoring infrastructure

laboratory network in the Caribbean was recognised in 2011 by the newly created Caribbean Public Health Agency (CARPHA) (Jack et al., 2011). Two declarations on Regional Laboratories arising out of CCAS workshops were presented to- and accepted by- the COHSOD Ministers of Health in 2012 and 2015 respectively. The first sub-regional network to be operationalized on the ground was the HIV Viral Load network between seven OECS member states and the Barbados National HIV Laboratory as hub (Landis et al., 2015). This research published by UWI with the Barbados Ministry of Health proved the principle that a quality assured laboratory network could be established and operated in the Caribbean region to deliver gold standard rt-PCR testing.

The capstone laboratory training event that took place the month before COVID-19 reached the Caribbean was a PCR testing activity in Washington DC in February 2020, co-ordinated by CARPHA and funded by PAHO. This training event prepared and equipped national public health laboratories across the Caribbean with the skills and the WHO approved test kits to meet the COVID-19 pandemic with a gold standard PCR test. The labs have never looked back. Around a dozen regional laboratories are delivering quality assured SARS-CoV-2 rt-PCR test results with turnaround times typically within 24 hours for tests conducted in-country and less than 36 hours for tests flown to another country. The laboratory lynchpin was therefore in place and ready to confront COVID-19 when the first confirmed case was recorded in a CARICOM territory on March 10th 2020.

The UWI COVID-19 Task Force: Informing the Caribbean's Response

The UWI COVID-19 Task Force, established on February 28th 2020 sought to replicate many of the paradigms established for the 2016 Zika Task Force, including the tagline: "Informing the Caribbean's Response". It was already evident at this point that the COVID-19 epidemic was playing out as two parallel epidemics: a viral epidemic and a misinformation epidemic (Landis, 2020a). The UWI COVID-19 Task Force therefore positioned an authoritative website as the fulcrum for its research and public outreach activities (www.uwi.edu/covid19). Four target audiences were served by the website: policy makers, healthcare professionals, journalists, and the public. The UWI COVID-19 Task Force is composed of twenty-one individuals spanning expertise in virology/ epidemiology/

laboratory diagnostics/ critical care/ pulmonology/ veterinary medicine/ tourism/ trade/ psychology/ psychiatry/ ethics/ gender/ development/ international relations/ youth advocacy/ continuing education/ and communication. It includes a UWI student representative and the University Registrar as an *ex officio* member (https://uwi.edu/covid19/taskforce).

The trust and *modus operandi* established in the Zika epidemic allowed the UWI COVID-19 Task Force to hit the ground running, with invitations issued to participate in CARICOM's decision making bodies even before the pandemic had reached the Caribbean, including Meetings of the Conference of Heads of Government, The CARICOM Council for Human and Social Development (COHSOD), the Eastern Caribbean Development Partners Group, and the Caribbean Disaster and Emergency Management Agency (CDEMA). UWI's role in CARICOM meetings coalesced under the pillar of information and research: to offer its academic and research expertise to help construct the evidence base for policy-makers to make informed decisions.

Among the priority initiatives for UWI in early March 2020 was that of conducting an epidemic scenario modelling exercise to estimate the impact of the epidemic on the healthcare systems of the Caribbean under three policy scenarios: Do Nothing, Act Early, Act Late. The modelling work was initiated on a bilateral basis between the George Alleyne Chronic Disease Research Centre (GA-CDRC) and the Barbados Government, and later replicated under the aegis of the UWI COVID-19 Task Force and at the behest of CDEMA for the entire English-speaking Caribbean. The data analytic products of the UWI were disseminated through CDEMA to CARICOM member countries, comparing the projections for COVID-19 patients needing hospitalization with the surge capacity in each country for hospital beds. These epidemic projections assisted policy-makers in estimating their procurement needs and may have influenced the decision of member states to opt for pre-emptive lockdowns, mostly before a single death had been recorded. In the case of Barbados, the decision by the Government to impose Non-pharmaceutical Interventions (NPIs) was specifically attributed to UWI's modelling research (Bradshaw, 2020).

The broadly synchronous actions among Caribbean countries to impose early border closures and movement restrictions helped contain the first wave of the pandemic (Murphy et al., 2020). The GA-CDRC at UWI compared twenty-two Caribbean countries with nine comparator countries chosen to reflect the variable sense of urgency and stringency observed in other parts of the world. These included other island states with small populations. The analysis demonstrated that Caribbean countries on average imposed border closures 23 days earlier than comparator countries, in-country movement restrictions 36 days earlier, and curbs on social gatherings 30 days earlier. These rapid policy actions effectively curtailed the first wave of COVID-19, with most Caribbean countries tracking the most successful Asian countries (notably Singapore and Vietnam) in blunting their epidemic growth curves. Geo-positioning data from smartphones confirmed that lockdowns in the Caribbean were very effective at restricting population movement in-country, thereby interrupting virus transmission chains.

The next priority for UWI was to analyze surveillance data to help Caribbean policymakers monitor the effectiveness of containment strategies and to identify any regional hotspots for cases and deaths. The GA-CDRC again took the lead to generate daily surveillance products for all member countries of CARICOM and the UK Overseas Territories. At the country level, epidemic growth curves were benchmarked against indicator countries such as New Zealand, Singapore and Iceland widely seen to have contained the epidemic. At the regional level, daily analyses provided an overview of the region's effectiveness in containing the epidemic. These data are shared publicly through the UWI COVID-19 Task Force website (www.uwi.edu/covid19/surveillance) following an express commitment to open access data sharing by CARICOM Health Ministers meeting at the COHSOD on April 22 2020.

It is worth mentioning that the GA-CDRC is a Unit of the Caribbean Institute for Health Research, that itself is a member of a distinguished group of research institutes at 'UWI Centre' that were established to serve the region as a whole (such as the UWI Seismic Research Centre, and the Institute for Sustainable Development). By sharing data in the role of a regional 'observatory', the UWI was therefore fulfilling its regional mandate while providing information for policymakers to guide the CARICOM response to COVID-19. The regional surveillance outputs of UWI remained critical as the Caribbean and other countries reopened their economies and borders in mid-2020 to allow tourism to resume.

The phase of the epidemic after border reopenings played out differently across the Caribbean, with some countries maintaining containment but others experiencing outbreaks of COVID-19. However, the overall picture remains one of broad control when compared to the regional scenarios witnessed in the Americas and the tourism source markets in Europe. Table 4.2 shows that COVID-19 death rates for Caribbean countries normalized per population size were typically one log below those reported for the USA and South America. For example, as of December 9th 2020, deaths per million in the USA and South America were 883 and 779 respectively, while in the European Union and Canada they were 667 and 348 respectively. By contrast, the most populous countries in the Caribbean – Cuba, Haiti and Dominican Republic – had death rates per million of 12, 20 and 217 respectively. Five CARICOM countries had recorded Zero deaths, eight CARICOM countries between 0 and 100, and five between 100 and 200 deaths per million. Only Belize and the Bahamas had experienced relatively high death rates of 460 and 414 respectively. Cognisant of the prevalence of mortality risk factors in the Caribbean, such as hypertension, obesity and diabetes, these Caribbean death rates provide evidence for the competent handling of COVID-19 from a public health perspective, especially when compared to regional neighbours in North and South America. But if saving lives is one thing, saving livelihoods has turned out to be an even greater challenge. The economic impact from COVID-19 was extreme due to the region's reliance on tourism as the principal economic driver.

Table 4.2 Professional HIV training workshops delivered by the Caribbean Cytometry & Analytica Society and co-organizing entities 2004–2018

Year	*Country*	*Organiser/ Co-organisers*
2004	Barbados	Barbados MOH
2005	Barbados	Barbados MOH
2006	Barbados	CCAS
2007	Barbados	CCAS
2008	Suriname	CCAS
2009	St. Kitts	CCAS-CHART-CDC
2010	Barbados	CCAS-CHART-CDC
2011	Jamaica	CCAS-CHART-CDC
2012	Aruba	CCAS-CHART-CMLF
2013	Bahamas	CCAS-CHART-CMLF
2014	St. Maarten	CCAS-CHART-CMLF
2015	St. Kitts	CCAS-CMLF
2016	Jamaica	CCAS
2017	Barbados	CCAS
2018	St. Lucia	CCAS

To assist Caribbean governments and decision-makers faced with a mix of medical, economic, social and other dilemmas posed by COVID-19, the ethicist on the UWI COVID-19 Task Force prepared a handbook entitled "*Ethics amidst COVID-19: A brief ethics handbook for Caribbean policymakers and leaders*" (https://www.uwi.edu/covid19/resources/resources-policymakers). The handbook outlined ethical principles of Respect – Inclusion – Responsibility (RIR) that called for alertness to the fact that in times of public health emergency situations human rights violations are often rife, especially of the most vulnerable, including migrants, women, children, the elderly, prisoners, the sick, etc. who are disproportionately affected. It emphasised: R-Respect the dignity of all persons (recognizing the equal inherent value of each person), including the right to life, which is oftentimes dependent on a right to healthcare; I-Inclusion of all, such that everyone is treated with fairness (Equity, Diversity and Inclusion); **R-Responsibility.** Individuals must take responsibility for their behaviour. These principles are as applicable in the Caribbean as any other part of the world, recognizing that different levels of responsibility inhere to different positions of power and practice. The principles of RIR encourage ethical commitment and action among persons who have the responsibility to impose policies that affect the lives of entire nations to discern the most appropriate measures to save lives and secure livelihoods in the time of COVID-19.

Managing a public health emergency shares many similarities with managing a weather-related emergency, from forecasting, to disaster planning, stay at home orders, safety and security, managing supply chain disruption, and above all consistency of public safety messaging. UWI has argued in published commentaries that the expertise and trust gained in managing annual weather-related disasters may partly be responsible for the comparative success of the Caribbean in containing the COVID-19 pandemic (Hambleton et al., 2020; Landis, 2020b). At the individual level, "stay at home" orders may have resonated with Caribbean people who are well used to such public advisories during the passage of tropical storms in the annual hurricane season. However, these factors alone do not adequately explain the regional collaborating mechanisms and coordinating actions that have helped control COVID-19 in a diverse region of low, medium and high-income countries. The theoretical construct of the WHO's Health Systems Framework is proposed to explain how the political, health and non-health entities appear to have combined to yield an effective regional pandemic response to COVID-19 (WHO, 2007).

Recapitulating the WHO's Health Systems Framework in a Regional Pandemic Response

Ordinarily, the WHO's Health Systems Framework (WHO-HSF) is used to strengthen a health system as a whole, not to map a specific disease response (WHO, 2007). However, the culture and history of CARICOM's collaborative response to natural disasters and other economic and health threats led us to consider that the multi-sectoral response to the coronavirus pandemic in fact recapitulated quite nicely the six building blocks of the WHO-HSF. As illustrated in Fig. 4.3, the first three building blocks of the WHO-HSF are within the health sector: Health Workforce, Medical Products and Technology, Service Delivery. The next three are in the non-health sector: Leadership and Governance, Healthcare Finance, Information and Research. In health policy literature, multi -sectoral responses are broadly viewed as actions which may involve the collaborative working of non-health sectors to promote the health of populations in the context of sustainable human development (Tangcharoensathien et al., 2017).

Strengthening of a health sector framework logically starts with Leadership and Governance. In the case of COVID-19, the time lag between the first phase of the epidemic in Asia in January and the second phase in Europe and North America in February gave the leadership of CARICOM a valuable window of time to assess or re-assess the strength of health systems and preparedness to the impending pandemic. Under the Prime Minister of Barbados, Mia Amore Mottley as Chair of CARICOM, the region sought to strengthen its health and non-health sectors to mount a multi-sectoral defence against COVID-19. Other functional units within the CARICOM organizational structure are also contained under Leadership and Governance, notably the CARICOM Council for Social and Human Development (COHSOD) which provides guidance and authority to the health sector agencies in the first three blocks of the WHO-HSF.

The lead role in the pandemic response mapping onto the first three health sector blocks fell to CARPHA. CARPHA was the regional public health agency charged with coordinating the regional health response with an intergovernmental mandate from CARICOM and guided by recommendations from the COHSOD – Health working group. In the COVID-19 response, CARPHA was responsible for: Assessment and identification of disease surveillance needs; Generation of health policy

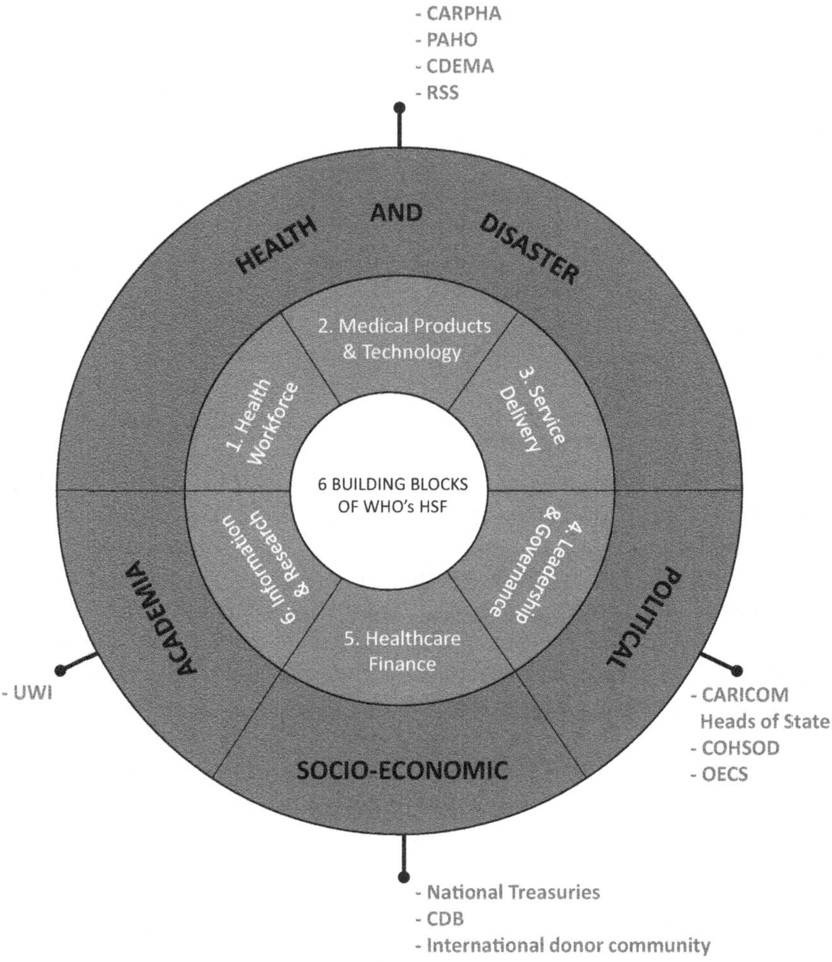

Abbreviations: The Caribbean Public Health Agency (CARPHA), The Pan American Health Organization (PAHO), Caribbean Disaster and Emergency Management Agency (CDEMA). Non-Health Sector Entities: Caribbean Community (CARICOM), Council for Social and Human Development (COHSOD), Caribbean Development Bank (CDB), The Organization of Eastern Caribbean States (OECS), Regional Security Service (RSS), The University of the West Indies (UWI).

Fig. 4.3 CARICOM's pandemic response mapped onto the WHO's Health Systems Framework

guidance to governments on International Health Regulation (IHR) protocols for ports of entry, guidance on the implementation of Non-pharmaceutical Interventions for the containment or mitigation of COVID-19 (including physical distancing, community containment, respiratory hygiene and the use of facial barrier equipment by the public) and co-ordination of laboratory testing. Other entities with major roles in these health sector blocks were PAHO, CDEMA and the Regional Security Service (RSS).

PAHO took the lead in: Procurement of COVID-19 test kits; Funding of training of CARICOM laboratories on WHO approved COVID-19 PCR kits; and Assessment of likely resource needs (human and physical plant infrastructure, such as estimation of Intensive Care Unit beds, ventilator capacity and Personal Protective Equipment regarding hospital readiness). UWI also shared their own modelling estimates with PAHO. The Caribbean Disaster and Emergency Management Agency (CDEMA) played the lead role for: Co-ordination of national, regional and international entities through the Regional Response Mechanism (RRM); Activation and coordination of the regional emergency response plan, including facilitation of regional surveillance and establishment of an Integrated Regional Logistics Mechanism comprising a hub with air and sea bridge; Dissemination of information including COVID-19 research products to member states; and Co-ordinating emergency airlift capacity with the RSS for transport of medical supplies and patient samples between islands while commercial airlift was suspended. The role of the RSS was crucial: flying up and down the island chain to maintain the flow of patient samples from smaller islands to the CARPHA reference laboratory in Trinidad and maintaining inter-island supplies of medical products and equipment at a time of major transport disruptions.

Healthcare finance was delivered through a grand coalition of grants and loans from the regional and international donor community and development banks, notably the Caribbean Development Bank to supplement the national treasuries of CARICOM member states. Although the WHO-HSF does not explicitly prioritize or weight the value of an individual block, from the CARICOM perspective of resource limitation, access to research and evidence to inform the COVID-19 response was critical to the development of protocols and policies. UWI was therefore invited by CARICOM to provide its research expertise – mapping onto the sixth building block 'Information and Research' – to help synthesize the evidence base upon which informed decisions could be taken. The

UWI used open access data to contextualise the evolving outbreak in the CARICOM with modelling studies and daily surveillance outputs, supplemented with peer reviewed publications. In this way the WHO- HSF system was effectively modified to suit a specific disease response, and the coordinated multi-sectoral actions of CARICOM that broadly map onto the six building blocks of the WHO-HSF are offered as a theoretical construct to understand the coherent and broadly successful response of the region to the COVID-19 pandemic. Indirectly that has highlighted the ability of a regional university to serve the needs of a complex archipelago of small island developing states by deploying its research expertise as a regional public good to assist CARICOM in developing a coherent response to COVID-19. The vision of those who framed the Grand Anse declaration 3 decades earlier had been vindicated.

References

Abayomi, E. A., & Landis, R. C. (2008). Flow Cytometry as the Spearhead for Delivering Sustainable and Versatile Laboratory Services to HIV-Burdened Health Care Systems of the Developing World: A Caribbean Model. *Cytometry B Clinical Cytometry, 74*(1), S80–S89.

Alemnji, G., Chase, M., Branch, S., Guevara, G., et al. (2018). Improving Laboratory Efficiency in the Caribbean to Attain the World Health Organization HIV Treat all Recommendations. *AIDS Research and Human Retroviruses, 34*(2), 132–139.

Alemnji, G. A., Branch, S., Best, A., Kalou, M., et al. (2012). Strengthening National Laboratory Health Systems in the Caribbean Region. *Global Public Health, 7*(6), 648–660.

Bradshaw, S. (2020, April 1). *COVID-19 Update Featuring Acting Prime Minister Bradshaw*. Prime Minister's Office. https://pmo.gov.bb/2020/04/01/covid-19-update-april-1-feat-acting-prime-minister-bradshaw/. Accessed 9 Sept 2020.

CARICOM. (1989). *Grand Anse Declaration and Work Programme for the Advancement of the Integration Movement*. CARICOM Secretariat, Turkeyen Georgetown. https://caricom.org/grand-anse-declaration-and-work-programme-for-the-advancement-of-the-integration-movement-july-1989-grand-anse-grenada/. Accessed 9 Sept 2020.

CARICOM – Who we are. (2020). Available from: https://caricom.org/our-community/who-we-are/. Accessed 9 Sept 2020.

Christie, C. D. C. (2016). Unravelling the Paediatric and Perinatal Zika Virus Epidemic Through Population-Based Research. *West Indian Medical Journal, 65*(1), 239–242.

Eurasia Group. (2020, November 25). *Ending the Covid-19 Pandemic: The Need for a Global Approach*. Eurasia Group. https://www.who.int/docs/default-source/coronaviruse/act-accelerator/2020-summary-analysis-of-ten-donor-countries-11_26_2020-v2.pdf?sfvrsn=6d6f630c_5&download=true. Accessed 10 Dec 2020.

Hambleton, I. R., Jeyaseelan, S., & Murphy, M. M. (2020). COVID-19 in the Caribbean Small Island Developing States: Lessons Learnt from Extreme Weather Events. *Lancet Global Health, 8*(9), e1114–e1115.

Jack, N., Gabastou, J. M., Kitson-Piggott, W., Landis, R. C., et al. (2011, November). Development of a Caribbean Laboratory Network to Support HIV/AIDS Programmes. Presented at *2011 Caribbean HIV Conference: Strengthening Evidence to Achieve Sustainable Action*. http://hivgateway.com/entry/96c3e917b490eb8d79dbfe6cfc55bc82/. Accessed 9 Sept 2020.

James-Powell, T., Brown, Y., Christie, C. D. C., Melbourne-Chambers, R., et al. (2017). Trends of Microcephaly and Severe Arthrogryposis in Three Urban Hospitals Following the Zika, Chikungunya and Dengue Fever Epidemics of 2016 in Jamaica. *West Indian Medical Journal, 66*(1), 11–19.

Landis, R. (2020a, January 28). The UWI Collaborative Model for Addressing a Regional Health Crisis. Presented in UWI*tv* webinar: *"Demystifying the Coronavirus (2019-NCOV)"*, The University of the West Indies. https://www.facebook.com/watch/live/?v=542819836322998&ref=watch_permalink. Accessed 9 Sept 2020.

Landis, R. (2020b, May 17). BBC World News Report on the Caribbean's COVID-19 Response Featuring Professor Landis. https://uwitv.org/intv/bbc-world-news-report-on-the-caribbeans-covid19-response. Accessed 9 Sept 2020.

Landis, R. C. (2007). The Caribbean: Riding the Dark Horse of HIV/AIDS Towards a Brighter Future. *Cytometry B Clinical Cytometry, 72*(2), 153–155.

Landis, R. C. (2019). Climate Change and Biosecurity – The Zika Lessons. *Journal of Eastern Caribbean Studies, 44*(1), 179–187.

Landis, R. C., Abayomi, E. A., Bain, B. C., Greene, E., et al. (2018). Shifting the HIV Paradigm from Care to Cure: Proceedings from the Caribbean Expert Summit in Barbados, August 2017. *AIDS Research and Human Retroviruses, 34*(7), 561–569.

Landis, R. C., Carmichael-Simmons, K., Hambleton, I. R., & Best, A. (2015). HIV Viral Load Trends in Six Eastern Caribbean Countries Utilizing a Regional Laboratory Referral Service: Implications for Treatment as Prevention. *PLoS One, 10*, e0125435.

Leonhardt, D., & Leatherby, L. (2020, June 10). Where the Virus Is Growing Most: Countries With 'Illiberal Populist' Leaders. *The New York Times*. https://www.nytimes.com/2020/06/02/briefing/coronavirus-populist-leaders.html. Accessed 9 Sept 2020.

Metzky, H., Matranga, C., Wohl, S., Schaffner, S., et al. (2017). Zika virus Evolution and Spread in the Americas. *Nature, 546*(7658), 411–415.

Murphy, M. M., Jeyaseelan, S. M., Howitt, C., Greaves, N., et al. (2020). COVID-19 Containment in the Caribbean: The Experience of Small Island Developing States. *Research in Globalization, 2*, 100019

Nkengasong, J. N., Nsubuga, P., Nwanyanwu, O., Gershy-Damet, G.-M., et al. (2010). Laboratory Systems and Services are Critical in Global Health: Time to End the Neglect? *American Journal of Clinical Pathology, 134*(3), 368–373.

Peter, T. F., Rotzm, P. D., Blair, D. H., Khine, A. A., et al. (2010). Impact of Laboratory Accreditation on Patient Care and the Health System. *American Journal of Clinical Pathology, 134*(4), 550–555.

Samms-Vaughan, M., Thame, M., Osmond, C., Hambleton, I. R., et al. (2006). Growth Curves for Normal Jamaican Neonates. *West Indian Medical Journal, 55*(6), 368–374.

Schneider, E. C. (2020). Failing the Test – The Tragic Data Gap Undermining the U.S. Pandemic Response. *New England Journal of Medicine, 383*(4), 299–302.

Tangcharoensathien, V., Srisookwatana, O., Pinprateep, P., & Posayanonda, T. (2017). Multisectoral Actions for Health: Challenges and Opportunities in Complex Policy Environments. *International Journal of Health Policy and Management, 6*(7), 359–363.

Thaman, K. H. (2007). The Role of Higher Education in Regional Development in Pacific Island Countries with Specific Reference to the University of the South Pacific, presented at the Regional Seminar *"Competition, Cooperation and Change in the Academic Profession: Shaping Higher Education's Contribution to Knowledge and Research"*, Hangzhou, China. UNESCO FORUM on Higher Education, Research and Knowledge. https://unesdoc.unesco.org/ark:/48223/pf0000157880. Accessed 9 Sept 2020.

Thannesberger, J., Rascovan, N., Eisenmann, A., Klymiuk, I., et al. (2020). Highly Sensitive Virome Characterization of *Aedes aegypti* and *Culex pipiens* Complex from Central Europe and the Caribbean Reveals Potential for Interspecies Viral Transmission. *Pathogens, 9*(9), E686.

UN News. (2020, August 6). *Global Cooperation Is Our Only Choice Against COVID-19, Says WHO Chief*. https://news.un.org/en/story/2020/08/1069702

United Nations. (2020, December 3). 31st Special Session of the General Assembly. "With More Than 1.5 Million Lives Lost to COVID-19, World Leaders in General Assembly Demand Urgent Action to Guarantee Equitable Distribution of Life-Saving Vaccines". https://www.un.org/press/en/2020/ga12293.doc.htm. Accessed 10 Dec 2020.

UWI. (2015, September 2). Press Release, The University of the West Indies: *UWI Sends Task Force to Assist Dominica After Tropical Storm Erika*, UWI St. Augustine Campus. https://sta.uwi.edu/news/releases/release.asp?id=1468. Accessed 9 Sept 2020.

UWI. (2017). The UWI Triple A Strategy 2017–2022: Revitalizing Caribbean Development. The University of the West Indies. https://www.uwi.edu/uop/sites/uop/files/Full%20plan.pdf. Accessed 9 Sept 2020.

Williams, C. R., Kestenbaum, J. G., & Meier, B. M. (2020). Populist Nationalism Threatens Health and Human Rights in the COVID-19 Response. *American Journal of Public Health, 110*(12), 1766–1768.

Wilson, K., Halabi, S., & Gostin, L. O. (2020). The International Health Regulations (2005), the Threat of Populism and the COVID-19 Pandemic. *Globalization and Health, 16*, 70.

World Health Organization. (2007). *Strengthening Health Systems to Improve Outcomes: WHO's Framework for Action.* https://www.who.int/healthsystems/strategy/everybodys_business.pdf?ua=1. Accessed 9 Sept 2020.

CHAPTER 5

Caribbean Economies and COVID-19: Impact and Prospects for 2021 and Beyond

Christine Clarke

Introduction

For perhaps the first time since the coining of the term small-island developing states (SIDS), there was a moment of hope that these fifty-two countries scattered around the globe would enjoy a positive outcome in the face of the most daunting crisis ever to sweep the globe. The very insularity and location that had previously been viewed as a disadvantage to the economic resilience and vitality of these nations, now offered benefits as the relative effectiveness of closing island borders became a positive, helping the SIDS to delay the advent of COVID-19 to their shores. For these countries, in which the health sector was not as strong as desired, these delays supported the readiness efforts inclusive of support in kind through the international development partners (IDPs). Despite this, it is important to note that in spite of the successful attempts at categorising

C. Clarke (✉)
University of the West Indies, Mona, Kingston, Jamaica
e-mail: christine.clarke03@uwimona.edu.jm

Y. Campbell, J. Connell (eds.), *COVID in the Islands: A comparative perspective on the Caribbean and the Pacific*,
https://doi.org/10.1007/978-981-16-5285-1_5

and studying SIDS, there are many nuances that impact and determine the experiences and outcome of any crisis on this widely varying group of countries.

The Caribbean is one of three major regional groupings of SIDS and home to twenty-two of the SIDS named by the United Nations. As a region, the Caribbean is home to diverse countries and populations on the basis of not only language and ethnicity, but also the structure of the various economies in terms of sectoral concentration such as tourism, natural resource exploitation, and manufacturing and this in turn impacts the flow of tax and other revenue flows to the governments; the degree of outmigration associated with the so-called brain drain; and the role of remittances as a major coping strategy for the population and source of foreign exchange for the central banks of the region. This chapter reviews the significance of the COVID-19 shock for the regional economies and reviews the currently assessed and potential future economic impacts to provides insights into where the headwinds at the tail-end of 2020 were likely to cause Caribbean economies to drift to in 2021.

Economic Context of the Caribbean

The International Monetary Fund (2013) indicates that Caribbean economies can be grouped into three categories: the micro-states of the Eastern Caribbean Currency Union, commodity exporters, and, service based – tourism – economies. Like other SIDS, Caribbean economies are small, highly dependent on international trade flows for domestic consumption, intermediate goods, and, end-markets for domestic production, and susceptible to economic shocks. They are also disproportionately exposed to disasters, globally. According to the United Nations Economic Commission for Latin America and the Caribbean (UNECLAC), the growth rate in the Caribbean is just under one-fifth that of other SIDS, registering at 0.8 percent in contrast to 4.7 percent for other SIDS for the period 2000–2017. Figure 5.1 illustrates the GDP growth trajectory of selected Caribbean countries between 2000 and 2018 (World Bank Group, 2020). Overall, trends in each country reflected declines in the rate of growth, with the exception of Guyana due to positive commodity prices and large gold discoveries, which outstripped the decline in traditional exports of agricultural products (United Nations Economic Commission on Latin America and the Caribbean, 2017).

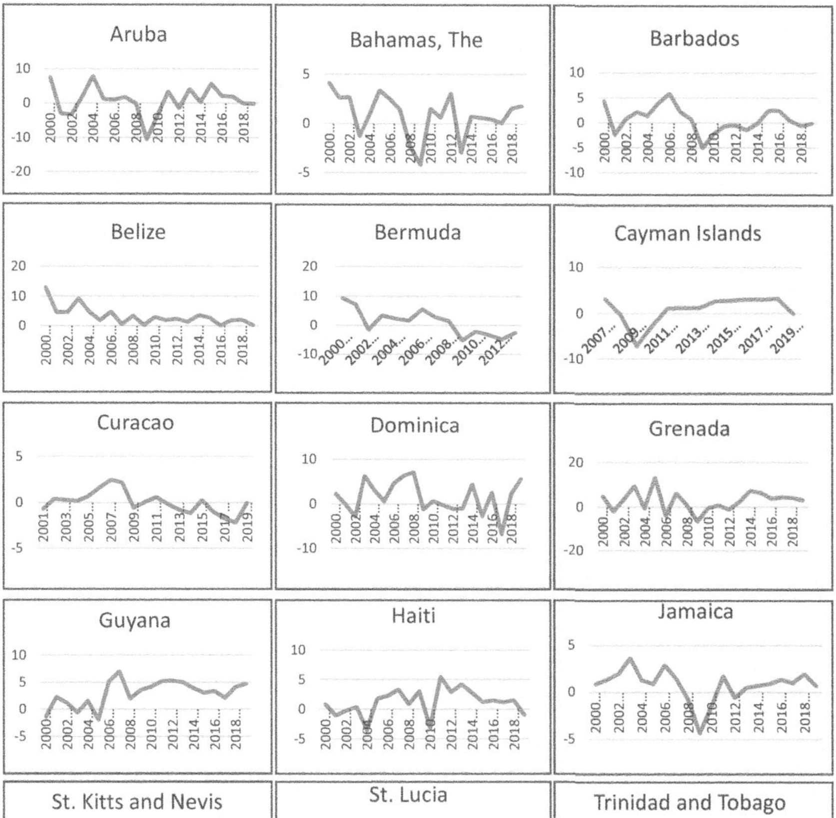

Fig. 5.1 Gross Domestic Product growth rates for selected Caribbean countries: 2000–2018

In real per-capita GDP terms, there were also wide differences within the Caribbean region from a low of USD$1715 in Haiti to a high of USD$70,363 in the Cayman Islands for the twenty-year period 2000 to 2019 (World Bank Group, 2020) (Fig. 5.2). The poor growth performance was associated with low levels of competitiveness in the majority of

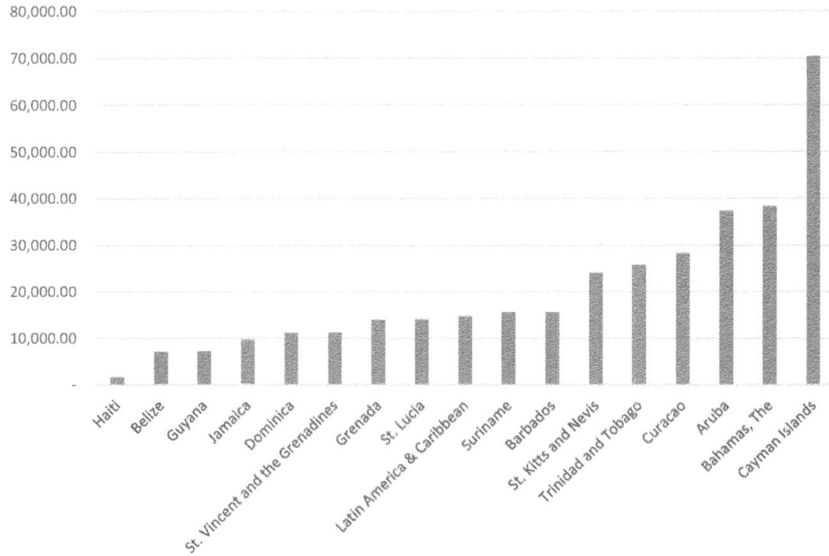

Fig. 5.2 Average Real GDP-per-capita

Caribbean states, catalogued in the successive Global Competitiveness Index (GCI) rankings published annually by the World Economic Forum. The GCI is based on 12 pillars that are derived from the factors in countries that, based on economic growth models, lead to productivity, growth and human development, and these pillars are aggregated into three groupings: enabling environment, human capital and markets. Many of the Caribbean countries typically experience bouts or continued levels of macroeconomic instability due to a combination of fiscal and monetary impulses, often associated with imbalances in the balance of payments. Only four Caribbean countries participated in the 2019 edition, with Barbados, Trinidad, Jamaica and Haiti receiving overall rankings of 77, 79, 80 and 138, respectively, out of a total 141 countries. Barbados received its best ranking in the area of ICT adoption (23rd) followed by their financial system (35th). Jamaica received a rank of 27 for its labour market and 33 in relation to its business dynamism due to the substantive reforms implemented during the period 2010 to present. Despite Jamaica's widely heralded success in implementing three IMF agreements, the first two of which were preceded by debt exchanges, and the agreements themselves seeking to entrench fiscal and debt consolidation as well as fiscal discipline, the country received a rank of 110 in the area of macroeconomic stability.

As a result of the poor growth outcomes driven by low levels of competitiveness, countries in the region have displayed disappointing trade statistics even when compared with other developing countries. UNECLAC (2019) noted that, whilst the exports generated by all least developed countries increased by just over thirteen percent, the increase in the Caribbean was 0.6 percent in the year just prior to the occurrence of the 2008/09 global financial sector crisis, whilst after the crisis the average annual export growth in the region declined to 0.36 percent in comparison to the global annual average of 1.45 percent. This reflects declining market share of the region's goods and services, and, declining market diversification. In turn, these two factors are apparent in the negative current account balances across the region.

The second notable characteristic of Caribbean SIDS is their comparatively high debt-to-GDP levels due to a variety of reasons. More than half of the countries in the region record debt-to-GDP ratios that exceed the sustainable level of sixty percent and the Caribbean was home to four of the twenty-five most highly indebted countries in the world in 2015, namely: Antigua and Barbuda, Barbados, Grenada and Jamaica, with total debt service payments across the region being equivalent to about one-quarter of the total revenue of governments (UNECLAC, 2019). In Jamaica, the ballooning of debt was associated with the resolution process associated with the domestic financial system during the late 1990s as well as changes in the terms and conditions offered to the Jamaican government during the decade of the 2000s (Clarke & Nelson, 2020). In Antigua and Barbuda, the explosion in debt was due to an accumulation of arrears and interest burdens of over thirty percent of government revenues in 2013 (Alleyne, 2010). These relatively high debt service costs compromised the ability of governments to effectively address development challenges and achieve development goals as it significantly reduced the fiscal space available to address the issues. Guyana had been able to arrest and reverse its high debt-to-GDP ratio through the Highly Indebted Poor Country Initiative (HIPC) and the Enhanced Highly Indebted Poor Country Initiative (EHIPC) which facilitated US$256 million in net present value terms to 1998 in debt relief provided largely by multilateral creditors (International Monetary Fund and International Development Association, 2000). Alongside the debt operation, Guyana was supported through Poverty Reduction Strategy Paper (PSRP) and various interim PSRP documents to support the social dimension of the society and the World Bank provided support for various structural reforms to address systemic challenges facing the country (Fig. 5.3).

The debt overhang across the region jeopardised the fiscal accounts of Caribbean economies and further limited the extent to which their governments could respond to the impacts of hazards in any substantive way. Many of these economies had to implement various liquidity management programmes (LMP) during the last decade in order to restructure their debt stock and the associated amortization schedules to restore sustainability in the debt dynamics. This would also redound to the benefit of government expenditure as the LMPs typically enabled governments to have access to more fiscal space from which to support increased infrastructure development and social sector spending under pro-poor, pro-growth mantras. The relatively poor quality and low capacity of national institutions across Caribbean economies has been the root cause of its various economic challenges and is borne out in the indicators of their

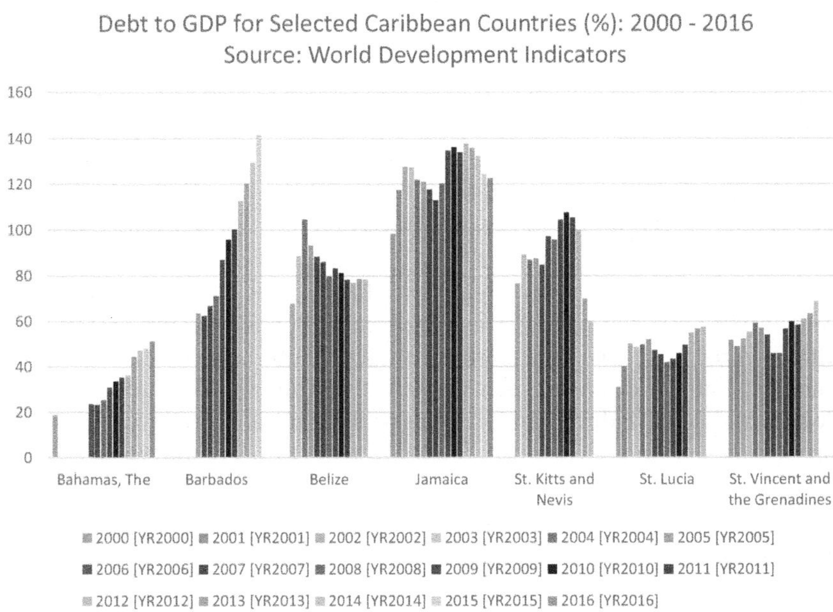

Fig. 5.3 Debt to GDP for selected Caribbean countries

economic performance over time (UNECLAC, 2019). Institutions, particularly those that controlled the various components of public financial management – especially relating to the estimation of the fiscal space and the prioritisation of budget line items – did not support the proper management of the budget process over the fiscal cycle through the necessary adjustments. Support was provided to Jamaica, St. Kitts and Barbados through IMF Standby Agreements; Jamaica in 2010, 2013 and 2016 (the latter on a precautionary basis); St. Kitts in 2011, and, Barbados more recently in 2018. Since then, debt dynamics have become more favourable and the countries were able to make provisions for the much needed capital expenditure which had increased by one percentage point between 2000 and 2015 (UNECLAC, 2019).

Economic Impact of COVID-19 on the Caribbean

Countries in the Caribbean had anticipated a turning of the tide for their economic performance in 2020. Jamaica, for example, had recently successfully completed two consecutive borrowing agreements with the IMF – without the country drawing down on the Special Drawing Rights available under the 2016 precautionary agreement – and was therefore looking to further consolidate the gains made whilst increasing capital expenditure. The question of continuing in a borrowing relationship with the IMF – possibly a precautionary one – was flatly rejected by the government as unnecessary. Barbados would have been well underway with the implementation of their IMF Extended Fund Facility (EFF) at the end of 2019. Along with the other countries in the region, however, generating and sustaining high levels of economic growth remained elusive. The region urgently needed to find the fillip that would support the efforts of the public and private sector to generate high rates of economic growth in a manner that was inclusive and broad-based thereby enabling increasing numbers of the population to enjoy higher standards of living through more secure livelihoods. The exception was Guyana, benefitting from the early boom phase of oil finds and favourable international conditions in the market for gold.

The prospects for the world economy in terms of growth were also relatively upbeat according to the IMF whose economic predictions at the beginning of 2020 indicated an increase in global GDP growth to 3.3 percent over the 2.9 percent recorded in 2019 and a further 3.4 percent for 2021 (International Monetary Fund, 2020c January). The downside risks were global trade trends associated with the trade relations between the United States and China, the outcome of the Brexit related issues and the need to see increased spending by the private sector. There was no mention of the emerging COVID-19 disease, already nascent in terms of its impact on nations much less the global economy at its January 29 publication date, as its forecasts for 2020 were against a backdrop of the downturn in manufacturing globally and the move by some key developed nations to reduce trade with others (Table 5.1).

COVID-19 has had a significant impact on the global economy and, despite the delay in its arrival to Caribbean SIDS, it wreaked havoc on livelihoods through the implementation of necessary prevention and control policies. What would have been thought of as a health crisis, has in fact matured to become the worst economic crisis ever observed globally and the Caribbean region has not been spared its impact. The extent of the impact and the ways in which it was felt in the Caribbean varied depending on the type of economy in terms of the concentration of economic activity in a particular sector. With respect to borders, there were mixed approaches across the region in terms of closures to international travellers, with the Cayman Islands remaining closed the longest – until October 1 – whilst the Bahamas and Barbados remained open throughout the period, even after the detection of their first cases. Clearance via a government website was required in Barbados, Dominica, Jamaica and St. Lucia and closures of entertainment activities and beaches were implemented across the region. Given these restrictions, the impact was greatest in countries and sectors that were service oriented. Service based economies and those with large informal sectors – both of which depend heavily on personal interactions for survival – such as tourism and wholesale and retail activities were most affected. Those economies that engaged in the business process by outsourcing models that were office-centric – were also hard hit during the period when lockdowns were implemented and Work From Home/Stay at Home orders were strictly enforced with the exception of essential workers.

The IMF growth forecast for Caribbean economies tabled in October 2019 was generally positive for the majority of the fifteen countries from the region that were included (International Monetary Fund, 2019). The Bahamas was the only country for which negative growth was forecasted for 2020 at the rate of -0.6 per cent whilst economic growth for Guyana was forecasted at a massive 85.6 percent, because of its booming resource sector. The rest of the region was expected to experience lower to moderate growth from 0.6 percent for Barbados to 4.9 percent for Dominica.

Table 5.1 Summarised COVID-19 Protocols implemented by Caribbean governments with implications for the economy

Country	*Border Closures*	*Work from Home Orders*	*Curfews*	*Beaches &/or other Entertainment Closed*	*Lockdown type activity*	*Entry Procedure/ Requirement*
Antigua & Barbuda	Open to international visitors		National			Assessment and monitoring
Bahamas	None		National		Weekend Lockdowns	
Barbados	None	March 28 – April 14			Restrictions on grocery store operations	Covid test either pre or on arrival Travellers required to complete online immigration/customs form (www.travelform.gov.bb) and submit 24 hours prior to travel
Belize	Closed to international travellers		State of Emergency related curfews		Fourteen day lockdown of Cayo District	
British Virgin Islands	Closed to international travellers		National	Beaches	Movement limited to districts	

(*continued*)

Table 5.1 (continued)

Country	*Border Closures*	*Work from Home Orders*	*Curfews*	*Beaches &/or other Entertainment Closed*	*Lockdown type activity*	*Entry Procedure/ Requirement*
Cayman Islands	Closed through Oct 1				Restrictions on public transportation Ban on private parties	
Dominica	Open to international travel		National curfews State of Emergency			Rapid testing required Clearance required prior to travel Fourteen day quarantine for persons with positive tests
Guyana			National curfews	Closure of bars, restaurants and other places of entertainment between 6 pm and 6 am		
Grenada			National Curfews		Grocery shopping by order of last name	Testing offered
Haiti			National Curfews		Road closures to prevent spread	

Jamaica	Closed: March 13 – May 31 for nationals and June 14 for international travellers.	March 13 – May 31	Ongoing national curfews with longer curfew hours during national holidays	All outdoor recreation closed Temporary reopening of entertainment activities; subsequently closed due to breaches	St. Catherine – in response to the detection of a workplace cluster Limits on size of public gatherings based on physical distancing requirements. Restrictions on public transportation re number of passengers Quarantines in multiple communities	Registration on Government Portal; travel subject to receipt of approval Pre-entry Negative COVID-19 test 2 to 5 days prior to arrival Quarantine on Arrival COVID-19 test on arrival
Saint Martin	Closed to US travellers			All outdoor recreation closed	National	COVID-19 test required 72 hours before departure
St. Kitts & Nevis	Closed to Tourism until October				Restrictions on public transportation re number of seats used	Negative COVID test taken within 72 hours of arrival

(*continued*)

Table 5.1 (continued)

Country	*Border Closures*	*Work from Home Orders*	*Curfews*	*Beaches &/or other Entertainment Closed*	*Lockdown type activity*	*Entry Procedure/ Requirement*
St. Lucia	Open		National curfews – various State of Emergency (April 24 – May 31)		One-week 24-hour shutdown (April 1–7) Private vehicles limited to carry 2 persons	Pre-arrival form to be completed and screening of arriving passengers Testing required for visitors showing symptoms on arrival Test results required within seven days of arrival
Trinidad and Tobago		Stay at home for April		Closure of bars and fast food restaurants		

Source: Various country websites

With the advent of the COVID-19 virus and the associated control policies to reduce the spread of the virus, the original growth projections had to be revised, a gradual process that was dependent on the identification of the approximate length of the crisis, the range of sectors impacted and the extent to which they were able to transition to digital or other alternative operational modalities.

The IMF's revised 2020 forecasts and its projections for 2021 pointedly illustrated the depth of the impact that the crisis was expected to have across the region, even in Guyana where growth prospects were estimated to decline fourfold to 26.2 percent (Table 5.2) (International Monetary Fund, 2020g). The remainder of the countries were expected to register declines in real GDP ranging from 4.0 percent in Haiti to 19.7 in Aruba. The larger islands – Jamaica, Barbados and Trinidad and Tobago, were expecting declines in real GDP of 8.6, 11.6, and 5.6 percent, respectively.

Table 5.2 Original and revised GDP growth forecasts for Caribbean countries

	Original Forecast for 2020	*Revised Forecast for 2020*	*Forecast for 2021*
Antigua and Barbuda	3.3	-17.3	4.7
Aruba	1.0	-19.7	9.0
The Bahamas	-0.6	-14.8	4.6
Barbados	0.6	-11.6	7.4
Belize	2.1	-16.0	8.0
Dominica	4.9	-8.8	3.3
Grenada	2.7	-11.8	3.0
Guyana	85.6	26.2	8.1
Haiti	1.2	-4.0	1.2
Jamaica	1.0	-8.6	3.6
St. Kitts and Nevis	3.5	-18.7	8.0
St. Lucia	3.2	-16.9	7.2
St. Vincent and the Grenadines	2.3	-7.0	3.7
Suriname	2.5	-13.1	1.5
Trinidad and Tobago	1.5	-5.6	2.6

Notes: Original Forecast from IMF (2019b) Revised Forecast for 2020 and Forecast for 2021 from IMF (2020g)

Source: International Monetary Fund (2019) *World Economic Outlook, October 2019 Global Manufacturing Downturn, Rising Trade Barriers* and International Monetary Fund (2020g) *World Economic Outlook, October 2020: A Long and Difficult Ascent*

The smaller islands – Antigua and Barbuda, the Bahamas, Grenada, and, St. Kitts and Nevis – were expected to experience larger declines of 17.3, 14.8, 11.8, and, 18.7 percent, respectively (Table 5.2). Given these large declines, and the anticipated identification of a safe, effective and reliable vaccine that would be distributed globally to arrest the spread of the disease, the IMF released 2021 growth forecasts that were quite robust; in some countries, such as Jamaica, these rates had not been recently observed. All fifteen countries included in the forecast were expected to register positive growth in real GDP above one percent. Aruba was expected to experience the largest growth rate at 9%, followed by Guyana with 8.1 percent, and, Belize and St. Kitts and Nevis expected to growth by 8% each. Barbados and St. Lucia were both expected to grow in the 7% range. Dominica, Grenada, Jamaica, and, St. Vincent and the Grenadines were expected to register growth in real GDP in the 3% range while the remaining countries were expected to growth by between 1 and 3%. It was a considerable turnaround in expectations.

Growth estimates were partly a function of the fiscal and monetary responses of selected Caribbean countries to counteract the impact of the COVID-19 on their respective populations based on the IMF Covid-19 policy response portal (Table 5.3) (International Monetary Fund, 2020f). Tourism dependent countries such as Jamaica, Barbados, Belize and The Bahamas used specific elements to support workers and employers in the sector. Barbados used public funds to reinforce tax relief operations in these countries, including extended deadlines for filing income and consumption tax, the deferral and in some cases reduction of taxes particularly the value-added taxes on disinfectants, cleaning supplies, protective gears and face masks to zero percent; and, pardons on income and value-added tax arrears for households and businesses. In Jamaica, under the COVID-19 Allocation of Resources for Employees (CARE) programme, compassionate grants were offered in addition to temporary cash grants to previously employed persons who were laid off after March 10. Under the CARE programme, self-employed persons also received income support. The public sector National Housing Trust provided support to households through a reduction in the interest rate on mortgages. In Jamaica, formal Parliamentary approval was secured to support the suspension of the country's fiscal rules until the 2021/22 fiscal year (Clarke N., 2020). In response to these fiscal measures, governments tabled emergency or supplementary budgets to account for additional or reallocated expenditure.

Table 5.3 Fiscal, Macro/Monetary and Balance of Payments interventions implemented across the Caribbean

Country	*Fiscal Policy Measure*	*Macro or Monetary Policy Measure*
Antigua and Barbuda	Increased health spending (0.5 percent of GDP); A 20 percent reduction in electricity costs to the public and fuel costs to fishermen for 90 days; One year investment incentives for home renovation and construction; Suspension of the common external tariff on food imports and all new tax measures announced in 2020 budget; and Expansion of social safety net programs. The reduction in the electricity bills, currently at 15 percent, has been extended through October 2020	Moratorium on loan repayment for an initial period up to 6 months, with a possible extension upon review; Waiver of late fees and charges to eligible customers during this period; and Targeted supervisory flexibility Increased credit line limits for governments (by reducing those for banks), Reduced discount rate from 6.5 percent to 2 percent
The Bahamas (As at September 23, 2020)	The government announced various support measures totalling B$121.7 million (1 percent of GDP), including B$15 million for health care, B$4 million for food programs B$15.9 million as income support for the self-employed, B$20 million to support business loans to SMEs with an additional B$5 million allocated to grants to assist with payroll expenses, B$60 million to provide tax deferrals and credits to companies with a minimum of 25 employees and annual sales of B$3 million that retain at least 80 percent of staff, and B$1.8 million to support to Family Islands (specifically to be used for any COVID-19 related expenditure).	A 3-month deferral against repayments on credit facilities for businesses and households that were negatively impacted by the pandemic Forbearance for borrowers in good standing before the onset of the pandemic **Exchange Rate/BOP** Suspension of exchange control approvals for domestic bank dividends The ceiling on the Bahamian open position on foreign exchange transactions has been relaxed to the maximum of 5 percent of Tier11 capital, Suspension of approval of applications to purchase foreign currency for transactions via the Investment Currency Market (ICM) and the Bahamas Depositary/Depository Receipt (BDR) program. Both programs fund external portfolio investments The CBOB has requested the National Insurance Board to repatriate some of its external assets, excluding any exposures to Bahamas and Caribbean domestic issuers

(*continued*)

Table 5.3 (continued)

Country	*Fiscal Policy Measure*	*Macro or Monetary Policy Measure*
Barbados (As at September 24, 2020)	Resources were provided to refurbish the hospital and clinics, build isolation centers, and provision critical medications and supplies Priority capital spending was increased and introduce social programs for displaced workers to mitigate the effects of COVID-19 on the economy including a ***Household Survival Program*** that provides minimum income for households unemployed due to COVID-19 and supplemental unemployment benefits though the National Insurance Scheme Income tax and VAT arrears will be pardoned to provide an infusion of liquidity to households and businesses Target for primary surplus in 2020 is 1 percent of GDP Tourism Loan facility implemented to provide urgent working capital and investment loans to hotels. There was also a deferral of the public wage-savings scheme (BOSS) to increase growth while the tourism sector recovers from the COVID shock A 12-month jobs program A tourism-sector stimulus and transformation package (available for up to two years)	Support for commercial banks and other deposit-taking include: Reduction in the: 1. Bank's discount rate from 7 percent to 2 percent; 2. securities ratio for from 17.5 percent to 5 percent 3. the 1.5 percent securities ratio for non-bank deposit taking licensees to 0; and, In addition the central bank affirmed support to entities in the system if needed

Belize	Fiscal stimulus amounting to BZ$25 million (about 1 percent of GDP) in 2020 to provide short term relief to employees affected by the crisis, especially those in the tourism sector So far, more than 40,000 applications for unemployment relief have been approved. The government has also introduced a bill to parliament that seeks to increase the maturity of treasury notes by an additional ten years and freeze the annual salary increment of public sector employees in 2020	Prudential measures implemented: Reduction of the statutory cash reserve requirements; Extended the time period to classify targeted non-performing loans in sectors such as restaurants, transportation and distribution companies, and other affected areas, from 3 months to 6 months; Encouraging domestic banks and credit unions to provide grace periods for servicing interest and/or principal of commercial and ancillary loans, as needed and where commercially viable; Reduced risk-weights for banks on loans in the tourism sector from 100 percent to 50 percent; and Reviews of the business continuity and cybersecurity plans of financial institutions' to ensure that an adequate level of financial services will be available to the public

(*continued*)

Table 5.3 (continued)

Country	*Fiscal Policy Measure*	*Macro or Monetary Policy Measure*
Dominica	Extension of the deadline for filing of personal and corporate income tax returns; Extension of three months for payment of corporate income tax; Waiving penalties for businesses that enter into payment plans within 6 months of the new payment deadline; Reduced corporate income tax rate (from 25 percent to 17 percent) to companies which commit to continue to employ at least 80 percent of their staffing as of January 1, 2020, for a period of 12 months Reduced import duty and the value-added tax charged on disinfectants, cleaning supplies, protective gears and face masks to zero percent in the; Increased funding to the Ministries of Health and Agriculture; Cash grants to approximately 2500 individual crop farmers, based on the size of the farmers holding; Implementation of multiple infrastructure projects with expenditure of up to US$100 million, with total additional investments in construction expected to amount to at least US$296.8 million; Pay to small contractors and merchants with amounts owed by the Government of EC$100,000 and less, utilizing the resources approved by the IMF under the Rapid Credit Facility (RCF); and Income support for the period April–June 2020 for heads of families and single persons who are currently unemployed	Moratorium on loan repayment for an initial period up to 6 months, with a possible extension upon review; Waiver of late fees and charges to eligible customers during this period; and Targeted supervisory flexibility Increased credit line limits for governments (by reducing those for banks), Reduced discount rate from 6.5 percent to 2 percent

Grenada	Measures include: Payroll support to the affected sectors (such as tourism) and individuals, Expansion of government employment programs; Credit support to small businesses; Increased health care spending, Reduced or deferred payment of some taxes	Moratorium on loan repayment for an initial period up to 6 months, with a possible extension upon review; Waiver of late fees and charges to eligible customers during this period; and Targeted supervisory flexibility Increased credit line limits for governments (by reducing those for banks), Reduced discount rate from 6.5 percent to 2 percent
Guyana	VAT waivers of VAT Waived duties on COVID-19 medical supplies and lab testing kits Tax deductions for all donations made by local businesses to staff and health institutions for the treatment of the virus VAT removed effective from: water and electricity April 01, 2020 to September 30, 2020; domestic air travel effective from April 08, 2020 to September 30, 2020; Extended deadlines for the filing of tax returns from April 30 to September 30, 2020 Expedited processing of VAT refunds for businesses and pay as you earn refunds to employees Relief grants provided to small businesses experiencing challenges to sustain operations and retain employees and for training and development Stimulus grants to assist farmers affected by the pandemic Resumption of the public assistance program - vouchers and packaged hampers for the coastland and the hinterland – after 2 months to facilitate a long-term strategic approach to Covid relief efforts Passage of an emergency budget, which includes funds for combating COVID-19, and revitalizing productive and infrastructure sectors, the health and education sectors account for approximately fifteen percent of GDP each	The Bank of Guyana (BoG) has extended the moratorium to allow banks to defer repayments and to classify affected accounts as non-performing, and the waiver on the regulatory treatment or condition for renegotiating loans under the supervisory guidelines to December 2020 Reserve requirements, liquid asset requirements for demand deposits, and, savings and time deposits have been lowered from 12 percent to 10 percent; 25 percent to 20 percent, and from 20 percent to 15 percent, respectively The commercial banks will provide short term financing for working capital at concessional rates of 5–6 percent, and reduce interest rates on consumer loans below G$10 million by 1–2 percent until December 2020. Other measures proposed to banks by the BoG include; the deferment of loan payments to assist customers in good standing, companies with liquidity requirements; and waiving or reducing fees and penalties for transactions with ATMs, POS, EFT, debit cards, loan processing, late payments on loans and special treatment on interest accrued during the moratorium period on outstanding balances below G$10 million The Bank of Guyana maintains an accommodative monetary stance

(*continued*)

Table 5.3 (continued)

Country	*Fiscal Policy Measure*	*Macro or Monetary Policy Measure*
Haiti	Boosting of some social programs Additional health care and security spending, and transfers to support workers and households, including supporting wage payments temporarily in some sectors are being considered	The central bank moved immediately to ease conditions in the financial system, including reducing the refinance and reference rates, lowering reserve requirements on domestic currency deposits, easing loan repayment obligations for three months, and suspending fees in the interbank payment system
Jamaica (As at July 14, 2020)	Taxes have been cut to approximately 0.6% of GDP, along with targeted measures for up to 0.5% of GDP to impede the effects of COVID19 A COVID19 Allocation of Resources for Employees (CARE) program was launched. The programme includes: 1. A Compassionate Grant to those who were unemployed or informally employed before the pandemic; 2. Temporary Cash Transfer to individuals who were previously employed who have been laid off or terminated since the pandemic by verifying unemployment since March 10, 2020; and 3. Grants to the self-employed persons whose regular earnings have been disrupted Fiscal Rules were suspended until the 2021/22 fiscal year A Supplementary Budget for Fiscal Year 2020/21 targeting a primary balance of 3.5% of GDP to account for the expected revenues shortfalls and necessary spending reallocations as a result of COVID19 The National Housing Trust (NHT) has announced mortgage rate cuts	The overnight policy rate remains at 0.5% The BOJ has taken actions to ensure uninterrupted system wide liquidity, with an estimated J$57B liquidity injection to date, and removal of limits on the amounts that deposit taking institutions can borrow overnight without being charged a penalty rate and a broadening of the range of acceptable repo collateral. The authorities encourage BOJ to conserve on capital by postponing dividends payments to shareholders reschedule loans and mortgages **Exchange Rate/BOP** The Bank of Jamaica has intervened in the FX market through limited sales of reserves via the B-FXITT auction mechanism, issuance of US$ linked notes and, repos of FX denominated Government of Jamaica bonds with banks and securities dealers

Source: International Monetary Fund (2020) ***Policy Responses to Covid-19***

The lockdown, quarantine and other emergency measures employed by governments across the region had significant economic impacts in the tourism dependent countries most of which are notable service economies. In the short-term, these impacts were more intense in countries where informality is relatively high, such as in Jamaica where it was earlier estimated to be around 40% of GDP (Torero et al. 2006). Other estimates indicate that the size of the informal sector varies considerably. For Jamaica the informal sector is likely to be between 35 and 44% of GDP while for The Bahamas, Barbados, Guyana, Suriname, and Trinidad and Tobago, the ranges are 20–30, 30–40, 29–33, 35–45, 26–33%, respectively (Peters, 2017). Business within the informal sector tend to operate through personal services and, in the context of COVID-19, controls over mobility meant severe limitations on the ability of informal entities to engage with customers and effectively maintain their livelihoods.

Worsening the impact on the informal sector are issues relating to financial inclusion, nuanced in its manifestations across the Caribbean. The region is generally still a cash economy due to a combination of structural and cultural factors. Structural issues include the high cost of maintaining accounts as well as the inconvenience of journeying to banks to access small amounts of cash to enable subsistence-type transactions to occur. This is the outcome of low levels of competition in domestic markets which tends to be associated with the regulatory and other frameworks that present barriers to entry for potential new entities. In Jamaica, for example, there is a surprisingly high rate of account holdings, the majority of which however are dormant. This jeopardised the ability of governments to reduce transaction costs associated with income support through small grants that were paid to vulnerable Jamaicans, because a large number of the beneficiaries did not have active or any bank accounts to receive the transfers digitally. The move to the digital space in the region has largely been through the existing banks and banking network which would necessitate the beneficiaries passing the account opening requirements – a somewhat challenging requirement in the context of the identification document gaps.

This digital divide was illuminated in the challenges faced by both businesses and consumers as they attempted to maintain their connections with each other and with government and other business services in the context of limited shopping time, absence of integrated payment mechanisms and e-commerce platforms to support low cost, low margin sales. Beyond the immediate economic impact, the long-term growth trajectory is likely to also be negatively impacted by the low levels of attendance in educational institutions when they transitioned to online and remote learning. The UN (United Nations, 2020) estimated that approximately 23.8 million children and youth would drop-out from school over and above the normal rates and the impact of this on both private and public sector vitality is likely to be more far-reaching than this numerical estimate. Labour markets in the Caribbean are heavily dependent on the ability of local talent to distinguish itself and to signal value to employers through the acquisition of tertiary education mixed with experience. The digital divide impacting school-aged children and young adults, manifested in both access to strong stable internet connections and to one-to-one access to appropriate devices in the home, meant the widening of existing learning and therefore earning gaps in the region. The distributional and equity implications are therefore stark compounding effects for the region in addition to the implications for economic growth.

Economic Support from the International Community

The international community played a supportive role in the response of Caribbean countries to the impact of the COVID-19 measures. The IMF mobilized US$1trillion to support countries around the globe as they contended with the COVID-19 pandemic. According to the IMF (2020d), eight Caribbean countries accessed these funds either through the Rapid Financing Instrument (RFI) or the Rapid Credit Facility (RCF) modalities whilst Belize benefitted from an augmentation of the EFF Agreement that it is currently engaged with the Fund (Table 5.4). The Bahamas and Jamaica accessed the RFI to the tune of US$250 m and US$520 m respectively – for a total of US$770 m. Dominica, Grenada, Haiti, St. Lucia and St. Vincent and the Grenadines accessed the RCF to the tune of US$14 m, US$22.4 m, US$111.6 m, US$29.2 m, and US$16 m respectively – a total of US$81.6 m. Barbados received an additional US$90.84 m under its existing EFF. Haiti benefitted from two allocations of US$5.61 m in two equal tranches from the Catastrophe Containment and Relief Trust in addition to access received under the RCF. Altogether therefore the Caribbean benefitted from a combined US$1065.26 m from the IMF, approximately 0.1 percent of the total made available.

Table 5.4 Caribbean Countries Accessing the International Monetary Fund Rapid Response Financing and World Bank Financing for their COVID-19 response

Country	*Type of Emergency Financing*	*Amount Approved in SDR million*	*Amount Approved in US$m*[2]	*Date of Approval*
Bahamas, The	RFI	82.4	250	June 1, 2020
Barbados	Augmentation of EFF	66	90.84	June 3, 2020
Dominica	RCF	10.28	14	April 28, 2020
Grenada	RCF	16.4	22.4	April 28, 2020
Haiti	RCF	81.9	111.6	April 17, 2020
Haiti (Debt Service Relief from the Catastrophe Containment and Relief Trust)	Catastrophe Containment and Relief Trust (CCRT) 1st Tranche	4.10	5.61	April 13, 2020
	Catastrophe Containment and Relief Trust (CCRT) 2nd Tranche	3.98	5.61	October 2, 2020
Jamaica	RFI	382.9	520	May 15, 2020
St. Lucia	RCF	21.4	29.2	April 28, 2020
St. Vincent and the Grenadines	RCF	11.7	16	May 20, 2020

Source: International Monetary Fund (2020d) *COVID-19 Financial Assistance and Debt Service Relief*

The multilaterals, notably the World Bank (WB) and the Inter-American Development Bank (IDB), and bilateral partners also played a role in mitigating the impact of COVID-19 in the Caribbean (Table 5.5). The WB provided much of its support to the countries that are members of the Organisation of Eastern Caribbean States (OECS) through the Regional Health Project (RHP). Through the RHP, the WB channelled almost half its US$108.712 m allocation to the COVID-19 response effort of the OECS – US$47.9 m or 44.1 percent with Dominica receiving the single largest allocation in the sub-region to the tune of US$18.4 m. Dominica

Table 5.5 Multilateral Support for Caribbean country responses to the COVID-19 pandemic

Country	*World Bank*	*Inter-American Development Bank*
Bahamas		Vulnerable Populations: US$19.5 m
Belize	Initial: US$12.4 m Additional: US$8 m	Immediate Public Health Response: US$6.2 m Vulnerable Populations: US$12 m
Dominica	Initial: US$6.6 m Additional: US$3 m, US$3.6 m and US$5.2 m	
Grenada	US$2.5 m Additional: US$2.5 m	
Haiti		Immediate Public Health Response: US$27 m Vulnerable Populations: US$60 m
St. Lucia	Initial: US$10.5 m Additional: US$5 m	
St. Vincent and the Grenadines	Initial: US$4.5 m Additional: US$4.5 m	
Suriname	US$412,000	
Trinidad and Tobago	US$20 m	Public Policy and Fiscal Management: US$100 m
TOTAL	US$88.3 m	US$224.7 m

Source: Inter-American Development Bank (2020) ***Operational Response to COVID-19*** and World Bank (2020) ***World Bank Response to Covid-19 (Coronavirus) in the Caribbean***

used US$5.1 m of its total allocation to strengthen its public health system through the procurement of drugs, equipment, and supplies for their laboratories as well as the retrofitting of facilities to provide appropriate isolation units for COVID-19 positive cases. The agriculture sector in Dominica – with particular emphasis on keeping local market logistics open – also benefitted from support as the Bank attempted to enable the authorities to avoid any threats to food security during the period (World Bank, 2020a). Dominica received two further allocations in June, to continue the support for the health sector (World Bank, 2020b) as well as the agricultural sector whilst also ensuring that some emphasis was placed on improving the country's resilience to climate-related hazards (World Bank,

2020c). Income support was also offered to the people of Dominica in the form of grants through further support from the WB, using previous employment, age and other criteria to determine eligibility (World Bank, 2020d). In Grenada, the US$5 m allocation, through two tranches, supported both the country's health sector and its capacity to respond to the demands placed on it by the COVID-19 pandemic. The funds were used to purchase critical supplies, protective gear for medical professionals, equipment to support the work of all medical personnel involved in the response effort, preparation of isolation units should they become necessary, as well as items to support the preparedness of workers at Grenada's ports (World Bank, 2020e). The St. Lucia authorities expended the US$15.5 m on strengthening the various dimensions of its health care system to anticipate, respond to, and cope with the number of potential and confirmed positive cases of COVID-19 on the island. Their efforts included the development of appropriate information campaigns for the general public to ramp up the level of awareness about the disease as well as the treatment protocols. The health facilities were also improved with the preparation of isolation facilities and the establishment of the appropriate mechanisms for laboratory testing (World Bank, 2020f). St. Vincent and the Grenadines used the US$9 m they received to enhance and improve the capacity of their health system to respond to the anticipated needs of the increased number of cases they expected to experience, like the other countries in the OECS that benefitted from the WB health sector support. The two equal disbursements were used to provide supplies and equipment for patient care and the protection of health sector staff who may come in contact with suspected and confirmed positive cases (World Bank, 2020g). The only other non-OECS recipients from the WB's allocation were Belize, Suriname and Trinidad and Tobago and the latter two countries also used their allocations of US$214,000, and US$20 m, respectively, to improve their health system's preparedness to deal with cases of the COVID-19 virus. The interventions implemented in Belize through the initial allocation of US$12.4 m were targeted to the granting of cash grants to vulnerable persons and families (World Bank, 2020h). Another allocation of US$8 m was used to support the country's agriculture sector and protect its food security (World Bank, 2020i).

The IDB provided US$224.7 million to four Caribbean countries, three of which – The Bahamas, Belize and Haiti – received allocations that were targeted to 'Vulnerable Populations' in these countries. The Bahamas redeployed US$19.5 m from a previously approved loan to support the provision of tax credits and tax deferrals in order to support continued employment in the industries most affected by the COVID-19 crisis (Inter-American Development Bank, 2020a). In Belize the government used the US$6.2 m loan to provide minimum income and employment to those citizens most affected by the economic dimensions of the crisis (Inter-American Development Bank, 2020b). In addition to the allocation for Vulnerable Populations, Belize converted an existing sustainable tourism loan to support their Immediate Public Health Response through the remaining balance of US$6.2 m. Through this the Government of Belize implemented a range of interventions to detect and interrupt the spread of COVID-19 and improve the capacity of the health system to provide essential care and properly manage waste products generated through the use of protective gear (Inter-American Development Bank, 2020c). Haiti received the largest proportion – approximately sixty-six percent – of the funds provided to support Vulnerable Populations in the Caribbean to the tune of US$60 m. These funds were to be channelled to supporting vulnerable groups to maintain their minimum income and their employment levels throughout the pandemic (Inter-American Development Bank, 2020d). The people of Haiti also benefitted from the support of the IDB for its health infrastructure through a loan of US$27 m. Trinidad and Tobago was the sole Caribbean country to receive allocations to meet the overall public response to COVID-19 through Public Policy and Fiscal Management support. The total amount of US$100 m committed to Trinidad and Tobago was the single largest made to all five recipient countries in the Caribbean representing just under one half of the total committed to the region. Trinidad used this to make extensive improvements to the capacity of its fiscal authorities to execute projects to strengthen the health sector's capacity to respond, enable the provision of counter-cyclical expenditure that could support vulnerable individuals and groups, as well as to facilitate the economic recovery (Inter-American Development Bank, 2020e).

Support was also provided to Caribbean countries by other international development partners. Entities within the United Nations' family invested in socio-economic impact assessment for several countries, including the United Nations Development Programme for Bahamas, Jamaica and Belize and UNICEF for children in Jamaica. UNECLAC has developed technical analyses of the early impact of the crisis in the region and the FAO has attempted to catalogue the food security and availability issues that emerged across the region. Bilateral partners provided in-kind support such as ventilators, protective gear, notably the renowned support and solidarity of Cuban doctors, to strengthen COVID-19 response across the region.

Policy Recommendations and Outlook

Broadly speaking, the impact of the crisis was devastating across the region as most countries are service-oriented and the majority of businesses, whether formal or informal, rely on in person interaction for their survival. Despite relatively high mobile phone penetration, the accessibility to high-speed internet was far below that required to drive economic activity. There is wide variation in rates of household broadband subscription, with Jamaica and Suriname at the bottom with subscriptions of 5 and 10 per 100 inhabitants in contrast to more than 20 in Trinidad and Tobago, The Bahamas, and the British Virgin Islands and more than 25 for Barbados (Fonseca-Hoeve et al., 2017). Consequently most businesses would not have developed digital platforms through which to deliver goods and services since the average consumer was unlikely to be connected digitally. In Latin America and the Caribbean generally even the population that is connected faced challenges with the quality of the connection and the cost of the service. Governments in the region fared no better than the private sector with only 7 per cent of the paperwork required by governments across the region being completed electronically; at least half the countries had neither cybersecurity strategies, digital agendas, and, only half of the region's population had access to a bank account or debit card (Ziegler et al., 2020).

The imperative to digitalise in the context of the modalities of the spread of COVID-19 and other potential pandemics or other crises requiring physical distancing regulations is beyond debate. The disparities between rural and urban rates of connectivity provide another dimension of the challenge as do the lower rates of mobile data uptake among the poor. Mobile data and fixed broadband will need to become far more accessible, affordable and of a higher quality in order to ensure that even after the crisis passes, learning, work, and trade can continue through this medium as it can be a powerful source of inclusiveness. Allied and integral to this has been the drive throughout the region to increase financial access which requires an integrated national and regional strategic framework to implement and ensure interconnectedness and effectiveness. Economic development across the region requires states to develop national strategies to improve access to government services and support the widespread use of standardised low-cost e-commerce platforms. These are public goods that will need significant government intervention as the private market is not likely to provide them adequately particularly in rural areas and at reasonable cost.

A thrust towards solving the various divides that have had far-reaching consequences for economies and societies, for example in access to education and health services, will have to balance these with efforts in relation to emerging blue and green economy orientations, as well as climate change. The robustness of the growth forecasts for 2020 and 2021 (Table 5.2) need to be interpreted in the context of the high degree of uncertainty regarding the global recovery, the effectiveness, safety, and uptake of the vaccine being initiated across the globe, and, the extent to which poor, vulnerable and marginalised groups can continue to cope within the strained economic environment. The glimmer of hope from the economic standpoint lies with the robustness of a major source of foreign exchange – remittances. During the early crisis period, the World Bank (2020j) had forecasted a global decline in remittances of 20% and a 19.3% percent decline for Latin America and the Caribbean. This sparked concern in relation to the social sector, as remittances have proved a major source of poverty alleviation, boosting the consumption capacity of the poor and vulnerable, and thus forced governments to consider the quality, scale and generosity of their social security and social protection systems in order to support these groups during the crisis. However, to the widespread benefit of Caribbean residents, remittances have been remarkably robust, defying expectations, registering growth across the region and in countries such as Jamaica increasing by as much as 30 and 40% in some months since April (CARICOM Business, 2020). This may be due to the resilience and critical nature of the jobs performed by the diaspora in the context of the crisis, supported by the existing networks of remittance offices and other forms of digital transfers.

As fiscal, monetary and national authorities move to design and implement their build-back stronger, better and more resilient plans, there will be a continued need to lean on the support of the international community financially as well as in the breadth of technical and development support from the IDPs more specifically. Collaboration across the various regional organisations through the CARICOM institutional infrastructure will prove useful in ensuring that regional public goods and externalities are addressed through appropriate state-led and where possible

IDP-supported interventions. Given the uncertainties in the pace that the global economy will rebound and the structure that it will take in the post-COVID eras, and lockdowns in the United Kingdom – a major tourism gateway for the Caribbean – it is even more necessary that the region pragmatically consider actions, investment and other forms of cooperation that would enable it to generate economic growth and development. Given the recent implementation of fiscal responsibility frameworks, governments across the region will be keen to support livelihoods, build better and more robust economies, while simultaneously balancing budgets to ensure that we leave no one behind.

References

Alleyne, T. (2010). *The Challenges of High Debt, Addressing the Debt Overhang, the Cases of Jamaica and Antigua and Barbuda*. Retrieved from International Monetary Fund Website: https://www.imf.org/external/np/seminars/eng/2010/carib/pdf/alleyne.pdf

CARICOM Business. (2020, September 7). *JAMAICA: REMITTANCES REBOUND IN JUNE*. Retrieved from CARICOM Business Website: https://caricom.org/jamaica-remittances-rebound-in-june-caricom-business/

Clarke, C., & Nelson, C. (2020). *Contextualising Jamaica's Relationship with the IMF*. Palgrave Macmillan.

Clarke, N. (2020, June 7). *COVID-19 and Suspension of Jamaica's Fiscal Rules*. Retrieved from The Jamaica Gleaner: http://jamaica-gleaner.com/article/commentary/20200607/nigel-clarke-covid-19-and-suspension-jamaicas-fiscal-rules#:~:text=JAMAICA'S%20FISCAL%20RESPONSIBILITY%20FRAMEWORK,per%20cent%20by%202027%2D28

Fonseca-Hoeve, B., Marius, M., Osepa, S., Coffin, J., & Kende, M. (2017). *Unleashing the Internet in the Caribbean – Stimulating Better Access in the Region*. Caribglobal Data Services, ICT Pulse Consulting, Internet Society.

Inter-American Development Bank. (2020a, August 9). *BH-L1037: Skills for Current and Future Jobs in the Bahamas*. Retrieved from Inter-American Development Bank Website: https://www.iadb.org/en/project/BH-L1037

Inter-American Development Bank. (2020b, June 15). *BL-L1034: Support to Safety Nets for Vulnerable Populations Affected by Coronavirus in Belize*. Retrieved from Inter-American Development Bank Website: https://www.iadb.org/en/project/BL-L1034

Inter-American Development Bank. (2020c, March 25). *Reformulation Proposal of the Sustainable Tourism Program II (BL-L1020) (3566/OC-BL) for the Financing of the Immediate Public Health Response to Contain and Control the Coronavirus and Mitigate Its Impact On Service Delivery In Belize*. Retrieved from Inter-American Development Bank Website: https://www.iadb.org/projects/document/EZSHARE-734982891-49?project=BL-L1020

Inter-American Development Bank. (2020d, July 8). *HA-L1145: Safety Nets for Vulnerable People Affected by Coronavirus in Haiti*. Retrieved from Inter-American Development Bank Website: https://www.iadb.org/en/project/HA-L1145

Inter-American Development Bank. (2020e, June 24). *Programme to Strengthen Fiscal Policy and Management in Response to the Health and Economic Crisis Caused By Covid-19 in Trinidad And Tobago*. Retrieved from Inter-American Development Bank Website: https://www.iadb.org/projects/document/EZSHARE-1196231878-11?project=TT-L1058

International Monetary Fund. (2013). *Caribbean Small States: Challenges of High Debt and Low Growth*. Internatonal Monetary Fund.

International Monetary Fund. (2019, October). *World Economic Outlook, October 2019*. Retrieved from International Monetary Fund Website: https://www.imf.org/en/Publications/WEO/Issues/2019/10/01/world-economic-outlook-october-2019#Chapter%201

International Monetary Fund. (2020c, December 3). *Antigua and Barbuda*. Retrieved from International Monetary Fund: https://www.imf.org/en/Countries/ATG

International Monetary Fund. (2020d, December 1). *COVID-19 Financial Assistance and Debt Service Relief*. Retrieved from International Monetary Fund Website: https://www.imf.org/en/Topics/imf-and-covid19/COVID-Lending-Tracker#WHD

International Monetary Fund. (2020e, January). *Tentative Stabilization, Sluggish Recovery? World Economic Outlook Update*. International Monetary Fund.

International Monetary Fund. (2020f). *Policy Responses to COVID-19*. Retrieved from International Monetary Fund Website: https://www.imf.org/en/Topics/imf-and-covid19/Policy-Responses-to-COVID-19

International Monetary Fund. (2020g, October). *World Economic Outlook, October 2020*. Retrieved from International Monetary Fund Website: https://www.imf.org/en/Publications/WEO/Issues/2020/09/30/world-economic-outlook-october-2020

International Monetary Fund and International Development Association. (2000). *Decision Point Document for the Enhanced Heavily Indebted Poor Countries (HIPC) Initiative*. International Monetary Fund.

Peters, A. (2017). *Estimating the Size of the Informal Economy in Caribbean States*. Inter-American Development Bank.

Torero, M., Robles, M., Hernandez, M., De la Roca, J., Webber, M., & Thomas, D. (2006). *The Informal Sector in Jamaica*. Inter-American Development Bank (IDB).

United Nations Economic Commission for Latin America and the Caribbean. (2019). Special Feature: The Caribbean Small States. In U. N. Caribbean (Ed.), *Latin American Economic Outlook 2019* (pp. 185–199). United Nations Economic Commission for Latin America and the Caribbean.

United Nations Economic Commission on Latin America and the Caribbean. (2017). Guyana. In U. N. Caribbean (Ed.), *Economic Survey of Latin America and the Caribbean*. United Nations Economic Commission on Latin America and the Caribbean.

United Nations (2020). Policy Brief: Education during COVID-19 and Beyond. Retrieved from United Nations Website https://www.un.org/development/desa/dspd/wp-content/uploads/sites/22/2020/08/sg_policy_brief_covid-19_and_education_august_2020.pdf

World Bank. (2020a, April 20). *World Bank to Strengthen Dominica's COVID-19 Response with US$6.6 Million*. Retrieved from World Bank Website: https://www.worldbank.org/en/news/press-release/2020/04/20/world-bank-to-strengthen-dominicas-covid-19-response-with-us66-million

World Bank. (2020b, June 26). *World Bank Provides US$15 Million Additional Financing for Eastern Caribbean Health Sector*. Retrieved from World Bank Website: https://www.worldbank.org/en/news/press-release/2020/06/26/world-bank-provides-us15-million-additional-financing-for-eastern-caribbean-health-sector

World Bank. (2020c, June 30). *World Bank Provides Additional Financing of $US16.4 Million for Agricultural Livelihoods, Food Security, and Climate Resilience in Dominica*. Retrieved from World Bank Website: https://www.worldbank.org/en/news/press-release/2020/06/30/world-bank-provides-additional-financing-of-us164-million-for-agricultural-livelihoods-food-security-and-climate-resilience-in-dominica

World Bank. (2020d, August 24). *World Bank Financing to Support Over 7,000 People in Dominica During COVID-19 Pandemic*. Retrieved from World Bank Website:https://www.worldbank.org/en/news/press-release/2020/08/24/world-bank-financing-to-support-over-7000-people-in-dominica-during-covid-19-pandemic

World Bank. (2020e, June 1). *World Bank Strengthens Grenada's COVID-19 Response with US$2.5 Million*. Retrieved from World Bank: https://www.worldbank.org/en/news/press-release/2020/06/01/world-bank-strengthens-grenadas-covid-19-response-with-us25-million

World Bank. (2020f, April 30). *World Bank Provides US$10.5 Million to Saint Lucia for COVID-19 Response*. Retrieved from World Bank: https://www.worldbank.org/en/news/press-release/2020/04/30/world-bank-provides-us105-million-to-saint-lucia-for-covid-19-response

World Bank. (2020g, April 21). *World Bank Provides $4.5 Million to Support Saint Vincent and the Grenadines' COVID-19 Emergency Response*. Retrieved from World Bank Website: https://www.worldbank.org/en/news/press-release/2020/04/21/world-bank-provides-45-million-to-support-saint-vincent-and-the-grenadines-covid-19-emergency-response

World Bank. (2020h, July 13). *World Bank Supports Programs to Assist over 13,000 Households in Belize in Response to COVID-19 Crisis*. Retrieved from World Bank Website: https://www.worldbank.org/en/news/press-release/2020/07/13/world-bank-supports-programs-to-assist-over-13000-households-in-belize-in-response-to-covid-19-crisis

World Bank. (2020i, August 27). *World Bank Provides US$8 Million to Strengthen Agriculture and Food Security in Belize*. Retrieved from World Bank Website: https://www.worldbank.org/en/news/press-release/2020/08/27/world-bank-strengthens-agriculture-food-security-belize?cid=lac_tt_caribbean_en_ext

World Bank. (2020j, April 22). *World Bank Predicts Sharpest Decline of Remittances in Recent History*. Retrieved from World Bank Website: https://www.worldbank.org/en/news/press-release/2020/04/22/world-bank-predicts-sharpest-decline-of-remittances-in-recent-history

World Bank Group. (2020, December 3). *World Development Indicators*. World Bank Group.

Ziegler, S., Segura, J. A., Bosio, M., & Camacho, K. (2020). *Rural Connectivity in Latin America and the Caribbean: A Bridge for Sustainable Development in a Time of Pandemic*. Inter-American Institute for Cooperation on Agriculture, Inter-American Development Bank, Microsoft.

CHAPTER 6

Did French Polynesia Cope with COVID-19? Intrinsic Vulnerabilities and Decreased Resilience

Charlotte Heinzlef and Damien Serre

Introduction

The COVID-19 pandemic, caused by the coronavirus SARS-CoV-2, emerged in China at the end of 2019, and an international public health emergency was declared by the World Health Organization (WHO) on January 30, 2020. The epidemic was declared a pandemic by the WHO on 11 March 2020, calling for essential protective measures to prevent saturation of intensive care units and to reinforce preventive hygiene (elimination of physical contact, end of crowds and large demonstrations as well as non-essential travel and movement, implementation of quarantine, etc.). The pandemic caused a series of cancellations of events around the world, the implementation by many countries of containment measures to curb the formation of new outbreaks of contagion and the closure of

C. Heinzlef (✉) • D. Serre
University of French Polynesia, Faaa, French Polynesia
e-mail: charlotte.heinzlef@upf.pf; damien.serre@upf.pf

Y. Campbell, J. Connell (eds.), *COVID in the Islands: A comparative perspective on the Caribbean and the Pacific*,
https://doi.org/10.1007/978-981-16-5285-1_6

international borders. It also affected social and economic instability due to the uncertainties and fears it brings to the global economy.

Small island states are structurally vulnerable to this pandemic, having few resources and limited care systems to respond to international health crises. French Polynesia, composed of 5 archipelagos (the Society Islands, including the main island of Tahiti, and the Tuamotus, Gambiers, Australs and Marquesas) comprising 118 islands, more than half inhabited, covering a land area of 4167 km^2 dispersed over 2.5 million km^2. It is occupied by about 280,000 inhabitants. COVID-19 was declared in Polynesia on March 10, 2020 with the first patient, a Member of Parliament who had returned from Paris. On March 20, lockdown was decreed on the same basis as in metropolitan France, but with a ban on the sale of alcohol in addition to other measures. After March 27, no more planes left or arrived in French Polynesia, cutting it off from the rest of the world. While the first cases were concentrated in Tahiti, in the urban area of Papeete, the 34th case appeared in the Tuamotu Islands on the atoll of Rangiroa, the first indication of spread to another archipelago. To control the epidemic, the local government then also suspended inter-island air links. The only planes authorized to fly were those that provided medical evacuations or transported medical equipment. With around 60 cases and no known patients showing signs of the disease, the authorities decided to end the lockdown on May 21. In order to boost an economy strongly linked to tourism, Polynesia reopened its international borders, effectively to all countries, on July 15th and also lifted all quarantine measures. Unfortunately, the number of COVID-19 cases then exploded (Fig. 6.1) reaching 3352 by mid-October, and was continuing to increase. By early December it had reached 16,000 cases, with 91 deaths, spread across all archipelagos, the most serious situation in the Pacific islands. Additional health staff had been flown in from France, and curfews restored to Tahiti and Moorea, in attempts to cope.

This significant increase in COVID-19 cases followed the reopening of international borders and raised questions for the health system. Indeed, with about sixty resuscitation beds equipped with artificial respirators in the main hospital, with about 1 person in 5 suffering from type 2 diabetes, 70% of the adult population overweight and 40% obese (Gatti et al., 2015; Reade, 2013), Polynesia was demonstrably a territory at risk to the spread of the virus. This structural vulnerability raised questions about the political decision to reopen the borders.

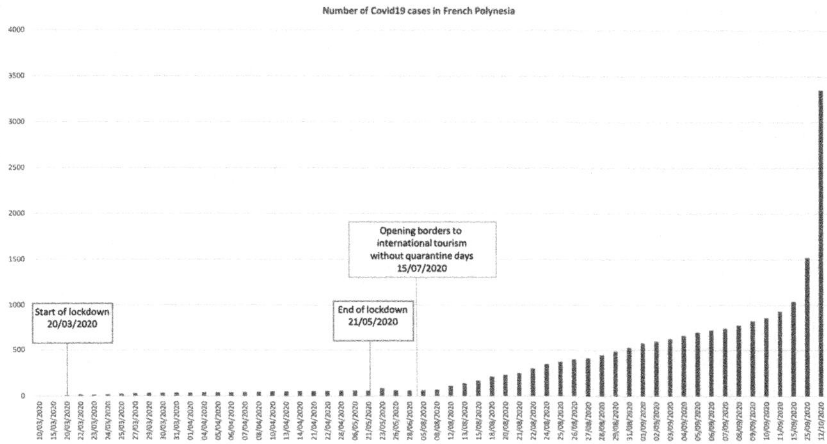

Fig. 6.1 Number of COVID-19 Cases in French Polynesia

This chapter aims to explore the intrinsic vulnerabilities of French Polynesia, vulnerabilities that can be defined in geographical, social, economic and spatial terms. These pre-existing vulnerabilities have conditioned the response of local authorities to the health threat. It is therefore necessary to analyse them in order to understand the spread and management of the virus in Polynesia. Finally, it is valuable to assess local resilience strategies in order to promote the implementation of strategies that encourage autonomy and effective local authority.

The Idea of Vulnerability

The concept of vulnerability has been widely studied and defined in the scientific literature, broadly regarded as the potential for harm or loss inherent in a person, thing or system (Burton et al., 2018; Wisner, 2016). According to the IPCC, vulnerability to a crisis is the degree to which a system is susceptible to, and unable to cope with, the adverse effects of crisis (Sharma & Ravindranath, 2019). Vulnerability may be defined as an internal risk factor of the person or system, that is exposed to a crisis and corresponds to its 'intrinsic predisposition to be affected or to be susceptible to damage. In other words, vulnerability represents the physical, economic, political or social susceptibility or predisposition of a community to damage in the case a destabilizing phenomenon of natural or anthropogenic origin' (Cardona, 2004). Vulnerability can therefore be understood as a result of a crisis (the amount of damage caused to a system by a particular event or hazard) or as a state (Birkmann, 2007; Cardona, 2004) that exists within a system before it is confronted with a particular hazard (Adger et al., 2004; Brooks, 2003; Burton, 2010; Cutter & Finch, 2008). One approach implies a post-crisis temporality, the other encourages a long-term analysis. We focus our analysis on this second dimension in order to understand the pre-COVID and containment vulnerabilities that determined responses.

Numerous vulnerability indices have been identified in the scientific literature (Adger et al., 2004; Cutter & Finch, 2008; Papathoma-Köhle et al., 2019; Spielman et al., 2020; Tate, 2012). We analyse the concept for French Polynesia through a methodology developed and tested in Avignon, France (Heinzlef et al., 2020, 2019), enabling a spatial decision-making system to support and encourage local actors to develop risk management strategies mitigating spatial and social vulnerability to achieve territorial resilience (Fig. 6.2; Table 6.1).

Vulnerabilities in French Polynesia

French Polynesia is first and foremost vulnerable to hazards because of its particular geography and its remoteness from other territories. French Polynesia is an overseas collectivity and part of the French Republic, yet it is about 18,000 kms from Paris. French Polynesia is entirely located in the southern hemisphere, in the centre of the Pacific Ocean, isolated from most of the world. It occupies an immense surface area of 5.5 million km^2, about the same area as the European continent. It consists of five archipelagos and 118 islands and atolls, a good half of which are inhabited and is mostly made up of water. Its geographical remoteness from any other territory as well as the small population and small land area make it vulnerable and dependent on external resources.

Social factors have been shown to be one of the most important causes in the vulnerability of populations at risk (Fatemi et al., 2017). Social vulnerability involves combinations of social, cultural, economic, political and institutional processes (Rufat et al., 2015; Tate, 2012) that condition socio-economic reactions in experiencing and recovering from crisis (Spielman et al., 2020). It refers to the characteristics of people in terms of their capacity to anticipate, cope with, resist and recover from the impact of a crisis (Wisner, 2016; Wisner et al., 2004).

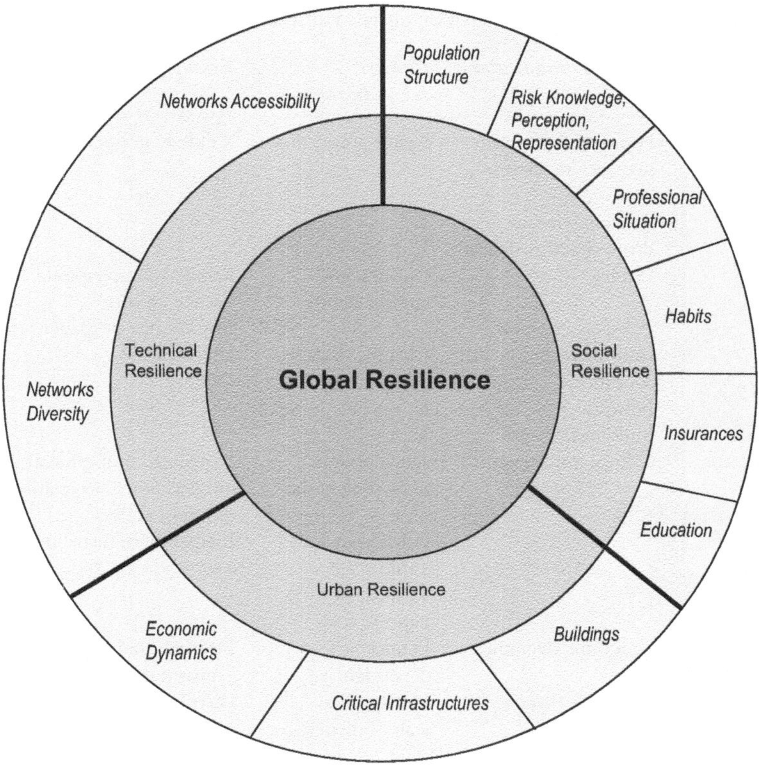

Fig. 6.2 Elements to analyse territorial resilience (Heinzlef et al., 2019)

- An aging population

People aged 60 years or more represent 12% of the Polynesian population, and that proportion is slowly increasing. Although the population of French Polynesia remains youthful (Fig. 6.3), with 31% under 20 years old and 8% over 65 years old, this aging trend tends to evolve, as a combination of fertility and migration (Fig. 6.4). Life expectancy is getting longer, the birth rate is falling, and thus the aging of the population is accelerating.

Older people are considered to be more at risk in the face of the COVID-19 crisis (Aronson, 2020; Colenda et al., 2020). Although all age groups are at risk for contracting the COVID-19 virus, older people face a significant risk of developing severe illness due to physiological changes

Table 6.1 Elements of French Polynesian vulnerability

Initial indicators	*Initial sub-indicator*	*Chapter sub-indicators*	*Examples*
Social indicator	Population structure (age, sex repartition, etc.)	Aging population	% elderly population
	Risk knowledge	X	
	Professional situation	X	
	Habits	Cultural and familial habits	Familial compositions Festive culture Importance of gatherings
	Insurances and health conditions	Chronic diseases	Diabetes Obesity
	Education	X	
Urban indicator	Buildings	X	
	Critical infrastructure	Metropolitan scale/Polynesian scale	Distribution of critical infrastructures in relation to different scales
		Polynesian scale/ Tahiti scale	Inequality of distribution
		Tahiti scale/ Papeete scale	
	Economic dynamics	Economic dependencies	International dependency Tourism dependency
Technical indicator	Networks diversity	Metropolitan scale/Polynesian scale	Distribution of networks in relation to different scales
		Polynesian scale/ Tahiti scale	Inequality of networks distribution
		Tahiti scale/ Papeete scale	
	Networks accessibility	Metropolitan scale/Polynesian scale	Distribution of networks in relation to different scales
		Polynesian scale/ Tahiti scale	Inequality of networks distribution
		Tahiti scale/ Papeete scale	

that come with ageing and existing underlying health conditions (World Health Organization, 2020). It is critical that older people can access health-care services during the pandemic for both emergency, ongoing and primary health care. Older people also have a significantly higher risk

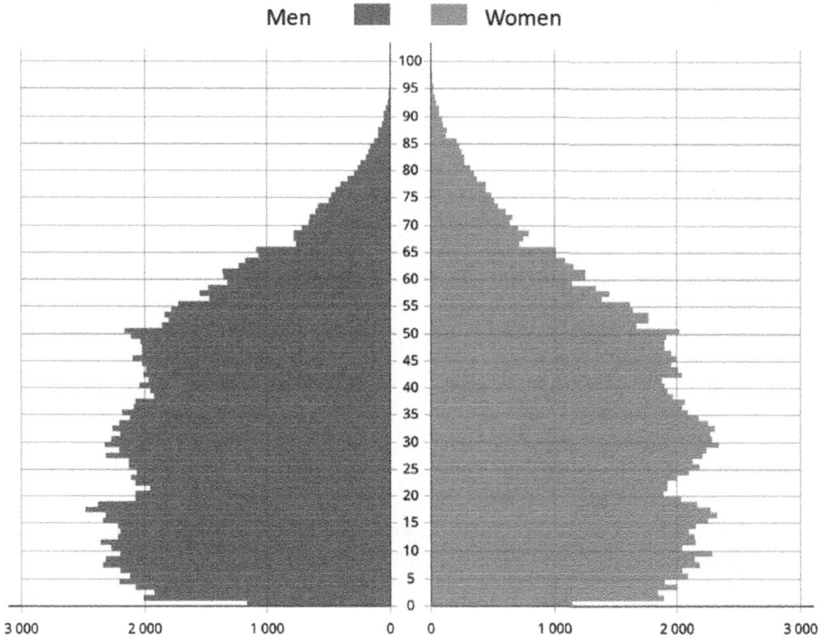

Fig. 6.3 2017 Age Pyramid. Source: Insee/Ispf – 2017 Census of Population

of death and serious illness from infection, with those over 80 years of age dying at a rate five times higher than the average. It is estimated that 66% of the global population aged 70 years and older have at least one underlying condition, putting them at increased risk of serious consequences from COVID-19 (United Nations, 2020). This vulnerability is expressed in statistics. While the median age of confirmed global cases of COVID-19 is 51 years (Fig. 6.5), the mortality rate for those over 80 years of age is five times higher than the average. More than 95% of deaths due to COVID-19 in Europe have been persons aged 60 years or older. In the United States, 80% of deaths were among adults aged 65 and over. In China, approximately 80% of deaths occurred among adults aged 60 or older (United Nations, 2020). Not surprisingly the first eleven deaths in Polynesia were *matahiopo*, people over 70 years of age, most therefore experiencing comorbidity factors.

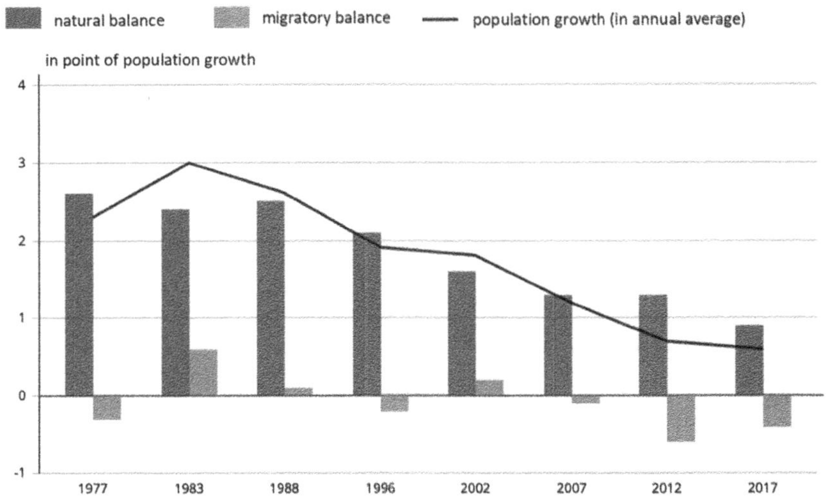

Fig. 6.4 Contributions to population growth (annual average), Source: Insee/ Ispf – Population censuses 1977–2017, Civil status of French Polynesia

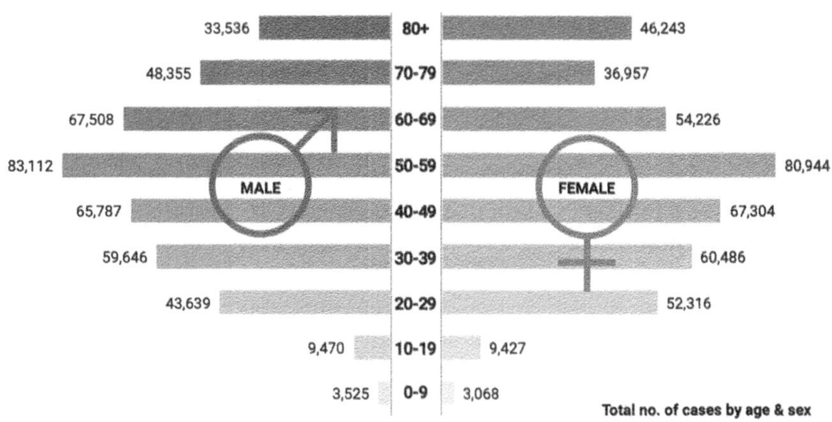

Fig. 6.5 Distribution by age and sex of confirmed COVID-19 cases (United Nations, 2020)

- A Polynesian population facing many chronic diseases

These comorbidity factors are important factors in the management, progression, complications and response to treatment of the initial illness. For COVID-19 they include diabetes, hypertension, obesity, cancer and lung problems. Indeed, some reports have shown that the vast majority of deaths occurred in people with pre-existing factors, such as these (e.g. Caballero et al., 2020; Wang et al., 2020; Yan et al., 2020). Among these factors, diabetes is a major risk factor for increased morbidity and mortality and diabetes patients were more likely to require management in the ICU with medical ventilation, and to die, compared with equally severe COVID-19 patients but without diabetes (Yan et al., 2020).

Diabetes is extremely prevalent in French Polynesia. The national Diabetes Atlas of 2017 estimated that the prevalence of type 2 diabetes in Polynesia is 22%, or more than 45,000 people aged 20–79 years old, 22,000 of whom are still undiagnosed (Niva, 2018). Type 2 diabetes is found in people who are obese; obese people are three times more likely to be diabetic than a non-obese person. Obesity is also prevalent in French Polynesia, and the rate of obesity has increased significantly over the last three decades (Abarca-Gómez et al., 2017). By far the greatest increase has been in the Pacific Islands. The obesity rate for children and adolescents in French Polynesia is now close to 30%. Obesity has been identified as a risk factor for hospitalisation, follow-up care, secondary effects and even mortality in patients with COVID-19 (Townsend et al., 2020). Indeed, obesity increases the risk of severe illness from COVID-19 being linked to impaired immune function and decreased lung capacity, so that having obesity may triple the risk of hospitalization due to a COVID-19 infection (Burström & Tao, 2020; Karaye & Horney, 2020).

- Cultural and familial habits impact in the spread of the virus

Spread of viruses such as COVID-19 is more likely to occur in places of close proximity, with significant contact between individuals, such as in homes, workplaces and schools (Jawad, 2020). High population densities lead to more face-to-face interactions, which create potential hotspots for the pandemic dissemination (Carozzi et al., 2020; Hamidi et al., 2020). As a result, international containment and border closures have been established to limit the interactions and spread of the virus. In Polynesia, lockdown lasted for over 2 months and the closure of international borders over 4 months. During lockdown the number of cases was kept below 70. However, long-standing habits and the way of life in Polynesia led, among other things, to an explosion of cases.

In French Polynesia, especially in disadvantaged families, the family brings together several generations under one roof. In ancestral Polynesia, the term *utuafare* designated an extended family community of five to twenty people who recognized each other through family ties, marriage or adoption. In this basic residential group, the person who has legal authority over the child (very often the progenitor) is not necessarily the person who takes care of the child. In 2012, it was estimated that 39% of children lived in such extended families (Guy & Ailincai, 2019). The containment and limitation of social interactions, in a narrow western sense, was therefore very difficult to implement in Polynesia. Consequently the president of French Polynesia, Edouard Fritch, on multiple occasions had to prohibit various public and private gatherings, and also meetings of family and friends. Social access to the Tahiti lagoon, a place of significance for historical and ancestral gatherings, was also banned. Ironically in October Fritch himself tested positive, and the virus spread to all archipelagos, believed to have been spread by politicians following a meeting in Papeete.

However, the weight of cultural tradition and lifestyles overcame these barriers and bans. In Polynesia, the "*bringue*", a traditional festival, reflected spontaneous festivity centred around a convivial meal. The party is an integral part of the local festivals, bringing together musicians and dancers, in intergenerational sharing and reunion. On 31st July, the "Piment Rouge" restaurant in Papeete, the capital, decided to relaunch these celebrations under the theme of the "Bayonne celebrations". Three days later at least one person who participated in this event had tested positive to COVID-19; another day later at least 71 positive people were counted, creating a cluster for the diffusion of COVID-19. This cluster then contributed to the explosion of cases in Polynesia, partly because any alert from the authorities and the restaurant's managers did not take place until a week after the party, after the people present had spread out, including from Tahiti to other islands. It also resulted in it spreading to France with the infection of a Mobile Gendarmerie Squadron from France returning there after a mission in Polynesia.

Economic Dependence on Tourism

Beyond the inherent vulnerabilities of the Polynesian population, economic realities led to decisions that were often undesirable in the context of COVID-19. Polynesia had managed to keep the number of cases with COVID-19 to less than 70 during the lockdown phase. The implementation of containment as well as the geographical location of French Polynesia, in the middle of the Pacific, and the many islands, limited both human interactions and the spread of the virus. However, this geographic location also led to hasty decisions.

Because of its isolation and fragmentation in five archipelagos, Polynesia has taken advantage of this geographical handicap to create the image of a dream and luxurious tourist destination (Kahn, 2011). Contemporary tourism in 2018 directly represented 8% of the GDP and 15% of salaried jobs in the commercial sector, partly the outcome of the 65 billion francs spent by foreign tourists. That does not include spending on international air transport estimated at 38 billion francs. Tourist spending mainly affects four branches of the economy: transport, whether by air, land and sea, the "hotel and restaurant" sector, trade and services offered to individuals such as excursions and leisure activities. However, tourism's value extends beyond this. Indirect effects must be accounted for, such as purchases of local products by the tourism sector. Thus, for every 10,000 francs spent by this sector, 120 go to the purchase of agricultural products, 650 to seafood products, 1100 to industrial products (including energy) and 1700 to services. In total, tourism's 'purchases represent 7% of local production of agricultural products and 18% of seafood products" (Institut de la statistique de la Polynésie Française, 2020). In addition, according to the ISPF, the impact of tourism is more important in remote archipelagos because "it represents, in remote islands, the only source of income for many families" (ibid). The 'tourist density' is very high on several islands, notably Moorea (7), Rangiroa (9.5), and Bora Bora (11) and most notably in Fakarava with 15 tourists per inhabitant. Such numbers emphasise the strong dependence of the population on tourism which 'is the main provider of employment as in Bora-Bora (56% of jobs), Rangiroa (31%) and to a lesser extent, the island of Moorea (27%)' (Institut de la statistique de la Polynésie Française, 2020).

French Polynesia is thus extremely dependent on the financial contribution that tourism represents, and this dependence has led to political decisions that are sometimes unsuited to a global health crisis. Between

March and July, only about ten flights operated between Paris and Papeete. Financed by France, they made it possible to repatriate Polynesians stranded in metropolitan France, to ensure medical evacuations and transport freight, particularly medical freight. All passengers arriving in Polynesia were subject to a 14-day quarantine. On July 3 however, the borders were reopened for French tourists, students, civil servants, and others, with an obligatory 7-day quarantine for new arrivals. However, 2 weeks later, quarantine was abolished to encourage the resumption of international tourism and make the territory attractive and more easily accessible. Thus, tourists from Europe and North America, the two main tourism markets, could travel to Polynesia without specific restrictions. The main reason for opening up was to save jobs, as declared by the Polynesian president Edouard Fritch, according to whom, nearly 19,000 Polynesians worked in tourism or had a job related to this sector. Nicole Bouteau, the local Minister of Tourism and Employment, declared that it was no longer a health emergency but an economic and social emergency. This decision helped to revive the tourism economy, as tourist numbers from both France and the United States grew (with 16,000 tourists from the United States and 7000 from France in 2020). But so did cases of COVID-19; numbers exploded and within 3 months cases had gone from about sixty during the lockdown to over 3000 cases and 11 deaths, and both numbers were still growing fast. Dependence on tourism was proving extremely costly.

Spatial Dependencies and Inequalities

Spatial inequality and political dependence further increased French Polynesian vulnerability, and at different scales: the metropolitan France/French Polynesia scale, Tahiti (main island)/French Polynesia and finally Papeete (the capital of Polynesia)/Tahiti.

- Metropolitan France / French Polynesia scale

Administratively, French Polynesia is an overseas collectivity of France. Power essentially resides in the hands of a Territorial Assembly elected by universal suffrage and possessing considerable power. The executive body is constituted by the government and presided over by the president of French Polynesia. However, sovereignty is ensured by the French State, represented locally by a High Commissioner of the Republic. Thus, France exercises authority over such issues as nationality (civic rights, electoral law, civil status records), guarantees of public liberties (justice, public prison service), foreign policy, immigration control, currency, university education, security and public order. Locally, within French Polynesia, the communes manage such matters as municipal police, roads and transport.

Although theoretically autonomous, Polynesia nevertheless remains extremely dependent on metropolitan France, especially politically and economically. That dependency was illustrated in the management of COVID-19, since it was essentially the High Commissioner who made the decisions on the measures to be put in place (Haut-commissaire de la République Française, 2020). As a representative of the French State, the High Commissioner thus followed metropolitan French decision-making and largely applied it to the Polynesian situation. Lockdown in France was declared on 17 March for about 6400 cases among 67 million inhabitants, and ended on 11 May. In Polynesia, lockdown was declared on 20 March for about 20 cases, and 280,000 inhabitants, and ended on 21 May. Many Polynesians criticized this decision as excessive, imposed on them effectively by a distant government unfamiliar with local circumstances, with one local citizen appealing to the Council of State, claiming that it was not proportionate to the importance of the risk of contamination according to the local circumstances (*La Dépêche de Tahiti*, 2020). This management of COVID-19, judged as "copy-pasted" by the locals, was the outcome of political dependence on the France.

Beyond political dependence, there is also an economic dependence that has influenced decisions regarding the management of COVID-19. At the end of the nuclear tests in French Polynesia, late in the twentieth century, France signed a "Progress Pact" with the territorial government to compensate for the loss of financial resources. It was initially planned for 10 years (1996–2006) but extended in 2010 to cover financial support. Direct financial transfers from metropolitan France concern the salaries of teachers, the payment of pensions to state employees living on the territory, justice and security. No taxes are collected by France in the territory. Therefore, the Polynesian debt to the French State was 83.45 billion Pacific francs (US$ 840 million) in 2019 (Viatge, 2019). This debt increased due to COVID-19, with French Polynesia planning to borrow 30 billion Pacific francs from France to boost the economy and 20 billion to support the social security fund. President Edouard Fritch observed that "borrowing means that tomorrow we will have to repay [...] today we are repaying nearly 12 billion a year because of the debts we had in the past and the outstanding amount is 80 billion" (*Polynésie la 1ère*, 2020). In order to repay this growing debt, it was considered necessary to revive the local economy, and thus tourism. This dependence therefore led to the reopening of borders without 14-day quarantine in order to revitalize the territory, the local economy and manage debt repayment. Indeed, attaching so much weight to the economy in crisis management was criticized by the local people, including labour unions, who called for a general strike. Union members considered that "health security is no longer properly assured" and that it had been "sacrificed at the expense of the economy of production [whereas] health and life have a fundamental role to play in achieving stronger and more job-rich economic growth" *(Outremers 360°*, 2020).

- Tahiti/ Archipelagos of French Polynesia scale

A further distinctive spatial vulnerability lies in the inequality of resources between Tahiti (and other Society Islands) and the more remote archipelagos. In terms of health care alone, but typical of education and other service provision, there are obvious inequalities in access. There are 5 hospitals in Polynesia: 2 in Tahiti, one on Moorea, 1 on Raiatea and finally 1 in the Marquesas Islands, the only one beyond the Society Islands (Direction de la Santé de Polynésie Française, 2020). Beyond hospitals, a parallel distribution of health practitioners highlights the inequality in health care resources. Thus, out of 514 practitioners, only 8 are in the Austral Islands, 14 in the Marquesas Islands, and 12 in the Tuamotu Islands (Gouvernement de la Polynésie Française, 2016). In the Tuamotu and Gambier archipelagos, where nearly 17,000 people live, only 29% have access to a doctor, 21% to a dentist, and none to a midwife. These numbers show that insularity and isolation are still little recognized and poorly reflected in the country's health strategies.

Isolation and insularity are compounded by transport connections, which accentuate health delivery deficiencies. An additional inequality comes from the air links between the different islands of Polynesia. Although 47 of the 118 islands have air transport connections, frequency and capacity vary enormously. Tahiti has many daily flights to the Society Islands several daily flights to some of the Tuamotus, one or two daily flights to the Marquesas Islands and a few flights per week to the Austral Islands. Such differences and disadvantages have contributed to gradual migration from the outer islands to Tahiti. However, isolation, insularity and remoteness suddenly became an asset in 2020, protecting many small islands from the spread of the virus. Late in 2020, many remote islands were still safe from the spread of the coronavirus, despite return migration from Tahiti to the islands, typically of pupils and students returning home.

- Papeete/Tahiti scale

Finally, spatial inequalities within Tahiti also occur. Tahiti is composed of the main island – Tahiti Nui – and a peninsula – Tahiti Iti. Papeete, the capital of French Polynesia, located on Tahiti Nui, hosts such institutions as the High Commission, the government of French Polynesia with its president, the Assembly of French Polynesia, as well as the Economic, Social and Cultural Council. It is also home to key infrastructure such as the main port and international airport of Papeete, fuel depots, the territorial hospital, private clinics and most of the industrial and economic infrastructures, as well as banking and financial institutions. Most formal employment and most economic opportunities are in Tahiti. More than two-thirds of the national population live in Tahiti, and two thirds of those live in Papeete, some at very high densities in overcrowded coastal suburbs such as Faaa, where social distancing presents a considerable challenge.

Conclusion

This chapter has highlighted the inherent fragility of the territory of French Polynesia and of Polynesian organisation with regard to the crisis management of COVID-19. It emphasises the more or less vulnerable components of the country, in order to raise awareness among local actors, and to make risk management evolve towards greater resilience. The critical elements are above all the underlying health conditions (notably co-morbidity), the spatial inequalities in service provision and Polynesian political and economic dependence (Fig. 6.6). The political decision to open the borders to tourism proved catastrophic. The social situation, which is extremely vulnerable to chronic diseases, endangers health resilience. Spatial inequalities, at different scales, weaken spatial and territorial responses to different crises. Lack of financial autonomy has resulted in multiple dependencies including on tourism and on financial assistance from France. Whether to set up strategies or management tools before the crisis or to revive the spatial, social and economic dynamics afterwards, the lack of financial autonomy questions the resilience of the territory.

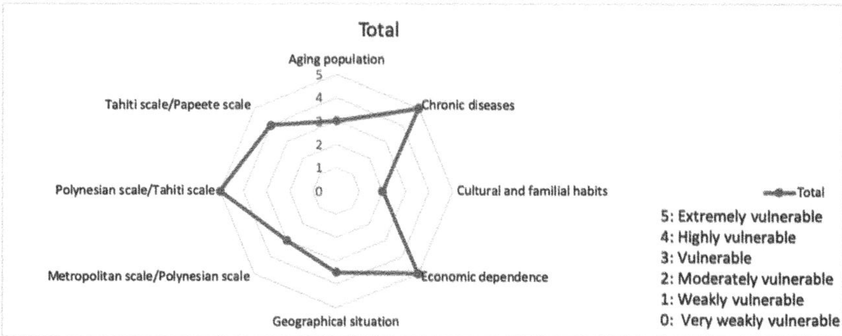

Fig. 6.6 Evaluation of vulnerability elements in French Polynesia

Certain actions and strategies can be put in place to mitigate these vulnerabilities and promote the implementation of resilience at different levels, but these have never been easy to achieve in such a vast territory as French Polynesia. Efforts to diversify from tourism and develop an export economy, whether from copra, noni juice, vanilla or pearls have never been sustained, especially in the face of a more urban and bureaucratic structure of employment, subsidised by France (Poirine, 2010). Rich sea floor mineral deposits offer some optimism for a future that is based more on local resources. As the population becomes more urban the prospects for decentralisation and greater autonomy for distant archipelagos weaken. Its intrinsic geography challenges the necessity for greater resilience.

We have emphasized the link between the intrinsic and organizational vulnerabilities of French Polynesia and the management of COVID-19. These vulnerabilities are inherent to the functioning of the territory. Whether geographical, social, economic or spatial, they have conditioned the response of the authorities to COVID-19 as well as the vulnerability of the populations and the uncontrolled explosion of cases. This – hopefully – one-off crisis illustrates the processes of dependency, social fragility, and their impact on responses to the current crisis. Faced with these long-term vulnerabilities, the development of resilience strategies is necessary. COVID-19 may be exceptional but French Polynesia regularly experiences cyclones. The concept of resilience can be defined as the abilities and capacities of a place and its people before, during and after a disruptive event in order to limit its negative impacts (Heinzlef et al., 2019). So defined in terms of planning for, adapting, absorbing, recovering from, learning and evolving, resilience must effectively be multidisciplinary and multi-sectoral.

At different scales, technical, urban, social, architectural, economic and political innovations need to relate to traditional risk management strategies. This injunction to innovation is adapted to the urban, economic, political, social and ecological complexity of the contemporary world and of French Polynesia. The practice of resilience is integrated with territorial, social, economic and political issues that might develop from local capacities and resources, to reduce vulnerabilities and promote the implementation of effective risk management strategies. To enable the development and implementation of these resilience strategies, a research project led by IRD and CNRS has been initiated to develop a tool to promote the use of the concept of resilience. The objective of an ongoing Pacific Island Long Term reSilience (ILOTS) project is to achieve the implementation of a resilience that will meet local needs in terms of scientific, technical, social and economic issues and extend into health issues (Heinzlef & Serre, 2019). That may be too late to mitigate the present deadly challenges of COVID-19 but the reduction in future vulnerability is crucial for multiple reasons.

Acknowledgements This project received financial support from the CNRS through the MITI interdisciplinary programs and from the IRD.

References

Abarca-Gómez, L., Abdeen, Z. A., Hamid, Z. A., Abu-Rmeileh, N. M., et al. (2017). Worldwide Trends in Body-Mass Index, Underweight, Overweight, and Obesity from 1975 to 2016: A Pooled Analysis of 2416 Population-Based Measurement Studies in 128·9 million Children, Adolescents, and Adults. *The Lancet, 390*, 2627–2642.

Adger, W. N., Brooks, N., Bentham, G., Agnew, M., & Eriksen, S. (2004). *New Indicators of Vulnerability and Adaptive Capacity.* Tyndall Centre for Climate Change Research.

Aronson, L. (2020, March 28). Ageism Is Making the Pandemic Worse. *The Atlantic.* https://www.theatlantic.com/culture/archive/2020/03/americas-ageism-crisis-is-helping-the-coronavirus/608905/

Birkmann, J. (2007). Risk and Vulnerability Indicators at Different Scales: Applicability, Usefulness and Policy Implications. *Environmental Hazards, 7*, 20–31.

Brooks, N. (2003). *Vulnerability, Risk and Adaptation: A Conceptual Framework* (Working Paper). Tyndall Centre for Climate Change Research.

Burström, B., & Tao, W. (2020). Social Determinants of Health and Inequalities in COVID-19. *European Journal of Public Health, 30*, 617–618.
Burton, C., Rufat, S., & Tate, E. (2018). Social Vulnerability. In S. Fuchs & T. Thaler (Eds.), *Vulnerability and Resilience to Natural Hazards* (pp. 53–81). Cambridge University Press.
Burton, C. G. (2010). Social Vulnerability and Hurricane Impact Modeling. *Natural Hazards Review, 11*, 58–68.
Caballero, A. E., Ceriello, A., Misra, A., Aschner, P., et al. (2020). COVID-19 in People Living with Diabetes: An International Consensus. *The Journal of Diabetic Complications, 34*, 107671.
Cardona, O. (2004). The Need for Rethinking the Concepts of Vulnerability and Risk from a Holistic Perspective: A Necessary Review and Criticism for Effective Risk Management. In G. Bankoff, G. Frerks, & D. Hilhorst (Eds.), *Mapping Vulnerability: Disasters* (pp. 334–363). Development and People, Earthscan.
Carozzi, F., Provenzano, S., & Roth, S. (2020). Urban Density and COVID-19. IZA Institute of Labor Economics 29.
Colenda, C. C., Reynolds, C. F., Applegate, W. B., Sloane, P. D., et al. (2020). COVID-19 Pandemic and Ageism: A Call for Humanitarian Care. *Journal of the American Medical Directors Association., 21*, 1005–1006.
Cutter, S. L., & Finch, C. (2008). Temporal and Spatial Changes in Social Vulnerability to Natural Hazards. *Proceedings of the National Academy of Sciences., 105*, 2301–2306.
Direction de la Santé de Polynésie Française. (2020). Direction de la Santé - Notre Organisation [WWW Document]. Direction Santé Polynésie Française. URL https://www.service-public.pf/dsp/notre-organisation/
Fatemi, F., Ardalan, A., Aguirre, B., Mansouri, N., & Mohammadfam, I. (2017). Social Vulnerability Indicators in Disasters: Findings from a Systematic Review. *International Journal of Disaster Risk Reduction, 22*, 219–227.
Gatti, C., Suhas, E., Côté, S., Laouan-Sidi, E., et al. (2015). Obesity and Metabolic Parameters in Adolescents: A School-Based Intervention Program in French Polynesia. *Journal of Adolescent Health, 56*, 174–180.
Gouvernement de la Polynésie Française. (2016). Schéma d'Organisation Sanitaire de la Polynésie Française. 2016–2021. Gouvernement de la Polynésie française.
Guy, E., & Ailincai, R. (2019). L'implication parentale en Polynésie française: le cas de deux écoles contrastées sur le plan socio-économique. Context. Didact. https://doi.org/10.4000/ced.1327.
Hamidi, S., Sabouri, S., & Ewing, R. (2020). Does Density Aggravate the COVID-19 Pandemic?: Early Findings and Lessons for Planners. *Journal of the American Planning Association, 86*(4), 495–509.
Haut-commissaire de la République Française. (2020). *Déclaration du Haut-Commissaire de la République au sujet du coronavirus.* Papeete.

Heinzlef, C., Becue, V., & Serre, D. (2019). Operationalizing Urban Resilience to Floods in Embanked Territories – Application in Avignon, Provence Alpes Côte d'azur Region. *Safety Science, 118*, 181–193.

Heinzlef, C., Becue, V., & Serre, D. (2020). A Spatial Decision Support System for Enhancing Resilience to Floods: Bridging Resilience Modelling and Geovisualization Techniques. *Natural Hazards and Earth System Sciences, 20*, 1049–1068.

Heinzlef, C., & Serre, D. (2019). Dérèglement climatique et gestion des risques en Polynésie française: conception d'un Observatoire de la résilience. *Cahiers d'Outre Mer, 280*, 531–563.

Institut de la statistique de la Polynésie Française. (2020). Points Conjoncture. Tourisme, Premier Trimestre 2020 (No. 40). ISPF.

Jawad, A. J. (2020). Effectiveness of Population Density as Natural Social Distancing in COVID-19 Spreading. *Ethics, Medicine and Public Health, 15*, 100556.

Kahn, M. (2011). *Tahiti. Beyond the Postcard. Power, Place and Everyday Life.* University of Washington Press.

Karaye, I. M., & Horney, J. A. (2020). The Impact of Social Vulnerability on COVID-19 in the U.S.: An Analysis of Spatially Varying Relationships. *American Journal of Preventive Medicine, 59*, 317–325.

La Dépêche de Tahiti. (2020). *Les pouvoirs du haut commissaire attaqués pendant le confinement.* Dépêche Tahiti. https://actu.fr/politique/les-pouvoirs-haut-commissaire-attaques-pendant-confinement_33342796.html

Niva, P. (2018). *Semaine polynésienne de lutte contre le diabète.* Ministère de la santé et de la prévention.

Outremers 360°. (2020). *Polynésie: un préavis de grève générale contre la gestion de la crise.* http://outremers360.com/politique/polynesie-un-preavis-de-greve-generale-contre-la-gestion-de-la-crise-covid/

Papathoma-Köhle, M., Cristofari, G., Wenk, M., & Fuchs, S. (2019). The Importance of Indicator Weights for Vulnerability Indices and Implications for Decision Making in Disaster Management. *International Journal of Disaster Risk Reduction, 36*, 101103.

Poirine, B. (2010). The Economy of French Polynesia: Past, Present and Future. *Pacific Economic Bulletin, 25*(1), 24–34.

Polynésie la 1ère. (2020). *Covid-19: le pays va emprunter 50 milliards.* Polynésie 1ère. https://la1ere.francetvinfo.fr/polynesie/tahiti/covid-19-pays-va-emprunter-50-milliards-835026.html

Reade, J. (2013). *Imported Foods, Obesity, and Health in French Polynesia.* Sea Education Association. https://www.sea.edu/spice_atlas/food_atlas/imported_foods_obesity_and_health_in_french_polynesia

Rufat, S., Tate, E., Burton, C. G., & Maroof, A. S. (2015). Social Vulnerability to Floods: Review of Case Studies and Implications for Measurement. *International Journal of Disaster Risk Reduction, 14*, 470–486.

Sharma, J., & Ravindranath, N. H. (2019). Applying IPCC 2014 Framework for Hazard-Specific Vulnerability Assessment Under Climate Change. *Environmental Research Communications, 1*, 051004.

Spielman, S. E., Tuccillo, J., Folch, D. C., Schweikert, A., et al. (2020). Evaluating Social Vulnerability Indicators: Criteria and Their Application to the Social Vulnerability Index. *Natural Hazards, 100*, 417–436.

Tate, E. (2012). Social Vulnerability Indices: A Comparative Assessment Using Uncertainty and Sensitivity Analysis. *Natural Hazards, 63*, 325–347.

Townsend, M. J., Kyle, T. K., & Stanford, F. C. (2020). Commentary: COVID-19 and Obesity: Exploring Biologic Vulnerabilities, Structural Disparities, and Weight Stigma. *Metabolism, 110*, 154316.

United Nations. (2020). *Policy Brief: The Impact of Covid-19 on Older Persons.*

Viatge, J.-P. (2019). La dette au plus bas en 2020. Tahiti Infos. https://www.tahiti-infos.com/La-dette-au-plus-bas-en-2020_a185869.html

Wang, D., Hu, B., Hu, C., Zhu, F., et al. (2020). Clinical Characteristics of 138 Hospitalized Patients with 2019 Novel Coronavirus – Infected Pneumonia in Wuhan, China. *Journal of the American Medical Association, 323*, 1061.

Wisner, B. (2016). Vulnerability as Concept, Model, Metric, and Tool. In *Oxford Research Encyclopedia of Natural Hazard Science*. Oxford University Press. https://doi.org/10.1093/acrefore/9780199389407.013.25

Wisner, B., Blaikie, P., Cannon, T., & Davis, I. (2004). *At Risk: Natural Hazards, People's Vulnerability and Disasters*. Routledge.

World Health Organization. (2020). *Health Care Considerations for Older People During COVID-19 Pandemic*. https://www.euro.who.int/en/health-topics/health-emergencies/coronavirus-covid-19/technical-guidance/health-care-considerations-for-older-people-during-covid-19-pandemic

Yan, Y., Yang, Y., Wang, F., Ren, H., et al. (2020). Clinical Characteristics and Outcomes of Patients with Severe Covid-19 with Diabetes. *BMJ Open Diabetes Research and Care, 8*, e001343.

CHAPTER 7

Covid-19 Management in New Caledonia and Wallis and Futuna: A Magnifying Glass for Local Political and Economic Issues

Olivier Hoffer

INTRODUCTION

With no COVID-19 related deaths, 39 confirmed cases and no community transmission recorded in early December 2020, COVID-19 management in the two French Pacific territories of New Caledonia and Wallis and Futuna appeared rather effective compared to the situation in other French overseas territories, including French Polynesia (Taglioni, 2020). However, anti-contagion COVID-19 policies provoked several controversies in both the radically different but closely linked territories, revealing their historical and institutional distinctiveness.

Beyond obvious contrasts in land and population size, New Caledonia and Wallis and Futuna political statuses are quite distinctive within an already diverse group of French overseas territories (David et al., 2016). New Caledonia is an archipelago of 18,575 km^2 and around 275,000 inhabitants whereas Wallis is a low-lying island of 78 km^2 and 8000

O. Hoffer (✉)
University of New Caledonia, Noumea, New Caledonia

Y. Campbell, J. Connell (eds.), *COVID in the Islands: A comparative perspective on the Caribbean and the Pacific*,
https://doi.org/10.1007/978-981-16-5285-1_7

inhabitants and Futuna is a mountainous island of 64 km² of 3200 inhabitants. New Caledonia has a complex population structure of indigenous Melanesians (Kanaks), now a minority, and more recent European and other migrants, most of whom live in the capital, Noumea. Wallis and Futuna is almost exclusively Polynesian, although the two islands have separate languages. While New Caledonia, in the wake of the Noumea Accord of 1998, became a *sui generis* French collectivity with considerable autonomy leading towards a process of self-determination, Wallis and Futuna was established in 1961 as a French overseas territory based on a highly unusual balance of power between the French Republic, the Catholic Church and the three customary kingdoms of Uvea in Wallis, and Sigave and Alo in Futuna (Soulé, 2005). In New Caledonia, the current division of power between the French Republic, the New Caledonia Government, the three Provinces and customary authorities originated from a succession of statuses following the indigenous Kanak struggle for independence in the 1980s, locally known as "the events" (David, 2019).

The self-determination process has dominated politics in New Caledonia for about 30 years (Connell & Aldrich, 2020; Connell, 2019 David, 2019). Most recently that led to a referendum on independence in November 2018 (in which 44% of voters supported independence) followed by a second referendum in September 2020, complicated by the COVID-19 pandemic, when rather more of the electorate – now 46% – voted for independence. To a lesser extent, the referendum also affected Wallis and Futuna, almost 2000 km from New Caledonia, though closely linked to it by the transportation network and through a significant diaspora mainly in Noumea, New Caledonia's capital city.

This chapter examines the implementation and impact of COVID-19 mitigation strategies in both territories, whose success mainly stemmed from islandness and remoteness, but also from specific institutional frameworks and processes. It then analyses the way in which COVID-19 management has been designed and sometimes used depending on local political agendas and economic issues.

Islandness, Remoteness and the Diffusion of COVID-19

Like many other island states in the pandemic, New Caledonia and Wallis and Futuna benefited from a relatively limited integration into the global transportation network, with preventive safety measures mainly conducted in international airports and port facilities. However, New Caledonia and Wallis and Futuna have very different transportation connectivities.

Tontouta international Airport, 52 km from Noumea, is New Caledonia's main gateway for overseas travellers, with 566,405 passengers being handled in 2019 (IEOM, 2020a). Four airlines (the national airline, Aircalin, and Qantas, Air New Zealand and Air Vanuatu) operate flights to 10 destinations in 5 countries and 2 French overseas territories. Although passenger traffic and flight frequency are relatively modest, New Caledonia is directly and daily connected to Tokyo and Sydney airports, which are among the busiest of the world, and closely connected to China, the epicenter of the pandemic. Daily Aircalin flights to Tokyo, and less frequently to Osaka, were mainly used by to reach France, connecting to Air France flights to Paris. Therefore, at the start of the pandemic, the risk of a rapid spread of the virus from Asia, Australia or Europe was not negligible.

Wallis and Futuna, on the other hand, is only accessible from New Caledonia, with two or three Aircalin flights per week. The Wallis airport handled only 32,204 passengers in 2019 (IEOM, 2020b), the vast majority of whom were either local residents or part of the Wallisian and Futunian diaspora (roughly twice the size of the resident Wallis and Futuna population) in New Caledonia. Futuna is particularly isolated (Gaillot, 2012), as this small mountainous island is only served by Point Vele Airport (handling 13,354 passengers in 2019), whose short runway enables only flights by a DHC-6 Twin Otter with a capacity of 10 passengers. In addition, flight cancellations occur when the wind does exceed 20 knots, or during heavy rainfall episodes. This quasi "hyper insularity" (Taglioni, 2011) considerably lessens the risk of COVID-19 infection in Futuna and, to a lesser extent, in Wallis. The same infection risk gradient applies to New Caledonia between the Mainland (Grande Terre) and the Loyalty Islands (Lifou, Mare, Tiga and Ouvea), Isle of Pines or Belep Islands, all connected primarily by similarly limited air services.

While islandness and remoteness are advantageous as long as these territories are COVID-free, their smallness potentially impeded regulation in the event of the virus arriving. Indeed, without significant French government subsidies, the few inhabitants of these two territories would not create enough tax revenue to provide efficient health services. New Caledonia received 1.48 billion Euros in 2018 (IEOM, 2020a), approximately 15% of its GDP. Wallis and Futuna received 143 million Euros in 2019 (IEOM, 2020b), including 40 million Euros for the Health Agency. The former amount represents more than 80% of its GDP. That has enabled the construction of health infrastructure that, in New Caledonia, has even been described as 'oversized for the population of the island' (Chauchat, 2018: 502). Although France's fund transfers have contributed to the construction of well-equipped hospitals that would be the envy of most small island Pacific countries (Saïdi, 2020a), most medical facilities are concentrated in Noumea, whereas other areas are relatively poorly equipped, and medical referrals to Australia continued. While 1426 beds are distributed in six hospitals in New Caledonia (all in the Grande Terre), all the 43 critical beds (20 resuscitation beds and 23 intensive care beds) are in a single Médipôle hospital in Noumea (IEOM, 2020a). Although the Médipôle medical team announced in March 2020 a possible extension of COVID-19 bed capacity to 102 beds in the event of a major outbreak (including 49 critical beds, at twice the rate of France itself) (Saïdi, 2020b), any uncontrolled spread of hundreds or thousands of COVID-19 cases could have quickly led to a health disaster, including any transportation issues for patients from remote places like the Loyalty Islands. As for Wallis and Futuna, two hospitals provide 51 beds (42 in Wallis and 9 in Futuna), including just one critical bed in Wallis Sia Hospital (IEOM, 2020b). The Wallis and Futuna medical team could not afford any virus diffusion.

Another element of vulnerability to COVID-19 lay in local health issues, especially the importance of comorbidity factors, due to a high prevalence of diabetes, hypertension and obesity. Most recent surveys showed high incidences: 38% of Caledonian adults and 70% of Wallisian and Futunian adults were obese, while 8% of Caledonian and 14% of Wallisian and Futuniasn were living with type 2 diabetes (ASS-NC, 2017).

The role of history was a further critical key in understanding the perception of the pandemic and thus the resultant management strategies. Deadly epidemics in the colonial era remain remarkably alive in Caledonian or Wallisian and Futunian memories. Repeated dysentery, smallpox, measles and flu outbreaks occurred in New Caledonia since the landing of James Cook in September 1774, destabilising Kanak societies and killing up to 80% of some tribes' populations (Sand, 1995; Kirch & Rallu, 2007). The last major outbreaks were the bubonic plague in 1899–1900 and 1912 (Cavert, 2016) and the Spanish Flu in 1918 (Shanks & Brundage, 2013). Outbreaks of leprosy were quite common until World War II in the Caledonian bush, motivating the relocation of certain Kanak villages by colonial authorities, like N'Dé in Greater Noumea's municipality of Païta in 1930. In Wallis Island, a typhoid outbreak in 1936 killed 5% of the population, despite zealous mitigation efforts of Dr. Jean-Joseph David, like the use of some islets in the lagoon as quarantine camps (Lachenal, 2017). The impact of these epidemics had not been forgotten.

Epidemic Management Controversies Despite Two (Almost) COVID-Free Territories

The historic and contemporary context of health care in New Caledonia partly explains the fact that Kanak customary authorities as early as mid-January had called for stronger measures to reinforced sanitary controls at Tontouta Airport. On 31 January 2020 the Customary Senate of New Caledonia declared the need to establish an immediate international travel ban. As soon as the first suspected COVID-19 case was declared on 6 February, and tested negative a few days later, customary councils of Loyalty Islands and Isle of Pines ordered a cruise ship ban (*Les Nouvelles Calédoniennes*, 2020b), despite the critical role of cruise tourism in the local economy. That first case was a New Caledonian woman, returning from China via Tokyo, who was put in quarantine in the Médipôle, along with 14 others who had been in close contact with her.

During the following weeks, an increasing number of politicians and doctors raised concerns about the New Caledonia government's strategy, that was seemingly based on waiting for the first confirmed case and then closing the borders. The Directorate for Health and Social Affairs (DASS) had however advocated an international travel ban on 10 March 2020. As Dr. Joël Kamblock pointed out on 17 March 2020: "We are one of the rare territories to be a step ahead. It's precious time. Waiting until the first case or the first hospitalization to take action would be too late. The first confirmed case means that many other infected people are already present. If we implement the same policy as in France, then our insularity will turn against us" (*Les Nouvelles Calédoniennes*, 2020a). Likewise, the former senator Simon Loueckhote called for the High Commissioner, the President of Government and the President of Loyalty Islands Province to establish a total lockdown of Ouvea Island.

The two first cases were finally diagnosed on 18 March, a Caledonian couple arriving from their honeymoon in Sydney, followed the day after by the announcement of anti-COVID-19 measures by Thierry Santa (President of the New Caledonia Government) and Laurent Prevost (High-Commissioner). These measures contained in the order Number

2020–4608 (https://juridoc.gouv.nc/JuriDoc/JdWebE.nsf/Juristart?openpage), included a ban on public gatherings with more than 20 people (including sport, cultural, religious and customary gatherings), closure of schools and the university, suspension of public transport (especially air and maritime connections to the Loyalty Islands), an international travel ban from 20 March (except for a reduced number of flights to Paris for freight delivery and other compelling reasons) and one flight every 10 days to Wallis. All arriving passengers were to quarantine for 2 weeks in designated Noumea hotels. A complete lockdown was planned from 25 March for at least 3 weeks. Other measures consisted of setting purchasing restrictions in grocery stores (such as 2 kg of rice and pasta per person).

Alongside the fact that approximately 2200 Caledonian people were suddenly stuck overseas, the communication of 3 new cases on 22 March 2020 sparked immediate and mixed reactions in the country. On the night of 22–23 March, a group of 60 Kanaks stoned entry hall windows of Tontouta Airport, as well as many parked cars and some airport shuttles on the road to Noumea. At dawn on 23 March, several filtering roadblocks were set up at the entry points of Kanak tribes in various parts of the Grande Terre, including Poindimie (Wagap, Napoémien), Kone (Neami), Pouembout, Bourail (Gouaro) and Yate, in order to ban contacts with non-tribal residents. Some tribal customary councils also ordered a curfew from 8 pm. All these spontaneous customary policies and activities were probably representative of the fearful memory of deadly epidemics brought by European sailors and settlers. More broadly, it also revealed the lack of understanding if not the disapproval of many Caledonians regarding the strategy of the Government, largely seen as political inertia.

Overall, the New Caledonian people accepted lockdown rules, especially the instruction to stay at home and only leave home with an exemption certificate to be completed and signed every day. Exemptions included commuting from and to work (only if the employer certified that the job could not be done at home), medical appointments that could not be postponed, essential family commitments (assisting vulnerable people and children), shopping for essential goods, and walking pets or exercising outdoors (alone or with family members) within 1 km of home for only 1 hour per day. The 1 km radius and the one-hour duration had been quickly added, a week after the initial order, after the police had observed an enormous number of people devoted to outdoor sporting activities. Fines of between 89,500 F.CFP and 447,493 F.CFP (respectively 892 and 4464 US$) could be handed out to lockdown rule breakers but, according to the police, these were relatively few. Perhaps surprisingly, unlike in other French overseas territories, including the Caribbean territories of Martinique and Guadeloupe and in Réunion and French Polynesia, masks were simply recommended in outdoor public spaces or in schools. Even though they were mandatory in public transport from 25 March to 15 June 2020, many bus users did not wear them. This complacency was tolerated by the Government because of the mask shortage and the high price of masks for the majority of bus users, who were often relatively impecunious.

As the last imported case of COVID-19 was confirmed on 2 April and no community transmissions had been detected, the Government decided to ease lockdown restrictions from 20 April 2020. With 18 imported cases, no deaths and only one patient admitted to intensive care, COVID-19 management was sufficiently successful to ease travel restrictions between New Caledonia and Wallis and Futuna. A quarantine free travel bubble was therefore established on 5 June 2020, with flight frequency increased to 2 flights per week. However, since early June, a few new imported cases were detected each month in New Caledonia, usually thanks to tests made just before the end of the 2 weeks hotel quarantine imposed on all arrivals. A total of 32 imported cases were registered between 18 March and 20 November 2020, New Caledonia being ranked as the 15th least infected country in the world (205 out of 220, just before Fiji) according to international data (https://www.worldometers.info/coronavirus/). The Government's strategy of early travel quarantine, fast and widespread testing (at twice the rate in France), contact isolation and case tracing had proved to be efficient (Saïdi, 2020b).

Until 16 October 2020, Wallis and Futuna was one of the last territories of the world to be labelled "COVID-free". A first and asymptomatic case arrived from France on 3 October and was tested positive at the end of the 2 weeks hotel quarantine, swiftly rising fear and even anger among the population. As a result, in the evening, a group of people claiming themselves to be a "Collective for the People's interests" blocked Wallis Airport for a few hours. Just before the blockage, the Administrator of Wallis and Futuna had denounced the uselessness of this action, which according to him was just "a way to release fear and insecurity", but also revealed "a lack of courtesy and cleverness" (https://la1ere.francetvinfo.fr/wallis-et-futuna-le-premier-cas-de-covid-19-entraine-un-blocage-de-l-aeroport-par-un-collectif-882622.html). With two more asymptomatic cases confirmed on 13 November and 24 November, the Wallisian and Futunian people and the authorities became increasingly divided about the way to avoid a spread of the virus.

COVID-19 Management Embroiled in Local Geopolitics

The evolution of COVID-19 management strategies in both Wallis and Futuna, and New Caledonia, demonstrated the complexity and peculiarity of local political issues. Both territories are characterised by complex and ambivalent political attitudes towards the French Government, alternatively seen as a potential threat to autonomy, an arbitrator of community and clan disputes or a support for development and welfare provision. The management of COVID-19 highlighted that very ambiguity, resulting in intricate political practices. It also revealed several political and cultural strata and fissures, usually less conspicuous in non-crisis times.

The first controversy was related to the closing of international borders, regarded as much too late by pro-independence leaders, as mentioned before, but also as part of a policy imposed by the French Government in collusion with the anti-independence leaders, especially Thierry Santa, the President of New Caledonia Government. Thus, Pierre Chanel Téin Tutugoro (of the pro-independence *Union Calédonienne* party) declared that the High Commissioner had encroached on local government prerogatives, by imposing a specific management mode. In a public letter issued on 18 May 2020, Daniel Goa (the leader of *Union Calédonienne*), denounced a breach of the Noumea Accord (that accorded a considerable measure of autonomy to New Caledonia) by the inclusion of New Caledonia in health emergency laws voted by the French Parliament on 23 March 2020 and 11 May 2020, since health affairs were the prerogative of the New Caledonia government. He then called for the dismissal of both the High Commissioner and the Commander of the French armed forces in New Caledonia, and the complete end to all flights from France in order to avoid new infections. Pro-independence leaders also denounced the arrival of numerous military personnel and raised concerns about the effectiveness of their 2 weeks quarantine in barracks.

Such assertions of French Government interference in COVID-19 management were surprising to many, as the COVID decision-making committee was composed in a collegial manner, including the Government's

President, the High Commissioner, the three Province Presidents and customary authorities. In a nutshell, this decision-making committee (working with an independent scientific council and a Government operational unit) seemed to epitomise and represent the diversity of New Caledonia's institutions (David et al., 2016; Saïdi, 2020b). Pro-independence leaders' attacks were therefore part of established and familiar political disputes, exacerbated in the lead-up to the 2020 referendum on independence. Both political camps accused each other of exploiting COVID-19 management in order to gain influence. Thus pro-independence leaders called for a postponement of the referendum, originally scheduled for 6 September 2020, to 25 October by invoking the disturbance of the referendum campaign caused by the virus. Most of the anti-independence leaders were opposed to any delay, and some of them speculated that the pro-independence parties were privately negotiating a rescheduling with the French Government. Eventually the French Prime Minister, Edouard Philippe, proposed the referendum for the 4 October 2020, a decision perceived by the pro-independence leaders as lacking impartiality. Though of minor importance, this episode symbolised the complexity of politics in the country.

The implementation of anti-COVID-19 measures also highlighted the functional evolution of New Caledonian institutions over the last 20 years, notably a shift towards territorial federalism. The three provinces, created after the Noumea Accord to give greater authority to what were the two Kanak dominated regions of the North and the Loyalty Islands, benefited from French financial support (notably in the fields of economy, education and environment), which enabled very different public policies and development strategies (David et al., 2016). The North and Loyalty Islands Provinces, led by pro-independence parties since their creation, have seen a growing influence of customary authorities, though that was limited until the COVID-19 crisis (Pantz & Robertson, 2018). On 19 March, the customary council of Lifou (Loyalty Islands) decided to ban all domestic flights and ferry services (except for emergency cases and for freight) 6 days before the official decision taken by the New Caledonian Government. A similar ban was also announced by the council of neighbouring Ouvea island 3 days later. In addition, customary authorities in Lifou enforced more severe restrictions than did government authorities, for instance a ban on alcohol sales, reduced shop opening hours, traffic controls, and the postponement of all weddings until 2021. Kanaks were rather more concerned about the impact of COVID-19 than European French.

The Customary authorities occasionally overshadowed provincial authorities, as on 9 April, when Jacques Lalié, the President of Loyalty Islands Province, called for ending lockdown in Lifou, citing the total absence of cases locally and the efficient management in the country. The following day, the President of the Customary Senate, Hippolyte Wakemi Sinewami-Htamumu, rejected the proposition, in stating that end of lockdown should not occur before the end of April. The President himself came from Mare, the third main island in the Loyalty Islands Province. What became a 3 weeks blockage of Air Calédonie (Aircal) flights by the Lifou Customary Council, in reaction to a slight price rise of 350 F.CFP (3.5 US$) due to COVID-19 related loss, was also indicative of the flexing of local Kanak authority, even though Aircal was largely owned by the New Caledonia Government (52%) and the Loyalty Islands Province (26%) and mainly served the Loyalty Islands.

In a wider sense, COVID-19 management provided an opportunity for all political institutions to emphasise their power and autonomy, which sometime fuelled speculation about a looming political partition whatever the result of the referendum would be. Thus the reopening of school date, scheduled by the New Caledonian Government for 22 April, was deferred until 4 May in North Province, citing the recommendation of the Customary Senate. This tendency towards federalism also included tourism promotion, with the withdrawal of both North and Loyalty Islands Provinces from the New Caledonia Tourism Point South (NCTPS) tourism recovery plan in May 2020.

In the same way, local and diverse approaches to COVID-19 management in Wallis and Futuna reignited debates about the level of dependence on both France and New Caledonia, as well as Futuna's dependence on Wallis. Indeed, the creation of the COVID-19 monitoring committee, including customary authorities, the Territorial Assembly, the High Administrator and the Catholic Church, proved an opportunity to address the old question of the modernisation of the territory's status. The extension of powers of the Territorial Assembly and the transfer of the Administrator's executive power to a new local authority, had been under discussion for more than 20 years, largely frustrated by the unwillingness of customary authorities to secede any power (Joissains & Sueur, 2014).

Many local leaders believed that a more autonomous status would offer a chance to reduce dependence on New Caledonia, especially by challenging Aircalin's monopoly on air transport, which led to overpriced flights (with a Wallis-Noumea return flight usually costing from 80,000 F. CFP (810 US$) to 120,000 F.CFP (1212 US$) and regular medicine shortages due to low freight capacity. (A proposal in 2019 to launch a local airline, WF Aviation, for the same reasons, to fly to New Caledonia, French Polynesia, and additionally Fiji, had come to nothing). However, in the COVID-19 context, the demand for better air connections became a divisive and paradoxical issue, regarding both the temptation to isolation and the needs for medical evacuations to Noumea. Being particularly concerned about the first COVID-19 case confirmed in Wallis in October, the Kingdom of Alo called for a reduction of flights to just once a month, while the Kingdom of Sigave asked for the repatriation to France of the infected person. Futuna's customary authorities subsequently requested that Futuna become a separate territory in order to end its dependence on Wallis.

COVID-19 Questioning Economic and Financial Dependence on France

While the international travel ban and the 3 weeks lockdown probably spared New Caledonia from a number of COVID-19 infections, it severely threatened an already slowing local economy (IEOM, 2020a). An economic recovery plan was consequently announced by the Government on 3 April. It was essentially based on partial unemployment benefits permitted for all companies whose turnover had decreased by at least 25% compared with the same month of 2019. These partial unemployment benefits were intended to guarantee 100% of minimum wages and 84% of average wages. Tax deferrals for businesses were also arranged.

The funding of this plan was concluded with the French Development Agency (AFD) on 13 May involving a substantial state loan of 28.6 billion F.CFP (286 million USD) with a repayment period of 25 years and 2 years deferred payment. This loan was also intended to ease the burden of repatriation flights to New Caledonia and the hotel quarantine process. All these measures led to the territory's budget skyrocketing from 66.5 billion F.CFP in 2019 to 93.7 billion F.CFP (665 million US$ to 937 million US$). What would rapidly be locally referred to as the "grand loan" sparked an ongoing controversy, as it questioned New Caledonia's financial dependence on France on the eve of a possible accession to full sovereignty (dependent on the referendum outcome).

Several political leaders, especially members of all the pro-independence parties (most under the umbrella of the *Front de Libération Nationale Kanak et Socialiste* (FLNKS)) but also *Calédonie ensemble* (a centrist anti-independence party), denounced the lack of transparency during the loan negotiation between the President of Government and the French Development Agency, whose conditions should have required further discussion in Congress. Thus, the 1.48% interest rate was regarded as staggeringly high while France itself could borrow at negative rates. Moreover, the loan agreement stated that New Caledonia committed to implementing "structural reforms to consolidate public finances", a major tax reform. Anti-independence leader, Philippe Gomès (*Calédonie ensemble*) disregarded the loan as "a total humiliation which will generate a complete loss of sovereignty in terms of taxes, social care healthcare" (*Les Nouvelles Calédoniennes*, 2020c). Others, like pro-independence leader Pierre-Chanel Tutugoro (FLNKS), felt "betrayed, deceived", pointing out the fact that "We are well aware that debt is a sort of planned reconquest. We have this feeling that in France's view, this debt issue is aimed at forcing us to do certain things and at controlling achievements, thanks to a monitoring committee composed by High Commission and the French Development Agency" (*Les Nouvelles Calédoniennes*, 2020c). Such debts would further threaten any move to a viable independence.

Although the Government would henceforth be embroiled in a budget controversy, the "grand loan" was probably necessary to avoid the collapse of the economy in such an important electoral year. It was particularly helpful for the tourism sector, which paradoxically saw new opportunities notwithstanding the international travel ban.

COVID-19 Restrictions Generating New Domestic Tourism and Development Opportunities

Border closure had meant huge revenue losses for New Caledonia's tourism industry, especially for the cruise ship sector. New Caledonia was a modest tourism destination which welcomed 130,458 international tourists and 343,962 cruise passengers in 2019. The sector accounted for 4% of the country's GDP in 2014, and employed 5387 workers (7% of the private sector workforce) in 2019 (IEOM, 2020a). However, tourism represented the second export sector behind nickel mining and its impact can be locally critical for some communities in the Loyalty Islands, the Isle of Pines and the Grande Terre's rural areas. Thus, 191,181 cruise passengers visited Lifou in 2019, bringing in an estimated 200 million F.CFP (2 million USD) in revenue in 2014, and maintaining approximately 200 jobs (Bouard et al., 2016). This is significant for an island population of only 9195 in 2019, although cruising benefited mainly just the northern part of Lifou, where cruise ships could anchor. In the same way, the Isle of Pines and Mare's economy increasingly rely on cruise ships sailing from Australia, whereas hotels and tribal accommodation in the rural Grande Terre largely depend on French visitors or newly arrived residents.

The tourism recovery plan unveiled on 4 May by South Province authorities was based on a strategy of enticing Caledonian people into visiting and rediscovering their own islands, by offering low-priced holiday packages. Calling for domestic "solidarity tourism", New Caledonia's authorities particularly sought to make the most of high-income households stuck within the country, particularly non-permanent residents working in the public sector who might be interested in discovering local places during their stay. (Many French civil servants are temporarily employed in New Caledonia under 2 years to 4 years term contracts and benefit from multiple financial advantages, notably French salaries multiplied by 1.73 or 1.94, plus a 10-month salary bonus every 2 years). Most work in the military and in education. Local civil servants also benefited from the wage indexation mechanism). Since 143,785 overseas trips of Caledonian residents occurred in 2019, rather more than international arrivals in the country, border closure suggested real domestic opportunities. Indeed since 2010, Caledonian people's stays in local accommodation surpassed those of international visitors, circumstances attributed to high-priced flights resulting from Aircalin's *de facto* monopoly (Blaise et al., 2016). An already important, domestic tourism boomed from May 2020. While the average occupancy rate for Noumea's hotels was only 59%, some accommodation in places like Bourail, La Foa and Isle of Pines, around 100 kms from Noumea, were in May 2020 fully booked until the end of 2020, if not the end of 2021 (*Les Nouvelles Calédoniennes*, 2020d). For several hotels facing budget deficits, like the Sheraton Deva in Bourail, COVID-19 proved a godsend, by attracting new guests and allowing the resort to finally become profitable. The Sheraton Deva opened in 2014, was the only five-star hotel outside Noumea, and until COVID-19 had an average occupancy rate of only 33%, and was on the verge of closing according to South Province authorities, one of the main shareholders. On the other coast of the Grande Terre, Lorenza M'Boueri, the head of

Thio's tourism office, summarised the situation: "If there were this many people all year long, there would be real tourism development". Optimism had been kindled.

Remarkably, seemingly against all odds, Wallis and Futuna's tourism sector also seemed set to make a profit from the quarantine free travel bubble with New Caledonia established in June 2020. Tourism was negligible in normal times, discouraged by intervening opportunities, high costs and effective invisibility, resulting in less than a hundred tourists a year (other than returning Wallisians and Futunans). It is one of the few places in the world without a single taxi. All that changed in 2020 as the number of Caledonian tourists increased significantly, according to hotel and accommodation owners, which further encouraged the territory to engage in creating the islands as a new niche tourism market and a more visible destination. That resulted in the launching of its tourism promotion website in September 2020 (https://www.wallis-futuna.travel/fr/) and its recent membership to SPTO (South Pacific Tourism Organisation) suggest. Out of adversity had come a new possibility.

Conclusion

The largely successful management and exclusion of COVID-19 in New Caledonia and Wallis and Futuna is indicative of how islandness, vulnerability and resilience are intertwined in island territories (Campbell, 2009; Baldacchino, 2014). Island resilience notably lies in the capacity to be food self-sufficient, highlighting the critical role of local food production in tribal and rural areas in New Caledonia and Wallis and Futuna, despite substantial food imports (much more evident in Noumea). The creation of new domestic tourism opportunities exemplify a part of island resilience, in fostering a partial revival of local living and consumption. However, while COVID-19 management was based on an international travel ban, which had valuable health consequences and revived domestic tourism, this had resulted in as many as 6000 Caledonian and Wallisian people still stuck overseas at the end of November 2020, and unable to return, despite repatriation flights.

In the meantime, the pandemic exemplified how global health issues and local political and cultural issues can be intertwined. In a crucial electoral year, supporters and opponents of independence used and manipulated the pandemic, in search of their own political objectives. Likewise indigenous Kanak cultural organisations took a lead in emphasising the need for decisive action, something of a contrast to their own traditionally more conservative role. Despite some controversies and political wrangling and manoeuvring, COVID-19 management proved quite efficient, perhaps even due to the diverse backgrounds and sometime conflicting approaches of all stakeholders. What might have seemed a paradoxical and indecisive approach typifies New Caledonia's political practices, based on constant consensus-building in a consociational democracy paradigm (David, 2019). It has even been suggested that not only is this is a more effective approach to health care management, but it is an improvement on the "Do Kamo" health system reform designed in collaboration with the WHO in 2018 (Saïdi, 2020a) by embracing both culture and political realities. It has at least offered some degree of optimism for the future.

References

ASS-NC (Agence sanitaire et sociale de Nouvelle-Calédonie). (2017). *Le Baromètre Santé Adulte 2015.* http://www.ass.nc/etudes-et-recherches/barometres-sante/barometre-sante-adulte-2015

Baldacchino, G. (2014). Small Island States: Vulnerable, Resilient, Doggedly Perseverant or Cleverly Opportunistic? *Études caribéennes* 27–28. http://journals.openedition.org/etudescaribeennes/6984, https://doi.org/10.4000/etudescaribeennes.6984.

Blaise, S., Sourisseau, J.-M., Hoffer, O., Bouard, S., et al. (2016). Des mineurs, des métallurgistes et des entrepreneurs au défi de la concurrence internationale. In S. Bouard, J.-M. Sourisseau, V. Geronimi, S. Blaise, & L. Ro'i (Eds.), *La Nouvelle-Calédonie face à son destin. Quel bilan à la veille de la consultation sur la pleine souveraineté?* (pp. 111–185). Éditions Karthala.

Bouard, S., Sourisseau, J.-M., Bellec, S., Hoffer, O., et al. (2016). Des stratégies de développement local volontaristes et différenciées. In S. Bouard, J.-M. Sourisseau, V. Geronimi, S. Blaise, & L. Ro'i (Eds.), *La Nouvelle-Calédonie face à son destin. Quel bilan à la veille de la consultation sur la pleine souveraineté?* (pp. 391–447). Éditions Karthala.

Campbell, J. (2009). Islandness: Vulnerability and Resilience in Oceania. *Shima: The International Journal of Research into Island Cultures, 3*(1), 85–97.

Cavert, W. (2016). At the Edge of an Empire: Plague, State and Identity in New Caledonia, 1899–1900. *The Journal of Pacific History, 51*(1), 1–20.

Chauchat, M. (2018). New Caledonia. *The Contemporary Pacific, 30*(2), 501–511.

Connell, J. (2019). Another Pause for Independence? The 2018 New Caledonia Referendum. *The Round Table, 108*(3), 241–258.

Connell, J., & Aldrich, R. (2020). *The Ends of Empire: The Last Colonies Revisited.* Palgrave Macmillan.

David, C. (2019). The Road to Sovereignty for New Caledonia? Analysis of the November 2018 Consultation on Self-Determination. *Small States and Territories, 2*(2), 141–156.

David, C., Sourisseau, J.-M., Gorohuna, S., & Le Meur, P.-Y. (2016). De Matignon à la consultation sur l'indépendance: une trajectoire politique et institutionnelle originale. In S. Bouard, J.-M. Sourisseau, V. Geronimi, S. Blaise, & L. Ro'i (Eds.), *La Nouvelle-Calédonie face à son destin. Quel bilan à la veille de la consultation sur la pleine souveraineté?* (pp. 23–67). Éditions Karthala.

Gaillot, M. (2012). Futuna. Un isolement pénalisant. *Journal de la Société des Océanistes, 135*(2), 265–268.

IEOM (Institut d'Émission d'Outre-Mer). (2020a). *Nouvelle-Calédonie: rapport annuel 2019*. https://www.ieom.fr/nouvelle-caledonie/publications/rapports-annuels/rapports-d-activite/article/rapport-annuel-2019-de-l-ieom-nouvelle-caledonie

IEOM (Institut d'Émission d'Outre-Mer). (2020b). *Wallis-et-Futuna: rapport annuel 2019*. https://www.ieom.fr/nouvelle-caledonie/publications/rapports-annuels/rapports-d-activite/article/rapport-annuel-2019-de-l-ieom-wallis-et-futuna

Joissains S., & Sueur J.-P. (2014). *Rapport d'information n°103 sur les îles Wallis et Futuna*. Sénat. https://www.senat.fr/rap/r14-103/r14-1030.html

Kirch, P. V., & Rallu, J.-L. (2007). *The Growth and Collapse of Pacific Island Societies: Archaeological and Demographic Perspectives*. University of Hawaii Press.

Lachenal, G. (2017). *Le Médecin qui voulut être roi: sur les traces d'une utopie coloniale*. Éditions du Seuil.

Les Nouvelles Calédoniennes. (2020a, March 23). *Covid-19: des mesures fortes attendues*. https://www.lnc.nc/article/nouvelle-caledonie/sante/covid-19-des-mesures-fortes-attendues

Les Nouvelles Calédoniennes. (2020b, May 7). *On a fait trop peu et trop tard face à l'épidémie de Covid-19 et on n'a pas fini d'en payer le prix*. https://www.lnc.nc/article/nouvelle-caledonie/politique/on-a-fait-trop-peu-et-trop-tard-face-a-l-epidemie-et-on-n-a-pas-fini-d-enpayer-le-prix

Les Nouvelles Calédoniennes. (2020c, September 4). *L'emprunt de 28 milliards enflamme le congrès*. https://www.lnc.nc/article/nouvelle-caledonie/politique/l-emprunt-de-28-milliards-enflamme-le-congres

Les Nouvelles Calédoniennes. (2020d, May 8). *Les Calédoniens appelés à sauver le tourisme*. https://www.lnc.nc/article/nouvelle-caledonie/societe/les-caledoniens-appeles-a-sauver-le-tourisme

Pantz, P.-C., & Robertson, S. (2018). *Exploring the Kanak Vote on the Eve of New Caledonia's Independence Referendum*. ANU Department of Pacific Affairs Discussion Paper No. 8

Saïdi, A. (2020a). *Comprendre la Nouvelle-Calédonie à travers son système de santé*. L'Harmattan.

Saïdi A. (2020b, May 15). *La stratégie de la Nouvelle-Calédonie face au Covid-19: un modèle pour les pays insulaires. Fondation Jean Jaurès*. https://jean-jaures.org/nos-productions/la-strategie-de-la-nouvelle-caledonie-face-au-covid-19-un-modele-pour-les-pays

Sand, C. (1995). *Le temps d'avant: La préhistoire de la Nouvelle-Calédonie*. L'Harmattan.

Shanks, G. D., & Brundage, J. F. (2013). Pacific Islands Which Escaped the 1918–1919 Influenza Pandemic and their Subsequent Mortality Experiences. *Epidemiology and Infection, 141*(2), 353–356.

Soulé, M. (2005). Relations coutume, État, Église à Wallis et Futuna 1837–1961. *Outre-mers, 92*(348–349), 117–125.

Taglioni, F. (2011). Insularity, Political Status and Small Insular Spaces. *Shima: The International Journal of Research into Island Cultures, 5*(2), 45–67.

Taglioni, F. (2020). La Covid-19 comme indicateur des spécificités sanitaires dans les outre-mers français: le cas de Mayotte. *Carnets de Recherches de l'océan Indien, 5*, 127–132.

CHAPTER 8

Double Jeopardy: Distance and Decentralisation in Tuvalu

John Connell

Introduction

Tuvalu is one of the smallest and most isolated nations in the world with a Polynesian population of about 10,600 people scattered across nine populated coral atolls, but with a growing population of more than 3000 in New Zealand. More than half the population usually live on the urbanised island of Funafuti. Tuvalu was one of the very few states to have remained free from COVID-19 late in 2020. It was however much affected, after a national State of Emergency was declared late in March, as transport links were cut and the nation moved towards greater self-reliance.

Beyond fisheries, local development opportunities are few. Tuvalu has long been dependent on international migration and remittances alongside aid, and has regularly sought new migration opportunities overseas especially in the context of potentially problematic rising sea levels (Connell, 2003; Mortreux & Barnett, 2009; Shen & Binns, 2012; Shen &

J. Connell (✉)
School of Geosciences, University of Sydney, Sydney, NSW, Australia
e-mail: john.connell@sydney.edu.au

Y. Campbell, J. Connell (eds.), *COVID in the Islands: A comparative perspective on the Caribbean and the Pacific*,
https://doi.org/10.1007/978-981-16-5285-1_8

Gemenne, 2011). Indeed Tuvalu has all too often been seen as the ultimate national 'canary in the coal mine' of climate change with the resilience and will of the local population doubted and the necessity for population resettlement seemingly imminent (Farbotko, 2010). The global pandemic of COVID-19 has brought double jeopardy: one more national challenge of external origin, alongside climate change. That has been met by reversing the flow of migration from urbanisation to decentralisation, seeking to revitalise agriculture and being the first SIDS to put in place a monthly Temporary Basic Income, to provide a financial lifeline for households. It also created an incentive for people to stay at home for the duration of the initial partial lockdown.

An Urban World?

Since Tuvalu became an independent state in 1979, separating from the Micronesian Gilbert Islands (Kiribati) in search of its Polynesian identity, it has been characterised by the growth of an urban bureaucracy and the transformation of the capital island of Funafuti into a small-scale version of classic urban primacy. An early urban bureaucratic development led to its being seen as one of the prototypes of MIRAB economies, characterised by migration, remittances, aid and bureaucracy (Bertram & Watters, 1985), quickly transformed for Tuvalu into MURAB (Munro, 1990; Besnier, 1995), though neither epithet was appreciated locally. However both Tuvalu and Kiribati were impoverished and aid donors perceived the need for non-traditional forms of assistance that might include migration opportunities (Connell, 1985). Nonetheless by 2017 Funafuti had grown to acquire a population of 6716, almost two thirds of the national population of 10,507. For the first time ever, between 2012 and 2017, that national population had also begun to decline because of emigration.

While economic growth has been generally positive because of large revenues from fishing licences, supported by substantial foreign aid and remittances, Tuvalu has the smallest GDP of any sovereign nation. The United Nations designates Tuvalu a least developed country (LDC) because of its limited potential for economic development, the absence of exploitable resources and its small size and vulnerability to external economic and environmental shocks. To offset these problems, at various times, Tuvalu has sought a number of distinctive and unusual development strategies, notably postage stamp sales and dotTV, that have had some intermittent success (Baldacchino & Mellor, 2015). The public sector makes up about two thirds of those formally employed. On the outer islands (every atoll except Funafuti) almost the entirety of employment is in the public service, including education and health, as they have become 'government islands'. Over the last census period every one of the outer islands has experienced a declining population with empty houses becoming evident.

On Funafuti population pressure on the tiny land area of 2.8 sq. km. has given it a population density of 2257 per sq.km- an extraordinarily high density that has placed enormous pressure on island resources, on land and in the lagoon. The overall population density in Tuvalu is around 410 people per sq. km. – the highest of any state in the Pacific. On Funafuti land has been reclaimed from the lagoon for residential use, little land remains in agricultural use (marked by the absence of the once key root crop, taro), access to local food is limited, with many households regularly experiencing food shortages and most being primarily dependent on imported foods of less nutritional value (McCubbin et al., 2017). Occasionally fresh foods are sold but Tuvalu is the only Pacific SIDS without a market. Funafuti has the only hospital in the country, with a small ICU. Outer islands only have access to nurses and patients are regularly referred to Funafuti and onwards for more specialised treatment. Dietary changes and greater sedentarism have contributed to high levels of diabetes, hypertension and cardiovascular diseases among Tuvaluans.

Even on the outer islands population densities remain high and many of the old characteristics of atoll agriculture, notably the swamp taro pits, have been abandoned as too demanding when wages and remittances make access to imported foods possible. Occasional droughts and cyclones are further deterrents. Exacerbating these problems are conflicts over land, that have grown alongside population growth. By the 1970s most atolls were experiencing a land shortage, land famine or even 'land crunch' (*oge manafa*) that was partly resolved by outmigration. Even on Nukulaelae, where claims on land were fewer than on any other atolls, it was 'well understood that in the hypothetical case of all non-resident Nukulaelae islanders returning to the atoll to live, the carrying capacity of the atoll's resources would be far exceeded' (Besnier, 1995: 28). Poverty exists, usually in large households with few wage earners, both in Funafuti and in the outer islands where, as the government has observed, 'people in the outer islands were increasingly reluctant to follow subsistence practices' (Tuvalu Ministry of Finance and Economic Development, 2014: 27). Less than a quarter of households had home gardens, having little land, while most faced challenges in getting planting materials should they have wanted to plant food gardens (Government of Tuvalu, 2013).

Only Funafuti has an airport so that most outer islands are served by a shipping service every two or three weeks, yet with solar panels and concrete houses, smart phones and wifi, the outer islands are no less 'modern' than Funafuti. However competing demands on the three inter-island vessels make even getting to Funafuti problematic, especially in the cyclone season. Transport issues have both disadvantaged Tuvalu and the outer islands.

Migration

Scarce resources, especially land, and occasional hazards have meant that Tuvaluans have invariably been mobile. In early times islanders were well able to be at least capable of envisaging the settlement of other islands and encountering new trade partners in alternative places. Tuvalu atolls became paired together, a long established relationship that may predate colonialism, but which was well established in the late nineteenth century; thus Nukulaelae is twinned with Vaitupu, and Nanumea with Nukufetau, a relationship of *fakasoa* (literally 'make friends') but also of competition (Besnier, 1995). In Nanumea the linkage reputedly began after Nukufetau sent a boat load of coconuts to Nanumea where serious drought existed (Chambers, 1983: 56). These ties consolidated land rights in other atolls, were useful responses to food shortages after cyclones, a local mobility that was a precursor of later migration: early examples of mobility as resilience.

As early as the first decade of the twentieth century British colonial officials had encouraged labour recruiting in the northern Tuvalu islands in order to counter perceived overpopulation (Chambers & Chambers, 2018). That was followed by migration to Nauru and Banaba (Ocean Island) to be involved in phosphate mining. Migration to both these islands was intendedly short-term, and residence there was only possible while working, but extended over several decades for some workers, being of particular importance from 1955 to the end of the century. After the Second World War, Kiribati and Tuvalu pioneered another form of international migration in the region. A Marine Training School was set up in 1968 in Tarawa (Gilbert Islands) to train both i-Kiribati and Tuvaluans as merchant seafarers for overseas shipping lines, mostly operating from Europe. A separate School was set up in Tuvalu after independence in

1979, training 60 seafarers a year. Initially the programs catered only for men but were later extended to women. At any one time, as many as 15% of the adult male population work abroad as seafarers, and their remittances have long been invaluable. Apart from Kiribati, no other SIDS had engaged in comparable contract labour schemes, and no other SIDS train their populations to be migrant workers. Both the Banaba and Nauru mines had closed by the end of the century and demand for Pacific islanders as shipping crew has continued to fall with greater global competition.

In due course that led to one more phase of temporary labour migration: with workers (initially mainly men) travelling to New Zealand and then Australia to participate for up to seven months in a year in temporary agricultural seasonal worker schemes that began in 2007. All recruitment went through the Ministry of Foreign Affairs and Labour (a significant ministerial conjunction of activities) from its 'work ready' pool of workers. That pool had been established from nominations by Island Councils, and screened by the Ministry according to previous migration history, character, health status, work ethic and basic English aptitude, with the final selection (both by the Ministry and by employers) intended to give reasonable and equitable representation to each island (Bedford et al., 2011). In every phase of temporary migration, the colonial and later the independent government of Tuvalu were firmly involved, but for these latter schemes, Tuvalu was disadvantaged by the tyranny of distance in comparison with workers from Tonga, Vanuatu and elsewhere.

Permanent Migration

Before the Second World War there were clear indications that population increases in some islands were putting too much pressure on local resources. Populations from some nearby islands in the Gilbert Islands (Kiribati) had already been resettled in unoccupied islands further north and the Ellice Islands (Tuvalu) followed suit. In the 1940s, a colonial official, D. G. Kennedy, suggested that the then main island of Vaitupu had limited agricultural potential, and the atoll faced 'probable overpopulation in the future', so that islanders might consider buying the unpopulated Fijian island of Kioa (White, 1965). The idea was enthusiastically received perhaps because it would make Vaitupu the most 'progressive' island in the archipelago. Kioa was purchased at auction in 1946 and settlement occurred in several phases (White, 1965; Koch, 1978). With a population of around 250 it was the first permanent overseas migration from Tuvalu. Vaitupuans remaining in Tuvalu have suggested that Kioa might be a possible relocation site for Tuvaluans in the event of unmanageable sea level rise (Stratford et al., 2013) indicative of the constant re-imagining of island migration possibilities.

Tuvalu (and its former British colonial partner Kiribati) were otherwise latecomers to more permanent international migration to metropolitan states, without migration corridors to the United Kingdom or elsewhere. From the 1960s onwards uneven development between the Pacific SIDS and metropolitan countries (the United States, New Zealand and Australia) resulted in high levels of migration, especially from the Polynesian PICs, with a growing diaspora emerging. So significant has international migration become that families created small-scale metaphorical 'transnational corporations of kin', by seemingly allocating family members to different countries, to diversify livelihood opportunities, alongside a 'culture of migration' where migration was normative, anticipated and unexceptional (Connell, 2008). The outcome has been a now falling population in the smallest independent Polynesian state.

This culture of migration was a 'safety valve' for high fertility rates yet is also accounted for them. At the household level, parents actively hoped to produce remittance earners. In the 1980s as one woman said of her only son: "One is not enough. If he goes away to work, there is no one to look after me here. If he stays and cares for me, no one earns any money overseas." Another woman recognized that her husband had been right to insist that they needed more children: "He said that if we had many children we might have a smart one who could go on to school and get good work. He will be our road to money and imported goods" (Chambers, 1986: 283–284). Migration spread and diminished risk. Remittances were crucial for atoll households and states and demands for them often substantially exceeded the ability of migrants to supply them, as in Nukulaelae where 'outbound letters are full of requests that exceed by far what most employed islanders can afford' (Besnier, 1995: 95). Letters to Nauru became long shopping lists. By 2012 at least half of all Tuvaluan households were receiving remittances from overseas, now stimulated by phone calls.

Little changed in the need for remittances, as belatedly Tuvaluans followed other Polynesians in settling overseas, seemingly permanently, mainly in New Zealand. Since 2002, New Zealand has offered a Pacific Access Category, which allocates places to several island states, dominated by Samoa (its former colony), but with 75 places a year for migrants from both Kiribati and Tuvalu. Effectively that was stratified by skill, so that the scheme constituted a skill drain from Tuvalu. Indeed the dominance of skills in migration streams moved the Speaker of the Tuvalu parliament to humorously suggest: 'we should focus on rugby, not only for the rewards and fame, but [since playing rugby] can also pave the way for permanent residence in New Zealand' (quoted in Shen & Binns, 2012: 77). Numbers grew in New Zealand (alongside much smaller numbers in Fiji and Australia, with about 600 and 220 respectively) so that by 2013 some 3537 Tuvaluans were counted in the 2013 census (though only 40% of these had been born in Tuvalu) and by the 2018 census that number had grown to 4653 (with 46% born in Tuvalu). Around a third of all Tuvaluans now live outside the country.

It was scarcely surprising, given the national preoccupation with migration, remittances and the existential threat of sea-level rise, that in 2014 the country should produce its own *Tuvalu National Migration Policy* (Tuvalu Ministry of Foreign Affairs, 2014) mainly directed to developing strategies for accessing superior labour migration opportunities overseas.

COVID-19 and Mobility Reversal

After more than a century of mobility, from the outer islands to Funafuti, and from Tuvalu to other parts of the world – whether temporary or permanent – COVID-19 dramatically swung these processes into reverse at two different scales. Funafuti residents were encouraged to 'return' to outer islands, partly because of some urban economic decline but mainly for safety from the virus (since Funafuti is effectively the only point of sea and air contact with other countries). Somewhat later overseas workers were repatriated to Tuvalu.

Tuvalu established a COVID-19 Health Taskforce as early as late January (during a separate emergency to deal with Cyclone Tino). Borders were first closed to possible arrivals from some high-risk countries in February, the last flight from Fiji landed on the 21 March and schools were closed. That disrupted educations since schools had no means of preparing alternative means of learning and many households had no access to e-learning or to reliable internet connections. Financial support was given to those in the tourism and transport sectors who lost employment. While tourism is trivial in Tuvalu, some 17 of the 40 tourist workers – most of whom were women – were made redundant. After the closure of the borders medical referrals were no longer possible putting additional pressure on the country's lone hospital. New regulations gave the government the authority to control the sale and price of essential commodities (crucial when households were so dependent on imports and shipping was likely to be restricted) and required the issuing of food vouchers. That followed a spate of panic buying as consumers rushed into shops to buy in bulk basic commodities such as sugar, rice, flour and biscuits, which led to the skyrocketing of prices (Del Castillo et al., 2020). The government put new emphasis on the distribution of agricultural planting stock and seedlings in an effort to revitalise agriculture. Landowners were encouraged to practise customary storage of foods, such as root crops and fish drying, preserving breadfruit and storing coconuts (Farbotko & Kitara, 2020), strategies and techniques that had generally disappeared, especially on Funafuti. Local leaders were empowered to consider rationing, were it to become necessary, to prevent hoarding and shortages.

Under the emergency regulations, the government acquired the power to order the compulsory relocation of people from Funafuti to their home islands and to prohibit movement of people to Funafuti. While that would protect people against the arrival of the virus, it was also thought that it would increase the labour supply for agriculture and fishing on outer islands (Farbotko & Kitara, 2020). However, that flew in the face of the rationale for migration to Funafuti, where few engaged in any kind of agriculture, so challenging recent history. Nevertheless, by early May, more than 1200 and perhaps as many as 1500 people, almost 15% of the resident national population, had voluntarily moved to the outer islands, where the population grew by 35%. While health security was the main reason for return, some looked forward to the prospect of spending time on home islands, eating local food and 'leaving the noise and pollution' of the capital (Kitara et al., 2020). The government allocated A$500,000 to each outer island to enable them to make necessary improvements in support of the returnees, such as expanding and servicing the clinics, and providing water tanks and tarpaulins for temporary accommodation.

Island councils (*faupule*) were free to choose how to best use the grant. Some islands used the funds to build 'getaway houses' or huts on remote land to settle in the event of COVID-19 reaching the atolls. Funafuti island council decided to build 55 huts for each of the 55 *mataniu* (the traditional household title system) on Funafala, an islet 15 kms south of Funafuti on the same atoll, that would continue the process of building up a community there, and is nothing other than 'settling there for good'. Vaitupu atoll, on the other hand, used the fund to build isolation wards, to accommodate those who might contract the virus, and improve access to water facilities.

Further complicating the movement of people to outer islands was the repatriation of overseas Tuvaluans, mainly stranded RSE workers from New Zealand, alongside students, observers on tuna fishing boats and medical referrals. Their return, and the impossibility of other workers taking the place of RSE workers, meant a small downturn in remittances. More than 200 people returned to Tuvalu, a considerable number to be quarantined on Funafuti, and who were only allowed to return after new medical equipment had arrived in the country (Kitara & Farbotko, 2020). The government also provided financial support to government officials and students stranded abroad.

Revitalising a semi-subsistence livelihood on outer islands was problematic. On every atoll water shortages were not unusual and saline intrusions hampered agriculture. Resettlement was made more difficult by Cyclone Tino in January that affected most islands, with hundreds of people temporarily evacuated from damaged homes, bananas and breadfruit trees uprooted in several places and salt-water intrusions damaging other plants. Migrant urban households would not be landless but might struggle to exercise land claims should they have been away for long and would be without actual food plots. Some early arrivals found it hard to gain family support and were reportedly turned away. Beyond pressure on agricultural land and fishing grounds, return migration put pressure on island schools, health clinics and shops. Not all those who returned stayed on the outer islands. Those with formal jobs in Funafuti filtered back early, and others returned temporarily for medical reasons and to restock supplies of foods and other goods.

At a different scale the government also encouraged movement of people from the main island of Funafuti to smaller islands around the lagoon, only one of which, Funafala, previously had a permanent resident population (a small village of about ten households without school or shop). Intriguingly, during the Second World War, Funafuti islanders had been relocated there when Funafuti itself was used by the American military. On smaller islands, such as Papaelise, bananas and even coconuts struggle on dry and salty land, but a small population had also moved there and were planting cabbages and other vegetables. Those who migrated to both these islands had to build traditional-style houses and transport their own water tanks, but were close enough to the capital to access services and employment (despite the high cost of local transport). Only Funafala was large enough to sustain a significant population, who created the first 'road' on the island, grandly named COVID19 Road, and established a community garden project.

During the relocation planning process and in the initial implementation stage, there were plans to replant root crops. Although *pulaka* (swamp taro) was the traditional food crop, sweet potatoes and cassava were encouraged since they could be in production more quickly. A few returnees did nevertheless plant *pulaka* suggesting an intention to stay for some time. The island councils had their own plans so that planting strategies varied. During this period people continued to eat staple imported foods, such as rice, flour, biscuits and chicken, and some returnees opened their own small stores on the home islands.

Key national initiatives involved distributing seedlings and other planting materials and supporting 'rapid gardening strategies'; landowners were encouraged to practice 'customary stockpiling' to create food buffers (ACIAR, 2020: 115) but that had never been realistic in the past and was no easier now. As one Tuvaluan pointed out: 'the ongoing challenges in making and sourcing compost for agriculture make it hard for us to grow staple foods' (quoted in ACIAR, 2020: 103). In any case planting material was in short supply since it had not been seen previously as essential. Even so the agricultural station on Funafuti, after largely being a token presence on the atoll, came into its own.

The outer islands of Tuvalu (and of Kiribati) had a more limited adaptive capacity because of both the distinct challenges of atoll agriculture (where the basic staples that are produced are neither easy to grow nor grow quickly) with a resultant transition to imported foods. Fishing is more feasible and thus more valuable in terms of a short-term response, but three of the atolls have no lagoons making fishing more challenging. On each of the atolls people now live in nucleated settlements where food gardens are more distant. Such distant parts of the atolls had distinctive value, as on Nanumea where some people had moved to Lakena, five kilometres from the main village, to 'temporarily escape an argument or an otherwise unpleasant situation on Nanumea proper … [and] minimize the potentially severe social strains that are inherent in the atoll's small size and limited resources' (Chambers, 1983: 11, 200). Even in good times, when atolls were cyclone and drought free, maintaining social and economic livelihoods could be challenging. Substantial and sudden return migration seemed guaranteed to exacerbate these difficulties and create new land crunches. Paradoxically, ensuring some flow of imported food, from Fiji and from Funafuti, was more crucial than ever.

Conclusion: Returning to the World?

Concerned over the potential consequences of the arrival of COVID-19 in a very small island state, unprepared for a pandemic, Tuvalu acted particularly swiftly and effectively – at least as fast as any other small state – to import PPE and other medical technology and to close its borders. After a century of gradual globalisation it abruptly and comprehensively cut itself off from the world. COVID-19 was kept out, repatriation was handled effectively and Tuvalu remained defiantly virus-free at the end of November.

Unusually and distinctively, migration became a tool of adaptation, both in government policy and in household behaviour. At three different scales, the pattern of migration was reversed; Tuvaluan workers returned from overseas, households moved 'back' to outer islands, and, on Funafuti at least, two small islands were populated by decentralisation from Funafuti. Gradually the capital city returned towards the status of merely 'first among equals' that it held prior to independence when the atolls all had much the same populations. That return sought a revitalised local resilience, where agriculture and fishing became revalued, but constituted a difficult adaptation to the past – challenging to sustain where agriculture was unusually difficult. Whether a new normal will have such a decentralised population seems improbable, yet it has provided a partial success story.

COVID-19 has proved a single relatively sharp and comprehensive shock, that Tuvalu has managed to cope with effectively, but global climate change remains the greatest ongoing existential threat to the existence of Tuvalu. In 2019 the Pacific Islands Forum, a key regional organisation aimed at coordinating economic and regional policies, was hosted in Tuvalu. The main agenda item was that of creating a comprehensive and collaborative response to the regional climate crisis, in a small island state that has gained unusual status through its global advocacy of climate change policies. COVID-19 has proved an unwelcome distraction from an 'old normal' but thus far that has seemingly been surmounted. A new normal will surely focus on climate change, public health care provision and on different and more diverse facets of resilience.

Acknowledgement I am grateful to Eliala Fihaki and Tapugao Falefou for their generous assistance.

References

ACIAR. (2020). COVID-19 and Food Systems in the Indo-Pacific: An Assessment of Vulnerabilities. Impacts and Opportunities for Action, ACIAR Technical Report No. 96.

Baldacchino, G., & Mellor, S. (2015). Tuvalu. Entrepreneurship and the Dot TV Phenomenon. In G. Baldacchino (Ed.), *Entrepreneurship in Small Island States and Territories* (pp. 268–281). Routledge.

Bedford, C., Bedford, R., & Ho, E. (2011). Engaging with New Zealand's Recognised Seasonal Employer Work Policy: The Case of Tuvalu. *Asian and Pacific Migration Journal, 19*(3), 421–445.

Bertram, G., & Watters, R. (1985). The MIRAB economy in South Pacific Microstates. *Pacific Viewpoint, 26*, 497–519.

Besnier, N. (1995). *Literacy, Emotion and Authority. Reading and Writing on a Polynesian Atoll.* Cambridge University Press.

Chambers, A. (1983). *Exchange and Social Organization in Nanumea, A Polynesian Atoll Society,* PhD thesis, University of California, Berkeley.

Chambers, A. (1986) *Reproduction in Nanumea (Tuvalu): An Ethnography of Fertility and Birth* (Working Paper 72). Department of Anthropology, University of Auckland.

Chambers, A., & Chambers, K. (2018). Engaging the World: Four Decades of Intensifying Change in Tuvalu. In J. Connell & H. Lee (Eds.), *Change and Continuity in the Pacific. Revisiting the Region* (pp. 17–36). Routledge.

Connell, J. (1985). Islands on the Poverty Line. *Pacific Viewpoint, 26*(2), 463–473.

Connell, J. (2003). Losing Ground? Tuvalu, the Greenhouse Effect and the Garbage Can. *Asia Pacific Viewpoint, 44*(2), 89–107.

Connell, J. (2008). Niue: Embracing a Culture of Migration. *Journal of Ethnic and Migration Studies, 34*(6), 1021–1040.

Del Castillo, N., Homasi, L., & Wainiqolo, I. (2020, December, 9–11). Leaving No One Behind: A Look at the Plight of the Vulnerable in Kiribati and Tuvalu Amid COVID-19. *Pacific Economic Monitor.*

Farbotko, C. (2010). Wishful Sinking: Disappearing Islands, Climate Refugees and Cosmopolitan Experimentation. *Asia Pacific Viewpoint, 51*(1), 47–60.

Farbotko, C., & Kitara, T. (2020, April 6). *How Is Tuvalu Securing Against COVID-19?* ANU DevPolicyBlog.

Government of Tuvalu. (2013). *Tuvalu 2012. Population and Housing Census.*

Kitara, T., & Farbotko, C. (2020, August 17). *How Tuvalu Is Doing Repatriation.* ANU DevPolicyBlog.

Kitara, T. Farbotko, C., Talia, M., Suliman, S., & Evans, C. (2020, August). *Reducing COVID-19 Risk Through Population Relocation and Closed Borders: Effects of Pandemic Emergency Measures in a Small Island State*. IOM Environmental Migration Portal.

Koch, K.-F. (1978). *Logs in the Current of the Sea*. ANU Press.

McCubbin, S., Pearce, T., Ford, J., & Smit, B. (2017). Social-Ecological Change and Implications for Food Security in Funafuti, Tuvalu. *Ecology and Society, 22*(1).

Mortreux, C., & Barnett, J. (*2009*). Climate Change, Migration and Adaptation in Funafuti, Tuvalu. *Global Environmental Change, 19, 105–112.*

Munro, D. (1990). Migration and the Shift to Dependence in Tuvalu. In J. Connell (Ed.), *Migration and Development in the South Pacific* (Pacific Research Monograph No. 24) (pp. 29–41). ANU National Centre for Development Studies.

Shen, S., & Binns, T. (2012). Pathways, Motivations and Challenges: Contemporary Tuvaluan Migration to New Zealand. *GeoJournal, 77*, 63–82.

Shen, S., & Gemenne, F. (2011). Contrasted Views on Environmental Change and Migration: The Case of Tuvaluan Migration to New Zealand. *International Migration, 49*, 224–242.

Stratford, E., Farbotko, C., & Lazrus, H. (2013). Tuvalu, Sovereignty and Climate Change: Considering *Fenua*, the Archipelago and Emigration. *Island Studies Journal, 8*(1), 67–83.

Tuvalu Ministry of Finance and Economic Development. (2014). *The Poverty Lines, the Incidence and Characteristics of Poverty in Tuvalu*. MFED Central Statistics Division.

Tuvalu Ministry of Foreign Affairs. (2014). *Tuvalu National Migration Policy*. MFATTEL.

White, G. (1965). *Kioa: An Ellice Community in Fiji*. Department of Anthropology, University of Oregon.

CHAPTER 9

Learning from the Past? Sovereign Space and Recreating Self-Reliance in Kosrae, Federated States of Micronesia

Ashley Meredith and David Fazzino

Through a series of closures, Kosrae, the smallest and easternmost State in the Federated States of Micronesia (FSM), became increasingly isolated from the outside world in order to prevent the entry of COVID-19. This chapter traces the continually evolving nature of government responses to the COVID-19 pandemic during 2020 while considering the economic and social implications on Kosraeans. The COVID-19 pandemic created the space for FSM, as a whole, and Kosrae in particular, to retain those aspects of globalisation deemed desirable to perpetuate while, at the same time, limiting other interactions determined to be too risky. Hence, FSM and its four component States exerted their sovereignty during this global

A. Meredith (✉)
National Archives of Culture and Historic Preservation,
Pohnpei, Federated States of Micronesia

D. Fazzino
Department of Anthropology, Bloomsburg University of Pennsylvania,
Bloomsburg, PA, USA

Y. Campbell, J. Connell (eds.), *COVID in the Islands: A comparative perspective on the Caribbean and the Pacific*,
https://doi.org/10.1007/978-981-16-5285-1_9

crisis while reflecting on the value of the intended renewal of the Compact of Free Association with the United States in 2023. The closure of the FSM border was however fraught with uncertainty due to the dramatic consequences it had for those who were residing off-island and were, even in December 2020, still unable to determine when they would be able to return to Kosrae.

We begin by providing an overview of Kosrae and its relationship to globalisation processes in the context of colonisation, missionary work, the introduction of diseases, development, and dependency. We do this in order to historically situate the contemporary responses to the COVID-19 pandemic in the context of a parallel history of globalisation and isolation. This draws on official FSM government accounts and reflections from contemporary Kosraeans. From here we consider the variety of impacts and restrictions on global flow of goods, people, and services that the COVID-19 pandemic has had on Kosrae, before examining the implications and opportunities that it presented for reclaiming subsistence-based practices in the face of growing uncertainty and precarity during the pandemic. Hence, this chapter reflects on official and unofficial responses to and preparation for COVID-19. It also considers the responses of the national FSM government in relation to the Kosrae State Government and how a shared understanding of federal responses to the COVID-19 pandemic emerged and solidified in order to maintain FSM's status as COVID-19 free. That success was incomplete, as negative impacts of the pandemic were experienced in Kosrae. After highlighting these we consider how COVID-19 has shaped continuing work towards community-based development and reinforced the popular trope of the importance of self-reliance and local ways. While not retreating from globalisation the response of FSM demonstrates that it is possible at a country level to carefully curate the level of global engagement desirable in an uncertain world, but that radical change is particularly difficult.

A Brief History of Kosrae

Kosrae, a single high island, is the easternmost of four States in the Federated States of Micronesia, a Small Island Developing State (SIDS) in the Caroline Islands (named after Carlos II of Spain in the sixteenth century) of the northern Pacific. Europeans were unsuccessful in establishing a stronghold on Kosrae until the mid-nineteenth-century and the arrival of an American Christian mission from Hawai'i. Coupled with more intense whaling operations in Kosrae, that led to a dramatic increase in death rates from diseases, reducing the population to a mere 300–400 people (Ritter, 1981; Hezel, 1983). The loss of lives coupled with missionary and whaler arrival altered traditional belief systems, practices, and governance. Distant Spanish tenure was short-lived and Germany bought the Caroline Islands and other Micronesian islands from Spain under the 1899 German-Spanish Treaty. The Germans began a process of economic development by planting export crops; on Kosrae these were bananas and coconuts (for copra). Germany lost control of the Caroline Islands to the Japanese in 1914 before World War I began. The Japanese held the islands through World War II and continued the work of the Germans by further developing Kosrae for economically valuable exports while also building military installations. The Japanese time in FSM, particularly during WWII, was marked by continued militarisation, disciplining of everyday life, and food rationing until the Japanese surrender, an event celebrated annually as Kosrae Liberation Day. Liberation Day events range from solemn occasions to a revival celebration of freedom including a focus on traditional skills and knowledge.

After World War II, the islands now known as the Federated States of Micronesia (FSM), along with other island groups (Republic of Palau, the Republic of the Marshall Islands, the Commonwealth of the Northern Marianas), were held as part of the United Nations Trust Territory of the Pacific Islands (TTPI), administered by the United States. The four central high islands and their respective outer island atolls (Yap, Chuuk, Pohnpei, and Kosrae) came together as a country and adopted a constitution in 1979. FSM secured its independence in 1986 and entered into a bilateral agreement with the United States, the Compact of Free Association (COFA). The FSM now has a population of just over 100,000 with Kosrae having around 6600. COFA, amended in 2003, will terminate in 2023 with current negotiations underway for a third phase. The key elements of COFA are that the US provides FSM with economic assistance (and eligibility for some US federal programs), defence, and visa-free migration to the US, while conversely the FSM allows US defence and other operating rights in the FSM, and denial of access to the FSM by other nations.

Any discussion of the COVID-19 pandemic must be situated in the context of these multiple globalisation processes related to connectivity, flows of goods, services, money, and ideas. These have had diverse and comprehensive consequences as Kosrae was dramatically and often adversely impacted by the successive imposition of four quite different outside colonising states (Peoples, 1985). These impacts are still being felt. The spread of the COVID-19 pandemic is the most prominent recent example of the costs associated with participation in a globalised world and it too had uneven impacts, just as did the spread of other diseases throughout history. That included diseases that decimated the population, upwards of 95% by some estimates upon contact, resulting in a loss of many traditional lifeways, practices, and belief systems. These have been primarily supplanted economically by participation in a mixed-subsistence economy and in community life through participation in organised religion and specifically the Congregational Church with which the majority of Kosraeans identify. These are subsumed within an overall dependency on foreign aid, wage (government) employment with nearly half of the workforce employed in the public sector. Incomes earned go into buying imported foods (Connell, 2015), that are desired but not essential, while expenditure on imported technologies used in agriculture and fisheries, has brought outboard motors, fibreglass boats, fishing lines and lures, gasoline, and "green machines" or weed eaters. These purchased goods replace historic and traditional agricultural, hunting, and fishing technologies and associated knowledge while increasing efficiency in terms of the amount of time spent procuring foods. While wage labour has taken time away from subsistence production, the public sector hours of the Kosraean work week allow for ample time to pursue subsistence pursuits particularly when assisted by time-saving technologies (automobiles to access upland farms and outboard motors to access fish). Beyond working in a more 'modern' context, there remains a desire for learning and utilising traditional knowledge and practices. Kosraeans have achieved a balance between outside resources and those found in their local communities.

Negotiating this balance is a constant process evident in the current work of the Kosrae Island Resource Management Authority (KIRMA), which seeks to reclaim some of the past of Kosrae through historic preservation, promotion of heritage-related activities, sustainable use of local resources, and the creation and development of the Mahkontowe Conservation Area (MCA) in Kosrae's interior. The MCA, created through legislation in 2018, is viewed by Kosrae governmental officials as a means to help ensure food sovereignty through the 'source to sink' promotion of healthy reefs while at the same time creating sites for activities geared towards local and international visitors alike, such as the building of rest houses in each of the four municipalities on the island to serve a dual function of meeting local needs through catering to international tourists and being used for subsistence practices. The design process and building for each of the four rest houses in 2019 and 2020 became a matter of community pride, highlighting the overall approach in Kosrae of balancing the local and the global through careful cultivation of international relationships that might build and maintain traditional knowledge. This proved a balancing act in seeking development not as an end in itself, but where projects and actions were judged on their merits to Kosraeans as a whole. That included both drawing from the past and innovative approaches in seeking grants from outside donors to build local capacity.

Between April–August 2020 KIRMA sponsored a consultation process with approximately 150 participants, composed of government officials and residents throughout Kosrae, to elicit collective perspectives on a variety of potential economic scenarios centred on developing an updated land use plan for Kosrae. Community consultations took the form of town hall style forums where community members commented on economies that were stagnant, contracting, and expanding. Although not explicitly sought out, mention of the impacts of COVID-19 was brought up throughout the community consultation process, and in informal conversations in various settings. Included in these discussions was an imagining of scenarios cloistered around hopes and fears centred on the closing, and potential reopening, of FSM, and what that might hold for the future of Kosrae.

Reduction, Acceleration, and Redirection of Global Flows

The early closure of the national FSM border in January 2020, and the absence of COVID-19 from FSM, meant that Kosraeans who remained on the island during the pandemic retained lifestyles untroubled and unencumbered by masks, social distancing, handwashing and sanitation protocols. Nevertheless COVID-19 impacted life on Kosrae in a variety of ways including a reduction in the flow of people, an increase in the flow of money, and concerns over the flow of goods. These manifested themselves in a variety of ways including a proliferation of preparatory exercises, distribution of PPE, distribution of medical equipment, legislative and policy efforts, and everyday sensibilities of space. Although the COVID-19 pandemic resulted in changes to flows in and out of Kosrae, life on the island, including dependence on outside sources of revenue to support island lifeways, remained largely unchanged. Here we discuss the reductions mandated in terms of mobility of people, then discuss the increased connectivity in terms of monetary flows before concluding with a consideration of the redirection of goods to maintain current lifestyles.

In the face of the uncertainties presented by the global diffusion of COVID-19, FSM maintained a precautionary approach by continuously increasing travel restrictions in early 2020 and maintaining these, despite a number of efforts to allow for limited and restricted access to Kosrae and the other States of FSM. Tables 9.1 and 9.2 highlight these actions. We

Table 9.1 FSM National and State Legislation regarding COVID-19

Date Adopted	*Order #*
Jan 31, 2020	The President's Public Health Emergency Declaration dated January 31, 2020
Feb 12, 2020	FSM C.R. 11-117, To further amend the President's Public Health Emergency Declaration dated January 31, 2020
Mar 11, 2020	FSM C.R. 21-129, To further amend the President's Public Health Emergency Declaration dated January 31, 2020, as amended on February 7, 2020
Mar 20, 2020	Pohnpei State Constitutional Emergency Order 20-01
Apr 17, 2020	FSM C.R. 21-138, To further amend the President's Public Health Emergency Declaration dated January 31, 2020, as amended on February 7, 2020, and further amended on March 11
May 30, 2020	FSM C.R. 21-149, To further amend the President's Public Health Emergency Declaration dated January 31, 2020, as amended on February 7, 2020, and further amended on March 11 and April 17, 2020
Jul 10, 2020	FSM C.R. 21-175, To further amend the President's Public Health Emergency Declaration dated January 31, 2020, as amended on February 7, 2020, and further amended on March 11, April 17, and May 30, 2020
Aug 20, 2020	FSM C.R. 21-185, To further amend the President's Public Health Emergency Declaration dated January 31, 2020, as amended on February 7, 2020, and further amended on March 11, April 17, May 30, and July 10, 2020
Nov 25, 2020	FSM C.R. 21-228, To further amend the President's Public Health Emergency Declaration dated January 31, 2020, as amended on February 7, 2020, and further amended on March 11, April 17, May 30, July 10, and August 20, 2020

Source: https://cfsm.gov.fm/ and https://pohnpeistate.gov.fm/

Table 9.2 FSM border closures

	Policy	*Date*
FSM National	Passengers from Wuhan, China not allowed to disembark	January 31, 2020
FSM National	Persons travelling directly or indirectly, either by air or sea transport, from anywhere in mainland China since January 6, 2020, are banned from entering FSM	February 15, 2020
FSM National	If a conflict emerges between the FSM National and FSM State Declarations and Emergency Orders, the Declaration by His Excellency President David W. Panuelo takes precedence	March 20, 2020
Yap State	Passengers allowed to disembark. But with 14-day quarantine	March 15, 2020
Yap State	No disembarkation on Yap	March 18, 2020
Chuuk State	No disembarkation on Chuuk	March 17, 2020
Pohnpei State	Pohnpei's Constitutional Emergency Order 20-01: only allowed quarantine for "Pohnpei residents", and "noncitizens who are employed by the Pohnpei State Government, the FSM National Government, diplomatic corps, businesses, international agencies and non-profit organizations" as well as "medical and technical personnel to assist with COVID-19, at the request of Pohnpei State." This also includes any medical or technical personnel requests by the FSM National Government	March 20, 2020
Kosrae State	No disembarkation	April 8, 2020

Source: FSM Public Information Office and Yap State News Briefs, 2020

provide a brief summary to highlight FSM's ability to effectively avoid the potential risk of infection by isolating itself from the rest of the world through taking swift early actions between January and March 2020. These actions began with the FSM President's initial Public Health Emergency Declaration on January 31, 2020, closing the border to anyone originating in China and mandating that anyone originating from another country with COVID-19 cases must have spent at least two weeks in a country without COVID-19 prior to arrival in the FSM.

As FSM and Kosrae continued to grapple with the economic and other implications of COVID-19, uncertainty of boundaries, including the jurisdictional boundaries between the national and state governments, as well as United Airlines procedures (the lone international airline flying to FSM), produced disruptions in even the potential of connectivity with the outside world through air travel. March was particularly disorienting between amendments to the President's Public Health Emergency Declaration and United Airlines' flight cancellations that occurred on a weekly and sometimes daily basis. There was uncertainty amongst Kosrae's residents about incoming passengers adhering to the quarantine in Kosrae and the actions required to monitor inter-island travel, as the FSM remained free of COVID-19 but maintained 14-day quarantines. Concerns over incoming passengers grew even between the FSM States as passengers leaving Kosrae on March 13, 2020 were placed in quarantine upon their arrival in Pohnpei. Confusion and uncertainty typified early weeks.

The focus shifted in May to work on coordinated efforts to build an effective quarantine program including facilities and protocols throughout the FSM. The potential efficacy of the quarantine program was continually contested even as the building of facilities and practice exercises led by a World Health Organization team, and the FSM National and Kosrae State Covid-19 Task Force continued.

While there was a decrease in the flow of people into Kosrae, there was a marked increase in the amount of funding coming into the FSM. Kosrae was no exception to feeling economic impacts from the pandemic, although the brunt was muted due to FSM's COFA connections with the US, China's continued competing interest and development money flowing from a variety of sources. Japan donated 382 million Yen (US$3.5 million) to FSM, for the procurement of medical equipment; China and the United States provided finance for PPEs, technical assistance, ventilators, and supplies, and Israel donated medical supplies. Specific to Kosrae, the US Department of the Interior approved US$1.5 million in funding for quarantine facilities and permitted a redirection of US$1.2 to build a quarantine facility.

Kosrae largely weathered the economic fallout of COVID-19 without dramatic alteration of the local way of life. However support from the FSM Government and US Government (via COFA) for some tourist businesses on the island was needed to maintain their viability as the complete elimination of tourist arrivals led to curtailing of operations, reduction of staff, and a decrease in the number of hours worked. Although annual tourism arrivals are typically under 1000 people, small relative to the global flow of tourists and to all tourist arrivals in FSM, direct support helped to maintain the viability of Kosrae's four hotels. None had to close although staff worked fewer hours. This external economic support made up for what would have been earned through government-sponsored travel as government employees usually received considerable reimbursement for travel through substantial government-mandated per diem rates, a standard boost to salaries. These were dramatically reduced in 2020.

In addition to support for the tourist sector, FSM also provided financial assistance to FSM citizens and families, stuck abroad. The US Pandemic Unemployment Assistance Program, applied for and implemented by the FSM, offered a one-time payment of $1000 to workers outside the formal sector, such as subsistence farmers, and to low-income families (who were encouraged to enhance their COVID-19 readiness by purchasing COVID-19 preventive items).

Overall, the COVID-19 pandemic resulted in an increase of government to government support for FSM as well as allowing for funds within FSM to be spent without restrictions. The pandemic raised questions over the long-term availability of remittances. Many Kosraeans are involved in the sending of remittances, estimated to be $600,000 annually (DevBizExperts LLC, 2019). These form an essential part of island incomes, needed for household goods, imported foods, cars, home repairs, and medical attention, amongst other expenses. COVID-19 disrupted remittance flow, although many feared that remittances would decrease as employment became increasingly precarious in the USA. However, as one Kosraean observed: "There seems to be more people receiving money through Western Union. The place is usually backed with cars and nowadays money at the Union usually runs out so you have to wait 24–58 hours to get your money'. At the same time, government assistance for Kosrae somewhat mitigated the loss of wages.

Along with a decrease in the amount of money able to be earned through formal employment on Kosrae, there were also delays in arrival of goods on which to spend this income. Nonetheless Kosraeans referred to a "business-as-usual scenario," where imported goods were still purchased at the same rate. Initial concerns on Kosrae centred on the possibility of a complete closure of transport, in effect isolating Kosrae from the outside world, which would have resulted in the decreased availability and increased cost of imported goods. What emerged was a continued and growing awareness of the precarity and wisdom of relying exclusively on imported goods, but also of completely dismissing global interconnections in fear of what might come. The initial and dramatic response limiting entry into FSM, coupled with uncertainty regarding the scope and effective management of COVID-19, carried with it the implied threat to a way of life which included the substantial incorporation of imported goods, including foods, and thus the knowledge, practices, and the social connections that were associated with these, including, but not limited to, subsistence practices, cuisine, and community activities. Any abrupt end to imports would necessitate reconsidering and reshaping the production, exchange, and consumption of both local and imported foods. Closing the borders led to initial questions about how and if foods and other goods could or should continue to be imported.

Concern over imports was particularly true in the early days of the pandemic, primarily that imported goods from the USA, a major locus of the COVID-19 pandemic, would have contaminated surfaces from contact with potential carriers in the USA. However such fears dissipated and food shortages never occurred with only bottled water (and toilet paper), that originated in the US, scarce for a short time. With delays in shipping, Kosraeans otherwise simply found substitutes for products previously consumed. The experience of finding substitutes or doing without staples for a period of time was not unknown to them, despite its absence in recent times.

Navigating Uncertainty

Alongside challenging mental and physical wellbeing and disrupting businesses and economies, COVID-19 introduced a sense of ambiguity regarding both the present and the future. Although there were and are many unknowns in relation to the virus itself, and the full extent of its impacts, there were some certain impacts. First, without accusations of "China virus", based on the occurrence of the documented initial outbreak, or "US virus" as a result of the gross negligence demonstrated by some (enough) members of the United States public and government (Kim et al., 2020), Kosraeans obviously recognised that this came from the outside. Second, unlike so many other changes emanating from beyond their shores, for multiple reasons, Kosraeans could see in advance that this plague was coming. Third, they were able to do something about it. On the island quarantine facilities were constructed in the Kosrae High School. Throughout 2020, Kosrae State, and FSM, delayed multiple times the disembarkation of passengers even as United Airlines flights continued to arrive and depart on Kosrae. Through their relative isolation, coupled with some luck, Kosraeans took deliberate actions to reduce any possibility of COVID-19 reaching the island. As global, and especially American, death counts mounted the desire to isolate themselves and prohibit entry grew. This was evident also at the national level, evident during discussions about the future of COFA, particularly as China demonstrated growing interest in Micronesia. The relationship between FSM, China and the USA allowed the FSM to be unusually selective about which aspects of globalisation and assistance they would accept, which even created a moment in history when the FSM President chastised and directed both the USA and China to end their squabbling over causation and work towards a solution (*Guam Daily Post*, 28 September 2020). The spread of COVID-19, and the inability of the rest of the world to effectively manage this, reminded islanders that outside practices, policies and procedures were not always the most efficacious or beneficial.

FSM remained COVID-19 free, as a result of its adherence to a strict policy on arrivals. The FSM placed a quarantine on the country, locking out those who might wish to enter, both citizens and non-citizens, while still permitting those who wish to leave to do so. The months following the prohibition of arrivals were marked by debates and discussions at national and state levels over the positive and negative effects from modifying the travel restrictions, which had caused some consternation for those separated from family members for months more than initially expected. December 5, 2020 was the fourth scheduled date for their return and was specifically designed for those who had been stuck overseas while seeking medical treatment. However the lack of medical doctors in FSM was one of the primary sticking points in preventing repatriation. That lengthy period was seen by some Kosraeans as both demoralizing and devastating while others supported Kosrae remaining COVID-19 free.

Concerns over lack of preparedness to effectively manage the virus trumped concerns associated with the suffering of those stranded overseas, whereas in Kosrae people were free from COVID-19, could participate fully in everyday life, practice subsistence on the land and seas, and continue local interactions with spouses, children, parents, and extended family members. Those who were "off-island" were away from their social support networks and often in contexts that now particularly limited their ability to live the lives that they desired. Frustrations increased as possible dates were continually pushed back a month, or two, at a time. Despite the problems for individuals and families stranded off-island, it remained unlikely that Kosrae would open its borders prior to the widespread distribution and recognised efficacy of a vaccine.

Kosrae adopted and retained a precautionary approach, seemingly analogous to Schrödinger's cat in that it was impossible to know if the cat was alive or dead in the context of the experiment (Thomson, 2020). On Kosrae caution was dictated by several events that were considered near misses. These included air passenger arrivals up to March 20, 2020, and the regular arrival of cargo ship crew and trans-shipment personnel; this was a stressful time waiting two to four weeks for any sign of COVID-19 on the island, since it could be asymptotic, prompting a heightened awareness at gatherings and legislative meetings. The outside world had become very much less alluring.

Seeing the Positive – Seeing the Island

The deceleration of globalisation, at least in some respects, created the space to consider and work on alternatives to dependence, building on pre-existing work, knowledge, and practices. Although the impacts of COVID-19 have in many ways been damaging everywhere, an exclusive focus on damage does little to highlight the agency of local communities. At a national level, swift action by the FSM government enabled the possibility that a viable vaccine would be widely circulated prior to any opening of the country to the virus.

Border closure helped to create spaces and opportunities for new forms of agency by potentially rekindling experiences and interests in tools and technologies associated with traditional cultural practices. In Kosrae, Utwe Municipality's celebration of Liberation Day in July 2020 included a rejuvenated Fahfah Festival where local artisans highlighted displays of traditional skills. Fahfah are Kosraean delicacies comprising pounded taro and coconut cream, or bananas. As a part of this event, several items were auctioned, including five canoes, which sold for between US$300 and US$550. Many of the winning bidders indicated that they would use them primarily for subsistence fishing. Local agriculture and fishing were being re-appreciated and revitalised.

The ban on arrivals necessarily extended to outside consultants prohibited from entering FSM. This reduced the availability of outside expertise (with consultants ranked below citizens for priority arrivals) and in Kosrae meant relying on local expertise, both lessening reliance on outsiders, and their perspectives on development, and resulting in more employment opportunities on the island.

We had the opportunity to work in a variety of roles to build capacity on the island. These included efforts by the Kosrae State Historic Preservation Office to develop ethnographic data collection protocols and capacities with the staff in oral history methodology on a variety of topics related to traditional knowledge including ancient routes, ethnozoology, ethnobotany, subsistence practices, navigation, funerals and burial practices, economic planning and development, and cultural identity in Kosraean culture (e.g. Meredith, 2020). Capacity building efforts were extended to high school students through holistic educational programming designed to empower youth and foster pride in place (Fazzino, 2020). These efforts at documentation were not merely about documentary process and practices, but also about sustaining the knowledge and everyday practices relevant to making a local living. Revitalising subsistence practices was part of the overall strategy of the State to ensure the health and wellbeing of Kosraeans. Kosrae State provided time for its employees to work in the formal economy and to pursue additional activities, including subsistence. Many older and middle-aged adults had memories of household work and farm work, either in the morning before work or school and in the evening afterwards, practices now being re-established. Other arrivals from outside Kosrae – a plethora of invasive and exotic species – had however reduced productivity of some lands (Cannon, 2015), but Kosraeans proved able to produce enough local food.

Ideas of the value of being local and of local self-reliance were familiar tropes on the island. There was a sense, amongst those in older generations, that Kosraean youth were too easily swayed by outside ideas which had led to changes in proper forms of respect and the loss of a work ethic. Many of these same elders however reported being enamoured with western foods while they were younger. For many this introduction to Western foods came when they spent the school week away from home eating primarily United States Department of Agriculture (USDA) food during Trust Territory administration times. During Trust Territory time, the lifeways of Kosraeans were further dramatically influenced by the introduction of goods and services, including electricity, indoor plumbing and bureaucratic employment.

Despite the opportunities to fundamentally rethink the way things are done, the food system ultimately remained largely unchanged as imports continued to find their way into Kosrae. There were no dramatic sustained price increases of imported goods nor a need or desire to systematically change eating habits. Both imported and local foods met nutritional and social needs. Pigs continued to be raised and fish caught for personal consumption as well as selling locally, but there was no need for or interest in, any overwhelming return to subsistence. Imported foods still took preference in local diets.

In short, in the face of ambiguity and uncertainty the overall response was to change relatively little, particularly when there was nothing that necessitated such a change. Stability and continuity were preferred where that was possible. Some changes were unavoidable; Kosraeans experienced an increase in activities associated with funerals since more funerary obligations were associated with the deaths of Kosraeans in the United States. Kosraeans continued to view wage employment as the key component of a suite of opportunities around which to build their lifeways. While there was some disruption of this due to COVID-19 related travel bans, this was partly mitigated by government assistance and by remittances. Benefits continued to come from outside Kosrae.

Concluding Thoughts – Departing and Arriving

Until April 8, 2020, Kosrae State hosted open borders, unlike the other FSM States. The closure of borders and relative isolation is not entirely new, as flights were rare during the Trust Territory Period and the SARS outbreak, but this was at a new scale and of a different duration. While FSM prepared to permit limited entry late in 2020, Kosraeans reflected on stories around the mid-nineteenth century when contact with 'western' influenza decimated their population to less than 400 people. In the context of the COVID-19 pandemic and the Compact, Kosraeans shifted toward self-reliance as the borders to the FSM started closing. As long as goods continued to arrive there seemed little reason to relax the border controls. As knowledge of the mortality experienced in the United States increased, so too did opposition to opening the border. Keeping people out of FSM and Kosrae became the primary coping mechanism. The language used in the FSM President's communications emphasised the importance of maintaining health and wellbeing by keeping COVID-19 out, even if this means that those caught outside FSM could not return. By contrast emigration continued as local people still left the island for a variety of reasons including seeking medical care in Hawai'i or reuniting with stranded family members overseas. This highlighted the relative lack of adequate local health care, and some facets of public health, where health benefits had also come from globalisation, albeit alongside crucial health costs.

As access to health care in Hawai'i and elsewhere indicates, Kosraeans have reaped some of the benefits of globalisation while still opting out of full integration with the global economy. Remembering history, Kosraeans recognised the potential impact of the COVID-19 pandemic and sought alternative means to assert sovereignty in a place free from it, with some realisation that local ways might continue to be foregrounded in development discussions and in everyday practice. Response to the COVID-19 pandemic created an opportunity to revalue local foods, reduce the incidence of NCDs and thus reliance on outside medical expertise to manage diseases. However a preference for the familiar, continuity and inertia, government assistance and government employment, remittances, and the operations of local stores, meant that incomes scarcely fell. One outcome was that diets still included a mixture of imported and local foods that was unlikely to reduce the burden of NCDs. Achieving a 'new normal' even in inadvertently fortuitous circumstances was exceptionally difficult: learning from the past and reshaping a much greater self-reliance proved too much of a challenge where multiple structures of dependence were long-established.

References

Cannon, P. (2015). *Forest Pathology in the Pacific (FSM, CNMI, Guam and Hawaii)* [Trip Report]. The Western Forestry Leadership Coalition. https://www.thewflc.org/sites/default/files/Forest%20Pathology%20in%20the%20FSM.pdf. Accessed 29 Nov 2020.

Connell, J. (2015). Food Security in the Island Pacific: Is Micronesia as Far Away as Ever? *Regional Environmental Change, 15*(7), 1299–1311.

DevBizExperts, LLC. (2019). *Kosrae State Strategic Development Plan (SDP) FY2020-FY2023.*

Fazzino, D. (2020). Holistic Pedagogical Approaches and Youth Empowerment. *NEOS, 12*(1), 10–11.

Guam Daily Post (2020, September 28). *FSM President Seeks Peace as US-China Squabbles Escalate.* https://www.postguam.com/the_globe/nation/fsm-president-seeks-peace-as-us-china-squabbles-escalate/article_eb19658a-006e-11eb-a861-a768ee1220c4.html

Hezel, F. (1983) *The First Taint of Civilization. A History of the Caroline and Marshall Islands in Pre-Colonial Days, 1521–1885.* University of Hawaii Press.

Kim, J., Reinke, A., Eldridge, E., & Grant, M. (2020). Between Georgia and Ohio: Constructing the Covid-19 Disaster in the United States. *Anthropology Today, 36*(4), 17–19.

Meredith, A. (2020). *An Intensive Ethnographic Survey on Traditional Routes in Kosraean Culture*. Kosrae State Historic Preservation Office. National Parks Service.

Peoples, J. (1985). *Island in Trust: Cultural Change and Dependence in a Micronesian Community*. Westview Press.

Ritter, P. L. (1981). Population of Kosrae at Contact. *Micronesica, 17*(1–2), 11–28.

Thomson, G. (2020). COVID-19: Leaving Lockdown – Of Schrodinger, Cats, Testing and Masks. *International Journal of Clinical Practice, 74*(8), e13519.

CHAPTER 10

Healthcare Denied? Covid-19 and Kiribati's Shrinking Transnational Space

John Marazita III

Introduction

The Republic of Kiribati comprises thirty-three mostly atoll islands. The low-lying islands span 3.2 million square kilometres of Pacific Ocean while straddling all four of the earth's hemispheres and several time zones. The small island state is spread over three groups of islands, from the densely populated Gilbert islands in the west to the sparsely populated Phoenix and Line Island groups in the east. With a population of just over 110 thousand and more than three thousand kilometres separating Tarawa and Kiritimati, great distances divide small, remote communities. No direct flights link the capital, South Tarawa, and the Line Islands group, requiring an international stopover in Nadi, Fiji. Remoteness and isolation hinder global economic linkages, including tourism, as few airlines serve Kiribati and access requires costly stopovers in other island states. Isolation and remoteness also impede inter-island travel. The limited domestic flight

J. Marazita III (✉)
University of Geneva, Geneva, Switzerland
e-mail: John.Marazita@etu.unige.ch

Y. Campbell, J. Connell (eds.), *COVID in the Islands: A comparative perspective on the Caribbean and the Pacific*,
https://doi.org/10.1007/978-981-16-5285-1_10

network offered by the local airline, Air Kiribati, is nonetheless invaluable to link outer-island communities in the Gilbert and Line islands groups that would otherwise necessitate arduous and sometimes dangerous ferry crossings (Air Kiribati, 2019).

Kiribati has few natural resources. Exhaustion of its phosphate deposits coincided with independence. Copra and fish, and even seaweed, are exports, but much of its national income comes from the sale of fishing licences and from international aid from the EU, Australia, New Zealand and elsewhere. Remittances, especially from overseas seamen, trained in the national Marine Training Centre, have long been valuable sources of income. By the 1980s, alongside its colonial partner, Tuvalu, Kiribati had been seen as a prototype of MIRAB where development was symbolised by the dominance of Migration, Remittances and Aid and thus of a bureaucratic structure (Bertram & Watters, 1985). Tourism is limited but Christmas Island (Kiritimati) in the Line Islands plays a role in drawing US tourists to fish the abundant waters (Watson et al., 2016). Because of limited domestic production potentiality, Kiribati imports many foods and manufactured items; with a gross national income of just $2986 per capita it is regarded as a least developed country and the poorest SIDS in the Pacific.

The Kiribati government aims to improve economic prosperity through its long-term Kiribati Vision 20 initiative (Kiribati Government, 2018). The government plans to build a resilient country that promotes tourism, local inclusion in the valuable fishing industry, supported by investment in transportation, including the recent purchase of Embraer E190-E2 jets capable of flying between the Gilbert and Line islands as well as on key international routes. The KV-20 represents a departure from previous policies of migration and climate adaptation – that celebrated the notion of migration with dignity – and toward economic growth and resilience based on local initiatives (Kiribati Government, 2018). As the COVID-19 pandemic spread quickly around the world, however, the government's ambitions and strategies were paused. Steady rural-urban migration over half a century has resulted in half the national population, and almost all national institutions, being concentrated in South Tarawa, where population density in some areas rivals that in London and Paris (Chauvin & Mathiesen, 2015; Connell, 2021).

In late March 2020, Kiribati once again reverted to an isolation unparalleled since becoming a British protectorate in the late nineteenth century (Nand, 2020). Following a deadly and highly scrutinised measles outbreak in the region, Pacific island states, beginning with Samoa, began shuttering borders to sequester the remote populations from the growing pandemic of COVID-19 (MacIntyre et al., 2020). Nearly overnight and with little notice, interstate mobility ended, families were separated, and Kiribati was unilaterally isolated from the rest of the world. Even government workers representing Kiribati's interests abroad in an official capacity at the time of the travel ban were left stranded in Fiji. Unlike other states in the region, only limited repatriations from COVID-19 free countries have been authorised, including from the Marshall Islands in October 2020 shortly before the first reported case occurred there (Kiribati Government, 2020a).

During the regional pandemic lockdown, media outlets in the Pacific region focused on reports of aid partner response, travel bubbles, dubious exemptions and demands for repatriation (Khosla & Pillay, 2020). As the pandemic extended beyond short term expectations, attention turned to the plight of migrant workers and the implications of their absence as part of Australian and New Zealand labour mobility schemes (Neef, 2020). While the governments of more populated and geographically less remote Pacific island states organised repatriation and chartered flights to facilitate the return of contract labour migrants, Kiribati's borders between March and October remained sealed. Unlike other Pacific island states, the Kiribati government has offered monetary assistance to internationally displaced citizens to offset hardship (Kiribati Government, 2020b). The plight of seafarers, however, has become sensitive as they finish contracts onboard cargo ships and face extended, cost-prohibitive displacement in final ports of call while lacking prospects of repatriation.

For most households in Kiribati, livelihoods and consumption have been little impacted by COVID-19. The government successfully secured continued access to cargo shipping routes and staple imports continued to arrive, and with the Kiribati government being much the largest employer, few jobs were affected. Ensuring access to medicine and healthcare heavily reliant on aviation transport, however, has been a pressing domestic concern for the Kiribati Ministry of Health and Medical Services (MHMS). Being a small country spread over a vast ocean, economies of scale do not support advanced domestic testing and treatment facilities (Oakes et al., 2016). Healthcare institutions in Australia, New Zealand, Fiji, and India are thus crucial components in both advanced testing and treatment. This chapter explores pre-COVID19 normalcy in the Kiribati healthcare system of the Gilbert Islands group and how the first six months of the closed-borders approach to containing the COVID-19 pandemic not only impacted access to healthcare but led to local adaptation as the transnational space of healthcare collapsed.

Healthcare in Kiribati

Since independence, Kiribati's remote location, post-colonial identity and urbanisation trends have compromised diets previously linked to sustainable livelihoods involving the labour-intensive activities of fishing and cultivation of local produce. Increased access to global food supply chains and shipping lines established through colonial networks facilitated a shift among the local people (i-Kiribati) away from labour-intensive food production and to a reliance on readily available, inexpensive, and effortlessly prepared imported foods. For urban households, white rice quickly replaced breadfruit and canned meats, chicken backs, and lamb flaps became staple meat sources (Cauchi et al., 2019). Expanding access to imported foods in the urban centre, and the outer islands, has driven rising rates of non-communicable diseases (NCD), obesity, and a stagnant life expectancy at birth of just over sixty years. Due to a lack of nutritional diversity, childhood malnutrition also plagues Kiribati (UNICEF, 2013).

Limited access to sanitation and urbanisation in South Tarawa that began during the colonial era have also led to illnesses associated with overpopulated conditions. With a high population density and only basic housing, personal space is limited. In addition to high rates of NCDs, Kiribati leads the Pacific region in rates of infectious diseases (Cavanaugh et al., 2015). Leprosy (15 cases/10,000 people), tuberculosis (30 cases/10,000 people) and hepatitis B (15–20% of the population) all burden the healthcare system and in some cases have prevented access of local workers into labour mobility schemes (Cavanaugh et al., 2015; WHO, 2015; Chambers et al., 2020). Infectious disease cases are especially prevalent in densely populated districts of South Tarawa, including Betio. Although the government has made strides to improve access to sanitation and clean water, occurrences of diarrhoea and respiratory infections continue to be a concern (Aretaake, 2019). Additionally, Kiribati has high maternal, infant, and neonatal mortality (Carter et al., 2016).

In 2000, the UN member states set out eight Millennium Development Goals (MDGs) relating to health, aiming to reduce global disparities and infectious disease rates by 2015. The MDGs were the precursor to the Sustainable Development Goals that have become the focus of the UN's work through 2030 (WHO, 2018b). By 2015, however, Kiribati had not attained any health-related MDGs (Kiribati Government, 2014).

The Kiribati government provides free access to healthcare through the administration of MHMS. Limited funding, however, restricts access to medical services and treatments on offer, and bureaucratic infighting often leads to poor working conditions for medical professionals. According to WHO (2013), Kiribati's healthcare system consists of only one tertiary-level hospital, three secondary-level hospitals, and primary-care clinics in outer-island communities. Of the four hospitals, three are in the Gilbert Islands group, centred on South Tarawa. While doctors staff hospitals equipped with essential medical equipment (secondary and tertiary care), outer-island clinics are staffed mainly by resident nurses, often with enhanced qualifications and midwifery training to treat routine illnesses (primary care). These clinics offer antibiotics, suturing, dressings, basic dental care and referral to the hospital system. They also arrange emergency treatment and transport when necessary. On the outer islands this transport is usually a boat. Advanced tertiary care is limited to South Tarawa. Patients residing in outer islands showing progressive illness symptoms are thus routinely transported to South Tarawa for further treatment (WHO, 2013).

The Tungaru Central Hospital in South Tarawa is the national centre for health and the country's only advanced, tertiary-level hospital, offering multiple wards and a surgical theatre. Although qualified nurses and doctors staff the hospital, there is limited capacity to perform some advanced diagnoses and treatments due to lack of equipment. Shortcomings in healthcare extend to a broad range of therapies, including dialysis, cancer (diagnosis and treatment), advanced surgeries and transplantations, and the handling of patients with rare blood types. For patients with advanced stages of cancer and other illnesses, domestic treatment consists of palliative care.

The Kiribati healthcare system is dependent on the mobility of patients, supplies, and training. As is the case in most Pacific island states where access to healthcare is limited, patients without adequate treatment domestically are outsourced by MHMS to foreign medical institutes in partner countries. Due to the cost-prohibitive nature of medical procedures, India has become an increasingly important destination hub for treating patients from Pacific island states, including Kiribati. Cancer and heart by-pass surgeries, for instance, that are untreatable domestically, are available in hospitals in Indian cities such as Hyderabad (WHO, 2018a). The ministry vets candidates for overseas medical referrals based on illness advancement and the patient's situation.

Before COVID-19 travel restrictions, international flight connections from the urban hub, South Tarawa, were already limited. However, these flights provided a crucial lifeline for extended care, leading to a dependency on access to foreign medical institutes for both testing and treatment. At the end of March 2020, however, international flights came to a halt. The flight closures began with Fiji Airways on 20 March and were followed closely by Solomon Airlines, Nauru Airlines, and Air Marshall Islands. As a regionwide lockdown began, the Kiribati President, Taneti Maamau, introduced the 'COVID-19 Preparedness and Response Regulations' through a State of Public Emergency proclamation (Kiribati Government, 2020a). The state of emergency led to a partial lockdown with schools closed, quarantine put in place, and restrictions put on spreading fake news. The government also postponed national elections for two weeks. The travel ban severely impacted both passenger travel and the flow of essential supplies.

The border closure also had implications for domestic transport. Firstly, the previously reliable flight connection between the Line and Gilbert Islands groups through in Fiji was severed. The east of the country was severed from the west. Secondly, unable to acquire spare parts from overseas, domestic airlines connecting the Gilbert Islands group became grounded as mechanical problems mounted. The travel ban thus catalysed a breakdown in both international and domestic mobility. Even newly elected members of parliament and government ministers were left stranded in their home islands. As international flights and shipping routes ground to a halt, outer islands that had grown dependent on South Tarawa for employment opportunities, education, and trade reverted to pre-colonial isolation.

Local Outcomes

The unprecedented closure of borders and uncontrollable pandemic gossip spreading through social media led to public anxiety over a possible COVID-19 outbreak. In March, the fear festered into chaos when a foreign-born fishing trawler employee was reportedly offloaded in the most densely populated district of South Tarawa, Betio. The patient, after being transported to the nearby Betio hospital, was rumoured to have COVID-like symptoms. As rumours spread across South Tarawa, the critically ill patient and contacting staff were isolated. Fear of the virus enveloped South Tarawa and parents and guardians rushed to collect children from schools. Much of the island went into lockdown and waited in suspense as a humanitarian flight ushered the, by that time deceased, patient's blood sample to New Zealand for further COVID-19 testing.

Much to Kiribati's relief, the patient tested negative for COVID-19, but the realisation that the community was ill-prepared for the outbreak became apparent, and MHMS began exploring contingency plans. Supported by limited supplies and without local testing capabilities, MHMS recruited medical staff as part of a COVID-19 task force. The early task force focused on procedural protocols for containing outbreaks with triage centres and quarantine facilities in line with the previously mentioned presidential proclamation. MHMS reportedly chose medical staff for the task force based partly on age and marital status. While these measures probably relieved some public sentiment, the healthcare system quickly turned its attention to more pressing matters in the hospital wards.

As international flights were grounded, so too were the flows of medicine and medical supplies. While Kiribati may never have had direct access to advanced medical facilities that have become commonplace in much of the world, the country's high rate of NCDs and infectious diseases required constant treatment and attention. Without the capacity to replenish stock, shortages began negatively impacting treatment and requiring MHMS to ration consumables. In the early days of the pandemic, Kiribati, like the rest of the world, faced shortages of masks, gloves, and hand sanitiser. During the first six months of the pandemic, the lack of international flights led to the prolonged scarcity of consumables and medical staff adapted practices to reduce consumable dependencies.

With border closures perpetually extended, the Kiribati hospitals began running out of crucial supplies and by September had severe shortages of medicine and testing equipment. Among the shortages were blood sugar testing kits (Glucostrips) and the antibiotic, Benzylpenicillin. Although it is too early to determine causality of hospitalisations impacted by healthcare deficiencies during the breakdown in transnational mobility, preliminary insights from medical professionals in Kiribati revealed a need for extended treatments of infections, and prolonged hospital stays due to a reliance on alternative antibiotics. As the duration of patient admissions increased, hospital wards began filling up, adding pressure to the already limited capacity.

More concerning than diminishing medical stock, however, was the sudden and complete disconnect of access to the global healthcare chain. Without reliable international transport, the Kiribati population became unilaterally isolated from advanced care previously supported by medical outsourcing. The travel ban stonewalled any patients awaiting referrals to overseas healthcare facilities. Patients requiring domestically unavailable dialysis, for instance, could no longer elect for private treatment in Fiji. Without testing capabilities, the disconnect severely impacted the diagnosis and treatment of cancer patients for whom even biopsy testing required outsourcing. In the first sixth months of the pandemic, barriers to accessing cancer treatment impacted every level of Kiribati society, from the outer island communities to urban members of parliament. Patients with an advanced illness that awaited referrals could only receive palliative care as travel bans became regularly renewed.

Beyond International Healthcare

Travel bans constrained not only international exchanges of people and supplies but also impacted domestic travel. As domestic flights halted due to mechanical failures, medical evacuations from outer islands to South Tarawa ceased. Inequalities in accessing healthcare grew between South Tarawa urbanites, having access to hospitals, and remote outer islands that became dependent on limited assistance from primary-care nurses. Except for facilities on the relatively large island of Tabiteuea, which has the only hospital outside Tarawa, outer island communities became detached from access to hospitalisation, medication, and advanced care. Just as South Tarawa had become dependent on connectivity, with international flights between Tarawa, Fiji, and elsewhere, so too had outer islands in the Gilbert Islands group become dependent on the urban incomes and livelihoods, markets and social networks of urbanised South Tarawa. With a lack of international flights to ferry spare parts, Air Kiribati's domestic fleet temporarily ceased operation.

Outer island communities, like Butaritari, no longer had markets for pumpkin harvests, and job seekers and students could no longer descend on South Tarawa in search of improved livelihoods and further education. The local graduates of the University of the South Pacific, Kiribati Institute of Technology, and Marine Training Centre could no longer hope for overseas job placements in Australia and New Zealand and onboard cargo ships plying the world's oceans. Prospective students could no longer receive education and further training in foreign institutes.

This chapter has so far recounted a unique narrative that may seem to be a metaphorical snapshot of a remote cluster of islands where 'time stood still.' This narrative, however, would not be accurate in singularly defining the first six months of Kiribati's COVID-19 period by the sudden and prolonged absence of connectivity. Long before colonial leadership and Christian missionaries reached the sandy Pacific atoll shores, the cluster of islands that now form the independent Republic of Kiribati, were by necessity largely self-sufficient island communities. The pre-colonial islands adapted to the constant threat of death and disease brought on by droughts and other hazards, by local exchange and occasionally local warfare. In a sense, the pandemic did not stop time but reversed time and resulted in communities adapting locally, if without the occasional violence of the past. Unlike the famines of their ancestors, however, pandemic-era Kiribati communities now confronted a modern vulnerability to both non-communicable and infectious diseases, such as cancer, without earlier precedents.

Adaptation to healthcare vulnerabilities brought on by the travel ban thus did not solely reflect on the MHMS. Instead, there was evidence of local agency. As travel restrictions cut off the import of medicine, communities in the Gilbert Islands turned to earlier generation remedies. In outer islands, local medicinal knowledge often varied between families and village practitioners. (Post-COVID-19 research in outer-islands of the Gilbert group will be necessary to fully understand the extent of outer-island adaptive capacity and its useful or detrimental impacts).

Evidence from South Tarawa, however, points toward the crucial role of local retreats as an alternative treatment to dwindling healthcare access. Although there have been no specific scientific studies to date on Kiribati traditional medicine, the purported success of 'detox' programs offered in South Tarawa has received increased local interest. During pre-COVID-19 normalcy, local practitioners trained in Fiji had begun locally adapted holistic therapy fused with Kiribati's traditional remedies. Local plants, often used in generational knowledge, such as pandanus, papaya, and noni, reduced local practitioners' need for more expensive imported medicinal ingredients, often influenced by rigid import restrictions.

Retreats, having less dependence on imports, could continue to provide extended naturopathic therapy during flight bans. The Kiribati Health Retreat Association (KHRA) manages the largest such retreat in South

Tarawa. It offers a range of services from specific detox programs for various organs to diet counselling and vegetarian meals, a rarity on atolls. KHRA, established in 2014, was a component of an Australian Department of Foreign Affairs and Trade (DFAT) funded food security program and is primarily youth-led (16–28 years of age). Selected volunteers receive professional training in naturopathic medicine at the Three Angels Missionary College in Fiji, including training in nutrition, exercise, massage, detox, hydrotherapy, and facials. Through various funding mechanisms, the retreat expanded to campuses in Kiritimati and has offered outreach programs in various outer island communities. Between 2014 and 2018, KHRA (2018) reported treating more than 1500 patients. Prospective patients sought treatment for a wide range of ailments. As of 2018, nearly half of the patients sought treatment for NCDs, but increasingly the retreat has been treating infectious diseases such as hepatitis, bone and joint ailments, menstrual problems, vision and hearing loss, fatigue, and inflammation.

Before COVID-19, the retreat's therapies complemented medical treatment and diagnosis. Retreats thus played a crucial role both during post-treatment therapy and as a long-term lifestyle change to improve health. Furthermore, hospital meal services were often inadequate, and patients' families with financial means often procured meals from KHRA to improve patient satisfaction. In the pre-COVID-19 normality, retreats played a niche position within the Kiribati society of preventive healthcare, long term lifestyle transformations, and improved diets and its popularity was disproportionately growing amongst female patients between forty and fifty years of age (KHRA, 2018).

Although the retreats often complemented medical treatment, they also had the capacity for in-patient, long term therapy. Kiribati retreats focus on the link between health and nutrition with most treatments being long-term. Anecdotal evidence has supported patient improvements for NCDs, hepatitis, obesity, cervical cancer, and liver and kidney illnesses (KHRA, 2018). Therapy for illnesses can often include juicing diets and purported natural antibiotics. Although there is no scientific evidence to support the reported results, there was growing local trust in retreat therapies as an alternative to modern medicine.

KHRA adapted to the pressures of COVID-19 flight bans by accelerating the substitution of imported commodities by locally grown crops. Among the commonly used local products were green coconut, pandanus, and the more recent agricultural introduction of cabbage. Juicing diets utilised local fruits, including the noni fruit, a staple home remedy for many ailments. These trends followed KHRA's previous work in raising awareness of often forgotten local, edible plants such as Pisonia (*Te Buka*) and other leafy greens.

As the Kiribati healthcare system faced shortages due to flight bans and ward occupancy, and patient frustration grew, retreats began attracting new clients seeking relief from NCDs and illnesses that otherwise could only receive palliative treatment in the hospitals. For patients unable to seek treatment in foreign medical institutions, the retreats offered hope, if not for a cure then for stability, until medical flights resumed. Unlike medical treatments provided by MHMS, however, in-patient treatment at retreats could be prohibitively expensive for Kiribati households often relying on little more than subsistence, but practitioners benefited.

Conclusion

COVID-19 has disrupted livelihoods and halted mobility worldwide. Kiribati, as of November 2020, remains one of the few virus free countries. Keeping the pandemic at bay required extreme measures from the government that effectively isolated the small island state from the rest of the world, and almost isolated many of the atolls. During the first six months of travel bans, the healthcare needs of the i-Kiribati that had become increasingly dependent on imports began to revert to traditional practices.

The Kiribati healthcare system has always struggled with providing free access to adequate healthcare with its citizens spread over many islands and three million square kilometres of ocean. Lacking economies of scale in healthcare, Kiribati forged a reliance on the mobility of patients, medications, and medical equipment in which the capital, South Tarawa, became a hub between limited outer island clinics and international global health providers. A new transnational space of healthcare was formed with nodes in countries such as Fiji, Australia, New Zealand, and India.

After travel bans paralysed the region and both cargo and passenger air traffic ground to a halt, this disruption acutely affected both the urban centre and the outer islands. More than six months after flights ceased, many households remain separated by international borders with little hope of being reunited before the new year. Migrant workers having finished overseas employment contracts remain in legal and social limbo. The travel ban was a sudden and unforeseen disruption to Kiribati's transnational healthcare strategy. As the last flight took off from Bonriki International Airport in March, the healthcare system began to brace for long-term adaptability to restrictions on medication and supplies. The public, in turn, lived with uncertainty and a lack of access to advanced testing and treatment. As flight bans were extended month after month, outer islands and local communities began accepting a new normal for healthcare. Local retreats practising naturopathic therapy infused with ancestral medical knowledge offered what was regarded as a trusted alternative to patients unable or unwilling to seek the limited treatments offered by MHMS in overcrowded hospitals and outer island clinics.

It is still too early to understand the full link between COVID-19 and healthcare in Kiribati, but interviews with KHRA staff and medical professionals on the healthcare frontlines suggest that Kiribati is resilient and adapting to the local health crisis by working across disciplines and enacting traditional medical knowledge to support gaps in healthcare. The travel ban, however, could have long term effects on the healthcare system, as opportunities for further training only available abroad remain out of reach for otherwise qualified local medical students.

As the first six months of isolation drew to a close, MHMS shifted focus. Firstly, increased humanitarian flights from New Zealand and Australia were beginning to replenish medicine and medical supplies. Secondly, the government was beginning repatriation flights leading MHMS to re-strategise the COVID-19 task force from a focus on perceived threats of an outbreak toward frameworks and regulations for quarantine. Reports from nurses during the first repatriation flight from the adjacent and then COVID-19-free Marshall Islands demonstrated a well-organised quarantine with arriving passengers hosted in a secluded location and attending medical staff quarantined nearby. As the repatriation was successful, the government was conservatively planning future flights, although quarantining medical personnel would stress the already overworked pool of medical professionals. As repatriations from Fiji, and humanitarian and medivac flights were scheduled to begin in November, a semblance of the former transnational healthcare system was gradually returning. With commercial flights grounded for the foreseeable future, however, post-COVID-19 Kiribati communities could become compelling field sites to study the long-term implications of healthcare disruptions on adaptive capacity, and whether future healthcare will become a more hybrid system of new and old therapies.

Acknowledgements This chapter is based on participant observation and narrative interviews with ministry officials, healthcare professionals, KHRA, and patients between April and October 2020. A special thanks to the private ward nurses, MHMS, Aroberte O'Connor and Pelenise Alofa for facilitating this research under such extraordinary times. *Te Mauri, Te Raoi, ao Te Tabomoa.*

References

Air Kiribati. (2019). *Air Kiribati – Our History.* Available at: http://www.airkiribati.com.ki/our-history. Accessed 24 Mar 2019.

Aretaake, R. (2019). Traditional Kiribati Beliefs About Environmental Issues and Its Impacts on Rural and Urban Communities. *Disaster Prevention and Management: An International Journal, 28*(1), 25–32.

Bertram, G., & Watters, R. (1985). The MIRAB Economy in South Pacific Microstates. *Pacific Viewpoint, 26*(2), 497–519.

Carter, K. L., Baiteke, T., Teea, T., Tabunga, T., et al. (2016). Mortality and Life Expectancy in Kiribati Based on Analysis of Reported Deaths. *Population Health Metrics, 14.*

Cauchi, J. P., Correa-Velez, I., & Bambrick, H. (2019). Climate Change, Food Security and Health in Kiribati: A Narrative Review of the Literature. *Global Health Action, 12*(1).

Cavanaugh, J. V., Kienene, T., Harley, D., et al. (2015). Effect of Diabetes on Tuberculosis Presentation and Outcomes in Kiribati. *Tropical Medicine and International Health, 20*(5), 643–649.

Chambers, S. T., Ioteba, N., Timeon, E., Rimon, E., et al. (2020). Surveillance of Leprosy in Kiribati, 1935–2017. *Emerging Infectious Diseases, 26*(5), 833–840.

Chauvin, R., & Mathiesen, K. (2015, March 17). The Pacific Islands Losing a Way of Life to Climate Change – In Pictures. *The Guardian.* Available at: https://www.theguardian.com/environment/gallery/2015/mar/17/pacific-islands-losing-way-of-life-to-climate-change-in-pictures. Accessed 19 Jan 2019.

Connell, J. (2021). Coral Islands, Climate Change and Distant Destinies? The View from Kiribati. In S. Moncada, L. Briguglio, H. Bambrick, I. Kelman, et al. (Eds.), *Small Island Developing States: Vulnerabilities and Resiliencies Under Change.* Springer.

Khosla, V., & Pillay, P. (2020). COVID-19 in the South Pacific: Science Communication, Facebook and "coconut wireless". *Journal of Science Communication, 19*(5), A07.

KHRA. (2018). *Kiribati Health Retreat Association REPORT 2014–2018, Republic of Kiribati.*

Kiribati Government. (2014). *Millennium Development Goals: Kiribati, Progress to June 2015.*

Kiribati Government. (2018). *Kiribati 20-Year Vision 2016–2036.*

Kiribati Government. (2020a). *COVID-19 Preparedness and Response Regulations 2020*. Office of te Beretitenti.

Kiribati Government. (2020b). *High Level Covid19 Preparation and Repatriation Committee Press Conference*.

MacIntyre, C. R., Costantino, V., & Heslop, D. J. (2020). The Potential Impact of a Recent Measles Epidemic on COVID-19 in Samoa. *BMC Infectious Diseases, 20*, 735.

Nand, E. (2020, March 21). Fiji Airways Suspends Flights, Implements Leave Without Pay. *Fijian Broadcasting Corporation*. Available at: https://www.fbcnews.com.fj/news/covid-19/fiji-airways-suspends-flights-implements-leave-without-pay/. Accessed 25 Nov 2020.

Neef, A. (2020). Legal and Social Protection for Migrant Farm Workers: Lessons from COVID-19. *Agriculture and Human Values, 37*, 641–642.

Oakes, R., Milan, A., & Campbell, J. (2016). *Kiribati: Climate Change and Migration – Relationships Between Household Vulnerability, Human Mobility and Climate Change*. United Nations University.

UNICEF. (2013). *Kiribati: Tracking Progress in Maternal and Child Survival*. UNICEF.

Watson, M. S., Claar, D. C., & Baum, J. K. (2016). Subsistence in Isolation: Fishing Dependence and Perceptions of Change on Kiritimati, the World's Largest Atoll. *Ocean and Coastal Management, 123*, 1–8.

WHO. (2013). *Republic of Kiribati, WHO proMIND: Profiles on Mental Health in Development*.

WHO. (2015). *Hepatitis B in Kiribati: An Island Epidemic*. Available at: https://www.who.int/westernpacific/news/feature-stories/detail/hepatitis-b-in-kiribati-an-island-epidemic. Accessed 25 Nov 2020.

WHO. (2018a). *Kiribati Health Financing System Assessment: Spend Better* (Report No: AUS0000154).

WHO. (2018b). *Millennium Development Goals*. Available at: https://www.who.int/news-room/fact-sheets/detail/millennium-development-goals-(mdgs). Accessed 25 Nov 2020.

CHAPTER 11

Small and Isolated Vava'u, Tonga: From Weakness to Strength During COVID-19?

Leody Cruzat Vainikolo

Introduction

The Kingdom of Tonga is a small Polynesian island state with a mainly agricultural domestic economy and much migration to New Zealand and Australia so that remittances make a significant contribution to both national development and household incomes. More than half of all ethnic Tongans live overseas, notably in New Zealand, but also in the United States and Australia. Tonga is an archipelago of four main island groups with 36 inhabited islands and a total land area of 747 sq kms but an Exclusive Economic Zone of about 720,000 sq kms. Nuku'alofa, the capital, is situated in Tongatapu, the largest island, which has about 70% of the national population of just over 100,000; its primacy has steadily been growing.

L. C. Vainikolo (✉)
Ministry of Agriculture, Food and Forests (MAFF), Nuku'alofa, Tonga

Y. Campbell, J. Connell (eds.), *COVID in the Islands: A comparative perspective on the Caribbean and the Pacific*,
https://doi.org/10.1007/978-981-16-5285-1_11

Vava'u, the second largest island group, some 300 kms to the north of Tongatapu, has about 14,000 people, just 14% of the national population. Neiafu, the urban centre of Vava'u, has around 5250 people. The remainder of the Vava'u population are scattered in small villages with economies centred on agriculture. Vava'u has also had a small but growing tourism economy with whale watching having become a popular activity in recent years (e.g Orams, 2013). With a rather warmer climate than Tongatapu, Vava'u has been a centre for the cultivation of vanilla, kava, root crops and various tropical fruits. Before COVID-19, Vava'u was often seen as somewhat peripheral to the rest of the country, with a small, rather isolated and declining population. This chapter examines how agriculture has changed under COVID-19 – despite its absence from the country – and the extent to which this has reshaped relations between Vava'u, Tongatapu and the wider world.

The Absence of COVID-19

At the end of 2020, Tonga remained one of the few countries without the COVID-19 virus. Like several other Pacific Island states Tonga closed its borders before COVID-19 could arrive. The last commercial airline flight to Tonga was on 21 March but subsequently several humanitarian flights brought Tongans back from New Zealand, Australia, the United States, Fiji and elsewhere. Some 846 Tongans had been repatriated from these countries by November 2020, but as many as 7000 overseas Tongans (many working on agricultural programmes in New Zealand and Australia) still sought repatriation, a measure of the number of Tongans employed in various overseas contexts. With limited local mobility, Tonga's own tiny domestic airline quickly became bankrupt, so cutting air links between the islands, but in November 2020 the Government established a new airline named Lulutai (Sea Eagle) to replace Real Tonga Airlines.

The Government declared a State of Emergency, which remained in place for the whole of 2020, and issued a National Lockdown Notice on 26 March 2020, which was extended many times. This imposed lockdown was accompanied by a curfew, that had been running for 9 months (March–December 2020), with the only change being curfew hours reduced from 10 hours (8:00 pm to 6:00 am) to 5 hours (12:00 midnight to 5:00 am). Alongside the night-time curfews, all activities and gatherings were not to exceed 50 people indoors (and 100 outdoors), with people having to comply with social distancing and Ministry of Health requirements for sanitation and hygiene. Church services were significantly reduced in numbers with this lockdown, a significant change in a nation where Christianity is extremely important and churchgoing almost universal. The Free Church of Tonga, which usually had church services on Mondays, Wednesdays, Fridays and Sundays with 3 services – at 5:00 am, 11:00 am and 3:00 pm – was reduced by the lockdown to a single hour-long service on Sundays.

The most obvious impact of the border closures was the loss of its tourism industry, that represented about a tenth of the national income (and rather more of Vavau's income), while contributing to the income of about a third of all Tongan households. Tourism collapsed in Vava'u and the whole of Tonga. Vava'u was the centre of seasonal whale watching from July to October with as many as 400 yachts alone coming to Vava'u during the season. Many guest houses and local restaurants that catered for these visitors closed down, and local shops faced much reduced business, although they remained in operation. However, with the borders closed, Vava'u eventually experienced some growth of domestic tourism, from Tongatapu and other islands, with airline tickets being booked out in November and many people also coming by ferry. A new niche in the tourism industry had opened.

Business decline quickly extended from tourism to the extent that over 60% of businesses in the country reduced the numbers of staff or their hours of work (Webb, 2020). The isolation resulting from border closures was however turned into something of an advantage, in providing an opportunity for the Government to review and renew the rules and regulations in giving licence permits to whale watching operators. Many whale watching operators were based overseas where money paid for whale watching remained, rather than being spent in Vava'u to support the local economy.

After an initial dip, remittances were largely maintained. The National Reserve Bank of Tonga Website reported in July 2020 that Despite expectations of a downturn due to the impacts of COVID-19, Tonga's remittance persistently remains at very high levels over the past few months to July 2020. Even without the annual festivities that usually attract remittance such as the church conferences, the Heilala Week Festival and the Agricultural Show, the inflow of remittances still increased in all categories and all major currencies. Higher family support in July 2020 indicates the willingness of relatives abroad to support their families despite the global economic downturn imposed by the COVID-19 pandemic. Total remittance receipts for

July 2020 recorded a new high record rising by $5.6 million (17.4%) from the previous month. This is the highest monthly remittance receipt since 2014 surpassing the remittance receipt recorded in May 2020 at $35.0 million and higher than that of July 2019 at $32.1 million.

While that reported the situation for the whole of the country there was no reason to suppose that this was not true of Vava'u. The diaspora proved invaluable.

Agriculture

The main islands of Tonga consist of mainly low and raised coral limestone overlaid by a volcanic ash soil layer. Surface water is absent and all agriculture is rain-fed. Temperature and rainfall increases northwards so that Vava'u has a much warmer climate than almost all the rest of Tonga; the island's warm climate and fertile soil are suitable for vanilla, pineapples and tropical fruits, and for a subsistence oriented agriculture dominated by various root crops. Vava'u is dominated by a landscape of agricultural production, both subsistence and commercial, long ago centred on coconuts (and copra), previously vanilla and now kava. Most households in Vava'u are semi-subsistence farmers growing local food and tending livestock.

Natural hazards can pose problems for agriculture and for broader livelihoods. Cyclones are regular occurrences. In 2014, category-5-cyclone Ian devastated the Ha'apai group and in 2018, the category-4-cyclone Gita devastated Tongatapu and 'Eua islands. Early in 2020 cyclone Harold caused much devastation on Tongatapu, but mainly spared Vava'u. That has been generally true in this century for Vava'u, where no major devastating cyclone or other natural disaster has occurred since 2001 where Vava'u was directly hit by Cyclone Waka with a sustained wind speed of 185 kms/hour. Agriculture was almost wiped out, including fruit trees and almost all the root crops. (During the finalisation of this chapter in December 2020 both Cyclones Zazu and Yasa affected the archipelago but neither was disastrous). El Niño droughts are relatively frequent, the last being in 2014. Earthquakes and off-shore volcanic eruption are more frequent. The 2009 tsunami killed nine people in the northern Niuas island group. According to the World Risk Index, Tonga ranks second only to Vanuatu in countries with a high propensity to natural disasters. The pandemic proved a different kind of disaster.

The agricultural economy is dominated by root crops, including the giant taro (*kape*), yams, other varieties of taro (*Talo Tonga* and *Talo Futuna*) and cassava. Traditionally, they are planted in sequence according to their importance with yam, highly prized for its traditional and cultural value, as the first crop followed by intercropping of *kape* and other root crops later in the cycle. For small islands without adequate facilities for food storage, perennial and annual root crops are highly appropriate. Root crops, effectively stored in the ground, represent food security. Farmers plant different types of root crops for food and for exchange and ceremonial reasons. Root crops serve as "money in the bank" to be harvested when needed and are particularly valuable for traditional occasions like weddings, funerals, birthdays, the celebration of national events and, above all, for gift exchanges. Vava'u islanders also have a range of livestock, especially pigs and chickens. The 2015 Tonga National Agricultural Census supported the island's proud claim to have more pigs than people – with some 25,067 pigs in Vava'u to a population of 13,738. There were almost as many chickens. Both proportions were higher than in other Tongan island groups.

Past Tongan diets were primarily composed of root crops, fish and occasional meat with few vegetables; the only traditional vegetables in the past were taro leaves and *pele* (H. manihot) leaves. Over time, with access to both money and imported food, and an unprecedented increase in the numbers of local shops (mainly established with remittances), there was a significant shift in eating habits, involving more processed and imported food. Many of these foods were devoid of nutrients and packed with high density calories, gradually contributing to an epidemic of non-communicable diseases (NCDs). Tonga has one of the highest rates of diet-related NCDs in the world, with poor food choices being a key factor (Evans et al., 2002, 2003). The incidence of type-two diabetes amongst the adult Tongan population increased from 5% in 1973 to 19% in 2012, and is roughly three times the global average (Lin et al., 2016), Consequently life expectancy in Tonga is presently declining, primarily due to NCDs. This transition was partly facilitated by the rise in household incomes because of remittance flows, and the consequent consumption of inadequate foods such as mutton flaps (Gewertz & Errington, 2010). Tonga has introduced various initiatives and policies aimed at promoting healthy eating and lifestyles, but there has been no effective dietary transition. One constraint to adopting a more healthy diet was that accessibility to affordable fresh fruits and vegetables in Tonga was often challenging and local marketed foods were expensive. However the Government distributed seedlings as part of its support for agriculture (see below) which eventually resulted in an extraordinary abundance of local vegetables at relatively low cost, which probably had nutritional benefits (as was intended).

The pandemic brought initial disruptions to transport flows, and therefore access to imported goods, while reduced local incomes (despite remittances being maintained) resulted in more people working on their farms, and planting and consuming local foods, as most households returned closer to subsistence. Coincidentally, many foods such as papaya, mango, Pacific lychee and pineapples, were all in season. Since many households were farming only a small portion of their land, with much being in fallow, many perennial and other fruit trees were growing naturally. Similarly more people returned to or took up fishing. Although fish and other marine products were expensive at the local market, people had the option not to buy but go and do their own fishing, and many chose to do this. The greater cost of fish in the local market was the outcome of both fewer commercial fishermen and few of them going out to fish for sale, as they were more likely to consume their own catch. Tuna, snapper and other export fish, cost as much as $12–15/kg locally. During the pandemic, the Government provided a subsidy on local fish where the Government bought the fish for $10/kg from commercial fishermen and the Fisheries Department of the Government then sold the fish for $7.00 per kg to the people. This was largely unsuccessful as commercial fishermen preferred to sell their fish to middlemen for the latter to resell the fish to people for $12–15/kg.

With lost employment (and reduced local incomes) people had more time and incentive to devote to subsistence activities. People even had more time to go to the bush than before with reduced time spent on social and church activities due to the restrictions imposed such as curfews and limited numbers of people in gatherings. That enabled resilience and a considerable degree of food security. Root crops eventually became relatively abundant, and available for marketing, within Vava'u if required and to the other main islands of Tongatapu, Ha'apai, and the Niuas. No official record of this local supply to the outer and main island exists but by mid-2020 weekly shipments were made on local vessels servicing these islands.

During the pandemic period, there were fewer traditional social events and feasting, and therefore reduced demand for money. Local food supply in the market initially declined, and became more expensive, as many farmers sought to either retain their root crops for their own food security, or gain better prices from regional sales. A significant change during 2020 was that there was also a growing demand for local root crops and vegetables from the overseas diaspora especially in the United States. That was the outcome of the context where many Pacific islanders in the United States were losing employment – as an indirect outcome of the pandemic – and were consequently seeking to buy cheaper food (rather than the more expensive root crops they had previously obtained from Mexico and elsewhere, whose prices were rising during the pandemic). That provided an unexpected and rather lucrative market and increased demand for local produce (as much as five containers a month) that was permissible to export to the USA; these included frozen root crops, frozen fruits and vegetables, such as frozen plantain, pineapples and *pele* leaves. Root crops were becoming valuable cash crops, perhaps tinged with nostalgia in an America in difficult times. It was much welcomed in Vava'u as the first time that exports had been marketed directly to the United States.

Cash Crops

Simultaneously households were also engaged in more obviously and more exclusively commercial agriculture in order to sustain local incomes. Vanilla has been a long-term perennial crop, strongly promoted by the Government, particularly in Vava'u where it grows well. It is an environmentally- friendly crop, with most vanilla plantations grown organically, and usually a high value crop that is not perishable once cured properly. It can be stored for many years without a special cooling facility and is suitable for shipment by boat and thus ideal for a small island. It was and is a key cash crop in Vava'u. However, fluctuating prices were at their lowest – around \$5.00/kg of green beans in the early 2000s – driving the farmers to destroy or neglect many of their vanilla plantations in favour of a shorter-term crop like kava. Although in the last 4 years, vanilla has commanded a very good price of about \$130–200/kg for green Vanilla beans, vanilla did not perform well (with limited flowering and a poor harvest) so that by 2020, many vanilla plantations were either neglected – two-thirds were destroyed – to give way to new kava plantations. However high prices and the availability of labour again resulted in a resurgence of vanilla planting.

Through the 2010s the kava price soared and many Vava'u farmers concentrated on kava, mainly for a local and regional market. From 2017, there was a significant shift in farming practices, particularly in the very small outer islands of Vava'u, with kava being planted first followed by yams, *kape* and other root crops. This followed the "Kava Fever" due to the significant increase in global demand and its price increasing almost eight-fold in 2018 from US$20 per kg. of dried kava powder to $150. Kava planting increased rapidly, with many secondary forests cleared for planting, and Vava'u recorded an increase in the area under kava from 805 acres in 2015 to 2640 acres in 2019. That continued during the COVID months. Large kava plantations were almost all intercropped with *kape*, which contributed to its becoming an even more important export crop. Ultimately the area under *kape* across Vava'u was greater than that under vanilla or kava.

As the Head of the Ministry of Agriculture, Food and Forests (MAFF) Extension Team in Vava'u, Mikaele Saipaia, observed late in 2020: "there is so much positive development at the farm level with more new crops planted, more new plantations established and better-looking farms which can be attributed to COVID-19 outcome. More people are farming". While no formal evidence exists, there was at least anecdotal indications that people were eating more local foods and that overall nutrition was benefiting.

While Vava'u is seemingly disadvantaged by distance, with some direct shipping services in the past being cancelled due to lack of economies of scale, and shipping companies experiencing losses, that gradually changed because of local produce being in demand overseas, particularly in the United States, a situation that did not exist before COVID-19, and because of the high prices for kava and vanilla. In many official documents and publications, being small and isolated like Vava'u appears a major challenge for development. Remote islands are costly to service, and it is expensive and unsustainable to establish modern infrastructure, including agricultural technology of various kinds. Capital too may be scarce and development bank loans harder to obtain. There are limited facilities such as coolers and blast freezers to store local produce like root crops for slow release in the market, and even short storage times can be problematic (Underhill et al., 2020). Vava'u benefited from the continuation of local shipping services with Tongatapu and other islands. Weekly services prevented any problem of market access (but now marked with strict regulations on loading and unloading cargo from overseas). At the same time agriculture had retained traditional practices and knowledge in all aspects of production and storage to maximize benefits from the local produce available without relying on modern technology like freezing and cooling, hence the island's primacy in *kape* production.

Under COVID one problem was reduced. Labour availability was always a challenge to crop production in the island, and had been aggravated in the previous decade by the Recognized Seasonal Employer Scheme (RSE) in New Zealand and the Seasonal Worker Programme (SWP) in Australia, since the younger and potentially more productive workers went overseas to work. Their perception was that if people were going to work in agriculture then it was much better to do it there for a 'proper wage. The consequence was that hitherto only overworked and older farmers, and younger men uninterested in farming, remained. Younger men and women preferred to work in offices and other kinds of work rather than farming. Under COVID-19 some of these jobs disappeared while potential migrants were forced to stay at home, effectively releasing labour – however reluctant – for agriculture.

Tonga Tackles COVID

Beyond individual and household initiatives, mainly in agriculture, the closure of the national border demanded a range of rapid responses from the government. Apart from the State of Emergency this also concentrated on supporting agriculture, and thus achieving greater national self-reliance. The Government approved a Cabinet Decision No 271 of 20 March 2020 on *"Food Security Preparedness Program for Corona-virus Pandemic Period"* which stressed the need for a greater focus on food crop production, with government support, that would encourage farmers to plough more land and plant more short-term food crops, particularly fast growing food crops like sweet potato, cassava, corn and various vegetables. Cassava could mature in 6 months and sweet potato varieties in 4 months.

Among specific recommendations the Government agreed to subsidise land clearance and preparation and aimed at the clearing of 350 acres in Vava'u and 1800 in Tongatapu. The MAFF advised Tongans to selectively use their domestic pigs, poultry and cattle, as sources of meat for family meals (rather than imports) and targeted the production and distribution of 30,000 broilers and 500 2-week old chickens, alongside some sheep, goats and pigs. MAFF also targeted the production and distribution of

30,000 papaya seedlings, 10,000 passionfruit seedlings, and 10,000 Soursops (*Apele Initia*). In the longer term it had a 3-year target production of 100,000 coconut seedlings, 70,000 breadfruit seedlings, 100,00 Sandalwood seedlings, 40,000 Kava seedlings, and 120,000 fruit tree seedlings (the last of these from Australia for distribution and for sale). MAFF also sought to produce and distribute vegetable seedlings and provide training for farmers, women, and youth groups on how to produce seeds from their own vegetable. It was a spectacularly ambitious program and never likely to succeed in its entirety.

In Vava'u the two parts of the program that got underway were the distribution of vegetable seedlings and chickens, and more extensive land clearance. Vegetable seedlings, with seeds sown at the MAFF Station, were given away to anyone interested in planting vegetables, between March and November, oriented at supporting a healthier community, and which would engage people in physical activity by taking care of their vegetables. The free seedlings eventually gave rise to more roadside fruit and vegetable stalls and sales of vegetables at relatively cheaper prices. Root crops took longer to be distributed and the produce of new plantings was only available later in the year. The MAFF also distributed broiler and layer chicks to households, restricted to 10 chicks per household, a program that was still going on in December 2020. Land clearance and preparation also went on, and a third (139 acres) of the anticipated area had been prepared by September, and more clearance was proposed for an eventual second phase. Farmers were given fuel allocations for their tractors to support clearance, although on the smaller outer islands, six of the nine villages had no tractors and were given farming tools in lieu of tractor fuel.

Local reception was positive, and the new developments were welcome, as noted by the District Officer of Neiafu, Ula Kaufusi: "in spite of all the difficulty that we are hearing and partly experiencing, so many good things are happening in agriculture with more and more people farming and planting more food crops for food security. Farmers are very aware of the need to plant more food in case we get longer lockdowns where there will be no other source of food but to rely on our own local food". The account of the Town Officer of Holeva, one of the more isolated villages in Vava'u, Mr. Uele Moala, was equally positive: "with the lockdown, there is more family time at night, and more time to go to the bush and do bush work like planting more crops and maintaining some of the crops planted earlier. These are some of the good results, aside from the assistance from the Government". Food crops took priority and projects aimed at longer-term plants, more evidently for cash crops, were deferred.

Conclusion

Some degree of remoteness from Tongatapu meant that Vava'u had always retained a relatively vibrant semi-subsistence economy. When COVID-19 hampered communications, and increased the cost of imported goods, that proved beneficial. It also meant – somewhat incidentally and equally unexpectedly – that Vava'u was able to respond to an unexpected demand for 'local' food from migrant Tongans in the United States. Thus, throughout the crisis, Vava'u was in the privileged position, compared with many other Pacific islands, of retaining food security, and having it boosted through government support. Meanwhile even the very small outer islands of Vava'u were benefiting from kava sales, and vanilla prices justified new plantings. Incidentally too, especially as seasonal workers were no longer able to migrate to Australia and New Zealand, labour was available for local work. Despite the loss of its tourist industry Vava'u survived during the pandemic period. Indeed it went from coping to thriving, to play a role as one of the main sources of local food for the whole of Tonga, remaining the main source for the giant taro (*kape*), and opening up new exports to the United States. By late 2020 it had responded both to increasing demand for local food and to the emerging market in the USA.

Agricultural revival, retention and growth became a virtue out of necessity. Through crucial if selective linkages to the world – not least remittances – Vava'u managed to be relatively unscathed by COVID-19. Remoteness, land availability and a resilient agricultural system all played a part. Vava'u, once more of an outlier, where population was falling, was able to support the main island and the capital city. A resilient COVID-19 population developed new opportunities. A diverse agricultural system, the retention of traditional knowledge and scope for fishing enabled it to survive and thrive. Food security and even diets probably benefited. During 2020 at least, the pandemic period provided an opportunity for a younger generation to be involved in local production, partly by default, as wage and salary employment contracted and, despite vital connections with the wider world, as Vava'u became more self-reliant.

The agricultural future of Vava'u seemed secured, with high prices and new markets, and the population had adjusted to new social contexts, where mobility was restricted. The experience of making a reasonable income in Vava'u may prove to be one brake on migration, especially if the prices of its three key commodities remain high (or if other niche products are developed) alongside a revival in tourism. Whether Vava'u will return to a more peripheral role, and its younger agricultural workers leave for better wages and excitement overseas, promoting higher incomes and thus the greater possibility of reverting to consuming imported foods remains to be seen. It may be that the reversion to 'tradition' really is a new normal, or, by contrast, that revival of tourism may prove even more valuable. Certainly, for many months, Vava'u like few other islands largely benefited from the reduction of certain distant ties and was fortunately able to benefit from being part of regional and international supply chains that continued to function.

References

Evans, M., Sinclair, R. C., Fusimalohi, C., Liava'a, V., & Freeman, M. (2003). Consumption of Traditional Versus Imported Foods in Tonga: Implications for Programs Designed to Reduce Diet-Related Non-communicable Diseases in Developing Countries. *Ecology of Food and Nutrition, 42*, 153–176.

Evans, M., Sinclair, R. C., Fusimalohi, C., & Liava'a, V. (2002). Diet, Health and the Nutrition Transition: Some Impacts of Economic and Socio-Economic Factors on Food Consumption Patterns in the Kingdom of Tonga. *Pacific Health Dialog, 9*, 309–315.

Gewertz, D., & Errington, F. (2010). *Cheap Meat. Flap Food Nations in the Pacific Islands.* University of California Press.

Lin, S., Hufanga, S., Linhart, C., Morrell, S., et al. (2016). Diabetes and Obesity Trends in Tonga Over 40 Years. *Asia Pacific Journal of Public Health, 28*, 475–485.

Orams, M. (2013). Economic Activity Derived from Whale-Based Tourism in Vava'u. *Tonga, Coastal Management, 41*, 481–500.

Underhill, S., Patolo, S., Zhou, Y., & Burkhart, S. (2020). The Agriculture-Nutrition-Income Nexus in Tonga: Is Postharvest Loss Undermining Horticulture Market Efficiency in Tonga? *Horticulturalae, 6*, 61.

Webb, J. (2020, December). Samoa and Tonga: Opportunities in the Storm. *Pacific Economic Monitor, 19-22.*

CHAPTER 12

The Bahamas: Tourism Policy Within a Pandemic

Michelle McLeod

The chapter examines tourism policy making actors and choices during the COVID-19 crisis in The Bahamas. As by far the most important sector in the economy, re-establishing tourism was anticipated to include public, private and non-government actors in influencing government policies. In 2019, The Bahamas received 7.25 million visitors, three quarters of whom were cruise ship passengers. They collectively enabled tourism to account for about 60% of national GDP (with only financial services otherwise being significant) and for direct tourism employment to represent about 60% of the workforce. With a national population of just 390,000 these tourist numbers represented one of the highest ratios of tourists to residents in the world. When the tourist system ground to an abrupt halt in March 2020, with the shutdown of major hotel establishments and cancellation of cruises, the effects of the global COVID-19 pandemic were devastating. For a small island state, still recovering from Hurricane Dorian in

M. McLeod (✉)
University of the West Indies (UWI), Mona Campus, Kingston, Jamaica
e-mail: michelle.mcleod@uwimona.edu.jm

Y. Campbell, J. Connell (eds.), *COVID in the Islands: A comparative perspective on the Caribbean and the Pacific*,
https://doi.org/10.1007/978-981-16-5285-1_12

2019, the pandemic proved a double whammy. Recovery was crucial and between May and November 2020 a suite of policies and statements followed aimed at enabling tourism to resume as safely and quickly as possible. The policies that will guide the tourism sector to recovery in The Bahamas are critical and the process of developing and enhancing such policies is important.

Policy making is an evolutionary process where guidelines exist to assist with adjustments that will improve outcomes. In a small island state context, with human and physical resource constraints, monitoring and responding to external shocks are particularly difficult. When these shocks are sudden, unparalleled and unexpected, uncertainty increases, and it is even more difficult to make decisions that will improve outcomes. Developing well-designed policies and practices that involved multiple stakeholders proved difficult.

In The Bahamas, tourism governance is similar to that in other Caribbean tourist destination, involving a number of local, national, regional and international agencies covering a range of concerns. In recent years there have been concerns about the region's tourism growth – that suggested overtourism in several places, especially associated with the cruise ship industry in Nassau (Nair, 2019), its susceptibility to climate change (Moore, 2010) and the broad relationship between climate resilience and sustainability in the context of the pandemic (Sheller, 2020). Several Caribbean island destinations are vulnerable to changes in temperature, and rising sea levels, both of which threaten the marine life of coral reefs. Projected sea level rise would threaten several Bahamas islands in the next few decades. Environmental vulnerability has been further intensified because of shifts in economic power, affecting tourism growth and development. The imposition of taxes on air transportation for long haul travel to the region and growing costs of airport security have threatened the already challenged tourism industry. However, travel restrictions and closing of borders, instituted by many Caribbean governments during the global pandemic, resulted in further and much more severe pressures on the tourism economy. COVID-19 did much more than intensify existing problems.

In normal times, a range of ongoing issues challenge tourism growth and development and in a small country with large tourism numbers, all of whom are from overseas, policy intervention to manage and sustain tourism activity is difficult. The arrival of COVID-19 provided an, albeit unwelcome, pause to reflect on the role and direction of the industry. That raised such issues as whether tourism would remain the same or change in a post-pandemic future and what that meant for sustainability, infrastructure provision and marketing. Are the tourism policies formulated appropriate to address the major tourism issues and challenges of tourism growth and revitalise development in small island states? Who are the actors most appropriately involved with tourism policy formulation? The chapter first provides an introduction to the Bahamas before examining the policies and the policy making process, and the implications of these in a time of uncertainty.

The Bahamas as a Tourist Destination

"It's Better in The Bahamas" has been the destination's long-lasting slogan used to tempt millions of visitors to the country every year. The Bahamas is an archipelago of 700 islands, about 30 of which are inhabited islands, with the main islands being New Providence, where the capital Nassau is located, Grand Bahama Island and Abaco (the two latter islands being badly affected by Hurricane Dorian). It has a population of about 390,000 and a GDP per capita of about US$27,500 in 2019, one of the highest in the Caribbean. Geographically, the Bahamas archipelago is not in the Caribbean, but off the coast of the United States, from where most of its tourists come. In 2019 some 80% of all tourists came from the United States (with 8% from Canada and a further 8% from Europe), most of whom were in search of a traditional holiday centred on beaches, rest and relaxation, and shopping in a pleasant climate (Bahamas Ministry of Tourism, 2020). As a former British colony, The Bahamas has historical and cultural ties with the rest of the Caribbean and is a member of CARICOM, the Caribbean economic community. Its tourist concerns are not dissimilar from those of several Caribbean SIDS.

As late as the 1940s, tourism in The Bahamas was seasonal, with tourism workers having to resort to other activities out of season, but from the 1950s The Bahamas was transformed from a high-end quality destination to a mass tourism resort, and numbers boomed. Two decades later tourism offered year-round employment, and minor fishing and farming activities disappeared. By the 1980s tourism dominated The Bahamas to the extent that the country could be described as a 'monoculture' with 'gross external dependence' and a 'convenient tourist suburb of south Florida' (Bethel, 1989: 133). As early as 1938 a government report had warned against over-concentration on tourism (Minnis et al., 2020: 33). The economic success of tourism was widely seen as having made The Bahamas overly dependent on tourism and on the United States, with infrastructure – and especially hotels – largely foreign-owned, and having contributed to excessive urbanisation, land alienation and various environmental problems. Tourism remains largely concentrated on Nassau, New Providence and Paradise Island with two major resorts, Atlantis with 3805 rooms on Paradise Island and Bahamar with 2200 rooms on New Providence Island. These two properties alone comprise one third of the country's room stock. Tourism only slowly embraced the smaller and more remote Out Islands (Rolle, 2015). The most rapid growth in tourism over the past four decades had been cruise tourism (Fig. 12.1) also centred in Nassau. By 2019 the ratio of cruise excursionists to stayover visitors was 3.3:1. While The Bahamas has cultural and scenic attractions, for many it is simply one stop in a cruise ship circuit rather than a distinctive destination in its own right. That in itself is a particular challenge to national tourism policy formation.

As early as the 1960s Gordon Lewis was arguing that the reliance of The Bahamas (and Bermuda) on luxury tourism is 'fundamentally parasitic [and] has incalculable consequences for the Bermudian-Bahamian

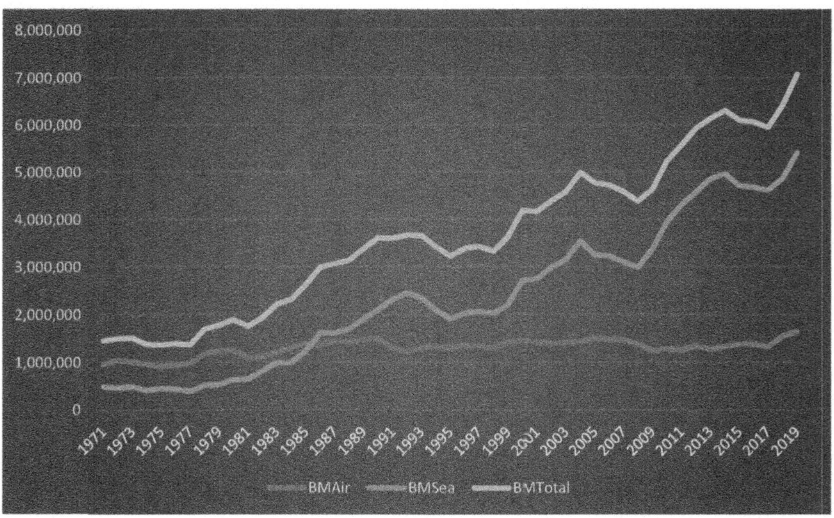

Fig. 12.1 The Bahamas visitor arrivals (1971–2019). (Source: The Bahamas Ministry of Tourism Visitor Statistics (Bahamas Ministry of Tourism, 2020))

way of life. It has made the ruling groups paranoid tourist worshippers. Public policies are shaped by consideration of the traffic's priorities' (1968: 327). While tourism has subsequently grown and diversified enormously, elements of this approach to the priority of tourism and its governance have scarcely changed. Much more recently it was readily evident that the relationship 'between tourism and sustainability has been little explored and even less implemented in The Bahamas, because the Government seems to be of the opinion that regulations and controls are unwarranted' (Rolle et al., 2020: 6). Planning and regulation were rarely evident as the industry rapidly grew.

The first confirmed case of COVID-19 in The Bahamas was on March 15th, 2020. The government then instituted an emergency order closing national borders and instituting health and safety measures advised by WHO. The country was locked down with the closure of hotels and resorts on the islands. The cruise ship business came to a halt with the immediate unemployment of straw market vendors and taxi drivers who depended on cruise ship business. Unemployment soon reached an estimated 40%. Social protection measures were introduced, and the National Insurance Board provided specially arranged benefits to workers who were affected by the shutdown of tourism. Without the lapse of much time, and following the immediate crisis, attention diverted to the economic need to restart tourism, and so to the need for tourism policy development.

Towards a Strategy?

A number of tourism policies emerged during the pandemic, over the period May to November, 2020, and were set out in online sources and media releases by the Caribbean Tourism Organization. It took just 2 months from the first case of COVID-19 in The Bahamas for new approaches and strategies to emerge. Secondary sources were utilised to review the nature of the policy decisions that were made that affected the tourism sector, some focused on the wider Caribbean region; most that are reviewed here focused directly on tourism but others centred on the economy of which tourism was the key part. These documents included: Caribbean countries reopening plans, initiatives post COVID-19 as of December 10th, 2020 (CARPHA, 2020); Bahamas Tourism Readiness & Recovery Plan for Re-entry into the Tourism Market (Government of the Commonwealth of The Bahamas, 2020); Contribution to the 2020/2021 Budget Debate (Office of the Prime Minister, 2020); Remarks by Prime

Minister Minnis on the occasion of the Commonwealth COVID19 Virtual Leaders Meeting (Minnis, 2020); Bahamas Minister of Tourism Presents Forward-Looking Tourism Plan (CTONews, 2020); Economic Recovery Committee Executive Summary Report (Economic Recovery Committee, 2020) and the Government of The Bahamas, Ministry of Health, Testing Protocol for Incoming Visitors Update #88. The documents were analysed to identify and evaluate the policy making process, determine the general themes and the tourism policies that were emerging and what they suggested that were distinctive or otherwise concerning tourism policy making during the pandemic.

Reopening plans started as early as May 2020 directed at business, travel and hospitality resumption. A consultation with a wide cross-section of the tourism sector including airlines, points of entry, taxi drivers, tour operators, hotels, vacation home rentals, retailers, straw market vendors, land and sea-based transport operators formed the basis of a reopening plan. An economic recovery committee was formed to address the overall recovery of the country and not only the tourism sector. Plans were made to reopen in July with testing requirements in place. An education event was held with cruise industry stakeholders. The pandemic continued but the number of cases were very low between March and July, and it was not until July that tourism promotion became an issue, as COVID-19 case numbers soared between July and November. In August, a travel advisory and issues of travel safety averted any attempt to reopen the country for tourism. By September, a Tourism Readiness and Recovery Plan had been drawn up, which formed the basis for activities to ensure the safe reopening for tourism, but numbers remined high.

In terms of a connection between data and policy making, a single document, the Prime Minister's address to the Commonwealth COVID-19 Virtual Leaders Meeting in June, 2020 (Minnis, 2020: 2) revealed any linkage between data and policy and that merely read 'no new cases for past ten days in Nassau, 41 days on Grand Bahamas, 35 days on Bimini and no cases on any of the remaining 23 major islands'. Based on this limited data alone, a decision was made to proceed with reopening the tourist economy. Other than this, scant evidence exists that data were utilised to make specific policy decisions. Information to feed into policy decisions were derived from the Caribbean Public Health Agency based in Trinidad and Tobago, and from a wide cross-section of stakeholders, and small businesses. Businesses were anxious to resume operations, but the absence of connections between data, opinion and policy decisions was evident.

As early as May, the Government of The Bahamas set out to resume commercial travel on July 1, by implementing point of entry best practices and reopening domestic borders. The Prime Minister argued that the policymaking would be innovative and progressive (Office of the Prime Minister, 2020). In the Prime Minister's Budget Debate the following phrases were noted: 'emergency policy recommendations', 'shape policy recommendations', comprehensive set of policies', 'curate and inform policy recommendations' but it was rarely specified or made clear what these might involve. Simply lifting restrictions and reopening the economy were the main issues to be resolved. It was evident that for the tourism sector to reopen, policies and procedures for all visitors to the Bahamas had to be instituted to ensure their safety. The greatest challenge was the unpredictable and evolving effects of the COVID-19 pandemic. By July, new testing requirements and an electronic Bahamas Health Visa application were instituted. Subsequently the goals and objectives for policy making shifted from one of travel safety to one of travel promotion. For the reopening of international commercial flights, a special offer rate of $229 on Bahamas Air and Balearia Caribbean, and The Bahama Paradise Cruise Line were promoted. A mobile app to assist with vacation planning was also launched.

However, even the limited success of tourism promotion activities during the pandemic contributed to another shut-down of the economy as the early reopening of tourism businesses resulted in a spike of COVID-19 cases. By August, the testing protocol was changed to a negative COVID-19 test, taken no more than 5 days prior to the date of arrival and quarantine for 14 days. This was changed again in October to add the COVID-19 Rapid Antigen Test on Day 5 of the visit. Such regulations naturally deterred potential tourists. After the spike in cases had fallen away at the end of the year the policy focus reverted to one of demand stimulation. A new extended stay programme and a "Vacation in Place" (VIP) programme were introduced (where visitors had to quarantine within the hotel for 14 days). By November, tourism promotion was once again on the agenda with announcements of hotel reopening, airport expansions and air credit. Elimination of the requirement to quarantine with follow up testing and monitoring while in the country followed.

How such practical decisions were made, and at what level and based on what information was rarely evident. Policy formation, in terms of practical decisions, seemed to remain detached from data and from the comparative experience of other island tourism destinations. Other than an immediate desire to secure border closure and the health of inhabitants of The Bahamas, the various policies revealed no clear targets or policy directions for tourism recovery. Planning for and implementation of tourism recovery in The Bahamas involved a multitude of organisations; while The Bahamas Ministry of Tourism and Aviation and The Bahamas Hotel and Tourism Association were the main implementing agencies, the Ministry of Health was a key strategic partner for the success of the recovery. Airlines were also key in the early days to restart tourism, including Southwest, JetBlue, Delta Airlines, United Airlines and American Airlines. Cruise Lines were only marginally involved.

The key decisions came from the Government, and their enthusiasm to get the tourist economy vibrant again was all too apparent. As the Minister of Tourism and Aviation, Dionisio D'Aguilar, declared in September: 'we cannot sit in a bubble and wait for everything to be perfect' while recommending that hotels on every island be opened within a month, despite case numbers mounting and the country no longer being able to advertise itself as a COVID-free destination. His view was that this was not necessarily problematic since most tourists will come from places with greater infection and deathrates. With tourism the basis of the economy 'we have to get real quickly' (quoted in Hartnell, 2020). Significantly the private sector was more reticent about opening up, hence the Minister suggested – after conceding that he did not have the power to force hotels to open:

> We'd love them to announce an opening date as quickly as possible, certainly in November before Thanksgiving. We've got to pull the trigger. We cannot behold the country to these two [the largest- Baha Mar and Atlantis] properties. They need scale. They have to look at COVID-19 conditions in their core markets to decide whether to pull the trigger. What happens in their core markets is a difficult call for them. Those [hotel openings] are private sector decisions. They drive it. We can only encourage them to do so. We feel the Government has to take the lead in this exercise and create the right environment for them, but ultimately it's their decision (ibid)

The dependence of the economy on external investment and decision-making was clearly evident. Ultimately the private sector failed to heed the Minister, the two largest resorts remained firmly closed and tourists failed to come in any numbers.

During the pandemic, tourism policy-making had taken on a new set of policy actors that did not have the sole interest – and not even the primary interest – of the recovery of the tourism sector. Indeed, the Caribbean Tourism Organization, the main body with managerial responsibility for tourism in the region adopted the approach of #thecaribbeanawaits. While the Bahamas Tourism Readiness & Recovery Plan for Re-entry into the Tourism Market involved wide consultations, consultations were generally few, in large part because health issues dominated the discussion at a national level, so that tourism policies and strategies were uninformed by debate and discussion and changed little. In a sense they were deferred until the pandemic was over. While the Prime Minister stated that innovative policies were to be developed (Minnis, 2020) the process by which this was to occur, and what that might entail, were never outlined. Clearly the pandemic brought on several issues that constrained the performance of the tourism sector – necessarily in terms of health – but also affected forward planning.

Changing health protocols delayed the restart of tourism, alongside the global loss of confidence in international tourism, that further constrained national policies to develop tourism in any form, however sensitive that might be to health concerns, sustainability or climate change. During a pandemic of this kind immediate tourism policy makes only limited sense, as long as the global climate is one of extreme uncertainty and humanitarian concern. In The Bahamas, rather than a focus on the recovery of the tourism sector, despite its massive national importance, public policy addressed the overall economic recovery of the country. While divergence from prioritising tourism policy-making may further delay the recovery of the tourism sector, as limited resources that can be placed into the tourism sector may be diverted elsewhere, that is scarcely surprising. Moreover, like many SIDS, The Bahamas has limited skilled human resources hence the ability to work in parallel on the future of the industry was also limited. Crucially that has also meant that the limited policy formation that did take place was directed to re-establishing the previous structure of tourism rather than taking the opportunity to contemplate let alone develop new directions.

Conclusion

From the onset of COVID-19, border closures prevented any significant incidence of COVID-19 and over time, a travel risk policy, with guidelines for health and safety protocols at all tourism touchpoints, seemed to have proved a promising follow-up. That was also logically directed at key markets where the pandemic curve had flattened and its impact was declining. However the opening up of the country, even with health protocols in place, resulted in a spike in the incidence of COVID-19 and a necessary reversion to 'hard' policies, so again briefly marginalising the tourism industry. Optimistic tourism promotion and practice had come all too soon despite precautionary health practices, but by September economic pressures for re-opening again took over.

While an evolving tourism policy making is critical to steer tourism activity to minimise the effects of a decline of tourist arrivals, or the problem of overtourism, or to steer towards sustainability in the face of climate change, the ability to do so during a pandemic – where market conditions were vastly different – was almost impossible. In the short-term the industry moved on as best it could. Unfortunately that also left little time or resources to consider the possible long-term future. Moreover what policies and practices that did exist seemed to exist in a vacuum and not be guided by local circumstances and data, or changes elsewhere in the region (or in the United States market). It was policy formation and practice as if nothing had substantially changed, and of seat-of-the-pants decisions over opening up. But the fluctuation of the pandemic made it inevitable that, even in The Bahamas, tourism would have to take a back seat to national development. While the Prime Minister had intimated that policy-making would be 'innovative and progressive' ever-changing circumstances left little time to consider what that might be.

An identifiable gap in policy formation was the lack of involvement of international agencies and organisations, and of the private sector. Airlines were marginally involved but hotel chains and cruise lines played no apparent role, despite their obvious substantial involvement (McLeod et al., 2018). Nonetheless without tourism policy-making evolving, and perhaps learning from wider CARICOM experience, effective tourism recovery in The Bahamas will be hampered and any future change of direction absent. Indeed both the dependence of The Bahamas on tourism, and the widespread enthusiasm to get back to normal, when so many depended on the industry, posed a long-term problem for it. There was an unmet need for more international consultation with airlines, tour operators and travel retailers to strengthen the tourism policies that would facilitate recovery, and with local workers and entrepreneurs over what that future tourism might look like. That certainly included the largest resorts. A further gap lies in linkages with climate change and sustainable development, pointing to a longstanding 'deficiency of comprehensive planning to address current and future impacts of climate change. This lack of planning places the country at high risk, particularly the tourism industry on which the country is financially dependent' (Thomas & Benjamin, 2020: 174). The absence of planning has long been so in The Bahamas – a striking absence and an anomaly where tourism simply is the economy. The irony is that the greatest opportunity to plan for the future came during the pandemic when tourism was at its lowest ebb. The Bahamas might even have proved a model for other Caribbean states and SIDS elsewhere considering their own touristic futures. In a planning vacuum it failed the test.

References

Bahamas Ministry of Tourism. (2020). *The Bahamas Ministry of Tourism Visitor Arrivals. 2020.* Ministry of Tourism from https://www.tourismtoday.com/services/statistics

Bethel, F. (1989). Tourism, Public Policy, and National Development in the Bahamas. In D. Collinwood & S. Dodge (Eds.), *Modern Bahamian Society* (pp. 129–138). Caribbean Books.

CARPHA. (2020). *Caribbean Countries Reopening Plans, Initiatives Post COVID-19 as of December 10th.* Caribbean Public Health Agency. Port-au-Prince.

CTONews. (2020). *Bahamas Minister of Tourism Presents Forward-Looking Tourism Plan.* Retrieved from https://www.onecaribbean.org/minister-of-tourism-presents-forward-looking-tourism-plan/

Economic Recovery Committee. (2020). *Economic Recovery Committee.* Office of the Prime Minister. Retrieved from https://opm.gov.bs/economic-recovery-committee-executive-summary-report-2020/

Government of the Commonwealth of The Bahamas. (2020). *Bahamas Tourism Readiness & Recovery Plan for Re-entry into the Tourism Market.* Government of the Commonwealth of The Bahamas.

Hartnell, N. (2020, September 8). Bahamas Must 'Get Real Over Tourism Re-Opening, *The Tribune.*

Lewis, G. (1968). *The Growth of the Modern West Indies.* Monthly Review Press.

McLeod, M., Chambers, D., & Airey, D. (2018). A Comparative Analysis of Tourism Policy Networks. In M. McLeod & R. Croes (Eds.), *Tourism Management in Warm-Water Island Destinations* (pp. 77–94). CABI.

Minnis, H. (2020). *Remarks by Prime Minister Minnis on the Occasion of the Commonwealth COVID19 Virtual Leaders Meeting*. Retrieved from https://www.thebahamaschamber.com/post/remarks-by-prime-minister-minnis-on-the-occasion-of-the-commonwealth-covid19-virtual-leaders-meeting

Minnis, J., Rolle, S., & Bethell-Bennett, I. (2020). The Impact of Tourism on Small Island Communities in The Bahamas: The Case of Abaco, Bimini and Exuma. In S. Rolle, J. Minnis, & I. Bethell-Bennett (Eds.), *Tourism Development, Governance and Sustainability in The Bahamas* (pp. 31–51). Routledge.

Moore, W. (2010). The Impact of Climate Change on Caribbean Tourism Demand. *Current Issues in Tourism, 13*(5), 495–505.

Nair, V. (2019, August 30). Overtourism and Cruise Tourism, *The World News*.

Office of the Prime Minister. (2020). *Contribution to the 2020/21 Budget Debate*. The Bahamas: Retrieved from https://opm.gov.bs/contribution-to-the-2020-21-budget-debate/

Rolle, S. (2015). The Bahamas: Individual Island Branding for Competitiveness in Archipelago Tourism. In G. Baldacchino (Ed.), *Archipelago Tourism. Policies and Practices* (pp. 162–179). Ashgate.

Rolle, S., Minnis, J., & Bethell-Bennett, I. (2020). *Tourism Development, Governance and Sustainability in The Bahamas*. Routledge.

Sheller, M. (2020). Reconstructing Tourism in the Caribbean: Connecting Pandemic Recovery, Climate Resilience and Sustainable Tourism Through Mobility Justice. *Journal of Sustainable Tourism*, 1–14.

Thomas, A., & Benjamin, L. (2020). Climate Change, Tourism and Sustainable Development in The Bahamas. In S. Rolle, J. Minnis, & I. Bethell-Bennett (Eds.), *Tourism Development, Governance and Sustainability in The Bahamas* (pp. 168–176). Routledge.

CHAPTER 13

An Industry in Crisis: How Vanuatu's Tourism Sector Is Seeking Economic Recovery

Anna Naupa, Sarah Mecartney, Liz Pechan, and Nick Howlett

Introduction

This chapter recounts and examines the events triggered by the COVID-19 pandemic that produced a steep decline in Vanuatu's tourism industry, the key economic sector during the preceding three decades. This necessitated

A. Naupa (✉)
Australia-Pacific Technical College, Port Vila, Vanuatu

S. Mecartney
Pacific Community (SPC), Nouméa, New Caledonia
e-mail: sarahm@spc.int

L. Pechan
The Havannah, Port Vila, Vanuatu
e-mail: liz@thehavannah.com

N. Howlett
Vanuatu Tourism Office, Port Vila, Vanuatu

Y. Campbell, J. Connell (eds.), *COVID in the Islands: A comparative perspective on the Caribbean and the Pacific*,
https://doi.org/10.1007/978-981-16-5285-1_13

a rapid policy shift by the Vanuatu Government towards COVID-safe business readiness and a programmatic response to build domestic tourism to revive Vanuatu's tourism industry during the pandemic. This chapter focuses on Vanuatu's experience in terms of impacts to tourism businesses, a new domestic tourism campaign and public-private policy shifts to support the resilience of an industry in crisis. It presents three cases that illustrate the response and recovery efforts at the business, sector and policy levels. Threaded through the cases is a common theme – the value of a public-private coalition-building approach to accelerate national economic recovery efforts, including through a reimagining of the industry to cope with the 'new normal'. The COVID-19 crisis in Vanuatu necessitated a public-private collaboration targeting economic recovery in both formal and informal sectors.

Formerly called the New Hebrides Condominium, the Republic of Vanuatu in the south-west Pacific gained independence from the colonial powers Great Britain and France in 1980. It has a population of approximately 285,000 people, speaking over 100 distinct languages spread across 83 islands, 65 of which are inhabited. Outside the five urban centres, rural economies depend primarily on subsistence gardening, market agriculture and small-scale tourism. Government services have limited reach beyond the centres but increased government and donor investment in infrastructure over the past 2 decades has gradually improved connectivity.

Prior to the COVID-19 pandemic, Vanuatu's tourism industry consisted of international visitation by air – or landed tourists – and cruise tourism from passenger ships carrying over 1000 passengers – or excursionists (Watt & Brenner, 2020: 345). Through the 2010s this had meant over 100,000 tourists a year. Nothing suggested a domestic tourism market. Until the pandemic, cruise ships called at between three to five islands: Aneityum (home to Mystery Island), Efate (where the capital Port Vila is located), Malekula, Pentecost (home to the famous *naghol* land-diving, which inspired the development of bungee jumping) and Espiritu Santo. The major non-cruise related tourism sites were Port Vila and Efate (with about 90% of tourism), Tanna (a major attraction being the accessible Yasur volcano) and Espiritu Santo (the largest island, with beach and dive tourism). Cruise tourism has high economic leakage and limited linkages within the tourism value chain compared to landed tourism (Watt & Brenner, 2020: 349). In 2019 total visitor expenditure was USD 171 million, primarily from landed tourism (NZTRI and Vanuatu Tourism Office, 2020; see Fig. 13.1). This figure was approximately sixty times the

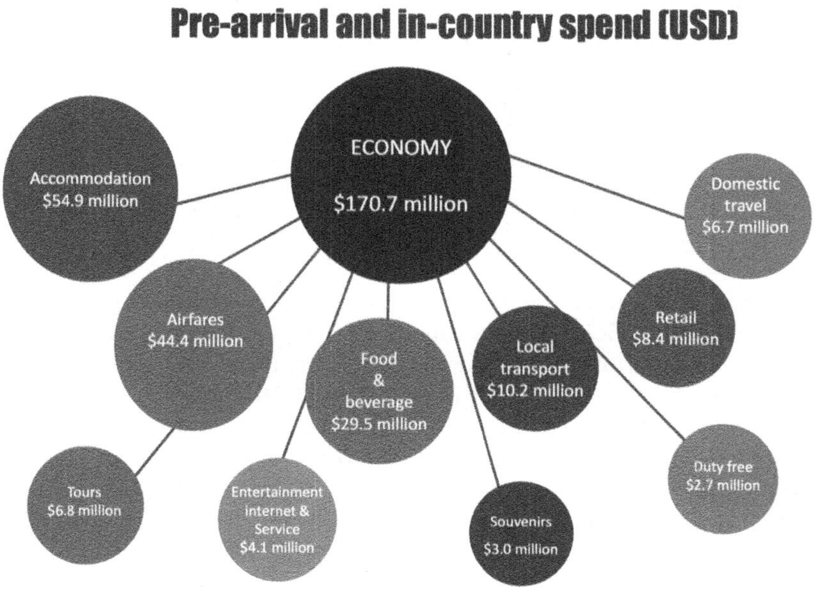

Fig. 13.1 Pre-pandemic dollars spent on tourism in Vanuatu (2019) (Source: Vanuatu International Visitor Survey 2019)

government's total 2019 budget for tourism development and marketing operations, and that total excluded indirect economic benefits through the tourism value chain.

Ranked by the World Bank's Risk Index as the most vulnerable country in the world (to natural disasters), Vanuatu is exposed to multiple natural hazards, including cyclones and earthquakes, which bring high risks and additional costs to doing business. The resilience of Vanuatu's tourism industry has been tested through annual cyclone seasons, including Cyclone Harold early in 2020, and immediately afterwards by the global COVID-19 pandemic. The interconnectedness of local and global systems and this influence on tourism resilience (Cheer & Lew, 2017) was dramatically demonstrated with the pandemic having a significant and probably long-lasting impact on the tourism sector. Studies on tourism resilience have typically focused on the environmental resilience of tourism systems, although increasingly there has been scholarship on how local tourism businesses build resilience to uncertain business environments (e.g. Orchiston et al., 2015). Common themes of "survival, adaptation and innovation" are key attributes of resilient tourism enterprises. Here we examine these themes through three case studies, highlighting the role of public-private collaboration in contributing to rebuilding and the enhanced resilience of the industry.

The Non-Arrival of the Pandemic and Vanuatu's Economic Context

The year 2020 marked the fortieth anniversary of Vanuatu's political independence, the nation's second category 5 tropical cyclone, and Vanuatu also graduating from the United Nations' Least Developed Country ranking on 4th December 2020. Above all, COVID-19 brought about unprecedented socio-economic challenges for Vanuatu that tested the resilience of both the country's economy and the role of economic links between Vanuatu and other countries during a period of prolonged border closures.

In 2019, Vanuatu reported 3.4% GDP growth, with a positive outlook for 2020. The Vanuatu National Statistics Office recorded robust tourist and visitor figures and there was a steady seasonal labour mobility flow to Australia and New Zealand, bringing in remittances. At the core of growth was tourism, with the ADB observing in 2019, as the industry was still recovering from Cyclone Pam in 2015, that 'the recovery of the tourism sector will be a crucial determinant of Vanuatu's medium-term economic prospects' (Dornan & Cain, 2019). As the largest sector in the domestic economy, the services sector was expected "to continue experiencing solid growth over the medium term, mainly from positive direct and indirect contributions from tourism-related services, including airport reconstruction and the *Shared Vision 2030*". The Vanuatu Government's budget policy priorities for 2020 included an emphasis on increasing national tourism development, consistent with efforts under the *National Sustainable Development Plan 2030* to improve business opportunities and the investment environment (Government of Vanuatu 2019). The Asian Development Bank (2019: 103) asserted that "growth is expected to remain stable in 2019 and 2020 as tourism benefits from the completion of major infrastructure projects. Inflation will ease further, and the current account will remain in surplus. With increased tourist arrivals, policies must ensure that benefits are broadly enjoyed and sustainable." However, by mid-2020 the reality was very different (Fig. 13.2).

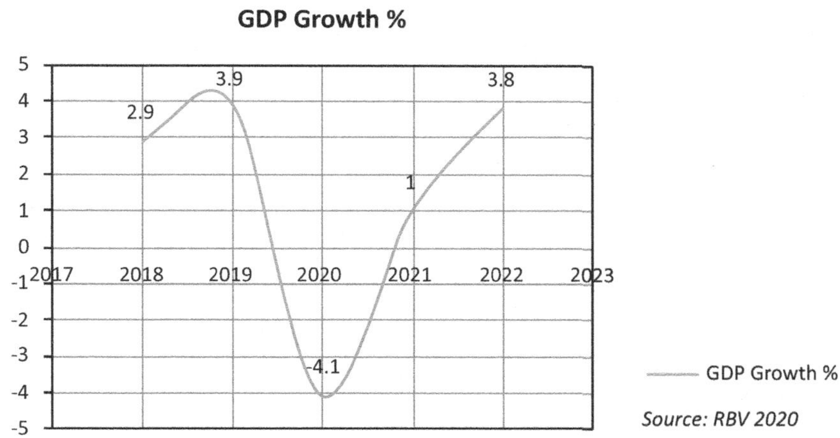

Fig. 13.2 Vanuatu's experienced and forecast GDP Growth 2018–2022 (%) (Source: Reserve Bank of Vanuatu)

On 11 March 2020, the novel coronavirus disease was declared a global health pandemic. Vanuatu, like other Pacific island nations, moved rapidly to close its borders to protect its citizens, announcing a State of Emergency on 26 March 2020, after national elections had been held a week earlier. By April 2020, the International Monetary Fund projected a contraction of -8.3% in real GDP growth. Before long, island nations with substantial tourism-based economies began to explore opportunities for diversifying their economies. In Vanuatu, the call for economic diversification was coupled with a renewed emphasis on the productive sectors (agriculture, fisheries, forestry and livestock), which appealed to populist narratives of subsistence-based, self-reliant, traditional lifestyles (Fig. 13.3).

From March 2020, Vanuatu focused on health protocols and border management. The growing crisis for businesses and for livelihoods was recognised early by the new government, which with donor support for the Tropical Cyclone Harold response, announced a VUV4 billion (USD40 million) economic stimulus package to support business continuity and employment protection for the five months to July 2020 (Table 13.1). The initial state of emergency was extended twice in 2020, on 10 April to include response efforts for TC Harold, and on 13 July to target COVID-readiness through to the end of the year, when a further 7-month extension went until 31 July 2021.

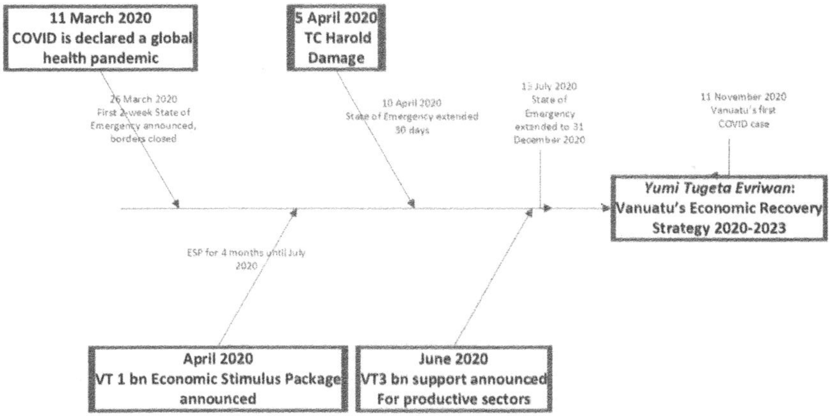

Fig. 13.3 Timeline of Vanuatu's responses to COVID-19

Several businesses complained about the lack of access to the Economic Stimulus Package (ESP). However, the Ministry of Finance and Economic Management noted that beyond the ESP, the series of other stimulus packages were designed so that "at some point, business communities could still directly or indirectly benefit from the stimulus package even if they are not qualified for the Employment Stabilization payment package" (*Vanuatu Daily Post*, May 9, 2020 updated May 18, 2020). The Vanuatu Chamber of Commerce and Industry (VCCI) also provided a platform for disseminating advice and information to businesses regarding support under the ESP. However, rather later, public holidays were extended enabling businesses to avoid salary implications.

The closure of international borders, combined with the unplanned restrictions on business operations and extended states of emergencies, produced a particularly significant downturn in tourism-related revenue and a general economic slow-down across all sectors. By April 2020 some 45% of tourism-related businesses had ceased trading, and a follow-up survey in August, with more businesses participating, showed that even fewer were still trading, with a considerable impact on employment and livelihoods (Table 13.2). Although the April and August surveys are not directly comparable, they provided a snapshot of industry experience at two point in time.

The compounding impact of both crises led to the development of the *Vanuatu Recovery Strategy 2020–2023*. As part of the 40th Independence

Table 13.1 Elements of Vanuatu's 2020 Economic Stimulus Package (ESP)

Vanuatu's VUV4 billion (USD40 million) economic stimulus package from April to July 2020 comprised of	
Employment stabilization payment package	Eligibility was restricted to business sectors severely or indirectly affected by the COVID-19 pandemic, including the tourism industry
School tuition fee subsidy	Registered schools were eligible for VUV 1,139,144,036 in addition to standard annual school grants. The Government further approved that each student would receive a tuition fee exemption of VUV42,000 for the 2020 academic year
Small and medium Enterprise Grant package	The Government committed VUV400 million (June–November) to support SMEs with an annual turnover of VUV200 million or less
Commodity subsidy	Targeting the agriculture sector, with a focus on kava, cocoa, coffee and copra exports, VUV300 million was made available, though details of eligibility are yet to be determined
Shipping subsidy	Government support to communities impacted by TC Harold including VUV100 million for sea freighting market produce from affected islands to Port Vila

In addition, the Vanuatu Government waived a number of taxes and charges to protect taxpayers from the financial impact of COVID-19. For example, Council of Ministers Decision No. 16 of 2020 agreed that road taxes would be waived from 1 April, 2020. Further business licence fees, other than for commercial banks, were also waived although licensees were still required to apply. Transport permit fees were also waived. The Vanuatu National Provident Fund also provided hardship loans, an interest-free withdrawal from a member's account for 6 months of up to a maximum of VUV100,000 (USD1000), after which the member had to choose to go on a repayment plan with interest or permanently withdraw the funds with a penalty. When the loan facility closed on 1 May, the VNPF had paid out about VUV1.5 billion (USD12.5 million). Source: https://www.imf.org/en/Topics/imf-and-covid19/Policy-Responses-to-COVID-19#V

Table 13.2 COVID-19 impact on tourism businesses in Vanuatu, April to August 2020

Tourism business status	*April 2020 (n = 267)*	*August 2020 (n = 348)*
Fully operational	55%	21%
Closed (indefinitely/ceased trading)	45%	3%
Closed (dormant, intending to reopen)	–	37%
Partially operational	–	39%
Net loss of employees	-2750	–2858

Source: Government of Vanuatu, 2020a

Anniversary Address on 30 July 2020, the Prime Minister announced the Government's intent to open a "Tamtam Travel Bubble" to mitigate further economic decline from border closure. A travel bubble, or travel corridor, was anticipated to support the safe and healthy passage of travellers between certain countries. Vanuatu's priority was its primary tourist markets and immediate neighbours: Australia, New Zealand and New Caledonia. By the end of 2020, the travel bubble concept had failed, because of COVID-19 in Australia and New Caledonia, limited interest in New Zealand and local fears of contamination. It was hoped that it might eventuate in 2021 coupled with Vanuatu's access to a COVID vaccine.

A Changed World: The Impacts of the Pandemic on Vanuatu's Tourism Industry

Tourism is deeply integrated throughout Vanuatu's economy. From March 2020, the tourism-related value stream became immediately evident with the sudden loss of jobs and livelihoods in both the formal and informal sectors, as tourism businesses closed. The most recent figures from 2017 pointed to a contribution of 30% to national GDP, with 11,000 people directly employed in tourism and an additional 19,000 employed indirectly (Vanuatu National Tourism Office 2019). After March 2020, the accommodation and food service sub-sectors suffered the most, with a reported decline of 79% in revenue (VCCI, 2020) and a significant reduction in full-time employment. Thousands lost jobs. Typically, some never found formal work again in 2020, and others, such as the former chef at an elite resort who worked 7 days a week selling donuts (*kato*) from a roadside stall (Graue, 2020). Typically those displaced had to develop their own creative strategies. Others who were self-employed in micro-businesses in the more tourism-oriented informal sector had to adopt similar strategies. Some traded from their homes where there were small markets; selling plants and home cooked food proved popular but most simply had to adjust by living on reduced incomes. Handicraft makers, who normally had a significant cruise ship market, lost almost all their sales.

Any quick rebound for Vanuatu's tourism industry was not anticipated and prolonged uncertainty around border closures remains. That initially led to strategies to construct domestic tourism, followed by rethinking long-term goals. The impact of COVID-19 on Vanuatu's tourism industry is illustrated by the following brief case studies of one locally-owned resort, the national destination marketing organisation's domestic tourism strategy and an industry-wide COVID response and recovery policy.

Case Study 1 – Business-Level Response: The Havannah Resort

The Havannah Resort on the main Efate island is an award-winning boutique resort (see: https://thehavannah.com). In recognition of its sustainable, indigenous-run business model, the resort was accredited as a National Geographic Unique Lodge of the World, a prestigious accolade that few resorts in the Pacific region have received. The resort's business model evolved considerably following Category 5 Tropical Cyclone Pam in March 2015, which prompted the resort's owners to incorporate resilience in its employment, supply chain, services and community engagement practices. This business model in turn helped to inform its response to the COVID-19 pandemic.

As with all tourism accommodation providers, The Havannah Resort was most affected by travel restrictions and the closure of Vanuatu's international borders in March 2020. This was compounded further by the uncertainty of the duration of the crisis, which had an adverse knock-on effect on local suppliers (agriculture and fisheries), community artisans (handicrafts and tours) and contributions to Government revenue (see Fig. 13.4).

Pre-pandemic, The Havannah Resort's business ecosystem stretched deep and wide with a long value chain and strong community ties that formed the basis of its business philosophy (Fig. 13.5). The pre-COVID supply chain was strongest on Efate Island, supporting over one hundred local suppliers that included security services, rubbish collection services, land and sea transport for staff and guests, handicrafts, tours, local food produce and plumbing services. The Resort had a strong focus on support for the local community, and supports local marine turtle conservation efforts and the Chief Roimata Domain, a UNESCO World Heritage Site. These formed common principles that underpinned its work and focus. The Havannah's pre-COVID supply chain was reduced by 95% due to COVID; the trickledown effect across the supply chain meant losses of income in multiple communities where The Havannah had links. While some rural suppliers with access to land and sea resources were able to 'pivot' and redirect towards subsistence farming and market gardening for

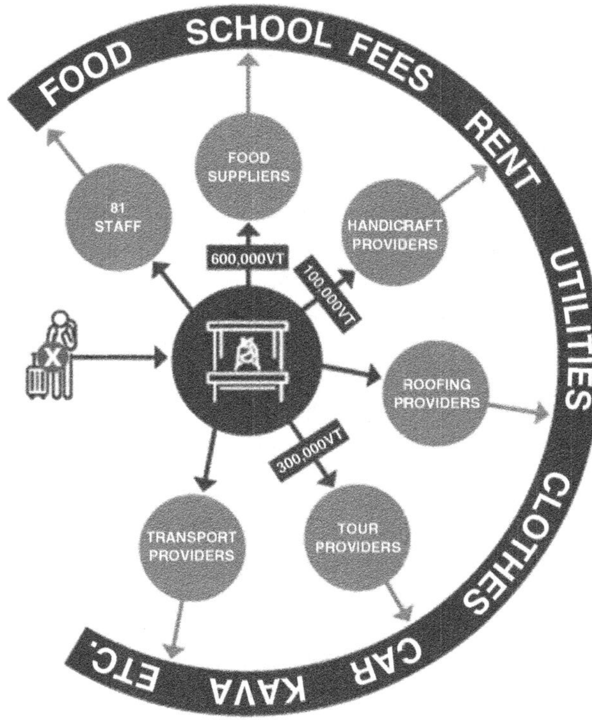

Fig. 13.4 The Havannah Resort's business ecosystem (Source: Vanuatu Economic Outlook Report September 2020)

income generation, having largely maintained these, others, particularly urban-based suppliers and resort employees had fewer alternative means of making a living. Since there was also limited investment by the government in tracking the impacts of the pandemic on businesses and on employment, there was no immediate evidence base on which to build a policy response.

While The Havannah Resort was able to secure access to ESP support for staff until July 2020, the decline in revenue meant that the majority of its more than 80 staff members had to be let go by October 2020 as the resort struggled to secure financial support for business continuity. The procedures used by The Havannah in laying off staff are illustrative of the challenges faced by both employers and employees in the crisis:

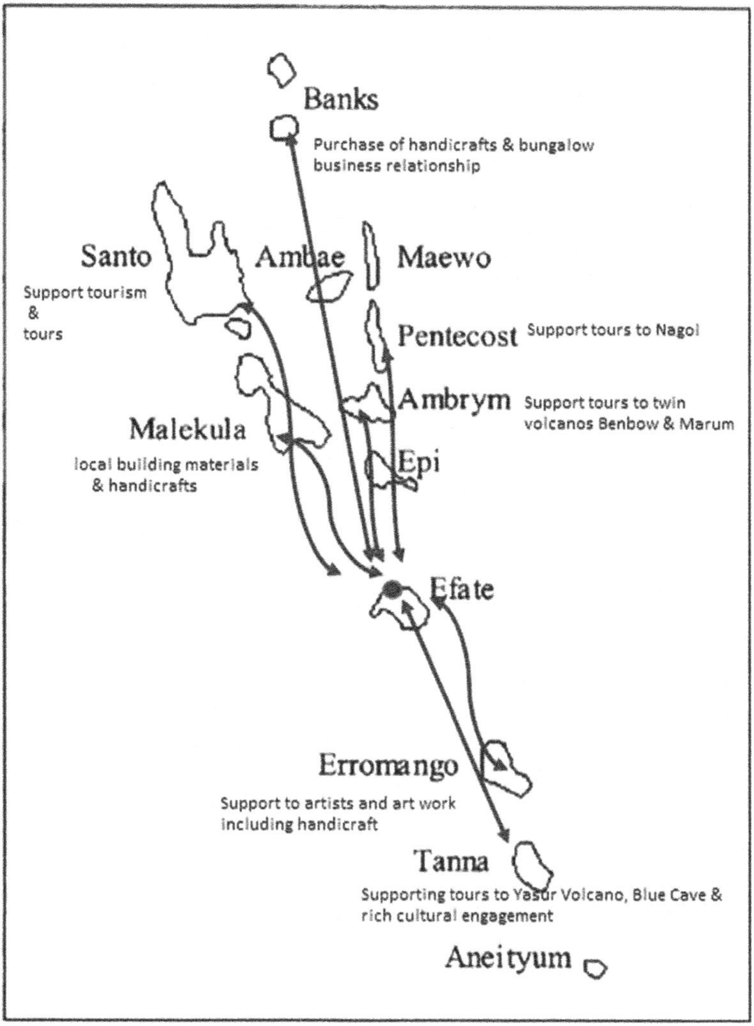

Fig. 13.5 The Havannah Resort's domestic value chain and business ecosystem pre-COVID

> The Havannah is about 40 kms from Port Vila and most of its employees came from four nearby villages with some from Port Vila. Employees were phased out as the severity and duration of the crisis became more apparent. Half the staff left in March but in July it was decided to terminate most remaining staff since it was not viable to keep them on even with existing reduced hours. Those first displaced had worked for less than three years and had lower termination payments so were the least costly to lay off. Staff who were displaced were offered the possibility of returning to the resort on given days for training and learning new digital and communications skills. None of those in nearby villages returned to full-time agriculture since the returns to agriculture had fallen as market prices fell. Some fished or provided boat transport to earn extra money. Some men were eventually re-hired for maintenance particularly of the grounds. Most of those who came from Vila returned there but few found fulltime employment. Many of those displaced were biding time and waiting for tourism to resume.

COVID-19 not only impacted the local supply chain and staff employment, but also meant that valuable community support programs through The Havannah Resort – such as the Pack for Purpose responsible tourism program (see https://www.packforapurpose.org/) with guests and sponsors providing essential supplies to schools in nearby villages; and resort staff skills sharing to assist and contribute to community development events – also ceased.

The different experiences for The Havannah post-disaster TC Pam in 2015, and in mid-crisis under COVID-19, caused the resort to re-think its own engagement in wider sector and policy level efforts as part of enhancing industry resilience (and therefore enterprise resilience). As Table 13.3 shows, the high impact of COVID-related job losses induced a much higher private sector engagement and government collaboration by the resort than in previous disasters. The trust that The Havannah Resort had built over the years through its local supply chain, community initiatives and staff support, positioned The Havannah Resort to represent the industry and community interests with a degree of legitimacy that few tourism businesses were able to achieve.

As a recognized 'voice' for tourism, The Havannah Resort – through indigenous co-owner Liz Pechan – was nominated for a range of private sector bodies and public-private committees (Fig. 13.6). Inclusion on these boards and committees allowed the industry's deep knowledge and experience to inform policy and shape the way forward. Liz Pechan led consultations for the Tourism Crisis Response and Recovery Plan (TCRRP)

Table 13.3 The Havannah Resort's disaster and crisis response experience, 2015 and 2020

	Disaster response – Cyclone Pam (2015)	*Crisis response – COVID-19 (2020)*
Job losses	Nil	90%
Staff skills transfer	100% during recovery period	Limited due to uncertain labour market
Period of recovery	4 months	Mar 2020 – unknown
Insured	Yes	No
Community support	High	Low
Government collaboration	Low	High
Private sector engagement	Low	High

for Product Readiness, and the creation of the Vanuatu Guidelines for Safe Business Operations, which resulted from close collaboration with the Ministry of Health, the Australia-Pacific Training Coalition, Department of Labour and Department of Tourism. A skills alliance was formed to deliver these guidelines across both formal and informal sectors. The collaboration with the Ministry of Health continued with work planned on future surveillance once border restrictions eased.

Case Study 2 – Sector Level Response: A New Domestic Tourism Campaign

With the closure of the borders, international tourism abruptly ceased in March 2020. Vanuatu's national tourism marketing agency, the Vanuatu Tourism Office (VTO), formulated a marketing response to stimulate domestic demand for tourism products to help support business continuity. The VTO established a domestic marketing program and designed a campaign targeting domestic consumers. In collaboration with the Department of Tourism, which was responsible for policy-making and product accreditation in the sector, the VTO conducted a National Tourism Business Impacts Survey on TC Harold and the COVID-19 Pandemic in April 2020, with a modified follow-up survey conducted again in August. These surveys predictably showed that businesses experienced significant decreases in revenue and that new marketing and product development were crucial. Prior to COVID-19, ni-Vanuatu

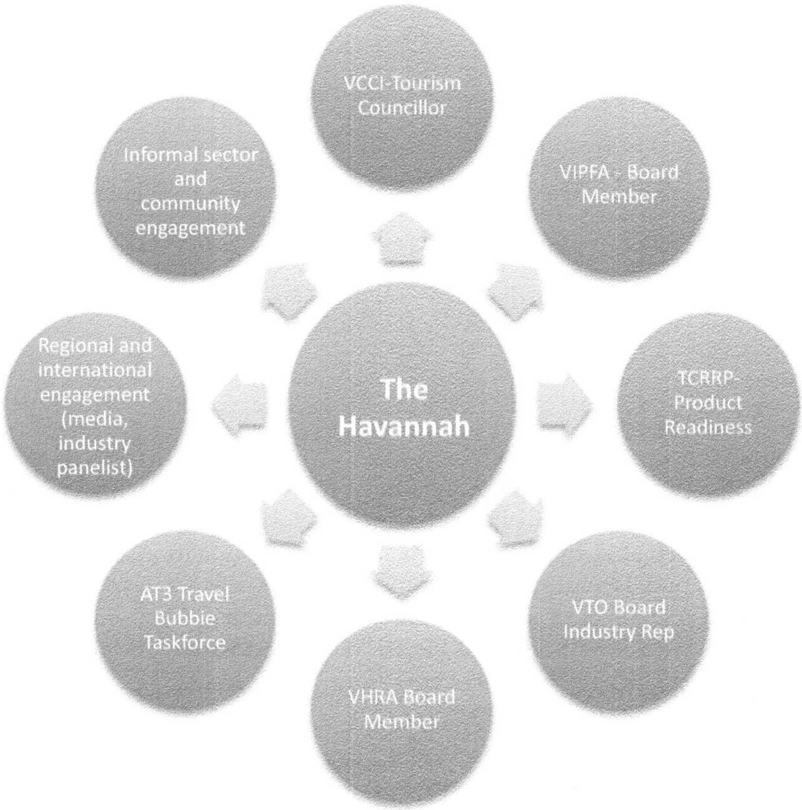

Fig. 13.6 The Havannah Resort's connections with boards, committees and communities underpinned effective public-private collaboration for tourism sector recovery efforts

participation in domestic tourism comprised mainly travel for work and visiting friends and family, hence VTO sought to stimulate demand for domestic tourism products by highlighting that tourism was not just for international visitors. The "*Sapotem lokol turisim*" ('Support local tourism' in the Bislama lingua franca), campaign was designed to stimulate new consumer behaviour and foster a more sustainable sector, involving a domestic tourism marketing strategy to assist businesses with marketing and packaging products for a domestic audience.

The first phase (mid-May to June 2020) consisted of an integrated marketing campaign using newspaper, Facebook and radio advertising to promote tourism businesses across the country. This included 105 promotional offers from tourism businesses. The strategies employed by businesses included offering group discounts, package deals, discounts, targeting new market segments within the domestic market, developing new partnerships with other domestic tourism businesses and shifting from full-time to casual staff employment. Businesses reported an average of 18% additional sales revenue as a result of the campaign.

A second phase (July to September 2020) of the domestic marketing campaign increased expenditure on Facebook advertising since this had generated the most effective response rate. VTO staff began a series of marketing workshops with industry in primarily rural areas to raise awareness about the COVID-19 situation on tourism, to support businesses to improve their own marketing efforts, by assisting them to participate in the campaign with offers, and to increase businesses' market visibility by assisting with updating Facebook profiles and Google My Business accounts so that consumers could easily locate them to make bookings. Thirty five tourist promotions, most in Efate, were put forward – mainly accommodation or restaurant providers – and most of those that succeeded in gaining extra trade were in Port Vila. One Port Vila resort that had previously oriented its marketing to couples from Australia switched to advertising itself as a family resort with discounted room rates and more budget food choices on the menu, and gained a consistently high local patronage.

During the second phase, VTO also conducted Vanuatu's first-ever survey of domestic tourism consumer preferences. A key finding was that there was no significant difference between expatriate and indigenous consumers in terms of willingness to spend on local tourism products. Indigenous consumers constituted the largest market segment and expressed an interest in participating in tours. The survey findings were complemented with focus group discussions, which showed that some indigenous consumers felt that tourism wasn't for '*man ples*' (local people), but was for international tourists only. In response VTO began publishing a series of guides for domestic consumers in Vanuatu's major lifestyle publication *Life & Style*, designed to educate and inform them about tourist attractions and tours in different islands or province. That continued into the third phase of the domestic marketing campaign (October onwards).

VTO developed image, video and social media content showing indigenous consumers enjoying tours and attractions, with a new catch-phrase and social media hashtag "*Sapotem lokol turisim*" and engaged local social media influencers to help the campaign gain visibility and presence. Potential domestic tourists who opted in to provide contact information as part of the consumer survey were provided with a VT3000 (USD30) voucher to spend at any accredited tourism business. VTO also developed a "Free Fun Bus" concept on Santo and Efate islands to provide free public transport to tourism attractions and businesses in rural areas beyond the urban centres of Port Vila and Luganville, to increasing local people's exposure to local tourism products (Table 13.4).

While domestic tourism could never deliver more than a fraction of the economic contribution that a revived operational international tourism industry would do, the effort put into the *Recovery Plan* process helped the industry operate, even as a shadow of itself, through a prolonged period of uncertainty, and provided a valuable measure of diversification.

Case Study 3 – Policy Level Response and Recovery?: COVID Safe Business Operations Guidelines

The 'voice' of the tourism sector during the crisis was coordinated through the Department of Tourism, which spearheaded the public-private development of the *Tourism Crisis Response and Recovery Plan* (TCRRP) for 2020–2023, and through the Vanuatu Chamber of Commerce and Industry's Tourism Councillor, who was also The Havannah Resort's co-owner. The industry responded swiftly to the immediate downturn in international visitor arrivals, working collaboratively across the public and private sectors, and produced the TCRRP by May 2020. The TCRRP focused on five priority areas, with Phase 1 until December 2020 and Phase 2 from 2021 to 2023, that reflected concern for a greater degree of personal security, involving safeguarding the health of citizens and visitors, COVID safe business operations in the industry, enabling effective international movement, marketing (to maintain Vanuatu as 'front-of-mind internationally', whilst also building a domestic tourism campaign) and greater community communication (https://tourism.gov.vu/images/DoT-Documents/Presentations/VTCRRP-Presentation.pdf).

The private sector, state-owned enterprises and community groups constituted an industry lobby group to influence tourism considerations as

Table 13.4 "Sapotem Lokol Turisim". Vanuatu's 2020 domestic tourism campaign in response to COVID

	Campaign activities	*Comments*
Phase 1 (May – June)	Integrated marketing campaign using newspaper, Facebook and radio advertising to promote product offerings from across the country	Businesses reported an average of 18% additional revenue as a result of the campaign
Phase 2 (July – Sept)	Increased expenditure on Facebook advertising and VTO series of marketing workshops with industry in primarily rural areas: (1) to raise awareness about the COVID-19 situation on tourism; (2) to support businesses to increase their marketing efforts, by assisting them to participate in the campaign (support to create offers); and, (3) to increase businesses' market visibility by providing assistance with Facebook profiles, Google MyBusiness accounts so that businesses could be located and contacted by domestic consumers	Facebook offers achieved market penetration of effectively all of Vanuatu's users of the platform over 18 years old; each user saw offer ads 60 times during the duration of Phase 2
Phase 3 (Oct – Onwards)	(1) VTO developed image and video content showing indigenous consumers enjoying tours and attractions. VTO engaged social media influencers to help the campaign gain visibility and presence in the market; (2) consumers who opted in to provide contact information as part of the consumer survey were provided with a VT3000 (USD30) voucher to spend in any accredited tourism business; (3) VTO working to develop local programming on television; (4) VTO developed the "Free Fun Bus" concept on Santo and Efate islands which provides free public transport to tourism attractions and businesses	Social media content showing indigenous consumers enjoying Vanuatu tourist attractions and businesses produced the highest engagement rates of any content ever published by the VTO – as high as 45%

part of Vanuatu's national economic recovery strategy. Private sector collaboration with the Vanuatu Government was recognised as essential for economic recovery, both in terms of protecting jobs, investments and ensuring business continuity. The VCCI lobby group continued to assist with the development of a phased approach that sought to balance public health and the economy.

The development of the *Vanuatu Guidelines for Safe Business Operations* (https://tourism.gov.vu/index.php/safe-business-operations-guidelines) to apply new health and safety measures at the business level in the context of COVID-19, was an important step in efforts to reshape the tourism industry response to the economic crisis. Developed through a multi-stakeholder collaboration between Government, the private sector, NGOs and training providers (including no less than the Department of Tourism, Department of Public Health, Vanuatu Chamber of Commerce and Industry, the Australia-Pacific Training Coalition, the Vanuatu Skills Partnership and World Vision Vanuatu), its guidelines required businesses to understand, prepare, take responsibility under a number of transmission scenarios, and carry out the recommended public health responses required by various scenarios. The six-partner coalition leveraged AUD 1.4 million (USD 1 million) to support the work. Of this, 50% was provided by the NGO sector, 25% by the private sector and the remaining contributions shared across government and training providers, demonstrating a strong public-private and community coalition-building. Launched in October 2020 by Prime Minister Bob Loughman, the SBO Guidelines were a hallmark achievement for Vanuatu, setting standards for safe operation during a pandemic not only for tourism-related businesses, but for all businesses.

Crisis Response: Lessons and Cautionary Tales

Prior to the declaration of the first COVID-related state of emergency, a National Novel Coronavirus (COVID-19) Taskforce was established on 27 January 2020 to monitor the evolving global health situation and to develop a preparatory plan for protecting Vanuatu from the spread of the coronavirus. It was chaired by the Ministry of Health and worked closely with the WHO, border agencies (Biosecurity, Immigration, Customs), airline, passenger and international shipping representatives and tourism representatives to respond to the WHO's declaration of a Public Health Emergency of International Concern. It issued a series of public advisories regarding cross-border travel and precautionary restrictions. The National COVID-19 Taskforce was dissolved within 3 months and its functions mainly devolved to the National Disaster Management Office (NDMO).

NDMO's original task to assist with economic recovery *after* the impact of COVID-19 became increasingly complex during the protracted crisis. COVID-19 was no traditional disaster with clear timeframes, and undertaking 'recovery' efforts whilst the crisis continued to play out required a more nuanced attention to state of emergency governance and recovery arrangements. As recognised in the *Yumi Evriwan Tugeta: Vanuatu Recovery Strategy 2020–2023,* launched in July 2020, "this compound disaster [of COVID-19 and TC Harold] has blurred the lines between response, recovery and disaster risk reduction (preparedness and prevention)" (Government of Vanuatu, 2020b: 1), with a "compound cost" from the loss and damage caused by both COVID-19 and TC Harold, as "with the rapid closure of international borders in March 2020, COVID-19 went from a health threat to an economic emergency" (op cit: 2). A key challenge was in navigating the dynamic state of the compound disaster, protracted crisis and the recovery strategies.

The tourism sector's own efforts for industry resilience provided both positive and cautionary lessons for Vanuatu's crisis recovery arrangements. The Department of Tourism's leadership of an active and effective public-private sector collaboration in the development of the TCRRP offered a valuable model that could be replicated across Government as part of the economic recovery. While consistency in Government commitment to public-private collaboration varied throughout the remainder of 2020, the efforts of the TCRRP, consolidated through the *Niufala Rod blong Tourism* forum held from 26 to 29 October 2020 highlighted the importance of genuine public-private collaboration for producing effective strategies. Significantly, while cruise ship tourism remained regarded as an important feature of Vanuatu's tourism mix, it did not play a major role in government planning and policy decisions. In terms of tourism recovery other sub-sectors were give priority as they represented a greater return to investment that came with potentially less risk.

The Havannah Resort used its pre-COVID model to inform COVID responses. Due to the prolonged hiatus caused by the pandemic, the resort had to adapt, with its co-owners becoming key lobbyists for public-private collaboration in relation to Vanuatu's economic recovery policy efforts. The resort management's informal-formal sector linkages provided a microcosm of the private/public, informal/formal, urban/provincial linkages that tied tourism businesses to communities throughout Vanuatu, and provided a voice to contribute to economic survival and recovery.

The COVID crisis produced a range of policy responses that were widely debated by the tourism industry. Chief amongst this was the Vanuatu Government's 'agriculture pivot' towards the productive sector, seen within the tourism sector as being at the expense of other economic sectors. The Department of Tourism, through the TCRRP and the October 2020 Tourism Forum, lobbied for support for tourism given its role in supporting rural livelihoods, and seized the opportunity to restructure the tourism industry around the new Vanuatu Sustainable Tourism Development Policy 2020–2030. Capitalising on the political rhetoric around reprioritising rural development during the crisis, the Department showcased the VSTDP as a vehicle for rural development through its facilitation of small-scale, sustainable rural tourism and cruise tourism. In this context, it pursued legislative change that would produce a diverting of industry-supported tourism marketing funds away from the Vanuatu Tourism Office. The unclear mandate for legislative reform was akin to 'rearranging deck chairs on the Titanic,' blurring policy attention on the fundamental issues facing the industry as a result of the COVID-19 pandemic and undermining the positive gains in public-private coalition building that the TCRRP and COVID Safe Business Operations Guidelines had achieved.

Conclusion: Policy, Politics and a Pandemic – Learning as we Go

Recognising the extraordinary situation that the COVID-19 pandemic had brought, in the wake of the devastating TC Harold, the Vanuatu Government faced massive challenges in salvaging and supporting the tourism industry, amidst other priorities, to formulating policies that required a broad assessment of alternative recovery responses and pathways for a resilient, long-term sustainable economy. The industry had effectively collapsed, with employment and linkages destroyed, despite valuable efforts at developing a domestic industry. The real-life experiences and needs of businesses and communities were therefore critical to identifying and developing strategies to support economic survival. Whilst early advances were made in terms of recognising the need to work together by leveraging private and community collaborative capacity, knowledge, and expertise, barriers to effective policy and its implementation remained at the end of 2020. These included varying levels of political commitment to supporting business continuity in a time of great uncertainty. The widely welcomed Economic Stimulus Package had provided crucial support for a four-month period to July 2020, but businesses subsequently experienced a sharp downturn, particularly in tourism and tourism-related sectors.

Despite the valuable outcomes of genuine public-private collaboration, as demonstrated by the TCRRP process, a mismatch persisted between public and private priorities under COVID-19. Human resources were inevitably strained. The COVID-19 crisis tested both public and private sector leadership skills to triage competing priorities and find opportunities in a severely constrained environment. Tourism resilience during the crisis extended beyond "survival, adaptation and innovation". Reimagining tourism required enhanced levels of coordination, and Vanuatu's development and implementation of the Tourism Crisis Response and Recovery Plan provided some critical insights into navigating a protracted period of uncertainty. The outcome of cooperation between the public and private sectors was evident in the development of the *COVID Safe Business Operations Guidelines* to safeguard workers, and the establishment of pandemic-ready standards that enable business continuity, targeting both the needs of the economy and livelihoods of Vanuatu citizens. Responding to the "compound disaster" described by the *Vanuatu Recovery Strategy 2020–2023* produced complex and dynamic arrangements that required agility and adaptability to navigate. For all the attempts at restructuring, continued border closures meant that Vanuatu's tourism sector remained in survival mode at the end of the year. Much the same was true elsewhere in the Pacific. Vanuatu's recovery efforts will depend on good management but most of all on the need for some opening up to overseas visitors, whether landed tourists in bubbles or cruise excursionists, but that scarcely seemed imminent. Plans were indefinitely postponed, but the new marketing slogan was firmly in place: 'We'll keep it beautiful for you'.

References

Asian Development Bank. (2019). Asian Development Bank Outlook. In *Strengthening Disaster Resilience*. Asian Development Bank.

Cheer, J., & Lew, A. (2017). Understanding Tourism Resilience: Adapting to Social, Political, and Economic Change. In J. Cheer & A. Lew (Eds.), *Tourism, Resilience, and Sustainability: Adapting to Social, Political and Economic Change* (pp. 3–17). Routledge.

Dornan, M., & Cain, T. (2019). Vanuatu and Cyclone Pam: An Update on Fiscal, Economic, and Development Impacts, *Pacific Economic Monitor*, May, 30–35.

Government of Vanuatu. (2019). *Half-Year Economic and Fiscal Update of 2019*. Ministry of Finance and Economic Management. https://doft.gov.vu/images/2019/Economic_And_Fiscal_Report/HYEFR_English_2019.pdf. Accessed 22 Nov 2020.

Government of Vanuatu. (2020a, April). *National Tourism Business Impacts Survey: TC Harold and COVID-19 Pandemic*. Department of Tourism and Vanuatu Tourism Office.

Government of Vanuatu. (2020b). *Yumi Evriwan Tugeta: Vanuatu Recovery Strategy 2020–2023.*

Graue, C. (2020, July 12). Vanuatu Feeling the Pinch as Coronavirus Pandemic Keeps Tourists Away, *ABC Pacific Beat.*

Orchiston, C., Prayag, G., & Brown, C. (2015). Organisational Resilience in the Tourism Sector. *Annals of Tourism Research, 56*, 128–163.

Vanuatu Chamber of Commerce and Industry. (2020). *Vanuatu Economic Outlook from a Private Sector Perspective.* https://vcci.vu/vanuatu-economic-outlook-report-from-a-private-sectors-perspective/

Vanuatu Tourism Office. (2019). *Towards 300,000: Sustainability, Partnership, Benefit for all: Vanuatu Tourism Market Development Plan 2030.*

Vanuatu Tourism Office. (2020). *Vanuatu International Visitor Survey 2019.* In Collaboration with the Auckland University of Technology, NZ MFAT, and the NZ Tourism Research Institute.

Watt, G., & Brenner, H. (2020). *Cruise Tourism in Vanuatu: Impacts and Issues.* The Council for Australasian Tourism and Hospitality Education 2020 Conference, 344–349.

CHAPTER 14

Vaccination for Vacation: Assessing the Resilience of the Jamaican Hospitality Industry in a Pandemic

Andrew J. Spencer and Diana E. Spencer

Introduction

In December 2019, waves of concerns hit the global media as it was reported that a serious outbreak of a new pneumonia was spreading through China. Other countries immediately enforced strict measures aimed at keeping the virus under control but that proved impossible. It quickly crossed borders and many countries responded by immediately closing their own borders to population movements; tourism was already collapsing and border closures provided a sudden, sharp and seemingly

A. J. Spencer (✉)
University of the West Indies, Kingston, Jamaica
e-mail: andrew.spencer@uwimona.edu.jm

D. E. Spencer
Caribbean Maritime University, Kingston, Jamaica

Y. Campbell, J. Connell (eds.), *COVID in the Islands: A comparative perspective on the Caribbean and the Pacific*,
https://doi.org/10.1007/978-981-16-5285-1_14

final shock. When Jamaica closed its border in March 2020 and cruise ships no longer entered its ports the industry entered into crisis mode.

In this century global tourism has been impacted by three deadly coronaviruses, SARS in 2003, MERS in 2015 and finally COVID-19. SARs and MERS resulted in no more than a 1.2% fall in international tourism arrivals (World Bank, 2020a, 2020b; Lee et al., 2012) of slight and localised consequence compared with COVID-19. Quite quickly the World Bank (2020b) announced the fastest growing recession since 1990. The report postulates that many economies were previously experiencing weaker and unfavourable growth before the pandemic which makes it extraordinarily difficult for these economics to rebound. Tourism was affected more than any other sector, alongside transport, and particularly so in many Small Island Developing States (SIDS) where it sometimes provided as much as 80% of the country's exports while contributing over 20% of their GDP (UNWTO, 2020). Many of the most affected SIDS, where tourism was effectively a monoculture, including Jamaica, were in the Caribbean. This dependence on tourism resulted in SIDS being some of the most volatile economies emerging from COVID-19 impact, although compared with many smaller Caribbean economies Jamaica had a degree of diversity.

Although many countries, including Jamaica, started re-opening their borders gradually at some point in 2010, their local economies continued to experience the negative impacts that the virus had placed on the operations and success of the economy. This chapter focuses on the considerable impacts of the coronavirus on Jamaica, which is particularly heavily dependent on tourism, then the principal contributor to the country's foreign exchange. Emphasis is placed on the impact of the reduction in tourism activity on small businesses, specifically hotels, visitor attractions, cruise ships and transportation. An assessment of the current protocols and measures employed will be addressed, as the government with the private sector has sought to counteract COVID's damaging effects on the industry. We explore the challenges and combination of solutions that will be required for the industry to have a meaningful survival, and which policies and practices may provide a degree of immunity to these shocks: what we have termed industry vaccination.

Jamaica: Heartbeat of the World

Jamaica, with a population of 2.8 million is the largest island in the English-Speaking Caribbean. Formerly with a plantation agricultural economy, it has gradually made a transition to tourism, mainly now centred on the north-west coast. Tourism and mining have been the leading sources of foreign exchange, but half the Jamaican economy relies on services, dominated by tourism. Jamaica continues to be affected by challenges due to its vulnerabilities relating to its natural disasters, negative external economic blows, high debt ratio, and crime. Collectively that has been a disadvantage since "countries with a more stable economy generate more tourists than those with economic woes" (Williams & Spencer, 2010: 362). While tourism began in the 1890s, it first boomed in the 1960s – and then slumped – with negative oversea press coverage, in its main market the United States. It grew again in the 1980s when cruise ship tourism became significant and by 1999 tourist numbers had reached 2 million. Two decades later 2019 arrivals were 4.3 million and their expenditure was US$3.2 billion; tourism accounted, directly and indirectly, for nearly 34% of Jamaica's GDP, with the sector growing at an annual rate of 9%. It had become the major driving forces of the national economy, and launched its new slogan at the start of 2020 – Heartbeat of the World – that hoped to capitalise on a number of topical factors in 2020 enhancing Jamaica's visibility on the international stage, including the April release of the latest James Bond movie "No Time To Die" – which was filmed on the island, the birthplace of the iconic fictional British spy. But tourism was about to collapse. The heart would soon need Intensive care.

The effective shutdown of the country had an immediate impact on the two main sources of foreign currency – tourism and remittances – that account for a third of the economy. Before COVID hit Jamaica had experienced ten successive years of tourism growth. Tourism collapsed, while inflows from remittances fell as there were contractions in the main economies that hosted the Jamaican diaspora, notably the United States. The combination of the fallout in tourism and the slowdown of remittances put pressure on the Jamaican dollar. Closed hotels, deserted beaches, empty docks, nightly curfews and a related slowdown in commercial and industrial activity marked the decline. Jamaica's Minister of Finance, Nigel Clarke, stated: "Government revenues have declined sharply, foreign-exchange inflows from tourism have dried up, foreign direct investment will decline, our external balance of payments will be negatively affected, economic output will contract significantly, and employment gains will be eroded (quoted in James, 2020). Workers in the tourism sector were most immediately and severely affected by the shutdown of the island's airports to passenger traffic. The closure of hotels left about 150,000 workers without jobs. The Prime Minister, Andrew Holness, declared that "the overall estimated loss to the economy from visitor's expenditure from stopover arrivals is J$107.6 billion [US714million]. Therefore, the nation is losing out on roughly J$400 million [US$2.6 million] per day". As in other tourism-oriented SIDS the ramifications were felt throughout the economy as tourism provided jobs in hotels, restaurants, transportation, stores, visitor attractions, alongside much indirect employment. We examine these impacts and linkages below.

An Industry in Crisis. Framing the Local Within the Global

Beyond deserted beaches and restaurants, the most visible impact was in the empty hotels. With the travel restrictions, closed borders, increased "stay home" campaigns and fears of contracting the coronavirus, the hotel occupancy levels experienced sudden shocks as the COVID pandemic swept through all nations. Room occupancy rates plummeted. The Planning Institute of Jamaica recorded that between January to June 2020 the services industry experienced an 11.5% decline, with a decline in hotels and restaurants due to pandemic restrictive measures of 50%, when compared to the same time period in 2019. Furthermore, as the crisis worsened and travel came to a virtual standstill, between April and June 2020 the services industry recorded a 21% decline while hotels and restaurants experienced 87.5% decline when compared to the same period in 2019. Most hospitality businesses in the accommodation sector operated a high-density model that required a 70% occupancy rate to break even. Even the few hotels which remained open only maintained an average of 25% occupancy. The level of unemployment among low and unskilled workers increased in parallel and the country's employment fell by more than 30%. Additionally, employees who were retained on staff experienced wage reductions of up 11%. Preliminary indications were that the impact on female employment was greatest since the majority of workers in hotels, restaurants and related activities were women. The immediate national consequence was the significant reduction in government revenue from the travel and tourism industry (Fig. 14.1). When compared to previous

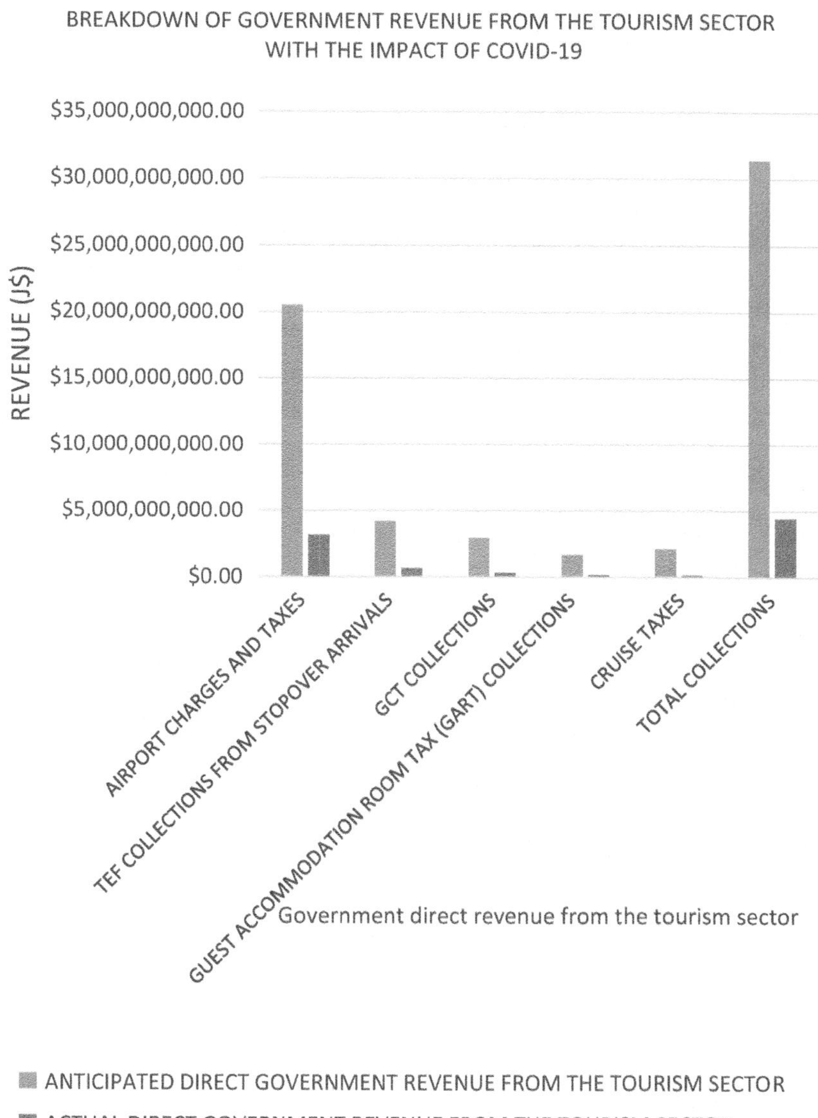

Fig. 14.1 Jamaica's government revenue from the tourism sector (Source: World Travel and Tourism Council, 2019)

years and Pre-COVID projections, its revenue had precipitously declined, wiping out earlier optimistic projections.

- *TEF- Tourism Enhancement Fund
- *GCT- Government Consumption Tax

Lives Versus Livelihoods

Universally COVID-19 presented an uncomfortable tension between the protection of lives from the virus and the ability to earn from the tourism industry, which by its very nature requires human interaction. Over 300,000 jobs have been lost with the lion share being from hotels, accounting for more than 40% of job loss. Around 170,000 of these represented direct job losses in the industry and the remainder were indirect. The Attractions sector inclusive of the larger players such as Chukka Caribbean, Dolphin Cove, Mystic Mountain and Dunn's River also saw massive numbers of employees being sent home. While not many Jamaicans were employed with Cruise Lines, the no sail order issued from April to October 2020, saw a major fall off in work for 3866 contract carriage (ground transport) operators and 949 craft vendors. For both these categories it was estimated that an even larger number operated informally.

The lack of any cruise ship arrivals between April and September meant no revenue whatsoever from the cruise industry – whether in employment and the related transport sector or through the taxation of cruise arrivals. That meant a sudden loss of US$1.2 Billion in taxes alone. Extending this analysis to visitor expenditure (whether in related accommodation, attractions, entertainment, food, transportation, shopping, etc.,) indicated a loss of US$53.3 Billion.

Before the crisis the tourism sector directly employed approximately 170,000 workers, representing 12.6% of the national labour force. After the borders closed in March 2020, most of the tourism sector was closed and between April and May approximately 90% of the staff were laid off. Approximately 153,000 workers were unemployed during this period. Significantly most of those who became unemployed were deemed unemployable – and so remained unemployed – a function of a combination of a lifetime commitment to their own particular niche in the tourism industry and low levels of skill (that were not in any case transferrable during a crisis that extended far beyond the tourism industry). By any standards these losses of incomes and employment were a major blow to the national economy and to the households involved in the tourism sector, especially since they could not have been predicted only 6 months earlier (and in direct contrast to the continued growth that was predicted). At the same time COVID-19 directly affected lives in Jamaica. By the end of May nine people had died, and there were 581 confirmed cases. This represented a combination of imported and local cases.

Ultimately it was decided to reopen Jamaica to the tourist industry; other Caribbean economies were beginning to do the same, and there was much debate about the need to balance lives and livelihoods. Since border closures were intended to prevent the spread of COVID-19 there was a real risk that Jamaica might now be exposed to a new pandemic threat.

The Reopening

Given the value of the tourism industry in Jamaica and its widespread linkages with other sectors, in June 2020, the Jamaican government re-opened the sector to visitors. It was among the first destinations in the Caribbean to re-open its border and doors to tourists. In an effort to recover from the debilitating effects of COVID-19 on the industry, the government argued that the re-opening was essential since it would potentially (or at least eventually) assist 350,000 workers who had been displaced, provide much needed revenue for tourism entities operating at a loss and contribute to the national budget. News of the reopening was met with concerns from the general population fearing the impact international travellers might have on the wellbeing of the local population.

In response to the citizens' growing concerns, the Government embarked on the creation of a "Covid Resilient Corridor" with its associated health and safety measures. In June 2020, a detailed protocol was drawn up to be used by hoteliers to accommodate their guests (Montevago, 2020). Hotels within the 'coronavirus-resilient corridor' were expected to follow the protocols and ensure that their guests understood the procedures established by the Ministry of Tourism. A corridor was designated along the north coast, from Negril in the west to Port Antonio in the east, and all tourists would remain there. Jamaica, like other Caribbean destinations, also provided isolation rooms for individuals to quarantine if they showed signs of the coronavirus. Jamaica Cares insurance was mandatory. Measures for testing visitors were introduced at airports, but cruise ship ports have remained closed. Cruise ships were perceived as too risky.

The Minister of Tourism, Hon. Edmund Bartlett, announced that reopening the sector would assist the displaced workers and the several tourism enterprises that were operating at a loss. As he pointed out:

> Tourism is big business – 80% of which is small business – the restaurants, craft vendors, tour and transportation operators, attractions, bars, duty-free shops. Due to tourism's transversal nature [sic] and the linkages with other productive sectors, it also stimulates agriculture, manufacturing and the creative economy. (*Travel Daily News*, 8 June 2020)

Based on the concerns of the general population about the implications of reopening tourism and its impact on the wellbeing of Jamaican citizens, the Minister contributed 10,000 masks to frontline industry workers. This initiative also contributed to a degree of economic sustainability by the creation of a cottage industry where the masks were made. With the assistance of the Tourism Product Development Company (TPDCo) and the Tourism Linkages Network, twenty-two small entrepreneurs were solicited to create these masks. Thereafter this cottage industry continued to expand as several micro, small and medium-sized enterprises sought to benefit financially in difficult economic times. In July there was a further relaxation of controls with a conditional reopening of small outdoor events, amusement parks and water attractions (which attracted their own set of regulations including mask wearing and sanitisation, rides and equipment sanitized after each us, and no sharing of towels, goggles or snorkels).

At the same time Jamaican hotels also concentrated on promoting staycations, by encouraging local citizens to take advantage of the low nightly rates that are now being offered (*The Gleaner*, 25 July 2020). This initiative was created by the Ministry of Tourism in July to increase occupancy rates as most hotels needed "an average of 60% occupancy to break-even, while the data is showing current occupancy rate at 10% to 35%" according to the preceding Jamaica Hotel and Tourism Association President Omar Robinson (*Observer*, 31 July 2020).

With the borders reopening in June to international travel, approximately 30% of workers returned to work in the sector between June and August. However they returned on a staggered basis, and there was still insufficient demand to have workers back full-time, so that many were employed at reduced hours and reduced wages.

Between mid-June and September the number of tourist arrivals was approximately 211,000, with numbers steadily growing (with no cruise visitors). American airlines were gradually increasing services to Jamaica, and 900 tourism related businesses (most of which had been certified as COVID-compliant) had restarted operations. This generated a revenue of $231 million. Compared with what had once been predicted, and the 2019 totals, it was almost trivial, but it was a solid start to a potential new normal, no COVID cases had occurred and it offered hope for a greater winter market.

Facing the Future

The government of Jamaica established a comprehensive, multi-phased tourism resilience program to establish and maintain a sustainable tourism product even during the pandemic. A Recovery Task Force was created in April 2020, consisting of major stakeholders from the private and public sectors. It was supported by two Working Teams – one for general tourism and another for cruise tourism – and a Secretariat, with a mandate to establish the strategic direction of the sector and a planned vision in a world learning to respond rapidly to a highly infectious virus. The Tourism Product Development Company (TPDCo) took the lead in creating tourism protocols that received the "Safe Travel" approval from the World Travel and Tourism Council (WTTC). These addressed issues of sanitization, wearing of face masks and protective equipment, the importance of physical distancing, digital enablement, real-time health monitoring and reporting, rapid response, and training.

Tourism workers and businesses that were affected by the pandemic were given grants, supported through a COVID Allocation of Resources for Employees (CARE) program (Ministry of Tourism, 2020). The Government quickly responded to the increasing unemployment with an innovative support programme with two key elements: Business Employee Support and Transfer of Cash (BEST Cash) providing temporary cash transfer to businesses based on the number of workers they retained; Supporting Employees with Transfer of Cash (SET Cash) providing temporary cash transfers to individuals who had lost employment after March 10 (the date of the first COVID case in Jamaica). The Government also provided the Jamaica Centre of Tourism Innovation (JCTI) online training programs for tourism workers to remain relevant. A Tourism Linkages Network was established to foster a deeper connection between the "ordinary Jamaican" and the "tourism product" to promote the role of the industry. Other initiatives included strengthening the Tourism Linkages

Policy, adopting a Tourism Environmental Stewardship Initiative (TESI) to highlight the environmental awareness of the sector, all geared towards strengthening the country's resilience.

To ensure that the industry met the enhanced health and safety standards required to reopen tourism, the Ministry developed its COVID-19 Tourism Health and Safety Protocols based on benchmarks from nearly twenty markets in the Caribbean and globally and from international health agencies. They covered everything from large and small hotels, guesthouses, attractions, beaches, transportation, shopping, social activities (restaurants and bars) and cruise ports, and were finally guided by a five-point recovery strategy, to meet international standards, provide training in all sectors, and develop a staggered approach to reopening. The core of this was the concentration of tourism in approved and regulated corridors. All of this was to be monitored with the Ministry of National Security intending to deploy over 140 TPDCo trained District Constables, to boost the surveillance.

The soft re-opening of the industry in Jamaica and the creation of the north coast resilience corridor sought to both generate demand while protecting the workers (enabling, if necessary, the Ministry of Health to be able to respond within a designated space). The promotion of a wide range of measures across the industry was required, covering a range of health and safety measures from social distancing to wearing masks that raised complex issues reflecting the need to balance safety and health considerations for both tourists and workers, maintain monitoring and surveillance while ensuring the degree of freedom and relaxation that would enable a pleasant holiday environment. Meanwhile the TPDCo was promoting a three-S branding "Safety, Security and Seamlessness": safety and security were oriented to obvious health, sanitation and relaxation issues, while seamlessness extolled the entire guest-host country interaction as a hassle-free experience.

The implementation and constant reviewing of cautionary and preventative measures were critical to making Jamaica safe and attractive. As the local and global COVID-19 context changed, especially in its major markets, Jamaica had to monitor and readjust its health and safety measures in light of constantly changing circumstances. These measures had to be communicated effectively to tourists: focusing on health and safety measures while reassuring tourists of the friction free nature of travel to and within Jamaica as a destination. To build this trust required messaging that balanced the positive, the believable and the actual. All such reflections, reconsiderations and evolving strategies needed to be developed quickly in a context where Jamaica was experiencing considerable national problems, irrespective of tourism, and where developing tourism had long been constrained by such challenges as the lack of financial resources, adequate infrastructure, social instability and lack of human expertise (Altinay et al., 2007).

The Final Analysis

As the world recovers from or at the very least tries to operate with COVID-19, tourism dependent economies like Jamaica must determine and map a path of recovery and revitalisation of the tourism industry. In light of the dwindling international demand for travel, igniting new interest will be shaped by familiar variables, about the merits, accessibility and cost of destinations, but also by perceptions of the destinations' health and safety measures. Promoting Jamaica as the Heartbeat of the World will be linked with promoting the implementation, practice and enforcement of sound health and safety measures to ensure that local hotels, restaurants, attractions and other tourism services are adequately prepared. That will require new strategies and marketing campaigns, linking local practices with global interests and concerns. At least late in 2020 tourist numbers were increasing, hotels and other attractions were reopening, employment was growing and there was a resurgent optimism and the possibility of vaccination was imminent. That at least suggested the possibility of the rebirth of an industry so crucial to national development that might support a new sustainable development. COVID-19 provided an opportunity to re-focus and emerge a stronger tourism destination. We are optimistic, that policy makers will finally accept that the "rinse and repeat" of decades of doing more of the same is over, yet any hope of a rebound will of necessity be a tale of flexibility, agility, innovation and resilience.

The future of the Jamaican tourism industry cannot be determined in Jamaica alone. While Jamaica was again promoting its own product, it was, as always, in competition with other Caribbean (and more distant) destinations, all also seeking to revive. Much depended on whether the United States-based cruise ship industry might recover and return, but, above all, on whether tourists were safe to travel. As Jamaica's extensive preparations to ensure the health of visitors – and of its own labour force – indicate, there are always risks in accepting visitors from overseas. Opening up too far, too fast and too early held dangers which thus far Jamaica had carefully avoided. But underlying future success are vaccinations and cures. Only the extensive acceptance of vaccinations will genuinely both enable vacations, restore confidence in the industry, and result in poverty reduction and a much-needed boost to the national economy.

References

Altinay, L., Var, T., Hines, S., & Hussain, K. (2007). Barriers to Sustainable Tourism Development in Jamaica. *Tourism Analysis, 12*, 1–13.

James, C. (2020). *COVID-19 Batters Jamaica's Tourism Sector.* Caribbean Investigative Journalism Network.

Lee, C. K., Song, H. J., Bendle, L., Kim, M. J., & Han, H. (2012). The impact of Non-pharmaceutical Interventions for 2009 H1N1 Influenza on Travel Intentions: A Model of Goal-Directed Behavior. *Tourism Management, 33*(1), 89–99.

Montevago, J. (2020, June 19). *Jamaica Takes First Step on Path to Post-COVID Tourism.* Travel Market Report. https://www.travelmarketreport.com/articles/Jamaica-Takes-First-Step-on-Path-to-Post-COVID-Tourism

Ministry of Tourism. (2020). *Building Tourism Resilience to Thrive in the 'New Normal'.* https://www.mot.gov.jm/speeches/building-tourism-resilience-thrive-%E2%80%9Cnew-normal%E2%80%9D

UNWTO. (2020, April 28). *100% of Global Destinations Now Have Covid-19 Travel Restrictions*, UNWTO Reports. Retrieved from https://www.unwto.org/news/covid-19-travel-restrictions

Williams, D., & Spencer, A. (2010). Advertising Expenditure and Tourist Arrival Figures: The Case of Jamaica. *Tourism and Hospitality Research Journal, 10*(4), 359–366.

World Bank. (2020a, April 4). *Air Transport, Passengers Carried.* Retrieved from https://data.worldbank.org/indicator/is.air.psgr

World Bank. (2020b, June 8). *The Global Economic Outlook During the COVID-19 Pandemic: A Changed World.* Retrieved from https://www.worldbank.org/en/news/feature/2020/06/08/the-global-economic-outlook-during-the-covid-19-pandemic-a-changed-world

World Travel and Tourism Council. (2019). *Economic Impact Report: Jamaica, World Travel and Tourism Council.*

CHAPTER 15

Tourism in a World of Disorder: A Return to the Vanua and Kinship with Nature in Fiji

Apisalome Movono and Regina Scheyvens

Introduction

In many Pacific island countries, tourism provides a suitable and often the main developmental driver for national economies (Harrison & Prasad, 2013; Movono et al., 2015). Although the impacts of tourism on Pacific countries and indigenous communities are well documented, few studies conceptualise tourism-dependent communities as complex and adaptive social and ecological systems (Movono et al., 2018). Consequently there is a void in understanding the complex ways Pacific islanders have responded and adapted to tourism and to the sudden changes in their circumstances, whether internally influenced by political means or through unstable global market forces. This chapter will contribute to this by showing how, over time, tourism and other shocks, including COVID-19, have inspired complex and transformational responses from participants.

A. Movono (✉) • R. Scheyvens
Institute of Development Studies, Massey University,
Palmerston North, New Zealand
e-mail: A.Movono@massey.ac.nz; r.a.scheyvens@massey.ac.nz

Y. Campbell, J. Connell (eds.), *COVID in the Islands: A comparative perspective on the Caribbean and the Pacific*,
https://doi.org/10.1007/978-981-16-5285-1_15

The adaptive responses of the people of Vatuolalai, an indigenous Fijian village, draw strength from their internal resources which act as safety nets during difficult times such as the COVID-19 pandemic.

The chapter first reviews the literature, focusing on sustainable livelihoods and complex systems theory, before outlining the approach taken, alongside essential background information to the area. This demonstates how the community has come full circle, initially weaning off their connections to culture and environment to focus on tourism livelihoods only to return to their traditional systems to cope with pressures from the pandemic and the closure of global travel. The discussion extends understanding of how indigenous Fijian knowledge, culture, and livelihood systems establish security for community members in the face of various shocks. Ultimately, we assert that a post-pandemic recovery must occur in a manner that considers indigenous kinship with nature as a cornerstone for building resilience in a world of increasing uncertainty and disorder.

Sustainable Livelihoods Approach and Complex Systems Theory

Chambers and Conway (1992), Chambers (1995), and later, Scoones (1998) initiated debates that aroused appreciation for the value of sustainable livelihoods research in rural development. Recently, livelihoods-based studies have expanded beyond their origins in human ecology to include applications in indigenous Pacific communities that deal with tourism (Movono et al., 2018). The Sustainable Livelihoods Approach (SLA) can guide enquiries about indigenous communities in a manner that focusses on the various livelihoods activities employed to sustain a living. SLA is people-centred, focused on community-level actions and considers the inherent capacities and knowledge systems of indigenous communities (Tao & Wall, 2009; Wu & Pearce, 2013). SLA is a means to analyse and change the lives of those who are disadvantaged by examining the context, resources, shocks, various forms of capital and activities required to attain a living. Such in-depth examinations are useful in suggesting ways forward for disadvantaged groups within specific communities facing challenging conditions (Wu & Pearce, 2013).

The concept of livelihoods as being more tangible than development emerges because it is easier to describe, discuss, observe, and quantify for rural indigenous communities, especially over an extended period.

The SLA promotes the use of adaptive strategies, loosely defined as adjustments that people make in their livelihood systems that allow difficult circumstances such as the global pandemic to be endured or overcome (Tao & Wall, 2009). Adaptive strategies are often based on local knowledge and developed by combining traditional and less traditional knowledge systems. Adaptation can also be examined if SLA is used within the conceptualisation of communities as socio-ecological systems (Bhandari, 2013). These cultural elements include indigenous knowledge, kinship systems, social capital and connections to nature which are resources that form the basis for traditional Pacific livelihoods (Movono et al., 2018).

Complex Adaptive Systems (CAS) theory is adopted here to complement the sustainable livelihoods approach, and compensate for its perceived weaknesses, providing an ideal theoretical framework to understand how livelihoods activities promote resilience building (Byrne, 1998; Shen et al., 2008). CAS justifies conceptualising communities as complex adaptive systems that are responsive to both gradual changes imposed by tourism and sudden perturbations such as COVID-19 (Morse et al., 2013). CAS is also suitable for studying evolving concepts and processes, such as community development and the multiple, interrelated, and interacting entities within an embedded system (Holland, 2006). CAS, a specific variation of systems thinking, provides a means of understanding how the structures, processes and resources within a community adapt to change (Byrne, 1998). In this way, specific societal functions such as livelihoods activities, customary resources, and adaptivity subject to tourism disturbance are considered in a more cohesive manner (Holland, 1992). CAS also has the potential to show how disorders at the macro, global socio-economic level can lead to very complex responses at the micro, communal level, or vice versa (Gunderson & Holling, 2002; Holland, 2006). Here this will be examined in the context of the Fijian village of Vatuolalai.

CAS theory is founded on the view that complex behaviour emerges as a result of interactions among internal system components and external influences as a response to change introduced into any part of the system (Holland, 1992; Holling, 2001; Morse et al., 2013). Elements of a complex adaptive system modify their behaviour and respond to various internal and external changes in the environment, and are characterised by panarchy (Gunderson & Holling, 2002; Coetzee et al., 2015). Panarchy is described as a nested set of adaptive cycles operating at discrete ranges of scale, emphasising vertical and horizontal structuring, multi-levelled linkages, and focusing on the processes and relationships between these corresponding elements (Allen et al., 2014; Coetzee et al., 2015). The concept of panarchy implies that processes, such as tourism development, and the pandemic which occur at the outer scale of the system, have the potential to affect the overall dynamics and responses of an indigenous community, and also vice versa (Folke et al., 2003; Gallopin, 2006). Panarchy is concerned with identifying emergent patterns, connections, and reactions of elements to processes or perturbations that are introduced to a system, but seldom used in the social sciences (Holland, 1992; Holling, 2001.). Both the SLA and CAS theory are used to support the aims of this study in understanding the complex changes that have occurred in Vatuolalai because of COVID-19.

The Research Area and Context

Fiji is an archipelago of more than 300 islands but with the majority of its population of over 900,000 on the main island of Viti Levu. The steady decline in sugar production over the past two decades along with limited industrial options, has seen tourism become the key driver of up to 40% of Fiji's GDP and its largest employer (Movono et al., 2015). Vatuolalai village, is on the Coral Coast, a two-hour drive from Nadi International Airport, which is Fiji's oldest tourism region where resort-based tourism was pioneered in the 1950s (Belt, Collins and Associates Ltd., 1973). The typical Fijian household has approximately six members, two or three of whom are employed at the nearby Naviti Resort or in their own, tourism-dependent businesses. Most of them experienced a loss of at least 50% of their tourism income during COVID-19 due to Fiji's borders being closed to international travel. Its thirty-two households increased by almost 100 people after COVID-19, to hover just under 330 people, the majority of whom were below the age of 35 and with a slightly higher percentage of females (54%) compared to males (46%). Over 92% of the population were involved in some form of tourism activity before COVID hit, indicating how dependant people of this community were on tourism and the extent to which they were then impacted by the pandemic (Movono and Dahles, 2017). In large part that was a consequence of the Naviti Resort, a relatively large-scale, all-inclusive resort, being built in 1972 on land leased from villagers, providing them with new opportunities and ways of life. Its fifty-year operation has also influenced their dependence on income and responses to other shocks such as the 2000 coup. That led to a 20% decline in tourism in Fiji, which reduced household incomes in Vatuolalai considerably resulting in complex social and ecological changes to which people have adapted in various ways (Movono, 2017). That was a partial precedent for the pandemic.

Approaches

The Fiji Vanua Research Framework was used as a central paradigm to guide the orientation and methodology of this long-term research project as it promotes the localisation of methods, decolonisation of research and focusses on the essence of Fijian society (Nabobo-Baba, 2008). Vanua is an all encompassing Fijian term which refers directly to the land, but includes the multiple socio-cultural and ecological features within a given area, suggesting its holistic and embedded nature. This approach also dovetails with SLA and CAS theory, which encourages a bottom-up approach, placing people, their culture and internal systems at the centre of analysis. The FVRF is ideal in guiding researchers to genuinely engage with participants, is adaptive during challenging circumstances and considerate of the holistic and interconnected nature of Fijian society.

The study employed a case study methodology for its suitability and applicability in conducting in-depth investigations, within a communal setting for an extended period of more than a decade. This allowed for the study of "social processes in their appropriate context", facilitating the incorporation of adaptive ethnography within qualitative research (Hartley, 1994: 208; Movono & Dahles, 2017). The lead researcher (an indigenous Fijian) immersed himself in village life engaging in long-term relationship building, and forging close bonds with the community, a process that began in 2009 (Movono et al., 2015). This approach built community members' trust and confidence, yielding a detailed understanding of their beliefs, motivations, and behaviour.

Relying predominantly on qualitative methods, participant observations and the modification of semi-structured interviews as conversations, or *talanoa*, were the primary tools for data collection. *Talanoa* refers to a process where people converse, share ideas and stories, and where rich communication and dialogue are established between two or more parties (Nabobo-Baba, 2008). Over the 11 years of this research, more than 100 *talanoa* were held with broad segments of the community, and specifically with individuals who were approached because of their particular knowledge. Such extensive engagement has yielded information about how the

Vatuolalai community has changed throughout their initial involvement in tourism to the shock hotel closures resulting from COVID-19. Most recently, as a result of the pandemic, conversations were held in groups and with individuals including youths, young mothers, special interest groups, and community leaders, using zoom and other video calling technologies. Because of restrictions on travel, a Research Associate conducted on-site *talanoa* in the case study community, providing another layer of rich data. Data from the study were analysed manually and incrementally throughout the study period because of the large volume and richness of qualitative data. Ethical considerations have resulted in the use of pseudonyms.

Adaptive Strategies: Looking Inwards

This study adds to the current SLA and CAS discourse by highlighting that the longitudinal responses of a complex and adaptive system to tourism promote adaptive behaviour, as people respond to challenging circumstances (Neely, 2015). Although not as sudden or as abrupt as natural disasters, tourism development is profound in its ability to stimulate ecological changes and spur further cultural and societal impacts that can hinder longer term sustainability (Movono et al., 2018; Paton, 2006; Folke et al., 2003). However, under COVID-19 people have moved quite rapidly from a tourism reliant livelihoods base to one that is now more reliant on traditional systems. Elina describes that:

> The main impact experienced is the loss of the primary source of income for our family as my husband and I (both tourism workers) are on leave without pay till further notice from different hotels where we work. This applies to the community that we lived in as more than 90% of breadwinners work in the tourism industry. We have since moved back to the village, cutting down on many expenses like rent, water bills, transportation cost, takeaways (due to the hectic work schedule), and day care bill…There is much anxiety around our economic well-being because many of us worry about where our income will come from even though we have resources available because we are so used to getting paid regularly. Many have combined resources to earn income, but we have reminded ourselves to support each other in all initiatives.

Elina thus pointed to a common trend where people were returning to their villages, resulting in an immediate increase in the population of almost 100 people. They then used their extended families to take care of children and share in the responsibilities of producing food and getting by. More importantly, the statement above also shows how people were anxious about their economic well-being, because they had become reliant on money earned from tourism work to sustain themselves. However, Elina added that,

> Being raised in the village has contributed to being resilient during this pandemic phase as we were taught to live on subsistence farming and fishing. We are now enjoying healthy lifestyles by making do with what we have and being content. We are currently engaged in traditional livelihood activities that were barely practised before COVID-19. Our family has now come up with ways to counter this loss of income by continuing farming and fishing for everyday living, selling food items needed daily like sugar, rice, dhal at my aunts' canteen, and other groceries at an affordable price. We are also selling Fijian delicacies for lunch at Sigatoka [the nearest town] Flea Market once a week.

This demonstrates the appreciation that Vatuolalai people had for their traditional knowledge and resources, which allowed them to again grow their own food and fish nearby. This provided them with a sense of security and sustenance. That nonetheless raised questions, at different scales, about the use of tourism as a tool by which long-term prosperity and sustainable development could be attained, mainly since the impacts of border closures resulted in people once primarily dependent on tourism being forced to fend for themselves. Such questions were hardly new (Rao, 2002; Harrison & Prasad, 2013).

The return to *vanua* was challenging since tourism had replaced traditional livelihood activities, along with the knowledge, skills and livelihood diversification benefits that accompany them (Scott, 1970; Ravuvu, 1987; Movono, 2017). Josefa, a village elder, reflected on the past five decades of tourism as a 'blessing and a curse' with COVID-19 now forcing people to 'appreciate' their internal systems, resulting in the community becoming more united in their motivations to cope with the pandemic. Josefa welcomes several such changes:

> This long-overdue pandemic has brought nothing wrong but has taught the people of Vatuolalai a lesson not to concentrate on tourism but to diversify, using traditional resources that are readily available. We need to think globally but act locally...I have noticed that a lot of villagers are beginning to

> utilise the availability of seafood, namely octopus, sea cucumbers, urchins, seagrapes which are healthy. COVID-19 has also increased our ability to sit in groups, [and] engage in strategic planning with village meetings now being well attended...numbers have tripled meaning that our communication network and the dissemination of information is perfect...The Church Minister has also been impressed by the increasing numbers of people present in all religious functions, showing that we are now becoming more united.

According to Josefa, the 'blessings' of COVID-19 thus extend beyond uses of traditional sources of sustenance and include improved participation in village governance, more time for community and spiritual activities, and a greater sense of social unity. As one traditional leader suggested: 'tourism must complement our way of life, rather than taking over'. The 'old' tourism model was increasingly seen as compromising family well-being: long hours of commuting to hotels or spending six weeks away from home on an island resort before having a week off seemed less than ideal for family and community life. By contrast some contemplated leaving tourism employment for something with better prospects and better wages.

Village food security efforts were being encouraged by leaders, especially as a way of motivating youths:

> Recently the youths and even the school children spend their weekends on farms with their families as a way of building food security and learning agricultural skills in line with our village plan. After conducting interviews with several women and youths, they talked about the drop in their income level or purchasing powers. Still, I heard no one complain about the consumption levels [indicating that they have enough to live on] ... The village youth group has started to work together and have planted more than 4000 taro plants and 200 plantains and bananas, which will enable the village to cater for all gatherings like funerals and marriage ceremonies. The village headman will assess individual family farming programs at the end of the year, and the winner receives two cows and a pig to encourage people to plant food to complement livelihood activities for the future.

Despite many difficulties, villagers continued to evolve and adapt by making deliberate attempts to "rescue" their way of life and improve their socio-cultural conditions under pandemic conditions. Village people have continually tried to diversify their interests to include, farming, livestock rearing and fishing to sustain life (Movono & Dahles, 2017), but now there was added urgency. Women, men and youths all reinvigorated their involvement in community projects, strengthening their communal ties and working together to build resilience and ensure survival. Thus, some participants noted that COVID-19 had been a 'blessing', giving them time to 'reflect on the realities in which they live' so that they could carve out a new way of life which would draw strength from their kinship and cultural systems. Despite financial hardships, people in Vatuolalai adapted by relearning skills and harnessing their various social, cultural and natural resources to sustain themselves during the global travel slowdown. That was embodied in a statement from Tako, a 70-year-old and respected male elder, saying that:

> I have witnessed many changes in my village, before tourism until now and I have observed that we continue to adapt to changing circumstances. I am happy now to see that people have come full circle, returning to the land, and respecting our culture, to me this shows our ability to endure just as we have for many years and the many challenges. We will continue to endure and will survive into the future.

Overall, the evidence indicates that COVID-19 is creating a plethora of social changes, capable of disrupting an unchanged post-COVID-19 social and economic order, having led people to reconnect with communal systems. Indigenous Fijian communities are novel, complex social and ecological systems that are multifaceted, interrelated, and with robust systems capable of supporting people through difficult economic times. This could provide valuable lessons for other tourism-related indigenous communities in the Pacific who must maintain their unique cultural identity while simultaneously pursuing economic development through tourism. Indeed, the retention of cultural identity is quite consistent with, and even beneficial for, tourism.

Conclusions: The Place of Vanua

Indigenous communities are not mere recipients and spectators in the face of COVID-19, but are conscious and active participants, aware of changes and deliberately adjusting their activities to cope with tourism-related stresses (Adger, 2000). Vatuolalai villagers have consistently adapted intuitively to the many changes experienced through time and, with COVID, became involved in yet another dynamic process of adjustment (Movono et al., 2018; Movono & Dahles, 2017; Movono, 2017). In many respects villagers have come full circle, from euphorically embracing tourism decades ago, and adapting to new, more modern lifestyles, to seeking ways to preserve their traditional livelihoods, many physically returning to the village and relying on past knowledge alongside customs and kinship in extended families.

Government policies rarely focus on the positive role of locally valued, indigenous systems, which in no small extent have helped people to cope. Tourism development strategies in Fiji, and elsewhere the Pacific, must expand planning and development approaches to involve communities and their socio-ecological systems as integral parts of the resilience discourse and practice (Movono et al., 2018). Attention to the complex nature of communities will allow for their situations and aspirations to be acknowledged and dynamic, and multi-levelled action taken to restore and re-establish lost elements of society or nature, where these can be of value. This could open up new opportunities for tourism to integrate perspectives in planning that complement the local adaptive strategies which sustained the people of Vatuolalai during difficult times. The people retain pride in their culture, traditions, and identity, which may enable increased synergies between public policy, tourism, and the community, thereby further improving consultation, harmonising interests, and taking specific collaborative actions to improve overall sustainability and resilience to future shocks. The people of Vatuolalai are aware of their customary resources, which were a source of strength and sustenance in the absence of tourism. In future, they will continue to adapt and react to the increasingly complex challenges, but where tourism development might be conducted to 'complement' the Fijian way of life and reduce vulnerabilities to future unforeseen perturbations. Ultimately, a post-pandemic recovery must occur in a manner that considers indigenous kinship with nature and each other as a cornerstone for building resilience in a world of increasing uncertainty and disorder.

References

Adger, W. N. (2000). Social and Ecological Resilience: Are They Related? *Progress in Human Geography, 24*, 347–364.

Allen, C. R., Angeler, D. G., Garmestani, A. S., Gunderson, L. H., & Holling, C. S. (2014). Panarchy: Theory and Application. *Ecosystems, 17*, 578–589.

Belt, Collins and Associates Ltd. (1973). *Tourism Development Program for Fiji*. United Nations Development Program/ International Bank for Reconstruction and Development.

Bhandari, P. B. (2013). Rural Livelihood Change? Household Capital, Community Resources and Livelihood Transition. *Journal of Rural Studies, 32*, 126–136.

Brondizio, E. S., Ostrom, E., & Young, O. R. (2009). Connectivity and the Governance of Social-Ecological Systems: The Role of Social Capital. *Annual Review of Environment and Resources, 34*, 253–278.

Byrne, D. (1998). *Complexity Theory and the Social Sciences: An Introduction*. Routledge.

Chambers, R. (1995). Poverty and Livelihoods: Whose Reality Counts? *Environment and Urbanisation, 7*(1), 173–204.

Chambers, R., & Conway, G. (1992). *Sustainable Rural Livelihoods. Practical Concepts for the 21st Century*. Institute of Development Studies IDS Discussion Paper 296.

Coetzee, C., Niekerk, D. V., & Raju, E. (2015). Disaster Resilience and Complex Adaptive Systems Theory. *Disaster Prevention and Management, 25*(2), 196–211.

Douglas, N., & Douglas, N. (1996). Social and Cultural Impacts of Tourism in the Pacific. In M. Hall & S. Page (Eds.), *Tourism in the Pacific: Issues and cases* (pp. 49–64). International Thompson Business Press.

Folke, C., Colding, J., & Berkes, F. (2003). Synthesis: Building Resilience and Adaptive Capacity in Social-ecological Systems. In B. Strarzomski (Ed.), *Navigating Social-ecological Systems; Building Resilience for Complexity and Change* (pp. 352–383). Cambridge University Press.

Gallopin, C. G. (2006). Linkages Between Vulnerability, Resilience, and Adaptive Capacity. *Global Environmental Change, 16*, 293–303.

Gunderson, L., & Holling, C. S. (Eds.). (2002). *Panarchy: Understanding Transformations in Human and Natural Systems*. Island Press.

Harrison, D. H., & Prasad, B. C. (2013). The Contribution of Tourism to the Development of Fiji and Other Pacific Island countries. In C. Tisdell (Ed.), *Handbook of Tourism and Tourism Economics: Analysis, New Applications, and Case Studies* (pp. 741–769). World Scientific.

Hartley, J. F. (1994). Case Studies in Organisational Research. In C. Cassell & G. Symon (Eds.), *Qualitative Methods in Organisational Research. A Practical Guide* (pp. 208–229). Sage Publications Ltd.

Holland, J. (1992). Complex Adaptive Systems Theory. *Daedalus, 121*(1), 17–30.

Holland, J. (2006). Studying Complex Adaptive Systems. *Journal of Systems Science and Complexity, 19*(1), 1–8.

Holling, C. S. (1973). Resilience and Stability of Ecosystems. *Annual Review of Ecology and Systematics, 4*, 1–3.

Holling, C. S. (2001). Understanding the Complexity of Economic, Ecological and Social Systems. *Annual Review of Ecology and Systematics, 4*, 1–23.

Morse, W. C., McLaughlin, W. J., Wulfhorst, J. D., & Harvey, C. (2013). Social-ecological Complex Adaptive Systems. *Urban Ecosystem, 16*, 53–77.

Movono, A. (2017). Conceptualising Destinations as a Vanua: An examination of the evolution and Resilience of a Fijian Social and Ecological System. In J. Cheer & A. Lew (Eds.), *Understanding Tourism Resilience: Adapting to Environmental Change* (pp. 304–320). Routledge.

Movono, A., & Dahles, H. (2017). Female Empowerment and Tourism: A Focus on Businesses in a Fijian Village. *Asia Pacific Journal of Tourism Research, 22*(6), 681–692.

Movono, A., Harrison, D., and Pratt, S. (2015). Adapting and Reacting to Tourism Development: A Tale of Two Villages on Fiji's Coral Coast. In S. Pratt. And D. Harrison (Eds.), *Tourism in Pacific Islands: Current Issues and Future Challenges*. Routledge, 100–114.

Movono, A., Dahles, H., & Becken, S. (2018). Fijian Culture and the Environment: A focus on the Ecological and Social Interconnectedness of Tourism Development. *Journal of Sustainable Tourism, 26*(3), 451–469.

Nabobo-Baba, U. (2008). Decolonising Framings in Pacific Research: Indigenous Fijian Vanua Research Framework as an Organic Response. *AlterNative: An International Journal of Indigenous Peoples, 4*(2), 140–154.

Neely, K. (2015). Complex Adaptive Systems as a Valid Framework for Understanding Community Level Development. *Development in Practice, 25*(6), 785–797.

Paton, D. (2006). Disaster Resilience: Building Capacity to Co-exist with Natural Hazards and Their Consequences. In *Disaster Resilience: An Integrated Approach* (pp. 40–67). Charles C Thomas Publisher Ltd.

Prosser, R. (1994). Societal Change and the Growth in Alternative Tourism. In E. Cater & G. Lowman (Eds.), *Ecotourism: A Sustainable Option* (pp. 19–37). Wiley.

Rao, M. (2002). Challenges and Issues for Tourism in the South Pacific Island State: The Case of the Fiji Islands. *Tourism Economics, 8*(4), 401–429.

Ravuvu, A. D. (1987). *The Fijian Ethos*. Institute for Pacific Studies, University of the South Pacific.

Scoones, I. (1998). *Sustainable Rural Livelihoods: Towards a Framework for Analysis*. U.K. Institute for Development Studies, University of Sussex.

Scoones, I. (2009). Livelihoods Perspectives and Rural Development. *The Journal of Peasant Studies, 36*(1), 171–196.

Scott, R. J. (1970). The Development of Tourism in Fiji Since 1923. *Transactions and Proceedings of the Fiji Society, 12*, 40–50.

Seruvakula, S. B. (2000). *Bula Vakavanua*. Institute for Pacific Studies, University of the South Pacific.

Shen, F., Hughey, K. F., & Simmons, D. G. (2008). Connecting the Sustainable Livelihoods Approach and Tourism: A Review of the Literature. *Journal of Hospitality and Tourism Management, 15*(1), 19–31.

Tao, T. C. H., & Wall, G. (2009). Livelihoods Approach Sustainability. *Asia Pacific Journal of Tourism Research, 14*(2), 137–352.

Wu, M. W., & Pearce, P. L. (2013). Host Tourism Aspirations as a Point of Departure for the Sustainable Livelihoods Approach. *Journal of Sustainable Tourism, 22*(3), 440–460.

CHAPTER 16

COVID-19 and Transnational Remittances in Samoa: Maintaining Family Ties in the Face of Crisis

Loïc Le Dé and Fotuosamoa Jody Jackson-Becerra

E iloa le tautai i aso e si'isi'i ai le tai po'o aso afā. (The test of a skilled navigator is in the midst of a storm)

Introduction

Remittances are the money and goods sent by migrants to their home country. In the last 30 years, remittances have reached substantial figures. In 1990 remittances sent to low- and medium-income countries reached US$ 31 billion while they represented US$ 554 billion in 2019 (World Bank, 2020). In this timeframe, the flow of remittances increased 18-fold. Nowadays, remittances represent almost four times the Official Development Assistance and amount to 30 times the funds directed to

L. Le Dé (✉) • F. J. Jackson-Becerra
Auckland University of Technology, Auckland, New Zealand
e-mail: loic.le.de@aut.ac.nz

Y. Campbell, J. Connell (eds.), *COVID in the Islands: A comparative perspective on the Caribbean and the Pacific*,
https://doi.org/10.1007/978-981-16-5285-1_16

Disaster Risk Reduction annually (Organization for Economic Co-operation & Development, 2019), even though the World Bank recognizes that these figures may only capture about 50% of the real flow of remittances. They are particularly important in small Pacific states, and especially in Samoa.

The highest remittance levels worldwide as percentage of Gross Domestic Product (GDP) are found in Small Island Developing States (SIDS), which receive on average double the remittances, as a share of GDP, of most 'developing' countries. In the Pacific Island region, Tonga and Samoa are the largest recipients of remittances with 40.7% and 17.3% as a share of GDP in 2019 respectively. About 90% of Tongan households receive remittances while 78% of Samoa households are remittance recipients (Takaneka et al., 2020). In the Pacific Island Countries (PICs) remittances are usually sent by family members in Australia, New Zealand and the United States of America (USA) who have migrated permanently, long-term or seasonally. Remittances often represent an important source of income critical to making a living and contributing to the fulfilment of basic needs such as purchasing food, upgrading housing, buying equipment and paying education fees. Remittances contribute to wealth creation, poverty alleviation, and thus play a significant role in countries like Samoa where formal social protection systems are largely absent (Brown et al., 2013).

Remittances play a particularly valuable role after emergencies, particularly in SIDS (Le Dé et al., 2013). Thus, during Cyclones Ofa and Valin 1990 and 1991 respectively, households in the Samoan village of Fusi received more remittances than in any non-disaster year (Muliaina, 2003). When a political coup occurred in Fiji in 2000, remittances increased by 43% (Prakash, 2009). More recently, a study of the five Samoan coastal villages impacted by the 2009 tsunami indicated that remittances were one of the main resources mobilised by households to overcome the disaster, offsetting the absence of formal insurance, limited savings and external aid support (Le Dé et al., 2015a). Remittances have thus often been described as a kind of self-insurance within extended families. Households receiving remittances tend to fare better in the face of crisis than non-recipients do. Overall, access to remittances implies a propensity to reduce vulnerability and a greater capacity to deal with natural hazards and disasters (Savage & Harvey, 2007).

The COVID-19 pandemic has potentially led to a remittance crisis both worldwide and in the PICs. In April 2020, the World Bank projected that remittances would fall by about 20% for the year, with a decline of 13% in the East Asia and Pacific region (World Bank, 2020). Both, long-term migrants and seasonal workers based in high-income countries have been badly impacted by the reduction of socio-economic activity and increased unemployment linked to COVID-19, potentially affecting their ability to assist their kin in time of disaster. This chapter focuses on Samoa, one of the main recipients of remittances worldwide as a share of GDP, and New Zealand, one of the main destinations for Samoan migration. It analyses the effects of COVID-19 on remittances sent by and to Samoans, including remitting behaviour and the mechanisms developed during challenging times.

Given the fast-evolving nature of the COVID-19 pandemic, we draw from a mix of secondary data including reports, peer reviewed articles and grey literature published in 2020. We also use primary data from *talanoa* conducted with Samoans in Auckland, New Zealand, to better understand how people and families were coping with the event. The concept of *talanoa* is recognised in many island nations across the Pacific region (Prescott, 2008). It is "a philosophy involving an open dialogue where people can speak from their hearts and where there are no preconceptions" (Halapua, 2003: 18). *Talanoa* were held between July and September 2020 both in person and virtually. The discussions were held in Samoan and translated into English.

A Short Background on Remittances and Disaster

The International Monetary Fund defines remittances as "household income from foreign economies arising mainly from the temporary or permanent movement of people to those economies. Remittances include cash and non-cash items that flow through formal channels, such as via electronic wire, or through informal channels, such as money or goods carried across borders"(IMF, 2009: 272). For many households in poor countries, transnational remittances constitute an important source of income contributing to the fulfilment of their daily needs notably in times of hardship. Remittances imply the migration of one or more household members, partly as a livelihood diversification strategy that offsets limited access to local resources. Those with weak entitlement to resources often use their social ties at transnational level. Migration and remittances are generally guided by different 'push' and 'pull' factors, such as the lack of employment or irregular wage-labour income, imperfect social security systems, absence of private insurance, land shortage, socio-cultural norms, family obligations, job opportunities abroad and favourable migration schemes. Such livelihood strategies may be intensified by deteriorating economic, environmental and political conditions, crises and disasters (Le Dé et al., 2013).

Unlike foreign investment, remittances are usually counter-cyclical as they tend to increase during crises. During crises remittances tend to increase substantially as migrants seek to assist their affected kin in survival and recovery, and in dealing with health treatments, getting food, clothing and other basic needs to cope. Moreover, remittances can be fast in reaching those affected, sometimes even before the arrival of international aid or government support. After the 2009 tsunami in Samoa, about 72% of remittances were sent within one week, even though there was a breakdown of communications for days (Le Dé et al., 2015a). Remittances may also be critical to re-start businesses, replace or offset losses such as cars or savings, repair or rebuild houses, pay school fees, or buy tools for agricultural production and productive activities.

Beyond their economic importance, remittances also constitute a social resource, as they involve people's social networks and are based on kinship relationships and notions of reciprocity (Levitt, 1998). Remittances may contribute to the maintenance of traditional customs, expression of cultural values, and to wellbeing and psychosocial recovery (Lee & Francis, 2009). Most studies on remittances are dominated by economic approaches and usually neglect these aspects, failing to integrate people's perceptions, feelings, and cultural meanings. After disasters, remittances may help those affected to feel supported, recover psychosocially and contribute to maintaining socio-cultural and religious practices (Leslie, 2000; Marsters et al. 2006). The role played by migrants in supporting their kin is critical. Several studies emphasised that remitting during disaster can have subsequent economic impacts on migrants such as reducing daily expenses, utilizing savings, selling assets and reallocating or requesting bank loans (Le Dé et al. 2015b). Remitting also has positive effects such as strengthening identity, social ties and culture, thus leading migrants to remit despite struggling economically. Most studies have focused on those receiving remittances, but very little attention has been paid to the perspectives of migrants.

Samoans, Transnational Ties and Remittances

Samoa is central to the Polynesian region of the Pacific and consists of two main islands, Savai'i and Upolu (Fig. 16.1). In 1962, Samoa gained independence after a period of colonisation initially by Germany between 1899 and 1915 and thereafter by New Zealand. The population of Samoa is about 200,000. The economy is centred around agriculture and fisheries with tourism of growing importance until the closure of the borders. Late in 2020 Samoa had no cases of COVID-19. As in many PICs, remittances play a significant role in the economy and represent a considerable source of income for most families. Processes of globalization, reduced agricultural productivity, employment opportunities overseas and migration schemes are factors all contributing to the development of a migration-remittances nexus (Connell, 2010). Samoans have mainly migrated to New Zealand, Australia, the USA and American Samoa, enabling the development of what was termed in the early 1980s the "transnational corporation of kin" as households effectively redistributed their human resources (Bertram & Watters, 1986). Although in the 1960s only 10% of Samoans lived abroad, nowadays, more Samoans live overseas than in the country itself. Thus, the Samoan community is the largest Pacific ethnic

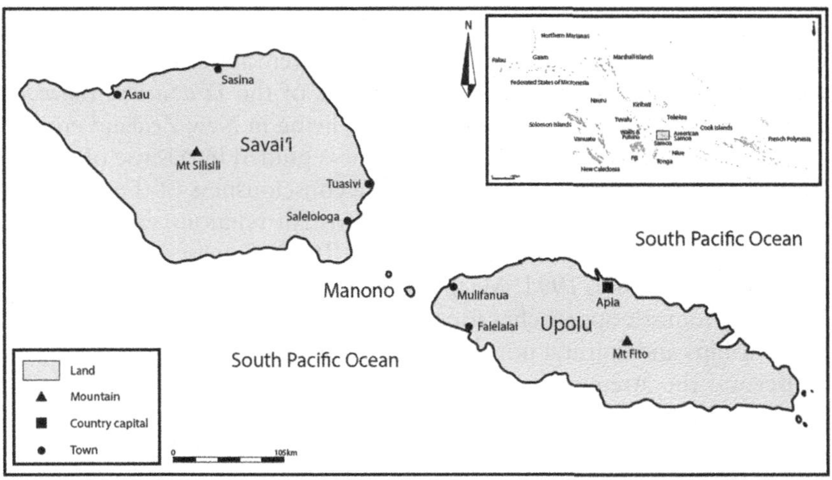

Fig. 16.1 Map of Samoa. (Source: Authors' own (2020))

group living in New Zealand, with 182,721 people identifying as Samoan at the 2018 New Zealand census.

The migration-remittances process became important from the mid-1980s and was interpreted as a way for Samoans (and others in PICs), to get maximum returns on transnational social capital (Bertram & Watters, 1986). The notion of MIRAB (Migration, Remittances and Bureaucracy) economies characterized remittances as a "comparative advantage" and migrants as "invisible exports" to high income countries; remittance behaviour was driven by economic motives of investment in the home country and asset accumulation (Poirine, 2006). In a study on Samoans and Tongans in Australia (Brown, 1997: 623) argued that "migrants are motivated to remit for reasons of self-interest, and in particular for asset accumulation and investment in their countries of origin". Subsequently Brown et al. (2014) concluded that Samoan migrants remitted because of financial obligation and family pressure.

Authors like Shankman (1976: 56) offered a softer viewpoint, arguing that Samoans were migrating to fulfil their obligations to the extended family (*aiga*) "without the immediate pressures and uncertainties of village life", and that in turn the family was encouraging migration to get some benefits, mainly remittances. Macpherson (1994) indicated the importance of social obligations towards the family as a core reason for remitting, rather than the more individualistic motives suggested by economists. More than a third of remittances are used for *fa'alavelave*, social events that encompass weddings, funerals, elections or the bestowing of chiefly (*matai*) titles, and other customary ceremonies. *Fa'alavelave* are traditional practices and customs at the core of the *fa'aSamoa (Samoan way)*. Recent research on Samoan migrants living in New Zealand emphasised that post-disaster remitting behaviour is guided by a sense of obligation or responsibility to assist the family, a consciousness of the economic difficulties experienced in Samoa, and Christian religious values (Le Dé et al., 2015b). Overall, insiders (Hau'ofa, 1993; Tamasese et al., 2010) and outsiders (James, 1991; Macpherson & Macpherson, 2009) have criticised economistic approaches for separating economic actions from social relationships and cultural norms within which remittances take place.

Beyond the attempts to understand the remitting practices of Pacific islanders, the literature has focused on issues linked to the very high reliance on remittances, questioning their sustainability and advancing the hypothesis of 'remittances decay'. This assumed that over time the number of migrants would decrease, Samoans based overseas would lose their ties with their country of origin and second generations would not remit. However, migration has continued, remittances have continued to increase, and Samoan migrants have proved reactive and effective in helping their kin especially during crises (Muliaina, 2003; Le Dé et al. 2015b).

The Impacts of COVID-19 on Samoans

Samoa declared a State of Emergency on the 21 March 2020 as a result of the COVID-19 pandemic. The State of Emergency meant that the movement of people was restricted, and public gatherings limited to five people. Events such as weddings, funerals and sporting competitions were affected, and all schools were closed. Travel between the two main islands of Savai'i and Upolu were reduced to only three sailings a week. Samoa also became one of the first countries in the world to close borders, preventing all travel to Samoa. Two days earlier, the New Zealand Government had closed its own borders to anyone who was not a New Zealand citizen or permanent resident. When these decisions were made, it had wider implications for Samoan migrants, many of whom were stranded in New Zealand while those in Samoa were unable to travel elsewhere for employment.

Samoans who have been in seasonal employment, such as the Recognised Seasonal Employer (RSE) scheme, were significantly affected by the COVID-19 pandemic when border closures came into effect. New Zealand's RSE programme allows up to 14,000 seasonal workers from the Pacific Islands to travel to New Zealand for up to seven months to work in the horticulture and viticulture industries (Bedford et al., 2020). The popularity of the RSE scheme among Samoan workers had long been evident in the number of applicants vying for limited job vacancies in each recruitment intake. The absence of new workers coming to New Zealand in 2020 led to industry leaders advocating for their return to New Zealand (Flaws, 2020). Up until November 2020, RSE workers had yet to be allowed back into New Zealand and based on the New Zealand Government's position, the immediate focus was more on ensuring New Zealanders, who had lost jobs due to COVID-19, had the chance to find new employment (Government of New Zealand, 2020). The RSE workers who were in New Zealand before the borders closed and unable to return were also affected. Their job contracts had ended, and many spent their savings that were meant to be sent home for their families to pay rents, school fees, buy food and meet other basic needs (Gilbertson, 2020;

Bailey, 2020). The same was true for those who are part of the similar Seasonal Worker Programme (SWP) in Australia (see Chap. 17).

Beyond the RSE scheme migrants, long-term Samoan migrants in New Zealand saw their ability to support their kin in Samoa significantly affected. The lockdown implemented in March/April 2020 meant that Samoans (and Pacific island migrants at large) bore the brunt of the COVID-19 outbreak through job losses and existing social and financial issues were further exacerbated by the slowing economy. In Auckland, where two thirds of Pacific people live, multiple lockdowns with the implementation of an alert system made it challenging for many households. Unemployment increased sharply with 37,000 more people losing their jobs, an increase of 32.5% since the June 2020 quarter (Stats NZ, 2020) The unemployment rate for Pacific people was higher than for all ethnic groups in New Zealand for the same timeframe. It was highest among Pacific people aged 15 to 24 years (15.3%), well above the national level of 10% for that age group (MBIE, 2020). The financial burden resulted in many high school students dropping out of studies to take up employment in order to support their families. In Auckland, the crisis exposed poverty levels of many Pacific families existing pre-COVID-19. In some parts of Auckland, it resulted in a food crisis (Foon, 2020) leading to concerns raised by Dr. Collin Tukuitonga over existing socio-economic conditions affecting Pacific people disproportionately (see Chap. 3; Ma'ia'i, 2020). COVID-19 had significant impacts on Samoan migrants, both on their own livelihoods and their ability to financially support their families and community in Samoa.

Although Samoa remained COVID-19-free, most Samoans' livelihoods were affected by the border closure and in-country restrictions. A study conducted between June and July 2020 indicated that the main income of 66% of Samoan households had declined and 50% had experienced at least one job loss as a result of the pandemic-related restrictions. Border restrictions and national measures to prevent the spread of the virus affected local communities in their ability to earn money for daily expenses, including food, health check-ups, water and electricity bills as well as school fees. By mid-2020, more than 71% of Samoan households declared that they had trouble servicing loans and debts (United Nations, 2020). The economic difficulties faced by many Samoans were particularly important since the country had barely recovered from a measles epidemic in December 2019, costing the lives of over 80 people and affecting the tourism sector which accounted for 25% of the GDP and where more than

12% of the workforce were employed. More than ever remittances were critical to counterbalance the socio-economic impacts of two crises on households in Samoa. Takaneka et al. (2020: 10) argued that "without continuous remittance flows, remittance-dependent households can fall into poverty or have difficulty meeting basic essential needs, as well as access education and health services".

Remittances in the Face of COVID-19

Although Samoan migrants faced strong socio-economic challenges, they continued sending remittances to Samoa to support families, churches, villages and, indirectly, the national economy. Data from the Central Bank of Samoa (2020) show that while remittances sent in March and April 2020 were much lower than for the same months in 2019, between April and May 2020 they increased considerably by 58% (Fig. 16.2). Remittance

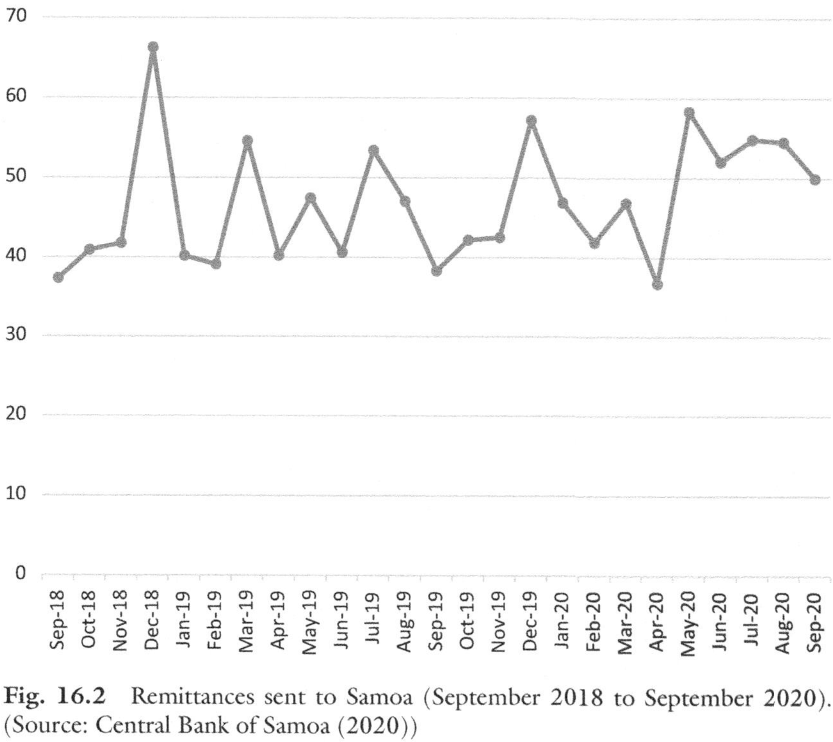

Fig. 16.2 Remittances sent to Samoa (September 2018 to September 2020). (Source: Central Bank of Samoa (2020))

levels in May 2020 were 23% higher than May 2019, reaching levels even higher than December 2019, known for being the most important month of the year for remittances. Overall the remittances sent from May to September 2020 were 19% greater than those sent for the same period in 2019. Thus, despite a decline in remittances in the early months of the pandemic (March–April), there was no sign of a sustained fall. Official records from the Central Bank of Samoa rather indicate that Samoan migrants were strongly supporting their families to overcome the COVID-19 crisis.

One explanation for this increase however was that, with borders closed and migrants unable to visit or return, more formal remittance-sending channels were used instead of physically bringing cash and goods. However, *talanoa* with Pacific people in Auckland suggest that the disruptions and challenges linked to COVID-19 had little impact on remitting behaviour or on familial and cultural ties to their motherland. Samoans in Auckland indicated that, despite the economic challenges faced, money and goods continued to be sent:

> Did COVID-19 stop the demands from our families? Absolutely not.
> COVID-19 made no difference, I still send money.

Talanoa participants emphasised that they still remitted because of the concern and care for families in Samoa. The following views are reflective of Samoans view on remitting during COVID-19 and the reasons why they assist their relatives:

> We continued to send because it's our love for the family.
> We don't think about ourselves but for our *aiga* (family) and we want to make sure they have the funds for their church donation and also to help out with maintaining our home and land.

Furthermore, Samoans highlighted the notion of *tautua* or cultural obligation to support and assist family members in Samoa:

> We always send money for the family in Samoa. Many of my family are here now but some are back home to look after our land and look after the family.
> We send our contribution because we serve our *aiga* and our village. It's our *tautua* (service) that our parents did, and we will continue this.
> Even though we are separated by the ocean but *tautua* continues. My children are successful here in New Zealand and we want them to continue their

> relationship to Samoa. So, we always send money for our *monotaga* (village contribution) and to help with the things that our family need

Samoans taking part in the *talanoa* explained that since the start of the COVID-19 pandemic there had been an increase of *e.talanoa*. These usually took place through Facetime (Facebook), Messenger, Skype and Zoom meetings connecting family members in New Zealand, Samoa and other places around the globe. Although these online discussions existed before COVID-19, they developed further and thrived after the start of the pandemic. In Australia Samoans and Tongans described these meetings as "virtual villages" (Enari & Faleolo, 2020: 116). Several Samoans labelled the COVID-19 period a time to "disengage to re-engage" and indicated that the COVID-19 isolation period was an opportunity to interact more with their family and help each other, including through remittances (Enari & Faleolo, 2020).

Samoans increasingly used different online tools not only to communicate but also to send remittances including via online shopping and services platforms (Srinivasan, 2020). The strict COVID-19 State of Emergency in Samoa meant the demand for online shopping increased considerably, as much as up to 500% (Siutaia, 2020). Online shopping and deliveries included groceries but also building materials, furniture and electrical appliances with the majority being sent to rural and remote areas. One of the remittance recipients in Samoa explained:

> It has been of great help to my family during these times of difficulties. [...] My daughter who lives in New Zealand placed the order and it was delivered the next day. She had ordered goods for daily use such as rice, soap, flour and other foodstuffs that we needed. (Siutaia, 2020)

In addition, Auckland-based companies developed a system of vouchers sent to mobile phones in Samoa so enabling households in Upolu or Savai'i to shop close to their homes. This enabled remittance recipients to buy essential needs, thus helping overcoming difficulties associated with both job losses and the restriction of movement in Samoa.

Quickly, after the COVID-19 pandemic started, governments, NGOs and international organisations both in New Zealand and Samoa developed several policies and actions to support Samoans facing the crisis, taking into consideration remittances and transnational support. The New Zealand government lowered fees for transferring remittances, enabling

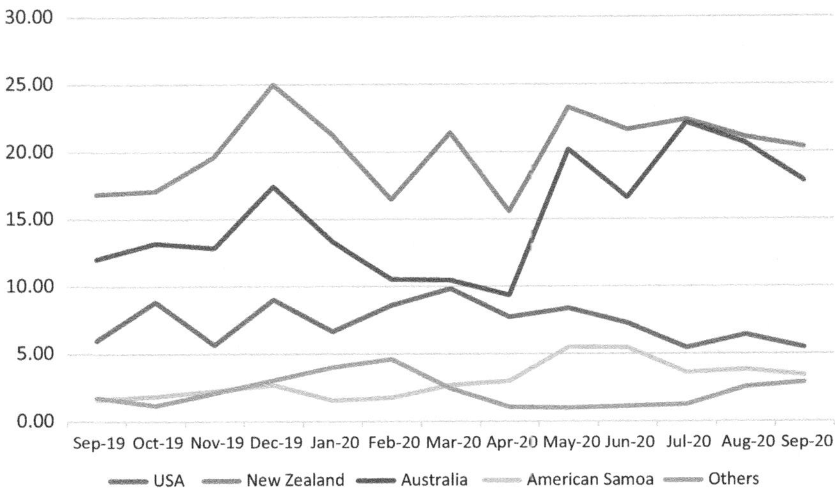

Fig. 16.3 Remittances sent to Samoa by Source (September 2019–September 2020). (Source: Central Bank of Samoa (2020))

Samoans to further assist their relatives back home (Immigration New Zealand, 2020). Different policies were developed to support the RSE workers in New Zealand, such as recognising their work as an 'essential service', providing an extension of working visas (Ackman and Taulealo, 2020) and including them as part of the Government's COVID-19 Economic Response Package. Migrant workers were also eligible for the national New Zealand wage subsidy if their employer was affected by the lockdown. In addition, RSE workers with little or no work were eligible for assistance to help meet basic needs through the Red Cross Hardship Fund. These different policies and programmes probably played a role in boosting Samoans' capacity to assist their kin in Samoa. The official records of remittances in Samoa indeed show that from May to September 2020, the increase of remittances came largely from New Zealand and Australia where similar policies were implemented (Fig. 16.3).

The Samoan government was also active in supporting the connections between Samoa and the diaspora. During the COVID-19 pandemic, the government of Samoa activated the Samoan Diaspora Unit developed during the measles outbreak in 2019 with the Prime Minister of Samoa acknowledging:

Samoans are proud people who always take pride in their heritage and culture just like most of our Pacific brother and sisters; Samoans living overseas tend to give support to national teams, family and church responsibilities and cultural obligations.

The Diaspora Unit aimed at facilitating the humanitarian contribution by Samoans abroad and in supporting their kin during crisis. Furthermore, measures by the Government of Samoa to reduce movement, public gatherings, and close borders removed financial pressures on remittance senders. The Government's intention was to lighten the burden (*fa'amāma avega*) for all families in Samoa, especially those with limited resources. Samoans in New Zealand have had to adhere to the strict rules affecting their ability to be physically present to support family members especially in cultural and social events like *fa'alavelave* (i.e. funerals, weddings, chief title change, anniversaries etc.), representing about 30% of remittances. Fa'alavelave are often hard to sustain economically but are part of Samoan culture and social organisation including *tautua*. Renowned Samoan leader and funeral director in Auckland, Tauanu'u Nick Bakulich, explained the lessening of financial pressures on families during COVID-19:

It is not uncommon for a Pasifika [Pacific islander ethnicity] family to spend over NZ$20,000 to farewell a loved one. This includes money for the funeral, the burial as well for the exchanging of gifts and the costs of catering for extended family over a two- to three-week period. [...] We've probably realised through COVID-19 that there are certain cultural practices that we don't have to put ourselves through. Some of those practices are things that have negatively impacted on us in a financial sense. (Latif, 2020)

Although there have been fewer large scale *fa'alavelave* taking place in Samoa since the start of the COVID-19 pandemic, remittance levels have kept increasing in significant proportions. This suggests that instead of being allocated to socio-cultural events, remittances were directed to meet basic needs and overcome the economic challenges faced by Samoans in the islands during the crisis. As the Christmas holiday season approached, the continued border closures meant that the usual influx of visiting family and friends to Samoa would not happen. This also meant a reduced financial burden to contribute to *fa'alavelave* ceremonies requiring financial and goods donations.

Disaster and Transnational Community Response: Beyond COVID-19

The COVID-19 pandemic had significant socio-economic impacts on Samoans in the islands and abroad. After dropping in March and April 2020 when borders were closed and lockdowns implemented, remittance levels increased noticeably and remained higher than usual. These remittance trends in Samoa were consistent with global trends, indicating the widespread support for kin during COVID-19. While at the beginning of 2020 the World Bank predicted a 20% remittances' fall (World Bank, 2020), it subsequently acknowledged that "remittances have held steady and, in some cases, even gone up" (Smith, 2020). This confirms that during disasters remittance levels increase substantially as migrants do all they can to assist their relatives affected by such events.

Despite the challenging situation, diasporic Samoans have demonstrated that socio-cultural connections between families and communities transcend borders and hardships. Historically, the Samoan diaspora has always been quick to assist their families affected by disasters, and often much more promptly than governmental and international agencies. For example, after the 2009 tsunami, Samoan families based overseas responded massively to support those affected through both individual and collective remittances (Le Dé et al., 2015a). The tsunami occurred when the 2008 financial crisis was still affecting high income countries around the globe, making it very difficult for Samoans based overseas to send remittances. Nevertheless, Samoan migrants mobilised to assist their kin, even though it sometimes meant contracting loans, getting into debt and placing themselves in a vulnerable position (Le Dé et al., 2015b). It is therefore not surprising that despite the difficulties faced during the COVID-19 pandemic, Samoans kept helping each other through remittances, even though this might result in short and long-term financial struggles.

Beyond the obvious financial assistance, the emotional and psychological support associated with remittances was crucial during the COVID-19 pandemic. Despite border closures and movement restrictions, Samoan communities (and Pacific Island communities at large) maintain strong linkages at the transnational level, exchange resources and information, and participate in social activities. Exchanges took place online through Zoom, Facetime or Messenger, highlighting both the strengths and creativity of Pacific islanders in overcoming the barriers, limitations and difficulties associated with the crisis. Samoans have remitted because of the love for their families, duty to help and religious values, which is in line with other studies on remittances and disaster (Tamasese et al. 2010; Thornton et al., 2010; Le Dé et al., 2015b). The sense of community and pride in nationhood (*loto nu'u*) is strong among Samoans whose values are based on *tautua*, which is service for one's family, village, church and nation and *fa'aaloalo* (respect), *fa'atuatua* (faith) and love (*alofa*).

Samoa, and PICs in general, are generally regarded as highly vulnerable because of their small size, remoteness, small economies and limited resources. However, Samoa has a long history of dealing with external shocks and stresses, and its populations have always displayed mechanisms to both cope with disasters and adapt to changes. These mechanisms are largely based on social ties, and norms of reciprocity and mutual help, which in a globalized world have taken the form of transnational remittances. The COVID-19 pandemic is one of the many disasters recurrently experienced in Samoa and another illustration of Samoans' capacities to proactively respond through transnational community ties. The impacts of COVID-19 are not over, and the socio-economic and health effects may last long. Samoans may struggle to recover and the efforts linked to remitting post crisis will surely create extra financial pressure on Samoan migrants. However, the COVID-19 has, once again, highlighted the resourcefulness of Samoans in the face of disaster, and that too will continue.

References

Ackman, M., & Taulealo, T. (2020, April 15). *COVID-19 and Pacific Labour*. Lowy Institute. https://www.lowyinstitute.org/the-interpreter/covid-19-and-pacific-labour

Bailey, R. (2020, July 21). *Stuck in New Zealand: The Experiences of Pacific Seasonal Workers* [Web Log Post]. Retrieved from https://devpolicy.org/stuck-in-new-zealand-the-experiences-of-pacific-seasonal-workers-20200721/

Bedford, R., Nunns, H., & Bedford, C. (2020, November 16). *Participation in the RSE Scheme: The Myth of Opportunity*. Development Policy Centre. . Available at: https://devpolicy.org/participation-in-the-rse-scheme-the-myth-of-opportunity-20201020-1/

Bertram, I. G., & Watters, R. (1986). The MIRAB process: Earlier Analyses in Context. *Pacific Viewpoint, 27*(1), 47–59.

Brown, R. P. C. (1997). Do Migrants' Remittances Decline Over Time? Evidence from Tongans and Western Samoans in Australia. *The Contemporary Pacific, 10*(1), 107–151.

Brown, R. P. C., Connell, J., & Jimenez-Soto, E. V. (2013). Migrants' Remittances, Poverty and Social Protection in the South Pacific: Fiji and Tonga. *Population, Space and Place, 20*(5), 434–454.

Brown, R.P.C., Leeves, G.D., and Prayaga, P. (2014) Sharing Norm Pressures and Community Remittances: Evidence from a Natural Disaster in the Pacific Islands, *Journal of Development Studies, 50*(3), 383–398.

Central Bank of Samoa. (2020). *Visitor Earnings and Remittances.* Retrieved from: https://www.cbs.gov.ws/index.php/statistics/tourism-earnings-and-remittance/

Connell, J. (2010). Pacific Islands in the Global Economy: Paradoxes of Migration and Culture. *Singapore Journal of Tropical Geography, 31*(1), 115–129.

Enari, D., & Faleolo, R. (2020). Pasifika Collective Well-being During the COVID-19 Crisis: Samoans and Tonga in Brisbane. *Journal of Indigenous Social Development, 9*(3), 110–126.

Flaws, B. (2020, September 12). 'This just cannot happen': $9.5 Billion at Risk as Horticulture Sector Struggles to Fill $25-an-hour Jobs. Stuff.co.nz. Available at: https://www.stuff.co.nz/business/farming/122741361/this-just-cannot-happen-95-billion-at-risk-as-horticulture-sector-struggles-to-fill-25anhour-jobs. Accessed 17 Oct 2020.

Foon, E. (2020). Pasifika Need Extra Help Dealing with Covid's Harsh Reality – Councillor. *Radio New Zealand.* Available at: https://www.rnz.co.nz/news/national/424205/pasifika-need-extra-help-dealing-with-covid-s-harsh-reality-councillor

Gilbertson G. M. (2020, June 25). *Thousands of Seasonal Workers Stranded in NZ as Work Runs Out*, Stuff.co.nz. Available at: https://www.stuff.co.nz/business/farming/agribusiness/121931976/thousands-of-seasonal-workers-stranded-in-nz-as-work-runs-out. Accessed 17 Oct 2020.

Government of New Zealand. (2020). *Seasonal Work Visa Available to More People.* Available at: https://www.beehive.govt.nz/release/seasonal-work-visa-available-more-people. Accessed 17 Oct 2020.

Halapua, S. (2003). *Walking the Knife-Edged Pathways to Peace.* Available at: http://www.pireport.org/articles/2003/07/08/. Accessed 07 Dec 2020.

Hau'ofa, E. (1993). Our Sea of Islands. In E. Waddell, V. Naidu, & E. Hau'ofa (Eds.), *A New Oceania: Rediscovering Our Sea of Islands* (pp. 2–16). University of the South Pacific School of Social and Economic Development.

Immigration New Zealand. (2020). *Key Updates for Recognised Seasonal Employers (RSEs) in New Zealand.* Available at: https://www.immigration.govt.nz/about-us/covid-19/recognised-seasonal-employers-rse-covid-19-information. Accessed 17 Oct 2020.

International Monetary Fund (IMF). (2009). *Balance of Payments and International Investment Position Manual Sixth Edition (BPM6)*. IMF.

James, K. (1991). Migration and Remittances: A Tongan Village Perspective. *Pacific Viewpoint, 32*(1), 1–23.

Latif, J. (2020, September 1). How Covid Is Upending Time-honoured Traditions for Māori and Pasifika families'. *The Spinoff.* Available at: https://thespinoff.co.nz/society/01-09-2020/how-covid-is-upending-time-honoured-traditions-for-maori-and-pasifika-families/. Accessed 20 Oct 2020.

Le Dé, L., Gaillard, J. C., & Friesen, W. (2013). Remittances and Disaster: A Review. *International Journal of Disaster Risk Reduction, 4*(1), 34–43.

Le Dé, L., Gaillard, J. C., Friesen, W., Pupualii, M., Brown, C., & Aupito, A. (2015b). Our Family Comes First: Migrants' Perspectives on Remittances in Disaster. *Migration and Development, 5*(1), 130–148.

Le Dé, L., Gaillard, J. C., Friesen, W., & Smith, F. M. (2015a). Remittances in the Face of Disasters: A Case Study of Rural Samoa. *Environment, Development and Sustainability, 17*(3), 653–672.

Lee, H., & Francis, S. T. (2009). Pacific Migration and Transnationalism: Historical Perspectives. In H. Lee & S. T. Francis (Eds.), *Migration and Transnationalism Pacific Perspective* (pp. 7–41). ANU E Press.

Leslie, H. Y. (2000). *Inventing Health: Tradition, Textiles and Maternal Obligation in the Kingdom of Tonga*. Unpublished PhD thesis, York University.

Levitt, P. (1998). Social Remittances: Migration Driven Local-level Forms of Cultural Diffusion. *International Migration Review, 32*(4), 926–948.

Ma'ia'i, L. (2020, August 18). 'Devastating impact': South Auckland's Pasifika Bear Brunt of New Covid-19 Outbreak. *The Guardian.* https://www.theguardian.com/world/2020/aug/19/a-devastating-impact-on-our-people-south-aucklands-pasifika-carry-the-weight-of-new-covid-19-outbreak

Macpherson, C. (1994). Changing Patterns of Commitment to Island Homelands: A Case Study of Western Samoa. *Pacific Studies, 17*(3), 83–116.

Macpherson, C., & Macpherson, L. (2009). Kinship and Transnationalism. In H. Lee & S. Tupai Francis (Eds.), *Migration and transnationalism: Pacific perspectives* (pp. 73–90). ANU E-press.

Marsters, E., Lewis, N., & Friesen, W. (2006). Pacific Flows: The Fluidity of Remittances in the Cook Islands. *Asia Pacific Viewpoint, 47*(1), 31–44.

Ministry of Business, Innovation and Employment. (2020). *Pacific Peoples Labour Market Trends.* Available at: https://www.mbie.govt.nz/dmsdocument/11845-pacific-peoples-in-the-labour-market-june-2020-quarter-pdf

Muliaina, T. (2003). Remittances, the Social System and Development in Samoa. In R. Iredale, C. Hawksley, & S. Castles (Eds.), *Migration in the Asia Pacific: Population, Settlement and Citizenship Issues* (pp. 259–272). Edward Elgar.

Organization for Economic Co-operation & Development (OECD). (2019). *Aid by DAC Members Increases in 2019 with More Aid to the Poorest Countries.* Retrieved from https://www.oecd.org/dac/financing-sustainable-development/development-finance-data/ODA-2019-detailed-summary.pdf

Poirine, B. (2006). Remittances Sent by a Growing Altruistic Diaspora: How Do They Grow Over Time? *Asia Pacific Viewpoint, 47*(1), 93–108.

Prakash, N. (2009). *The Development Impact of Workers' Remittances in Fiji.* Thesis Presented in Partial Fulfilment of the Requirements for the Degree of Master of Arts at Massey University.

Prescott, S. M. (2008). Using talanoa in Pacific Business Research in New Zealand: Experiences with Tongan Entrepreneurs. *AlterNative: An International Journal of Indigenous Scholarship, 4*(1), 127–148.

Savage, K., & Harvey, P. (2007). *Remittances During Crises: Implications for Humanitarian Response.* Overseas Development Institute, Humanitarian Policy Group Report 25.

Shankman, P. (1976). *Migration and Underdevelopment. The Case of Western Samoa.* Westview Press.

Siutaia, H. (2020). S.O.E Increases Online Shopping Orders by 500 Per cent. Available at: https://www.samoaobserver.ws/category/article/61798. Accessed 6 Dec 2020.

Smith, S. (2020). *The Great Remittance Mystery.* (Online). Npr.org. Available: https://www.npr.org/2020/10/23/927249563/the-great-remittance-mystery. Accessed 28 Sept 2020.

Srinivasan, P. (2020). *Samoans Around the World Buy Groceries Online to Support Families Back Home During COVID 19 Emergency.* [online] Available at: https://www.abc.net.au/radio-australia/programs/pacificbeat/pacific-online-groceries/12152644. Accessed 6 Dec 2020.

Stats New Zealand. (2020). *Unemployment Rate Hits 5.3 Percent Due to COVID-19.* Available at: https://www.stats.govt.nz/news/unemployment-rate-hits-5-3-percent-due-to-covid-19#:~:text=In%20the%20September%20 2020%20quarter,since%20the%20June%202020%20quarter. Accessed 19 Oct 2020.

Takaneka, A. K., Gaspar, R., Villafuerte, J., & Narayanan, B. (2020). *COVID-19 Impact on International Migration, Remittances, and Recipient Households in Developing Asia.* Asian Development Bank Briefs No. 148.

Tamasese, K., Parsons, T., Sullivan, G., & Waldegrave, C. (2010). *A Qualitative Study into Pacific Perspectives on Cultural Obligations and Volunteering.* Pacific Section and The Family Centre Social Policy Research Unit.

Thornton, A., Kerslake, M., and Binns, T. (2010). Alienation and obligation. Religion and social change in Samoa. *Asia Pacific Viewpoint, 51*(1), 1–16.

United Nations (2020). *COVID-19 Socio-economic Response Plan.* Available at: https://unsdg.un.org/resources/covid-19-socio-economic-response-plan-samoa. Accessed 28 Sept 2020.

World Bank. (2020, April 22). *World Bank Predicts Sharpest Decline of Remittances in Recent History* [Press Release]. Available at: https://www.worldbank.org/en/news/press-release/2020/04/22/world-bank-predicts-sharpest-decline-of-remittances-in-recent-history. Accessed 10 Oct 2020.

CHAPTER 17

Pacific Labour Mobility on Pause: Consequences of Temporary Immobility During the Pandemic

Kirstie Petrou, Olivia Dun, Carol Farbotko, and Taukiei Kitara

Introduction

At the beginning of 2020, labour mobility of Pacific Islanders to Australia and New Zealand was steadily increasing. In 2018–19 approximately 11,000 Pacific Islanders travelled to New Zealand through the Recognised Seasonal Employer (RSE) scheme to meet seasonal horticultural labour needs. Australia's analogous Seasonal Worker Program (SWP) had over 12,000 participants in 2018–19, a number that was projected to grow as

K. Petrou (✉)
Flinders University, Bedford Park, SA, Australia
e-mail: kirstie.petrou@utas.edu.au

O. Dun
School of Geography, University of Melbourne, Parkville, VIC, Australia
e-mail: olivia.dun@unimelb.edu.au

Y. Campbell, J. Connell (eds.), *COVID in the Islands: A comparative perspective on the Caribbean and the Pacific*,
https://doi.org/10.1007/978-981-16-5285-1_17

places in the scheme had recently been uncapped (Lawton, 2019). Australia's newly launched Pacific Labour Scheme (PLS) was taking off, and provides Pacific Islanders with employment opportunities in industries such as meat processing and aged care, on two to three year-long contracts. There was a general sense that participation in these three Pacific labour mobility (PLM) schemes would only continue to expand, and Pacific Islanders were keenly seeking out these opportunities. Yet by early 2020, the looming coronavirus pandemic brought PLM to a sudden and quite unexpected halt. By the end of March, and with little warning, Australia and New Zealand had closed their borders to international arrivals to prevent the spread of COVID-19. In Pacific Island Countries (PICs) too, borders swiftly closed and participation in the regional labour mobility schemes was suspended. PLM was no longer possible, and COVID-19 travel restrictions meant thousands of Pacific seasonal workers found themselves either stuck in Australia and New Zealand, or unable to travel to participate in the schemes as planned. It is the experiences of these seasonal workers that provide the focus of this chapter.

Border closures, both international and subnational, have had various implications for home and host countries involved in PLM. In Australia and New Zealand, international seasonal workers – Pacific Islanders, backpackers and others – account for a significant proportion of the horticultural labour force (Dufty et al., 2019; Bedford, 2020a). A report commissioned by Horticulture Innovation Australia predicted that in 2020–21, COVID-19 travel restrictions within and beyond Australia would lead to a horticultural labour shortfall of approximately 26,000 jobs. New Zealand is similarly facing significant labour shortages due to the pandemic. In addition, state border closures within Australia meant international seasonal workers still in Australia faced difficulties travelling interstate to access work. Suddenly, the importance of international seasonal workers was brought into sharp relief.

C. Farbotko
University of the Sunshine Coast, Sippy Downs, QLD, Australia
e-mail: carol.farbotko@csiro.au

T. Kitara
Griffith University, Brisbane, QLD, Australia
e-mail: john.connell@sydney.edu.au; taukiei@gmail.com

For participating PICs, the impacts of border closures were felt at various scales. PLM can provide an important source of income at the national level. For example, in Tonga, one of the largest labour supplying nations, earnings from PLM contribute more to Gross Domestic Product (GDP) than aid and trade combined (Howes & Orton, 2020). Individuals and households too had come to rely upon income from PLM. COVID-19 travel restrictions meant that a significant source of employment had – for the time being – disappeared for some households, with implications for household wellbeing. Drawing on interviews with Pacific diaspora members in Australia and New Zealand who are connected to PLM workers in varied ways and with SWP workers from Solomon Islands and Kiribati, this chapter examines what the pandemic meant for current and potential PLM workers. We argue that, while COVID-19 presented opportunities for a small number of Pacific seasonal workers to earn more than expected, it also exacerbated longstanding issues around access to enough work, isolation from local communities and the social costs of family separation.

A Brief History of Contemporary Pacific Labour Mobility

Interest in establishing the contemporary PLM programmes was sparked by the 2006 UN High-Level Dialogue on International Migration and Development which examined the advantages of temporary and circular migration (Bedford et al., 2017). The New Zealand RSE scheme was established shortly thereafter, in 2007, and responded to long-running labour shortages in the New Zealand horticulture sector. Modelled on the RSE, Australia established the Pacific Seasonal Worker Pilot Scheme (PSWPS), which operated between August 2008 and 30 June 2012, subsequently becoming the SWP. Both the RSE scheme and SWP aim to fill seasonal labour shortages in industries including horticulture (Australia and New Zealand), viticulture (New Zealand), agriculture, and, to a lesser extent, tourism and accommodation (Australia). Following the success of the SWP, in 2018, Australia introduced the PLS to enable rural and regional employers to engage workers for up to three years in roles not filled by Australians. Unlike the SWP and RSE, PLS employment does not need to be seasonal, and common industries include aged care, agriculture, hospitality and tourism, aquaculture, and meat processing. All three schemes aim to contribute to economic development in labour sending countries through workers sending remittances and gaining new skills, knowledge and employment experience. Participation in the schemes had been increasing each year (Lawton, 2019) until the COVID-19 pandemic.

The RSE, SWP and PLS are open to the Pacific Island countries of Fiji, Kiribati, Nauru, Papua New Guinea (PNG), Samoa, Solomon Islands, Tonga, Tuvalu and Vanuatu. While they are culturally diverse, these small island states have high domestic unemployment levels and growing youth dominated populations. The PLM schemes provide a valuable opportunity to access employment and earn much higher incomes over a short period than would be possible in PICs. Under normal circumstances, SWP contracts last for 6–9 months, while RSE contracts are 7–9 months duration within any 11-month period. However, during the COVID-19 pandemic, travel restrictions meant many workers were unable to return home, and their stay in Australia and New Zealand exceeded these time frames.

Temporary labour mobility schemes are often promoted as a 'triple (sometimes quadruple) win', as they aim to benefit employers, the workers themselves, and their communities/countries of origin. Service providers in host countries, such as transport operators and accommodation providers, are sometimes cited as a fourth beneficiary (Underhill-Sem & Marsters, 2017). Despite these 'wins', the PLM schemes have been periodically plagued with negative issues common to temporary labour mobility programmes (Costa & Martin, 2018), such as inadequate accommodation, insufficient work, lack of access to permanent residency, restrictions on freedom of movement, vulnerability to exploitation, and worker injuries and deaths (Underhill-Sem & Marsters, 2017; Petrou & Connell, 2018; Stead, 2019). While gaps in PLM research remain, existing analysis of PLM also reveals positive outcomes such as the ability to save money and send remittances, poverty alleviation, education, housing and small business investments, and financially supporting extended family members (World Bank, 2018; Bedford et al., 2020).

Pacific Labour Mobility and Immobility: COVID-19-Induced Changes

The following tables provide a rough timeline of changes brought about by the COVID-19 pandemic (Table 17.1) that impacted some of the PLM workers who we spoke with, and details around repatriation

Table 17.1 Timeline of COVID-19 pandemic impacts on PLM in Australia and New Zealand, March–November 2020

Date	*Events impacting PLM workers*
18th March 2020	RSE workers are entitled to New Zealand government funding if they fall sick, are required to self-isolate or if they cannot work because the employer's business is being affected by lockdown measures
20th March 2020	Australia and New Zealand close country borders to all non-residents
24th March 2020	South Australia closes state borders
2nd April 2020	New Zealand extends RSE visas until 25th September 2020
4th April 2020	Australia announces that SWP workers already in Australia will be allowed to continue working in agriculture
May 2020	7000 SWP workers remain in Australia and 9300 RSE workers remain in new Zealand[1]
8th July 2020	New South Wales-Victoria border closes. 'International seasonal workers' are considered 'high-risk' and are not eligible for a permit to cross the border, although permanent residents and citizens undertaking seasonal work are eligible. Restriction challenged by farmers. New Zealand announces that stranded RSE workers who did not have a current RSE employment agreement may work part-time (minimum 15 hours per week) in any industry. Qualifying workers will be issued a limited visa that is valid until 30th October 2020.
18th August 2020	New Zealand extends RSE visas until 31st December 2020
3rd September 2020	First group of 160 ni-Vanuatu SWP workers arrive in Australia since border closures, for mango picking trial in Northern Territory
9th October 2020	151 Tongan SWP workers arrive in Queensland. Australian government confirms that all 10 SWP countries had accepted an invitation to re-start SWP recruitment.
13th October 2020	Second group of 160 ni-Vanuatu SWP workers arrive for mango picking in Northern Territory
15th October 2020	Western Australia announces plan for mobilizing seasonal workers from Pacific Islands and Timor-Leste
16th October 2020	Government of Samoa announces pool of 2000 workers ready to work in New Zealand.
27th November 2020	172 Fijians and 175 Solomon islanders arrive in Brisbane to participate in the SWP and PLS. New Zealand announces that 2000 RSE workers from the Pacific will arrive in the country in early 2021

Source: Doan et al. (2020) and IOM (2020)

opportunities for 'stranded' workers (Table 17.2). This summarises the indicative impacts that restricted worker mobility. Given the constantly shifting situation, information in these tables is not necessarily complete.

Methodology

Data for this chapter come from two different sources. Between June and August 2020 Kirstie conducted semi-structured interviews over the phone and internet with 24 Pacific Islanders living in Australia and New Zealand. These diaspora members were in contact with RSE, SWP and/or, to a lesser extent, PLS workers, and were either involved in providing pastoral care to workers; had relatives working in the schemes; and/or, sat on seasonal worker advisory boards. Some diaspora members had been in limited contact with seasonal workers since the pandemic began, while others were involved in providing practical support to workers either in person, over the phone or via social media. Diaspora members represented all of the PICs that participate in the PLM schemes except for Nauru. Interviewees were located in four Australian states (Victoria, New South Wales, Queensland and South Australia), Auckland and Dunedin (New Zealand). Themes covered in interviews included the perceived social and economic impacts of the pandemic on PLM workers.

Data gleaned from the diaspora interviews are not firsthand accounts from PLM participants themselves; however, this is in some ways beneficial. PLM workers are often reluctant to voice negative opinions about the schemes for fear that this might jeopardise their own participation or that of their community (Bedford et al., 2020; Tazreiter et al., 2016; Smith, 2016). Pressure also comes from home countries and within the PICs themselves, officials have occasionally blacklisted entire villages when workers have behaved 'badly'. Indeed, one Samoan informant told Kirstie 'Every time an RSE worker speaks, they know that their reputation and their village and their family would be affected'; many are reluctant to complain. In this respect, including the opinions and observations of diaspora members, who are unlikely to face the same barriers to voicing opinions about the schemes, provides an invaluable perspective.

To complement these diaspora interviews, we draw upon interviews that Olivia and Taukiei conducted with three i-Kiribati SWP participants who had been in Australia throughout the COVID-19 pandemic (interviewed in June and November 2020). Olivia additionally interviewed the Australian-based Labour Mobility Coordinator for Solomon Islands and

Table 17.2 Repatriation opportunities and conditions for returning to PICs as at October 2020

PIC	Border status and flight availability, as at October 2020	Repatriation possible?	Health requirements to enter country
Fiji	Closed to non-Fijian citizens except for entry by sea. One international flight per week between Fiji and Auckland limited to 150 passengers.	Yes	Fijian citizens are encouraged to return a negative COVID test prior to repatriation. Fourteen days quarantine required upon arrival.
Kiribati	Borders closed to all passengers until December 2020 with exceptions for cargo, fuel, medical, food and humanitarian supplies. Currently no commercial pathways are available for return to Kiribati.	No	When borders reopen fourteen days quarantine will be required. Kiribati has 20 rooms available for quarantine/treatment and 60 beds available for group repatriation.
Papua New Guinea (PNG)	Reduced commercial flights available from Brisbane and Auckland, but it is difficult to secure a seat for the Auckland flight.	Yes	Proof required of negative COVID-19 test completed within seven days prior to travel. PNG government funded quarantine is available for RSE workers.
Samoa	Borders closed except to returning citizens and residents. A one-off repatriation flight from New Zealand occurred on 18th September 2020 (170 RSE workers). Further repatriation opportunities expected in October.	Limited	Upon arrival hard copy evidence of a negative COVID-19 test result dated within three days of departure for Samoa and a letter from a GP confirming no flu-like symptoms must be presented. Returnees must spend fourteen days in managed isolation.
Solomon Islands	Borders closed. No regular scheduled commercial flights in operation between Honiara, New Zealand and Australia.	Limited	Two negative COVID-19 test results (two and ten days prior to arrival) must be presented prior to departure for Solomon Islands. Medical declaration form must be completed five days prior to departure. Health screening occurs on arrival in Honiara and fourteen days quarantine required. Three negative COVID-19 tests required before leaving quarantine.

(*continued*)

Table 17.2 (continued)

Tonga	Borders closed to non-Tongan citizens/permanent residents. Most international flights were cancelled. Limited repatriation opportunities for RSE/SWP workers via charter and commercial flights.	Limited	Quarantine capacity is approx. 150 people. Must obtain a negative COVID-19 test four working days prior to departure. Fourteen days quarantine is required upon arrival in Tonga followed by one week of self-isolation.
Tuvalu	Borders closed to all vessels and aircraft, with exceptions for cargo, fuel, medical, food and humanitarian supplies, and other public interest grounds. Some flights specifically for repatriation of Tuvaluan citizens were permitted, however no commercial flights were available.	Limited	Quarantine capacity is 90 people. Fourteen days isolation required upon arrival, pre-departure testing and possibly pre-departure isolation will be required.
Vanuatu	All ports of entry closed. All flights require exemptions. Fortnightly freight flights from New Zealand can accommodate up to 80 passengers. Repatriation has included New Zealand military charter flights.	Limited	Must spend fourteen days in government approved quarantine upon arrival.

four Solomon Islander SWP participants who were due to be mobilised to Australia in May 2020. These interviews were conducted in Australia in August and October 2019, and remotely in April and May 2020. We combine this data with the diaspora interviews to provide different viewpoints and qualitative data insights about the impact of the COVID-19 pandemic on PLM workers. Our data is not intended to provide a representative sample of all PLM workers' experiences during the pandemic, rather it represents a rich, qualitative account that explores some of the key issues.

Impacts on Workers

Employment Impacts: Prospective Workers

Unsurprisingly, in 2020, the inability to travel (both internationally and subnationally) created issues around access to the PLM schemes. Due to travel restrictions, workers who were meant to arrive in Australia and New Zealand after late March 2020 were unable to depart as planned. The four Solomon Islander SWP workers interviewed were due to travel to Australia in May 2020. All four were eager to participate in the SWP again as soon as it was possible. As Ramsey explained: '*I was very desperate to return back [to* Australia*]...I was looking, if only they could allow us to return...I would be very happy...to go back to work [in* Australia*]*'. They wanted to return as the benefits of participating in the SWP had already helped them to improve their lives (Dun et al., 2020) and they were eager to continue with their plans (building a new house, establishing a commercial farm, starting a new small business) that additional income earned in Australia would enable.

The four Solomon Islander workers were in their first year of SWP participation in 2019. When interviewed in April and May 2020, they had already begun looking for other means to earn an income in Solomon Islands. Their pre-existing social networks and increased assets and skills gained from their time in the SWP during 2019 were helping them to adjust to the new circumstances in Solomon Islands as is demonstrated by the experience of the following three workers:

> Benjamin had begun a new job search stating that '*due to this sick[ness] COVID-19, Solomon Island every company locked down too. That's the big problem*'. He had recently successfully secured a job as a driver for a new hydropower project in Honiara based on his prior experience working as a driver explaining '*I've got the forklift licence from* Australia'. Benjamin had obtained this licence in 2019 to do citrus packing shed work in the Sunraysia region and emphasised '*It's helped so much to me, because I saw the licence, and the [hydropower] company see and they asked me, "Where did you get this licence?" So I said, "I've been working in* Australia *last year, I do forklift there." So the licence helped so much.*'
> Cathy, a young single mother, when asked if she could possibly go back to working in a job she held prior to participating in the SWP, explained "*I don't know because some of the employees have been [made] redundant and [are] going back to the province.*" Cathy had applied to her Australian superannuation company to access her funds so that she could use them to start her own business, a retail shop to sell "*general goods like* rice, *biscuits and all this. Just a small shop. That's what I'm thinking to do with my* money *when I receive it from the superannuation fund if I'm not going back – we are not going back to* Australia".

> Since the COVID-19 situation emerged, Pauline said she '*Just stay at home and…just listen out if anything happens*'. She had been looking after her four children and explained she was teaching them '*every evening*…I *bought them some [learning] material*' as their school had not provided any material. Pauline asserted '*I really miss. I'm not happy*' when she found out she could not return to participate in her second SWP season in Australia. She was waiting to see how the COVID-19 situation evolved and planned, if unable to return to Australia, to rely on the new taxi business she had started with the car she had been able to purchase with her 2019 SWP earnings.

While some 'cancelled' workers successfully drew upon skills and savings from previous PLM work, others experienced difficulties securing employment in home countries. Julia lived in New Zealand and had been in contact with Tongan RSE workers, and observed that some Tongan workers who could not travel were unable to find alternate work in Tonga:

> *So they [cancelled workers] talked about how difficult it is for them to try and gain employment in Tonga because their only work experience is seasonal work, so it's hard for them to try and pick up another employment in Tonga because of that. So a lot of them have been planning for when they're here and they had financially planned that while being here in NZ they could cover a, b, c of their expenses. They can't do that now.*

While PLM provides an important source of income for Pacific Islanders, some cancelled workers successfully engaged in other livelihood activities. For the Solomon Islanders who were only in their first couple of years of PLM participation, prior PLM work provided some extra human and financial capital that acted as a buffer to transition to other work as the COVID-19 pandemic took hold (with social networks from employment prior to their PLM participation still intact). However, they were still keen to return to Australia to work as soon as permitted. Our limited data indicate that these 'positive' impacts were not uniform. For some 'cancelled' Tongan workers, the inability to participate in PLM had negative financial repercussions and, as we discuss below, plans and project were consequently put on hold.

Employment Impacts: Current Workers

One of the longstanding difficulties around PLM schemes has been the potential for fluctuating hours of employment. Seasonal work is, by its very nature, unpredictable and Pacific Islanders have long expressed frustration when weather or other conditions mean they are working fewer hours than expected (e.g. Holani, 2017; Petrou & Connell, 2018). Many PLM workers who travelled to meet 'summer season' (roughly November–April) labour requirements found themselves also working the 'winter season' (roughly May–October), a time of year with quite different labour needs. In New Zealand, due to limited repatriation flights during the pandemic, there was an estimated oversupply of more than 3000 RSE workers during the 2020 winter season (Bedford, 2020b). In order to combat these issues, both Australia and New Zealand allowed the (controlled) movement of workers to new employers and regions.

There was some variability in seasonal workers' ability to access work during the pandemic, and while some were fully employed, others were not (IOM, 2020). The diaspora members who Kirstie spoke with stated that for the most part, seasonal workers were still working. However, access to 'enough' work varied. Richard lived in Auckland and reported that to prevent redundancies, i-Kiribati RSE workers along with their non-RSE co-workers were working four- rather than five-day weeks, with a resultant drop in income. Diaspora members also observed that some PLM workers had been unable to work or faced reduced hours when packing and warehousing facilities were closed due to COVID-19 restrictions. In both Australia and New Zealand, the slow bureaucratic processes required to approve workers' movement to a new employer meant some workers faced several weeks with no work. Many PLM workers relied on their savings to get through these periods of no or reduced work and, as we discuss in the next section, this raised concerns over workers' ability to save money.

In Australia, state border closures further impacted workers' access to employment. In July, 'international seasonal workers' were banned from crossing the New South Wales-Victoria border which dissects an important horticultural region. Many PLM workers were employed in this region and regularly travelled across the Victoria-New South Wales and Victoria-South Australia borders to access different farms. With little warning, some workers who were accommodated on the 'wrong' side of the border lost access to their workplace and their employment. This was the case for the i-Kiribati workers who explained they had 2–3 weeks without work and spent that time just '*waiting and sleeping all day, all night*' until their employer could find them work on farms located within the state where they were accommodated. Accommodation shortages in the region meant moving to new lodgings was often difficult. Immediately prior to the closure, one labour contractor had planned to move 800 workers across the border to access more work, but the plans had to be put on hold. Local informants expressed frustration at the lack of consultation on this decision and its impacts on the region.

Financial Impacts

As a result of fluctuating work, the financial impacts of COVID-19 also varied. Solomon Islander SWP workers who were unable to travel as intended needed to readjust their finances. For some this meant stopping planned housing construction or business investments which had knock-on financial impacts:

> Cathy who had been building a new house from her 2019 SWP earnings explained '*We finished with the digging for the post and then when we wanted to start putting up the post, the COVID-19 suddenly broke out so we have to stop there…I'll have to tell my carpenters "we have to wait. Stop there until I earn good* money… *[again] and then we start building the house again"*.

> Ramsey, who was in his first year of SWP participation in 2019, had been actively preparing to start two poultry farms since returning to the Solomon Islands in November 2019. He hoped that one day these farms would be '*very huge*' whereby he was '*going to need more hands on the job…[and] to recruit …relatives just staying back at the home village….[and] help them to get jobs when they work in the poultry farm.*' Ramsey had begun negotiating to pay for the use of a hectare of land in Western Province and had established a partnership with a nephew on a different area of land (just under one hectare) in Guadalcanal province. In addition, he had been researching market prospects and viability, making business plans and was just about to purchase 250 roosters and 250 layers [hens] with his SWP earnings to get the farms started when the Covid-19 restrictions started to unfold in the Solomon Islands. If the restrictions on travel to Australia were to continue, Ramsey still planned to try to start his poultry farm near Honiara, but not the one in Western Province, and was in the process of seeking casual employment at the time of interview.

In contrast, diaspora members indicated that a few seasonal workers in Australia and New Zealand had earned far more money than they would have done in a 'regular' season. For example, a few RSE workers chose to stay and work on 'the farms' when others (RSE workers and Kiwis) were not available or interested in undertaking further work. However this was probably true of only a minority of workers, and additional quantitative surveys also indicate that most workers experienced an overall reduction in earnings (Doan et al., 2020). Indeed, the i-Kiribati SWP workers explained '*Because we lose a lot of weeks, that means we lose a lot of* money'. They, along with some diaspora members, mentioned that government financial support would have been invaluable at the time. Since the SWP is intended as both a development initiative, and a means of ensuring Australian horticultural labour needs are met, these expectations of government support during such unprecedented 'emergency' times was not misguided or unreasonable.

In Auckland, some Tongan and Samoan RSE workers stayed with family members while they awaited repatriation. Diaspora members raised concerns that the workers were meanwhile using up their savings by contributing to household expenses and purchasing consumer items. Nancy observed:

> *Because when they are in the regions, they don't get to go into The Warehouse [department store] or to like JB Hi-Fi or like go into Vodafone and see the fancy stuff. So their* money *just goes to the dairy [convenience store] or to the local thing. By coming to Auckland, they will end up spending. And you know, it's very strict, RSE for* Samoa *is no* alcohol, *pretty much no sex, no* alcohol, *no life. You just work and home and prayers and work and that's it for 6 months or however long. So when you take away all those rules and they're in Auckland or Wellington, there is, some are most likely going to go home with zero dollars. Especially the young ones. The older ones tend to be more responsible and you know like if they have wives or families back home they send that* money *more diligently [...] I saw a few, you know one gets like the fancy earphones or the headsets, and then you see everyone with like headsets, and I'm like 'Oh shit, that's $300'.*

In New Zealand, PLM workers were unable to access hardship payments from the government. More informally, in both Australia and New Zealand, some employers helped cover workers' weekly expenses. For example the i-Kiribati SWP workers were unable to send remittances home and received a cash advance (to be repaid) from their labour hire contractor to cover expenses during the period of no work. Additionally, one of the workers explained '*luckily they give us a chance to claim our superannuation…because – you know that superannuation you [normally] have to claim it when you back at [out of] the country, that's your time to claim it back, but now they give us the chance to claim it early and we lucky we already got it*'. Even under such circumstances however, diaspora members expressed concern that many workers' savings were dwindling. Ben, who lives in rural Victoria, described the situation

> *They continue to rack up bills in terms of rental bills and food. Like they have to get a minimum payment from their host employer which is round about $100 I hear just to sustain themselves, but that obviously goes on to the deductions that they'll have to work off later, so that throws the whole thing into a bit of a tail-spin […] what we've noticed is more that they're worried […] it's the fact that they've really got no income, that they're being supported by their host employer.*

The Australian-based Solomon Islands Labour Mobility Coordinator revealed that for those Solomon Islander SWP workers still in Australia, pressure to send financial remittances home had increased since the COVID-19 situation took hold. Family members in Solomon Islands had been laid off work and unemployment had increased. More broadly, diaspora members observed that reduced hours meant some seasonal workers were remitting less. In some cases, this created stress and tension with families in PICs who rely on remittances for basic needs such as food, health care and school fees. Indeed, workers were sometimes trying to meet multiple financial commitments at home:

> *They support their extended family with school fees, you know because the same issue back home is* unemployment *and lack of opportunities. So basically these people [RSE workers], as much as they would like to save, they are cultured to send* money *every week because they support myriad of things happening within their families. So if they've got loans, that's something with, that's something [you] could say is another factor that contributes to their hardship now with losing their hours and having an obligation to make whatever payments in a week or month.*

Nonetheless, although workers were remitting less, many were remitting a higher proportion of their wages than they had pre-COVID (Doan et al., 2020). To do so, they cut daily expenses by eating less or skipping meals. While the financial impacts of the pandemic on workers and their families inevitably varied, COVID-19 caused an overall reduction in earnings and financial security.

Community Assistance

Pacific diasporas can provide an important source of support for PLM participants (Nishitani & Lee, 2019), and, during the pandemic, diaspora members as well as local communities provided invaluable assistance and support. Some diaspora communities collected and distributed warm clothes for workers who had not expected to spend winter away from homes where 'winter' was a quite different phenomenon. Others provided food relief, one of the most commonly needed forms of support. In regional Victoria, Ben reported

> *I know certainly in the Solomon Island community, when seasonal workers were here we were taking it in turns of giving, you know taking pots of* rice *around to some of the workers who were staying in a motel, and boxes of bread that we'd managed to get from local bakeries at the end of the day, cooking up meals that used to supplement their food when they were not working.*

Sometimes, this support spanned geographical divides. In Victoria, Tongan and Samoan community members living in Melbourne and Geelong had been doing their best to support workers during extended periods of lockdown. Melanie described how Tongans in Melbourne were supporting workers in rural Victoria:

> *There's a lot of Tongan seasonal workers out there and a lot of families who are connected…to Melbourne metropolitan area, and they were just saying about how their living conditions are not good. There's barely any food and that a lot of support needs to be done. …we were meant to …go down to the rural areas to send some food packages; some Coles vouchers, clothes, warm clothes for the seasonal workers. And we were going to do it on the weekend just to send it out to all of them. But because of the* lockdown, *we were unable to go. So we've just been trying to just keep them in* communication *with us, just to…keep us in the loop of how everything is going, [and asking] "What can we supply that's online for you guys to be able to access?"*

Social distancing imposed during the pandemic could make providing practical assistance difficult. However, online options, such as electronic food vouchers, provided avenues for remaining connected and helping one another. Several seasonal workers reached out to diaspora members they had 'met' on Facebook for assistance when they faced challenges in the workplace. In some ways, COVID-19 thus increased the already important role of social media in keeping workers connected and helping them navigate life in Australia and New Zealand (Petrou & Connell, 2019). Help and assistance from diaspora members – both in person and over social media – was invaluable during this time.

Isolation and Missing Families

Isolation from local communities is an oft reported issue with seasonal worker schemes, particularly when workers are employed or housed in geographically remote areas (Petrou & Connell, 2018). For PLM workers, the pandemic intensified feelings of isolation as coronavirus restrictions meant some workers could not leave their workplaces. Silas observed that for ni-Vanuatu SWP workers in Victoria, 'life is very monotonous: *slip girap wok kakae slip* [sleep, get up, work, eat, sleep]'. Feelings of isolation and monotony were further compounded when workers relocated to different areas and social distancing restricted their contact with communities in the new locations. Sam described the situation of Samoan SWP workers in Victoria:

> *And they say they have enough work, particularly with picking oranges at the moment, I think it's that time of the season, and yeah. So, but they're really really struggling because they've been away from home and it's like a* lockdown *where they are because they can't really go out, they can't do anything really, just work […] I think at this stage it's more a spiritual and psychological support that they need […] it's actually missing the interaction and some sort of, you know, as part of the* community, *that's what they're missing.*

Aside from workers being worried in general about the COVID-19 pandemic and their family members possibly contracting coronavirus, the stress of extended family separation was particularly difficult when family members at home became unwell or died. For ni-Vanuatu workers, Cyclone Harold, which hit the archipelago in April 2020, destroying infrastructure and homes, caused extra strain. Edna explained:

So that was a, that was a doubly maybe triply major blow for those guys who were so far away from their families in Vanuatu. *And us, we know [there was tele]*communication *breakdown. And so it created major problems for those, for the seasonal workers who were in Central Otago […] when Cyclone Harold struck, they were still here in* New Zealand. *And at the same time COVID hit as well, which saw the whole world closed borders […] Our people who were working in Central Otago who were seasonal workers, were about to finish off their season, the seasonal work. And they were ready to actually go home. But they weren't able to go home because the borders were closed. And so it created a major havoc for them in terms of physical effects, the financial, the mental part of everything.*

Diaspora members emphasised that money was not enough to alleviate the stress of the situation

So as much as we can put the value dollar into what's happening in terms of work and stuff I think we should also be talking more about the mental state *right now.*

Right at the moment I think most people there, they don't worry about money *at the moment, they worry to go back to their family. But anyway, can't do it because Tongan borders closing.*

These sentiments were echoed by Tekiau, an i-Kiribati SWP worker 'stuck' in Australia

[I feel] Kind of hopeless I would say. Because you never knew that – what if you got infected by this COVID-19. Each and every night you have to think. What you're thinking of was your families back home: "What if I got infected, am I able to see my family again and my loved ones".

Diaspora members believed that the stress of family separation was compounded by the lack of clarity about what the future might hold:

It goes beyond, like I said, the financial losses, there's also the social and the mental implications of being away from home […] one of the consistent things we've heard from all of the team leaders is 'We want to go home, please let us know now' […] the New Zealand *government had extended everyone's visa until September, so one of the questions they've asked is "what happens when it comes to September and we're still stuck here?" You know, "what will the implications be?" So I think for them, a lot of it's the uncertainty, the not knowing […] And they've said it too, like we've had meetings with Tongan officials,*

> *they've [RSE workers] come on and they've said 'Look just, we'd rather get a solid answer. You know, like if we're going to be here until next year, just tell us. That's better than us, the false hope of thinking we're going to go home and we're not.'*

Many workers found themselves in a difficult situation. Nonetheless, and even as workers benefited from extra pastoral care and support during this time, it is important to emphasise that seasonal workers are resilient:

> *But you know, I guess the* resilience *in some of these people, this type of isolation is not foreign to them so they kinda take it in their stride and… It's actually quite humbling to see some of them, you know these people come into a scenario and actually be themselves OK, well while the system doesn't look that great, but I think in themselves they seem to be OK.*

PLM During a Global Pandemic: An Opportunity for Recognising and Revaluing Pacific Islander Labour

COVID-19 has created a constantly shifting and often difficult experience of labour mobility for Pacific Islander seasonal workers in Australia and New Zealand, while simultaneously highlighting their essential contribution to food production in Australia. Some workers benefitted economically from extended seasons, but many faced financial losses and emotional stress during this time. Indeed, COVID-19 has exacerbated existing and longstanding issues around access to enough work, isolation from local communities and family separation. At the beginning of the pandemic, Haley et al. (2020: 36) suggested that migrant farm workers in Canada would face heightened vulnerabilities as a result of COVID-19, and argued that the pandemic 'exposes systemic problems facing migrant workers, which have long been the target of criticism among experts working in the field'. In many ways, this rings true for the experiences of Pacific Islander labour migrants. At the same time, the pandemic emphasised the important role that local communities, Pacific diasporas and diaspora civil society groups often play in supporting seasonal workers and filling the gaps left by 'official' pastoral care programs that may be ill-equipped and slow to adapt to certain situations or issues – such as a global pandemic.

The issues we have highlighted here are neither uniform nor ubiquitous – many workers experienced no problems and enjoyed their time working in Australia or New Zealand. Nonetheless, various issues recur with enough frequency in the literature (and the media) to suggest that they constitute a concerning pattern. COVID-19 provides an opportunity to increase dialogue on, and address some of these longstanding issues, as a matter of policy. Diaspora civil society groups, for example, do not only have the capacity to be reactive, supporting workers after problems occurred. Rather, increasing the participation of civil society groups in policy dialogues will be crucial in developing proactive, comprehensive, practical and equitable solutions. This is especially important given the development objectives of labour mobility, some of which were undermined – in the short-term at least – by cessation of mobility. In October 2020 the Australian government announced A$9 million towards improving worker wellbeing in the PLM schemes. This provides an excellent opportunity to think about how the wellbeing of workers can be better supported, both in times of global pandemic and in 'normal' life. Certainly as a formally managed migration programme with development objectives built in, there is an extra onus on both host and recipient governments involved in PLM to make sure workers are properly protected, supported and taken care of so that the basic objectives of PLM are not undermined.

After much media speculation and lobbying from Australian and New Zealand farmers and agribusinesses, the pandemic-induced labour mobility 'halt' discussed in this chapter evolved into a 'pause'. By November 2020, and after a successful trial of ni-Vanuatu mango pickers in the Northern Territory, all the SWP labour sending countries had agreed to recommence their participation in the scheme. Soon afterwards more workers arrived in Australia from Tonga, Fiji and Solomon Islands. New Zealand was set to follow with plans to restart the RSE in early 2021. The pandemic had demonstrated just how important the PLM schemes had become in the region for Australian and New Zealand farmers and agribusinesses. PLM restarted even before a COVID-19 vaccine has been developed, before many Australian citizens themselves were able to return to Australia, and before Pacific Island international tourism – an important contributor to GDP in many PICs – had resumed. The value of Pacific Islander labour migration to Australia and New Zealand is clear. In New Zealand, policy had already started to better reflect this value, with an increase to the hourly horticultural rates for RSE and domestic workers alike: an important example of policy improvements that can ensure that PLM schemes move closer towards a true 'win' for all involved.

Acknowledgements Olivia's and Taukiei's interviews form a part of the 'Transformative human mobilities in a changing climate' (LP170101136) project funded by the Australian Research Council. Kirstie's data was collected as part of short-term consultancy for The World Bank's study of the impacts of COVID-19 on Pacific labour mobility and remittances. The World Bank research was partially funded by DFAT. The findings, analyses and interpretations presented in this paper do not necessarily reflect the views of the World Bank or DFAT. All authors thank the participants who kindly agreed to be interviewed for these respective studies.

References

Bedford, C. (2020a, October 02). New Zealand's Seasonal Labour Shortage, and How to Solve It. *DevPolicy Blog*. https://devpolicy.org/new-zealands-seasonal-labour-shortage-and-how-to-solve-it-20201002/

Bedford, C. (2020b, July 09). RSE COVID-19 Responses, Supporting Seasonal Workers in New Zealand. *DevPolicy Blog*, https://devpolicy.org/rse-covid-19-responses-supporting-seasonal-workers-in-new-zealand-20200709-2/

Bedford, C., Bedford, R., & Nunns, H. (2020). *RSE Impact Study: Pacific Stream Report*. Wellington.

Bedford, R., Bedford, C., Wall, J., & Young, M. (2017). Managed Temporary Labour Migration of Pacific Islanders to Australia and New Zealand in the Early 21st century. *Australian Geographer, 48*(1), 37–57.

Costa, D., & Martin, P. (2018). *Temporary Labor Migration Programs: Governance, Migrant Worker Rights, and Recommendations for the UN Global Compact for Migration*. Economic Policy Institute.

Doan, D., Dornan, M., Parsons, K., Petrou, K., & Yi, S. (2020). *Pacific Labor Mobility, Migration and Remittances in Times of COVID-19: Interim Report*. World Bank Group.

Dufty, N., Martin, P., & Zhao, S. (2019). *Demand for Farm Workers: ABARES Farm Survey Results 2018. ABARES Research Report 19.10*. Australian Bureau of Agricultural and Resource Economics and Sciences (ABARES).

Dun, O., McMichael, C., McNamara, K., & Farbotko, C. (2020). Investing in Home: Development Outcomes and Climate Change Adaptation for Seasonal Workers Living Between Solomon Islands and Australia. *Migration and Development*. https://doi.org/10.1080/21632324.2020.1837535

Haley, E., Caxaj, S., George, G., Hennebry, J. L., Martell, E., & McLaughlin, J. (2020). Migrant Farmworkers Face Heightened Vulnerabilities During COVID-19. *Journal of Agriculture, Food Systems, and Community Development, 9*(3), 35–39.

Holani, A. (2017). *Enhancing the Development Impact of Labour Mobility in Pacific Island Countries: The Role of the Seasonal Worker Program and the PACER Plus*. Unpublished Doctoral Thesis. University of Adelaide. https://digital.library.adelaide.edu.au/dspace/bitstream/2440/114064/2/02whole.pdf

Howes, S., & Orton, B. (2020, January 21). For Tonga, Australian Labour Mobility More Important Than Aid and Trade Combined. *DevPolicy Blog*. https://devpolicy.org/for-tonga-australian-labour-mobility-more-important-than-aid-and-trade-combined-20200121/

IOM. (2020). *Rapid Assessment of the Socioeconomic Impacts of COVID-19 on Labour Mobility in the Pacific Region*. International Organization for Migration (IOM).

Lawton, H. (2019, July 25). Australia's Seasonal Worker Program Now Bigger Than NZ's. *ANUDevPolicy Blog*. https://devpolicy.org/australias-seasonal-worker-program-now-bigger-than-nzs-20190725/

Nishitani, M., & Lee, H. (2019). Becoming 'Overstayers': The Coloniality of Citizenship and the Resilience of Pacific Farm Workers. In V. Stead & J. Altman (Eds.), *Labour Lines and Colonial Power: Indigenous and Pacific Islander Labour Mobility in Australia* (pp. 159–184). ANU Press.

Petrou, K., & Connell, J. (2018). "We don't feel free at all": Temporary ni-Vanuatu Workers in the Riverina, Australia. *Rural Society, 27*(1), 66–79.

Petrou, K., & Connell, J. (2019). Overcoming Precarity? Social Media, Agency and ni-Vanuatu Seasonal Workers in Australia. *Journal of Australian Political Economy, 84*, 116–146.

Smith, R. (2016). *The Goal of the Good House: Seasonal Work and Seeking a Good Life in Lamen and Lamen Bay, Epi, Vanuatu*. Unpublished Doctoral Thesis. Manchester: The University of Manchester. https://www.research.manchester.ac.uk/portal/en/theses/the-goal-of-the-good-house-seasonal-work-and-seeking-a-good-life-in-lamen-and-lamen-bay-epi-vanuatu(f21bc829-0c6b-4c55-a419-1572fff5fa1e).html

Stead, V. (2019). Money Trees, Development Dreams and Colonial Legacies in Contemporary Pasifika Horticultural Labour. In V. Stead & J. Altman (Eds.), *Labour Lines and Colonial Power: Indigenous and Pacific Islander Labour Mobility in Australia* (pp. 133–159). ANU Press.

Tazreiter, C, Weber, L, Pickering, M, Segrave, M and McKernan, H. (2016). *Fluid Security in the Asia Pacific*, Macmillan.

Underhill-Sem, Y., & Marsters, E. (2017). *Labour Mobility in the Pacific: A Systematic Literature Review of Development Impacts*. New Zealand Institute for Pacific Research (NZIPR).

World Bank. (2018). *Maximizing the Development Impacts from Temporary Migration: Recommendations for Australia's Seasonal Worker Programme*. World Bank.

CHAPTER 18

From Face to Face to Meeting in Space: The Impact of COVID-19 on the Events Industry in Trinidad and Tobago

Leslie-Ann Jordan and Tisha Jack

> Event professionals are problem-solvers. We make the impossible look easy, and we do it under circumstances that would make most people slink away to weep in a dark corner. But none of us can dig out of this hole alone. (M. Adelman, Event Safety Alliance)

Introduction

The novel coronavirus pandemic (COVID-19) has caused significant disruptions to all forms of economic activities, but especially international travel and tourism, since the first case was reported in China in December

L.-A. Jordan (✉)
Department of Management Studies, University of the West Indies (UWI), St. Augustine, Trinidad and Tobago
e-mail: Leslie-Ann.Jordan-Miller@sta.uwi.edu

T. Jack
TCJ Events, Port of Spain, Trinidad and Tobago
e-mail: tisha@tcjmanagement.com

Y. Campbell, J. Connell (eds.), *COVID in the Islands: A comparative perspective on the Caribbean and the Pacific*,
https://doi.org/10.1007/978-981-16-5285-1_18

2019. In Trinidad the first case was reported on March 12th, 2020, the day after the WHO declared COVID-19 a pandemic. By November 2020, Trinidad and Tobago (T&T) had reported 5904 positive coronavirus cases and 111 deaths.

Trinidad and Tobago's's economy is one of the strongest in the Caribbean region due to its heavy reliance on oil and gas (Roberts 2020). Oil and tourism are two of the main contributors to Trinidad and Tobago's GDP, with a 45% and 7.6% share, respectively, in 2018 (Oxford Business Group, 2020b). The pandemic had a significant impact with the Prime Minister, Dr. Keith Rowley, stating that the local economy had shrunk by 10% (Caribbean National Weekly, 2020). Citing the collapse in world oil prices and the wider negative impact of the pandemic, the country's Finance Minister, Colm Imbert, said that the Government anticipated a 2020 budget deficit of TT$15.5bn (US$2.3bn), three times the original estimate.

Although tourism in T&T has been less of a priority for economic growth and development than in other Caribbean states, it has been consistently earmarked as an engine of growth to help diversify the economy (Lewis & Jordan, 2008). However, over the past five years, there has been a noticeable decline in tourist arrivals at an average rate of 5.4%, with a small increase of 3.8% recorded in 2019 (Fig. 18.1). The events industry has been a significant niche market for T&T, but has shared in this decline, exactly paralleling the overall decline, and culminating in 59,471 business and convention visitors in 2019, accounting for 15% of all stopovers (Ministry of Tourism, 2020) (Fig. 18.2).

The chapter analyses the impact of the pandemic on the Meetings, Incentives, Conferences and Exhibitions (MICE) sector in T&T, with specific exploration of some of the strategies that have began to be implemented in order to respond, revitalise and transform the industry. According to the International Congress and Convention Association (ICCA) (2020), the industry chose not to use the 'MICE Market' label and instead say the 'Meetings Industry', which encompasses all the above. As Senkhane (2020) noted, "The Meetings Industry consists of a broad range of organizers, suppliers and facilities engaged in the development and delivery of meetings, conferences, exhibitions and other related events which are held in order to achieve a range of professional, business, cultural or academic objectives". Others have recommended the use of the term 'Events Industry' as an umbrella term for the vast scope of meetings and events and this is used here. This study mainly used secondary data, including industry publications, newspaper articles, and reputable websites.

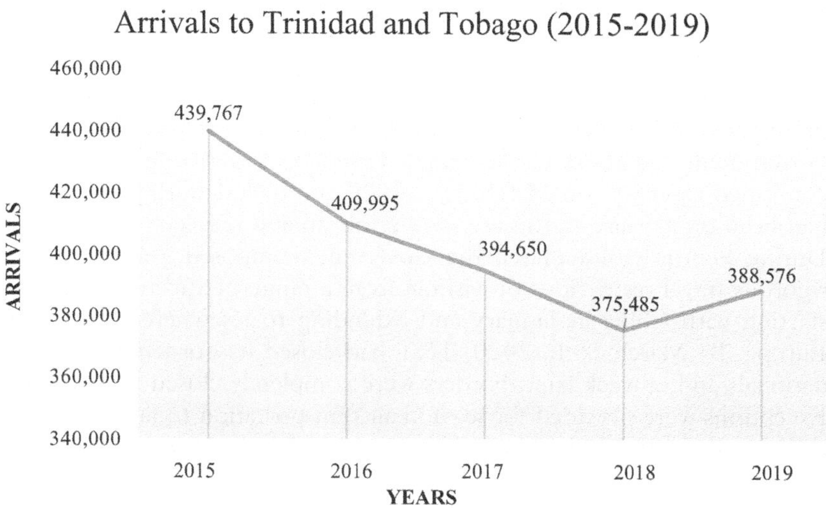

Fig. 18.1 Arrivals to Trinidad and Tobago (2015–2019). (Source: T&T Ministry of Tourism, 2020)

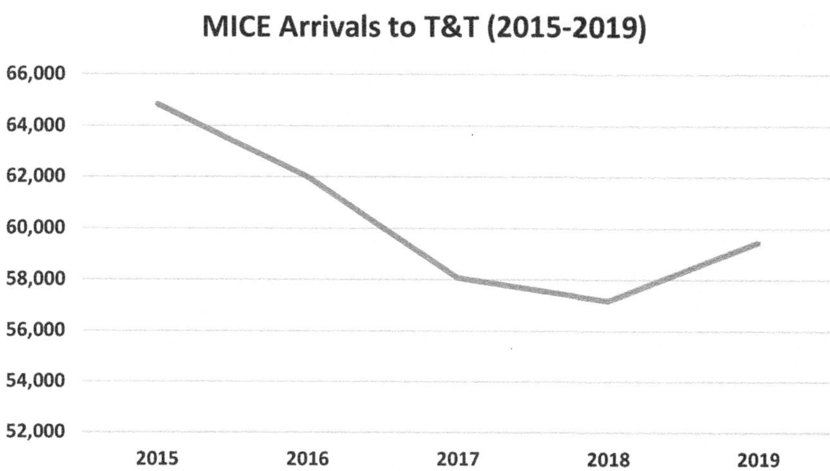

Fig. 18.2 MICE Arrivals to T&T (2015–2019). (Source: T&T Ministry of Tourism, 2020)

Trinidad and Tobago's Response to COVID-19

Trinidad and Tobago was first directly affected by the arrival of the virus in mid-March and all early cases were imported as people returned from holiday, overseas education, or business trips, some in transit. Despite the Government's advice to restrict travel, Trinidad's largest single increase in confirmed cases was on March 21st, 2020 and included 40 persons who had been on a cruise to the French Caribbean and returned to Trinidad. During February and March the Government imposed gradually more rigorous travel restrictions on visitors from a range of affected countries, starting with China in January and extending to elsewhere in Asia and Europe. By March 16th, 2020, T&T had closed its borders to all non-nationals and, a week later, borders were completely closed to everyone. Exceptions were provided for sea and air transportation to facilitate trade and ensure a sustainable supply of food and goods, including pharmaceutical and medical goods required for protecting the population. Concurrently, the Government directed the closure of public places and limited mass gatherings to no more than 25 people initially, encouraged new arrivals to self-isolate, sensitized citizens to social distancing, increased the resources available to the Ministry of Health to enhance its ability to respond and expanded testing facilities for citizens with have symptoms of the virus.

To prevent the community spread of the virus, on March 31st, the Government imposed a Stay at Home Order for Non-Essential Services. Further lockdown measures included practicing social and physical distancing with public gatherings reduced to ten people. The cruise ship season was also brought to a premature halt with the closing of the country's seaports to all cruise ships, while citizens were advised to cease all non-essential travel. As the country began experiencing more cases and potential exposure to confirmed cases grew, schools and tertiary education facilities were closed. Bars and restaurants were ordered to suspend in-house dining, cinemas and beaches were off limits and citizens were encouraged to work from home if possible.

These restrictions effectively brought the events industry to a grinding halt as conferences, meetings, seminars, weddings and other similar events, that depended on human assembly, were postponed or cancelled. The consequences of the pandemic saw the shutdown of hospitality and tourism operations, as well as many of the sectors that supported the events industry, for the better part of 2020. International, regional and even domestic tourism also ground to a halt in order to prevent the spread of the disease.

COVID-19's Impact on Trinidad and Tobago's Events Industry

Perhaps not surprisingly the events industry has been one of the hardest hit industries during the global pandemic (Madray, 2020; Rwigema, 2020) as a component of the tourism industry that, in some cases, can be replaced by online activities. Its demise has resulted in the loss of economic and socio-cultural benefits in many destinations including the generation of foreign exchange, revenue generation, creation of employment, stimulation of other economic sectors and fostering of local culture and pride. In 2018, business events hosted more than 1.5 billion participants across more than 180 countries, and the events industry generated more than US$1.07 trillion of direct spending, in planning and producing business events, related travel, and spending by exhibitors. The industry also created 10.3 million direct jobs globally and generated $621.4 billion of direct GDP (Events Industry Council and Oxford Economics, 2018).

In many small island developing states (SIDS), the events industry has been critical to economic growth and sustainability given the plethora of activities that it supports, such as accommodation, food and beverage, entertainment, transportation, tour operators, travel agents and the retail sector. However, the events industry is volatile and can be affected by natural weather conditions such as hurricanes, and by earthquakes, seasons and holidays, wars and terrorist attacks, economic shocks and global health crises such as the SARS epidemic, bird flu and now COVID-19. The pandemic has caused hundreds of sporting events, conferences, weddings and other events worldwide to be cancelled, postponed or transformed into "digital-only" affairs or minute versions of their original version based on the health regulations that guide public gatherings. By late in 2020, as many as 83 million potential attendees had been forced to change their plans. Global sporting events such as the Euro 2020 football championship and the Summer Olympics in Japan, were postponed until 2021. Expo 2020, to be held in Dubai in October 2020 was rescheduled for a year later and the ITB Berlin, the world's largest tourism trade fair, was cancelled.

Despite the declining numbers, the Ministry of Tourism (2020: 34) acknowledged that "Events and festivals can fulfil important tourism functions, generating tourism business growth directly and indirectly while serving as "brand builders" and tactical levers to offset seasonal events". More specifically, the events industry has always been a significant niche market for T&T, evidenced by its tagline – "*Trinidad: Where the World Meets!*" (Tourism Trinidad Limited, 2020). The destination recorded 57,183 business and convention stopover visitors in 2018, accounting for 15% of all stopovers. Additionally, Trinidad is well known as the home of Carnival, which saw visitor expenditure total almost TT$400 m (US$59.1 m) and attract 35,560 tourists in 2018 (Oxford Business Group, 2020a). When T&T was forced to close its borders to international travel, this had a sudden and immediate negative impact on its strategy to position itself as the conference and convention capital of the Caribbean, as well as the home of Carnival.

Events Cancellations and Postponement: Economic Impacts

In May 2020, the Caribbean Association of Event Professionals (CAEP), conducted a survey of 170 event professionals in T&T in order to determine the state of the industry. Its conclusions recognised the significant impact that the pandemic was having on the industry; some 30% of the event businesses surveyed, had cancelled all their events for 2020 although, ambitiously, another 30% predicted that about 90% of their planned events were still going to take place (CAEP, 2020, p. 25). According to the CAEP (2020: 23), the hardest hit people in the events industry were the lower, skills-based workers who "…rely heavily on the cash-based income that they are paid immediately after an event, in order to survive. While many usually find alternative jobs during the event industry's slow periods, those too have become scarce to non-existent". Of those surveyed, 46% had either been laid off or furloughed while 22% had remained working in a traditional brick and mortar establishment (CAEP, 2020: 24). Furthermore, 60% of events had reduced event budgets for 2020 while 7% had increased event budgets for 2020 (CAEP, 2020: 24). This dire situation was exacerbated later in the year, when the Prime Minister announced that the T&T Carnival 2021 would be cancelled, only the third time in the history of the event that this has happened. The event attracts thousands of visitors every year and generated more than US$3 million in 2019 (AP News 2020). However, since March 2020, both large and small events of all types were affected due to the pandemic and the vast majority of cancellations came following confirmation of the first case of the virus in March 2020 (see Table 18.1).

According to UN T&T (2020: 25), "The cancellation of major festivals, signature events and holiday weekend packages as a result of COVID-19 restrictions negatively affected a significant share of the working population and resulted in overall loss of economic activity within the country".

Table 18.1 Sample of events cancelled in Trinidad and Tobago due to COVID-19

Type of event	*Name of event*
Music festivals, concerts and theatrical productions	Tobago jazz festival TT music festival Jazz artists on the greens Dil-E-Nadan's everybody loves Raymond Vaughnette Bigford's up on the hill concert Milwaz productions' school shows of the tempest and the taming of the shrew National Philharmonic Orchestra's concert presentation of Spirit of a people: Dance tan tan!
Cultural festivals	Point Fortin borough anniversary Easter goat & crab race festival Second annual film and folklore festival
Religious events	Phagwa celebrations at the Aranguez Savannah, Children's Phagwa SWAHA International's Annual Cultural Show 2020 Spiritual Baptist shouter liberation day Celebration
Conferences	AMCHAM TT's annual Women's leadership conference UWI international conference on emerging trends in engineering and technology (IConETech-2020) UWI seismic research Centre's (UWI-SRC) volcano Hazard mapping network building session
Networking and educational events	Maternity road show UWI's academic advising and mentoring in higher education symposium TT transparency Institute's town hall meeting on gender-based violence Rotary Club of Central Port of Spain's model UN session teen together networking Creative TT's workshop CARIRI ideation workshop

Sources: T&T Newsday (2020a, 2020b)

Likewise, the Trinidad and Tobago Services of Coalition Industries (TTSCI) and the Trinidad and Tobago Manufacturers Association (TTMA) Report (2020: 12) stated: "One sector that has been negatively impacted is the arts, entertainment & recreation sector known as the creative industry. Many businesses experienced cancellations in performances, sporting events and training have been halted, workers have been laid off and companies are struggling to survive". This is a significant impact as this very active entertainment sub-sector, which supports the events industry, is labour intensive and contributed roughly 7% of GDP annually (UN Trinidad and Tobago, 2020).

Unfortunately, T&T no longer has a public sector organization specifically dedicated to the events industry, like a Meetings and Conventions Bureau. The absence of such a Bureau meant that there was no central organisation responsible for continuously supporting the local events industry by summarising and interpreting information about COVID-19 and its impacts, as part of their industry support program. It is therefore difficult to estimate the total number of events affected and cancelled by COVID-19 and the consequent comprehensive direct and indirect economic loss. However, in order to fill this knowledge gap, the Caribbean Association of Event Professionals (CAEP) was formed in 2019 to provide greater regional support for the event management profession.

Pivot or Perish: Strategies for Managing Events

The essence of the events industry is having persons gather for one reason or another: the very act that health authorities advised against as the world continues its fight against the COVID-19 virus. Industries worldwide continued to champion re-opening, but since they were unable to demonstrate how they could do so safely, calls to re-open remained unanswered. Event professionals were forced to pivot and devise alternatives in order to remain viable.

Emergence of Virtual Events

In order to continue hosting events during the pandemic, there had to be a transformation of face to face events. Consequently, two main options emerged: virtual events and hybrid events. Virtual events maintained planned events on the calendar. The most pertinent question asked was how could a virtual event mirror the experience of the face to face event?

With limited knowledge, many event professionals turned to what they knew and relied on familiar video conferencing software such as Zoom, Microsoft Team, Google Meet and Skype to conduct training sessions, seminars, media and product launches. It was the easiest solution for both the event host and attendees, but it soon saw the emergence of "zoom fatigue".

To create an experience beyond what this software offered, some event planners conducted research on available event platforms that could help achieve this. Platforms that offered networking capabilities (video and voice calls), attendee engagement, and even opportunities for fun interactions were chosen options. Some new platforms emerged, and existing ones upgraded their product/service offers to meet the growing demand. Popular event management software company CVENT became inundated with requests and subsequently had a rigid screening process for clients. Being traditionally known for event registration and payment systems, they launched a new Virtual Attendee Hub to meet the growing demand. Other platforms included GoToWebinar, Accelevents, Hubilo, Remo Conference and Whova.

Virtual events necessarily abandoned such event-related sectors as caterers, live entertainers, and the onsite staff. Some downstream companies were able to pivot their business, as in the case of Macknun Events, an event rentals and event staffing service provider, where the entire business relied on live events; its manager, "in a short time ... managed to innovate her Cocktail Delivery Service, which allowed her to keep her business going through these trying times" (CAEP, 2020: 16). Some event managers perceived some benefits to transitioning face to face events to virtual ones. For example, Priya Marajh, the Vice-President Advocacy and Member Engagement, of the Energy Chamber of Trinidad and Tobago, stated that "we will be able to reach a wider audience because of the platform. Also, because the registration cost is lower, we are hoping to get smaller operators or companies/persons who may not have attended before". In terms of marketing the event, she explained that they did "more online marketing, particularly on social platforms like: Linkedin, Facebook and Instagram, as well as direct email marketing". Finally, the process for planning the event also changed. Marajh explained that "we planned this virtual conference in about two months, as opposed to a "face to face" conference, which takes a much longer time to plan. The planning was all done in-house, so we didn't use a corporate event planner" (CAEP, 2020: 13).

To restructure the Carnival 2021, both the Trinbago Unified Calypsonians' Organisation (TUCO) and the National Carnival Commission (NCC) planned to organise virtual or digital events that could be monetized (Fig. 18.3). NCC's CEO, Colin Lucas, said the Commission's virtual events would mirror and showcase the three major elements of Carnival: music, mas and pan. Lucas stated that the Commission was not going to use the term "virtual" "because it implied a level of interactivity. He said it was really digital platforms being used to showcase different aspects of Carnival" (T&T Newsday, 2020c). Similarly, Pan Trinbago stated that they would also be hosting virtual events and there are plans to organize a pay-per-view event called 'The Sweetest Pan in the World' (T&T Newsday, 2020c).

Despite these initiatives, the idea of going virtual was new to some and not an option for others. According to CAEP (2020: 23), although some businesses managed to pivot to virtual events, the challenge was that "...this only offers a small percentage (approximately 10%) of the income regularly generated for the live version of the same event". However, the survey recorded that 47% of event professionals planned to offer virtual events as an option for their events with only 9% advising that they have no plans to make events virtual in 2021. Clearly event professionals saw little choice but to seek to develop virtual events or go out of business.

Fig. 18.3 Virtual Trinidad and Tobago carnival events 2021. (Source: Loop TT (2020) and CNC3 News)

Emergence of Micro Events

One of the main strategies that was implemented by stakeholders in T&T to navigate the impacts of the pandemic, was executing micro events, particularly for weddings and other social events. With restrictions on gatherings limiting the number of persons allowed, events saw fifty, twenty-five and even ten people in attendance. On a positive note, these smaller numbers saw a lot more attention being paid to the details and design of the events to still create an ambience that was enjoyed by the host and the fewer but specially selected guests in attendance. That continued to be an option for those who could still host events, without the technology driven version juxtaposed to it. For those who still wished to have more than the legally allowed number of guests in attendance, providing a stream of the event was the only alternative. For private social events, this was as user friendly as setting up a Zoom link, but for more intricate events or those that desired more of a production for their online attendees, enlisting the services of a video production company was required. Some new employment thus emerged.

A necessary transformation in the events industry was the introduction of streaming events. Streaming was not widely used before March 2020, but, to still allow intended guests to participate in and 'attend' events like funerals, weddings and annual general meetings, event planners and hosts set up streams for viewing. Videographers met the demand by providing this as an add-on service to their existing packages and it became their 'pivot' during the period. Live streaming provided a lifeline for the entertainment industry and provided an escape for persons with the physical inability to participate. Social media platforms, such as Facebook and Instagram, became stages for entertainers to still reach their fans. With this in mind, in March 2020, entertainment and music moguls Swiss Beatz and Timbaland created now internationally known Verzuz Battles that brought together some of the best in the music industry in a friendly battle of their biggest hits on Instagram Lives. Caribbean Dancehall icons Beenie Man and Bounty Killer's episode in May 2020 proved to be a pivotal one in the series. According to Billboard, it was "easily the most exciting and entertaining yet, as well as the first to delve into dancehall reggae." The battle drew 500,000 viewers and both artistes recorded their highest single-day streaming the day after the battles – Beenie Man saw a 188% increase over the two days with Bounty Killer receiving a 291% surge (Billboard, 2020). In T&T, similar events were conducted to accommodate the situation where residents were no longer able to attend concerts and similar shows. For example, Digicel contracted various artistes to perform live from their homes in their *Cameo Concert Series* and included jazz singer Vaughnette Bigford, calypsonian Preedy and reggae singer Isasha (Loop TT, 2020). Similarly, Ubersoca kept its fans entertained with its Online Quarantine Cooler Couch Fete featuring different soca artistes and DJs. Over 8000 users tuned in from 29 different countries from the Caribbean and all over the world to participate in the first one that took place in March 2020 (Loop TT, 2020). This demonstrated that live streaming could be a viable option for events in the entertainment industry and indeed was necessary, as the pandemic was prolonged, but it raised new questions about how income and employment might be generated in these changing circumstances.

While in-person events were irreplaceable, hybrid models presented unique opportunities to connect both in-person and remote/online attendees from multiple locations and have them enjoy an interactive experience. The international events industry began moving towards this format for meetings, conferences and seminars, since it was a great way to gradually re-introduce in-person events as the world adjusted to living with the COVID-19 virus. They allowed for physical distancing for those gathered in-person and still gave distant attendees access to the content. Additionally, they allowed planners and hosts to protect revenue generated. Should people no longer be able to attend in-person due to a spike in COVID-19 cases, or unable to travel or if there was a change in the health restrictions, having a virtual option of an event provided a viable plan B. As this evolves, event platforms will be the online venue for remote attendees and should the event host/planner choose to, they also provide the "meeting point" for these two groups of attendees to connect.

A question that underpinned some events developments was "Is this live?". Some event hosts and planners opted for pre-recorded event, particularly where there were considerable logistics and coordination challenges involved in the production. For example, T&T Soca Artiste, Kes the Band, launched their 2020 album with a pre-recorded concert premiered live on their Youtube channel. Also, Caribbean corporate giant ANSA McAL Limited brought together some of the Caribbean region's best music artistes for their fundraising initiative "One Yard". The virtual benefit concert allowed viewers to "travel" to Guyana, Trinidad, Jamaica, Barbados and St. Vincent for entertainment by performances from various soca and reggae artistes which was all pre-recorded. This option for events of this scale and nature allowed the event host and planners to mitigate all risks associated with doing such events live, and gave them full control of the quality of the content to be received by viewers and of their experience.

Moving Forwards, Moving Outwards

COVID-19 raised basic questions about safety and thus the location and physical structure of events, questions that had never been needed in the past. As the possibility of staging events became more likely later in the year, outdoor venues, where feasible, were the suggested option as air circulation in indoor venues pose a higher risk of infections occurring. The downside to this option for planners was the risk of the weather, that could present cost challenges, alongside physical distancing. There was an increased demand for larger event venues in order to adhere to the 6 feet apart rule. Venues also revised their packages with tiered pricing according to the number of guests in attendance. Those that were able to quickly invest in required health protocols such as sanitization stations, temperature checks on entry, frequent cleaning of high traffic areas and touch points, were most sought after and booked.

The pandemic further resulted in greater emphasis being placed on developing standard protocols and legally enforceable regulations for the industry stakeholders to follow; strict safety regulations and a hybrid approach to events was the suggested strategy. Regardless of the format of the event, the number one priority for any event and its stakeholders became the need to communicate safety! Potential participants based decisions to attend events on their level of comfort and the safety of an environment. An Event Safety Alliance Reopening Guide (2020) outlined recommendations for the industry as a guide to reopening safely. These recommendations spanned attendee education (sharing information prior to the event on the safety protocols in place for the event, reminders during the event about practices expected by attendees), worker safety (event staff properly attired in personal protective gear, on top of safety practices and a designated role of someone specifically tasked with implementing and enforcing infection mitigating activities) and sanitisation of the venue: (cleaning and disinfecting event spaces before, during and after the event).

Mandatory wearing of masks by attendees, when food and drink was not being consumed was also considered, but, while it was important to ensure standardisation of practices by stakeholders in the industry, the nature of events varied so much that establishing and enforcing standards was unusually difficult. The challenge was to devise and implement rigorous and effective practices without destroying the ambience and enjoyment of events of whatever kind.

Beyond the events planners, recommendations and protocols extended to the various suppliers who made things happen in their respective fields. It was therefore necessary for all stakeholders to explore how they could support events in these different formats. Thus caterers had to consider how food could be served safely at events (perhaps preparing packaged meals) and what to do at tea breaks. Videographers and audio technicians had to provide a higher quality of coverage as virtual and hybrid events became more common. Decorators were in greater demand as the set itself became more important. Viewing events from a different eye was required by photographers. In every facet of event staging new structures and approaches had to be developed. Underpinning all these changes, and covering advertising, ticketing and accommodation, was the greater role of technology. 'Bandwidth' and 'streaming' were two words that everyone in the industry was having to become familiar with. Pre-recorded content will form at least part of live agendas. It can bring an exciting element to how events are executed and allows event planners to become more internationally focused, but may conceivably disadvantage some older or more 'traditional' events participants. If nothing else videography and technology had brought new horizons.

Despite innovative and creative use of technology, that could not counter the negative economic impacts from the cancellation and postponement of events. Event revenue generation, employment, sponsorship opportunities and consumer spending all significantly declined through 2020. Those most affected have been the small and medium sized entrepreneurs that provide support services to the industry such as caterers, event decorators, brand marketers, rental companies, service staff, florists, craftsmen and even event venues. There were fewer positive impacts, as the industry was forced to transform and innovate simply to remain in existence. Event planners quickly upgraded their technical skills to learn the benefits of online platforms and software required to host online events and engage distant, virtual participants. Virtual events may cost less to host than conventional events, but their earning potential will now be a balance of lower attendance fees and the ability to host more attendees. As countries like Trinidad and Tobago get closer to winning the battle against COVID-19, event planners will again have to readjust to accommodate two audiences simultaneously – virtual and in person. Given the importance of the events industry to Trinidad and Tobago, a more detailed evaluation of the impact of the COVID-19 shock would be critical to a better understanding the extent of the economic and socio-cultural impact of the pandemic, and to establish what have been the most successful responses that will again allow the events industry to become a valuable source of foreign exchange, income, job opportunities, economic stimulation and cultural expression.

References

AP News. (2020, September 29). *Trinidad Cancels Famed Carnival for 2021 Due to Pandemic.* https://apnews.com/article/virus-outbreak-carnivals-caribbean-pandemics-puerto-rico-46493288832ce5223c975f65bc2bfce1

Billboard. (2020). *Beenie Man vs. Bounty Killer in 'Verzuz' Battle of Dancehall Legends: See Billboard's Scorecard and Winner for the Showdown.* https://www.billboard.com/articles/columns/hip-hop/9389721/beenie-man-bounty-killer-verzuz-battle-scorecard

Caribbean Association of Event Professionals (CAEP). (2020). *Analysis of the Local Events Industry.* CAEP.

Caribbean National Weekly. (2020). *PM Keith Rowley Says Trinidad and Tobago's Economy Has Shrunk by 10% Due to COVID.* https://www.caribbeannational-weekly.com/caribbean-breaking-news-featured/pm-keith-rowley-say-trinidad-and-tobagos-economy-has-shrunk-by-10-due-to-covid/

Event Safety Alliance Reopening Guide. (2020). https://static1.squarespace.com/static/5aec979d3e2d09db8bcad475/t/5eb86f694a67d30048528163/1589145456606/2020-05-11+Event+Safety+Alliance+Reopening+Guide.pdf

Events Industry Council and Oxford Economics. (2018). *Global Economic Significance of Business Events.* Microsoft PowerPoint - OE-EIC Global Meetings Significance (FINAL) 2018-11-09-2018.pptx (eventscouncil.org)

International Congress and Convention Association (ICCA). (2020). *Definition of "MICE".* https://www.iccaworld.org/aeps/aeitem.cfm?aeid=29

Lewis, A., & Jordan, L. (2008). Tourism in Trinidad and Tobago: Carving a Niche in a Petroleum-based Economy. *International Journal of Tourism Research, 10*, 247–257.

Loop TT (2020, March 28). *Take Your Mind Off COVID-19 with These Virtual Concert Series.* Take your mind off COVID-19 with these virtual concert series | Loop News (looptt.com).

Madray, J. S. (2020). The Impact of COVID-19 on the Event Management Industry. *International Journal of Engineering Applied Sciences and Technology, 5*(3), 533–535.

Oxford Business Group. (2020a). *Trinidad and Tobago Taps Business and Events, Sports and Ecotourism*. Available on: https://oxfordbusinessgroup.com/overview/standing-out-country-seeks-elevate-its-position-competitive-region-focusing-business-and-events

Oxford Business Group. (2020b). *Banking in Trinidad and Tobago: Will Covid-19 Spur a Shift to Fintech?* https://oxfordbusinessgroup.com/news/banking-trinidad-and-tobago-will-covid-19-spur-shift-fintech

Roberts, L. (2020, October 2). Saving Tobago's Economy Post COVID-19. *Caribbean Insight*. https://mycaribbeaninsight.com/saving-tobagos-economy-post-covid-19/

Rwigema, P. (2020). Impact of Covid-19 Pandemic to Meetings, Incentives, Conferences and Exhibitions (MICE) Tourism in Rwanda. *The Strategic Journal of Business & Change Management, 7*(3), 395–409.

Senkhane, M. (2020, April 17). MICE Definitions – What Is MICE? https://theplanner.guru/2018/04/17/mice-definitions-what-is-mice/

Tourism Trinidad Limited. (2020, May). *Talk Tourism*, Issue 02. https://static1.squarespace.com/static/5c3f8af136099b4b446a18a4/t/5eb2c4f25c1a04720609c50b/1588774137464/FINAL+TALK+TOURISM+Issue+02.pdf

Trinidad and Tobago Coalition of Service Industries (TTCSI and the Trinidad and Tobago Manufacturing Association (TTMA). (2020). *Economic Impact Assessment of COVID-19 on Services & Manufacturing Sectors*. https://www.ttcsi.org/wp-content/uploads/2020/05/EIA2020-TTCSI-TTMA-Final-VG-May5-2020.pdf

Trinidad and Tobago Ministry of Tourism. (2020). *Revised National Tourism Policy (Draft) Trinidad and Tobago*. http://www.tourism.gov.tt/Portals/0/Documents/Press%20Releases/Revised%20Draft%20National%20Tourism%20Policy%20(2020-2030).pdf?ver=q3x37qPEOiT9ReRaCEsgBg%3D%3D

Trinidad and Tobago Newsday. (2020a, March 12). Events Cancelled Over Covid19 Fears. https://newsday.co.tt/2020/03/12/events-cancelled-over-covid19-fears/

Trinidad and Tobago Newsday. (2020b, March 13). *TT, Regional Events Cancelled Over Covid19*. https://newsday.co.tt/2020/03/13/tt-regional-events-cancelled-over-covid19/

Trinidad and Tobago Newsday. (2020c, November 27). *Virtual Events in Train for Carnival 2021*. Virtual events in train for Carnival 2021 (newsday.co.tt).

United Nations Trinidad and Tobago (2020). Socio-economic Response to COVID-19: Assessment and Framework. https://reliefweb.int/sites/reliefweb.int/files/resources/TTO_Socioeconomic-Response-Plan_2020.pdf

CHAPTER 19

Challenging the Idea of Work: Assessing the Impact of COVID-19 on the Future of Work, Health and Well-Being of the Jamaican Workforce

Kenisha V. Nelson, Ijah Thompson, and Trevor Riley

Introduction

The novel coronavirus (COVID-19) pandemic is one of the most significant health and economic threats countries have faced in recent history. After the declaration of this health emergency, most governments across the globe enforced strict measures to reduce the risk of transmission of the

K. V. Nelson (✉)
University of Technology, Kingston, Jamaica
e-mail: KNelson@utech.edu.jm

I. Thompson
University of the West Indies, Kingston, Jamaica
e-mail: hjlewis@utech.edu.jm

T. Riley
Shipping Association of Jamaica (SAJ), Kingston, Jamaica
e-mail: KNelson@utech.edu.jm

Y. Campbell, J. Connell (eds.), *COVID in the Islands: A comparative perspective on the Caribbean and the Pacific*,
https://doi.org/10.1007/978-981-16-5285-1_19

virus including social distancing, travel restrictions and border closures, stay at home orders and curfews, bans on large gatherings, and business and school hour restrictions or closures (Gostin & Wiley, 2020). The resultant decline in economic activities has had a dramatic impact on businesses and workers. In response, businesses and industries also took unconventional measures to remain viable during this period and to improve their chances of recovering after the pandemic. Abrupt changes in work arrangements resulted in employees having to make significant adjustments to their modes of work. At the same time, the devastating effects of COVID-19 on business resulted in many workers losing their jobs or being vulnerable to lay-offs, reduced hours, and reduced income (ILO, 2020a).

For those who remained employed, work either changed or intensified depending on the industry. Frontline workers, especially healthcare and emergency response workers tasked with the safety of others, were under enormous strain, due to increased workloads, long working hours, and personal safety concerns (ILO, 2020a). Workers in industries that involved the manufacturing and delivery of essential goods, transportation, and security and safety of the population also faced similar situations as their jobs required frequent contact with the public. For workers who had to physically be at work, there was continued concern over health and safety practices due to the increased risk of contracting the virus in the workplace, especially for those with existing health concerns. Others who could work from home however were psychologically vulnerable due to increased risks from isolation and loneliness due to reduced social interactions, the struggle to balance home and work responsibilities, and challenges related to learning new technology or communication tools (ILO, 2020a). The drastic changes accompanied with the many uncertainties about the evolution of the crisis severely affected economic security and increased occupational stress (APA, 2020)

Safety and health at work were more relevant than they became in managing the COVID-19 pandemic. New work presented new challenges and exacerbated existing ones. Psychosocial risks had to be addressed along with infection control as, if left unmanaged, they would lead to severe physical and mental health problems which would leave the workforce ill placed to subsequently rebound. The accumulation of stress and burnout as a result of poor working arrangements, lower job security, and blurred boundaries between work and home life, have a significant impact on health which in turn impact productivity, performance, job engagement, increased absenteeism, and presenteeism in the workplace (ILO, 2020a). In this chapter, we provide an analysis of the impact of the COVID-19 pandemic on the health and well-being of workers in Jamaica, within the context of the country's legislative environment and the implications for sustainable development. The analysis is informed by the review of primary and secondary sources including the extant literature, global and local policy documents, and the use of statistics from local and international organisations such as the Statistical Institute of Jamaica (STATIN), Caribbean Policy Research Institute (CAPRI), and the United Nations (UN). We argue that COVID-19 presents unique opportunities to examine the gaps in occupational health and safety policy and practices in Jamaica, including the need for more emphasis on the psychosocial hazards of work and the impact on the health and well-being of workers. The chapter will provide discussions on five occupational themes that were amplified by COVID-19: leadership/management practices, remote work, digitisation, gender inequities, and economic and psychosocial impact. We conclude with a call to strengthen the policy and legislative framework to advance the health and well-being of workers to meet future challenges.

A Conceptual Framework

Definitions of a healthy workplace have evolved greatly in parallel to the nature of work itself. It has shifted from an almost exclusive focus on the physical work environment (i.e., physical, chemical, biological, and ergonomic hazards), to more broadly include health practice factors (lifestyle); psychosocial factors (work organization and workplace culture); and a link to the community (WHO, 2010). Moreover, with the emergence of new occupational hazards over the years, increased focus has been on assessing the indicators of work-related stress and well-being and understanding the mechanisms by which job characteristics affect the health of workers (Brauchli et al., 2015). While a pandemic is extreme and unique, there are existing occupational health frameworks that can provide an understanding of work-related stress during a crisis and inform strategies to respond to the needs of workers (Sinclair et al., 2020). The discussions in this chapter draw upon the well-established job-demands-resources model which provides a broad and flexible conceptual framework for understanding the increased pressures and risks associated with changes within the work environment that can adversely impact the health and well-being of workers during a pandemic.

The job-demands-resource (JDR) model assumes that an imbalance between negative (job demands) and positive (job resources) job characteristics result in adverse physical and psychological health consequences. The model suggests that the workplace consists of various stressors whether organisational (e.g., workload, time pressure), physical (e.g. loud noise, heat), or social (e.g. poor supervision) which have physical and psychological consequences (Schaufeli & Taris, 2014). By contrast, resources are health-protecting factors and can counter job demand, lower levels of strain, and promote greater work engagement. Job resources consist of those positive organisational, physical, social, and psychological factors that may "(a) be functional in achieving work goals, (b) reduce job demands and the associated physiological and psychological costs, and (c) stimulate personal growth and development." (Demerouti et al., 2001: 501). Job resources include control, participation in decision making, task variety, and social support (e.g., support from supervisors and colleagues). Research supports the JDR in predicting both positive and negative well-being and associated outcomes including burnout, job performance, absenteeism, and work engagement (Demerouti et al., 2001; Schaufeli & Taris, 2014). The analysis is anchored in the JDR model, underpinned by the principle that increased demands during a time of crisis, accompanied by limited resources, will adversely impact workers' health and well-being.

The Jamaican Context

Economic Impact

The first case of COVID-19 arrived in Jamaica on March 10th. Three days later, the Prime Minister declared Jamaica a disaster area and enforced The Disaster Risk Management Act. On the 12th of March, the Government of Jamaica ordered the closure of all schools, and subsequently closed Jamaica's borders to all arrivals. The Government also implemented social distancing orders, restrictions on public gatherings, and curfew hours which resulted in most business operations closing by 5 pm. The economic and labour market fallout was severe, with service industries like hotels and tourism being the most affected. Workers were confronted with significant strain associated with uncertainty, employment status, and financial instability (Ricketts, 2020). As of July 2020, estimates from the Statistical Institute of Jamaica (STATIN, 2020a) showed a 9% decline in the employed labour force compared to January 2020. A COVID-19 survey also conducted by STATIN between April and June 2020, showed that approximately 47% of respondents reported a reduction in their income and 13% percent had severe or complete loss of income. Thirteen percent also indicated that their workplace implemented reduced hours with pay reduction and 7% said their place of work instituted days off without pay (STATIN, 2020b). A study by UNICEF in collaboration with the Caribbean Policy Research Institute (CAPRI) showed that COVID-19 effects were felt severely among households, with 80% of households suffering some income loss. The average income lost was 46% and one in two households only had enough income or savings to sustain them two weeks or less (UNICEF, 2020).

Remote Work

Independent of the financial fallout, new occupational risk factors emerged for workers who remained employed. As organisations adapted to the challenges brought on by the pandemic, they had to adjust quickly to issues never faced before, notably transitioning their workforce to working remotely. Remote work and working from home, while a significant adjustment for business, was critical to sustaining their viability. Several businesses quickly transitioned their workforce to working from home arrangements, at least in the early days of the pandemic. A survey conducted by the Jamaican Manufacturers and Exporters Association (JMEA) in August 2020 showed that from a sample of 20% of its members, more than half (51%) of companies had at least 10% of their employees working remotely since the pandemic (JMEA, 2020). Surveys also conducted by CAPRI, STATIN, and the Jamaica Productivity Centre in the Ministry of Labour and Social Security found respectively that 14, 15, and 30 percent of businesses had staff working from home (CAPRI, 2020; Jamaica Productivity Centre, 2020; STATIN, 2020a). Moreover a quarter (26%) of respondents to the Jamaica Productivity Centre survey indicated their intention to maintain working from home arrangements post COVID-19.

Remote work is not new and before COVID-19 it represented an important area of discussion for the future of work. However, the scale and timeframe for making this adjustment for workers and businesses in the context of a pandemic was dramatic. While remote work can help to reduce the potential for contagion and has other obvious benefits such as reduced travel time (and cost) to work, some challenges may cause stress and impact mental health. Thus, many workers required to work from home had to adjust to coordinating intensified family and work demands. Since schools were closed, workers had to balance working from home with child-care duties and supervision, creating additional stress. The CAPRI online poll showed that among participants who had school-aged children, approximately 30% indicated that they were getting less than 50% of their work done while working from home. It was also a challenge for healthcare, emergency, and other essential workers who had to be at work and work longer hours, and make supervision arrangements for their children who were at home. Work-family conflict defined as "a form of inter-role conflict in which the role pressures from work and family domains are mutually incompatible" (Greenhaus & Beutell, 1985) was associated with a variety of adverse health consequences (Allen et al., 2000).

Some people did not have conducive facilities to work effectively from home, while others struggled to remain motivated due to distractions and limited social support. Several challenges with remote work were noted in the JMEA survey, including computer and technology problems, and internet capacity at worker's homes (JMEA, 2020). A report generated by the Private Sector Organisation of Jamaica (PSOJ) in May 2020, which was based on feedback from its members, acknowledged that some of the early challenges that managers and workers faced with remote work included reduced productivity, problems with workers staying motivated, and increased distractions when working from home (PSOJ, 2020). However, elsewhere, studies have shown that employees who work from home tend to work longer hours and that the work extends into evenings and on weekends, creating an intensification of work (Eurofound and the ILO, 2017). The PSOJ (2020) report also highlighted that workers missed the social interactions within the workplace. Social isolation and quarantine measures, although necessary to limit the spread of the virus, were linked to adverse health-related outcomes (Brooks et al., 2020; Tuzovic & Kabadayi, 2020).

Technology and Digitisation

Many companies in Jamaica had to move their businesses online swiftly or improve their online capabilities. In the context of work and the workplace, the ILO refers to digitisation as "the presence and use of cloud computing and scheduling tools, as well as web-based applications across different platforms to facilitate remote access and collaborative work." Digitisation in the workplace was already driving changes in the way work is organised, skill requirements, labour standards, and well-being of workers (ILO, 2020b: 8). Digital transformation has been seen as the driver of the fourth industrial revolution and small island states must capitalise on opportunities that digitisation offers to increase global competitiveness that will drive growth and sustainable development. COVID-19 comprehensively accelerated the challenges and opportunities associated with the adoption of digital technologies.

On the one hand, COVID-19 has offered opportunities to re-imagine business operations and in some instances has strengthened the resilience and viability of some businesses. Moreover, if adopted effectively, digital technologies can foster positive work changes including increased productivity, enhanced communication, and increased work flexibility (Barnett et al., 2011). On the other hand, digital technology which involves the use of the internet, computers, or mobile phones assumes access to these devices and internet connectivity. In Jamaica, the telecommunication network infrastructure remains critically underdeveloped which results in connectivity issues that lead to loss of productivity and frustration among employees. Internet connectivity and technology problems were among the primary challenges for workers in the manufacturing industry.

The circumstances of COVID-19 have also highlighted digital divides and deficiencies in digital readiness, particularly in developing nations. The rapid switch to the use of digital or online tools means that workers must quickly get familiar with new or different technologies and communication devices. This may mean initially working longer hours to adapt to this new reality (ILO, 2020b). One sector where the impact was most apparent was education. Schools and colleges at all levels had to make a swift transition to online learning in efforts to reduce physical contact in the classroom. Educators had to improve their technological literacy quickly, adapt curricula to fit into online platforms, and manoeuvre distance learning sessions with students. Confusion and stress for teachers were identified as two of as many as 13 adverse consequences of school closures by UNESCO (2020) and was associated with transitions to online learning. Distance learning in small island developing states like Jamaica is exacerbated by poor technology infrastructure, particularly in rural areas, and inadequate access to technology tools within the school systems, and a shortage of skilled human resources to remedy these problems.

Gender Inequalities

The world of work is gendered, and the COVID-19 pandemic has exacerbated existing social and economic gender inequities (ILO, 2020b). Women take on responsibilities such as childcare and domestic duties, even in dual-career couples (Shockley & Shen, 2016). These kinds of inequities became even more apparent during the pandemic as women in the workforce faced the brunt of the psycho-social pressures of balancing intensified work demands and home-based responsibilities such as childcare and homeschooling. This was supported by data from the CARPI-Bluedot online poll which showed that mothers (51%) or other female relatives (23%) were primarily responsible for supervising the children while they are at home (CAPRI, 2020). Women are also disproportionately represented in key health and social sectors, and therefore were at increased risk for exposure to the virus and the increased demands associated with being on the frontline. WHO estimated that, globally, approximately 70% of jobs in the human health, service, and support sectors are held by women (Boniol et al., 2019) and STATIN data point to women being overrepresented in these industries in Jamaica (STATIN, 2020b). The Labour Force Survey and the UNICEF survey showed that women in employment and female-headed households were more severely affected by the pandemic, inducing a gendered economic crisis, compared to men (STATIN, 2020b; UNICEF, 2020).

Leadership and Management

The pandemic has been a test for leadership in organisations across the world. Research shows that strong and effective leadership is important particularly in times of crisis and can in turn help to mitigate stress for employees (Dirani et al., 2020). During a pandemic, employers are faced with multiple challenges, including trying to maintain their economic viability by ensuring their business processes are minimally affected. Good leadership, however, will also make the health and safety of workers a priority, to mitigate the physical risks and psychosocial stressors generated by the pandemic. Workplace measures implemented to respond to COVID-19 may support the mental health of employees and maintain work performance, as appears to have occurred in Japan (Sasaki et al., 2020). Equally, leadership approaches that involve the communication of realistic expectations, setting achievable deadlines, and offering adequate support to employees in the context of a crisis will also help to reduce employees' experience of occupational stress (Dirani et al., 2020).

Like other countries, the Government of Jamaica through the Ministry of Health and Wellness, issued early directives that required organisations to develop and implement comprehensive health and safety practices to prevent contagion of the virus within workplace settings. By May 2020, the Ministry of Health and Wellness had issued workplace guidelines and protocols which encouraged businesses to establish and implement policies and practices such as social distancing in the workspace, making adjustments to work arrangements to include flexible work, improving ventilation systems in buildings, regular cleaning and sanitisation of surfaces, promoting proper respiratory and handwashing hygiene, provision of facilities for sick employees and engaging in risk communication, and training and education among workers. However, the STATIN COVID-19 survey showed that the most common measures implemented in job settings, as reported by employees were simply little more than increased sanitisation and cleaning (58%) and enhanced screening measures (21%). Fourteen percent of workers indicated that their place of work implemented other safety measures and protocols, but a similar number said that their business did not implement any safety measures (STATIN, 2020a).

To ensure business continuity and consistent performance while protecting the health and well-being of employees, there is an obvious need for agility in management practices to match new realities (Dirani et al., 2020). Some organisational leadership struggled with the idea of remote work before COVID-19, having ingrained conventional ideas that productive work is done within physical and controlled office settings. Such presumptions about work affected the ability of some organisations to adapt quickly to the changing culture of work during the pandemic. Furthermore, with this attitude, organisational leaders even increased monitoring and pressure on workers to be productive while at home (Kniffin et al., 2020). Adapting to work from home arrangements will see success when there are structures, well-defined functions, and trust in place (ILO, 2020b). As the survey of member companies of the JMEA showed, there were challenges with communication and coordination for those employees working from home (JMEA, 2020). Similarly, the Jamaica Productivity Centre found that mindset and leadership were two of the factors reported as inhibiting transitioning to alternative work arrangements (Jamaica Productivity Centre, 2020). During crises, effective leaders can reduce the level of stress of their employees through support and communication. The best strategy for managing teleworking is through the process of Management by Results, where management and employees agree on reasonable measures of productivity output (Sorenson, 2016).

Situation Analysis of Occupational Health in Jamaica in the Context of Sustainable Development

A healthy working population is necessary for the economic sustainability of any society (WHO, 1995). In September 2015, Jamaica adopted the United Nations Agenda for Sustainable Development which is integrated into its 21-year *Vision 2030 Jamaica National Development Plan* (PIOJ, 2009). *Vision 2030 Jamaica* is a strategic framework for the development of national policy and programmatic initiatives that will drive the government's commitment to sustainable development goals (SDGs). Specifically, promoting "a healthy and stable population" is aligned with Goal 3 of the 2030 agenda for sustainable development, to "ensure healthy lives and promote well-being for all ages" (PIOJ, 2019).

Demonstrating health as a priority, in May 2019 the Ministry of Health was rebranded as The Ministry of Health and Wellness, signalling a greater focus on well-being among the population through health education and promotion. This reflected the WHO's conceptualisation of health as "a state of complete physical, mental and social well-being and not just the absence of disease or infirmary." The Ministry increased its focus on the prevention of non-communicable diseases through its 10-year strategic plan by seeking to strengthen health systems and multi-sectoral actions and partnerships. Among its promotion of healthy lifestyle initiatives is the Jamaica Moves Programme for schools, communities, and workplaces. There has also been increased emphasis on mental health, particularly promotions to reduce the stigma against mental illness, increase help-seeking behaviours, and expand services to treat people with psychological difficulties (PIOJ, 2015). In light of this, progress is being made with the implementation of the Mental Health Action Plan 2013–2020, in which community mental health services are critical components (PIOJ, 2015).

The Ministry of Health and Wellness (MOHW), as the primary driver of the country's health plan, has broadly acknowledged that the environment in which people live, work, learn, and play all impact their health. Moreover, the Minister believes that the social determinants of health are the remit of many non-traditional partners, including employment organisations, and as such has called for greater support from employers in promoting health and wellness in the workplace, as it not only benefits individual employees but increases productivity. Collaborative efforts with these social partners will therefore be critical to design policies and practices within their specific settings to promote healthy environments through "healthy lifestyle programmes in the special settings – schools, communities, and workplaces, utilizing the 5 pillars of health promotion" (MOHW, 2019).

With the growing recognition that COVID-19 was having a deleterious effect on mental health, the Government through the MOHW in October 2020 implemented a COVID-19 Mental Health Response Programme to address anticipated or emergent mental health issues among the population. Although the programme was meant to serve the wider population, the Minister acknowledged specific groups such as healthcare workers who endured significant strain due to increased work demands and long work hours and who were at increased risk for mental health issues as a group of primary concern. However, it is unclear what measures have been put in place by individual organisations to mitigate the increased psychosocial pressures associated with COVID-19 disruptions of work life in other industries.

While the MOHW has the overall mandate for promoting the health and well-being of the population, the Ministry of Labour and Social Security governs policies and practices within the labour market, including monitoring and enforcing the requirements of occupational safety and health standards (OSH) for the protection of workers. Establishing strong labour and legislative framework is important for achieving SDG 8, to promote inclusive and sustainable growth, employment and decent work for all. However, the environment in Jamaica around workplace health and safety remains in its legislative infancy. Although Jamaica is a signatory to the ILO convention on health and safety, a long-awaited Occupational Health and Safety (OSH) bill was yet to become law. The OSH Act will replace the longstanding Factories Act of 1943, which was limited in its consideration for standards for psychosocial hazards, however the bill did not improve significantly on specifically addressing psychosocial risk factors within the workplace.

Other legislative instruments have also fallen short. In 1996, a Labour Market Reform committee recommended that flexible work arrangements form a fundamental part of labour market reform in Jamaica, recognising that flexibility in the workforce enhances global competitiveness, efficiency, and productivity. Following a series of stakeholder consultations, The Employment (Flexible Work Arrangement) (Miscellaneous Provisions) Act was adopted in 2014. However, its legal provisions are yet to be fully implemented and adopted widely across industries and work settings, though there was much support from consultations with stakeholders who acknowledged the benefit of flexible work arrangements (Hibbert, 2020). Despite consideration of telecommuting within the provision, it lacked work from home specific guidelines within the labour laws. In fact, during the pandemic, the Disaster Risk Management Orders (under the Disaster Risk Management Act), which called for employers to allow their employees to work from home if the conditions of the job allow it was the only legal framework for enacting remote work. Even so, that Order was replaced by one which only encouraged rather than mandated remote work for employees under 65 years of age.

Taking time off from work is now mandated under the Disaster Risk Management Act for workers with flu-like symptoms to avert the risk of infection spread to other workers. Employees are also mandated to stay away from work if they are in a quarantine zone or if they have come in contact with someone suspected or confirmed of having COVID-19. However, there are no formal provisions or amendments in law for sick leave or other COVID-19 induced absence to protect employees. Furthermore, contract workers who are not paid when not at work may well try to hide illnesses in order to prevent losing income. Therefore, it is imperative that compensation policies be reviewed to protect workers and ensure that there is no pressure to work while sick, and to consider how work from home is treated in the context of ill-health (Kniffin et al., 2020). Other states, such as the USA, have amended Acts like the Families First Coronavirus Response Act that "requires certain employers to provide their employees with paid sick leave or expanded family and medical leave for specified reasons related to COVID-19" (U.S. Department of Labor, 2020). This local gap in amending the laws to protect employees is likely to encourage presenteeism (people going to work when ill), and foster situations that promote the further spread of COVID-19 among unsupported workers. This sets up a greater likelihood of the health and well-being of some workers being negatively impacted by this absence of legislative support.

Conclusion

The national strategic goals are intricately intertwined and interdependent, and if achieved will increase the chances for sustainable development. To realise macroeconomic stability, the Government of Jamaica must improve the efficiency of its labour market by building and maintaining its human capital which must be guided by protective principles. A healthy population is needed to drive productivity and thus national development. The impact of COVID-19 has been dramatic, and it will no doubt change the way people work in fundamental ways. COVID-19 has accelerated discourses and initiatives towards considerations of the future of work and amplified areas that are of concern, but developing appropriate and effective legislation has been slow. Increased use of technology, learning new skills quickly, balancing home and work life, the need to adapt to new types of work, work inequities, and ineffective leadership increase the psychosocial pressures associated with work and have slowed innovation.

COVID-19 had had different effects on particular industries. Some organisations have been more easily able to restructure operations and the labour force to make permanent changes brought on by benefits associated with COVID-19 such as increased use of technology and remote or virtual work. While there have been benefits of remote work before COVID-19, it was typically considered within the context of employee preference and mutual agreement between employer and employee. It is therefore difficult to discern the value of remote work, particularly mandatory work from home for the employer and employee during COVID-19 (Kniffin et al., 2020). It is also evident that for many, work conditions have deteriorated which will contribute to a greater risk of burnout as a result of prolonged stress. To be able to cope with work demands during and also post COVID-19, employees need tangible, social, and psychological resources (Kniffin et al., 2020). That will vary for different groups of workers, and according to gender. Leadership and management practices need to quickly adjust to meet these challenges adequately.

In Jamaica, and other small island states, where knowledge of the particular challenges of occupational health and safety is limited, ongoing research will be critical to informing policy, legislative, and programmatic initiatives. Local legislative provisions aligned with international conventions and practices will be essential to bolster a climate of safer and healthier workplaces. Presently, Jamaica strives to catch up with ILO and other international recommendations. It is generally emphasised that the quality of human capital is one of the main causes of the country's longstanding challenge with productivity within the workforce (PIOJ, 2015), but there is little follow-up in recognising the constraints of poor health and well-being of workers as a determinant of low productivity levels. If the current lethargy persists in the development of a robust legislative, policy, and programmatic framework regarding psychosocial health and well-being in work settings, specific objectives and targets for Vision 2030 Jamaica may fall short. Both the government and the private sector have much work to do to enhance the policy and support framework for protecting the physical and psychological well-being of workers. COVID-19 has effectively offered, even demanded, rethinking; the opportunity needs to be taken.

References

Allen, T., Herst, D., Bruck, C., & Sutton, M. (2000). Consequences Associated with Work-to-Family Conflict: A Review and Agenda for Future Research. *Journal of Occupational Health, 5*, 278–308.

APA. (2020). Stress in America™ 2020: Stress in the Time of COVID-19. https://www.Apa.Org, May, 2019–2021. https://www.apa.org/news/press/releases/stress/2020/report%0Ahttp://files/461/report.html

Barnett, K., Chiveralls, K., & Moretti, C. (2011). Stress, Work and Technology Across the Life Cycle. *Journal of Management Research and Analysis, 1*(January), 1–21. https://www.researchgate.net/publication/283470896

Boniol, M., McIsaac, M., Xu, L., Wuliji, T., et al. (2019). *Gender Equity in the Health Workforce: Analysis of 104 Countries.* WHO Health Workforce Working Paper No, 1.

Brauchli, R., Jenny, G., Füllemann, D., & Bauer, G. (2015). Towards a Job Demands-Resources Health Model: Empirical Testing with Generalizable Indicators of Job Demands, Job Resources, and Comprehensive Health Outcomes. *BioMed Research International, 2015.* https://doi.org/10.1155/2015/959621

Brooks, S., Webster, R., Smith, L., Woodland, L., et al. (2020). The Psychological Impact of Quarantine and How to Reduce It: Rapid review of the Evidence. *The Lancet, 395*(10227), 912–920. https://doi.org/10.1016/S0140-6736(20)30460-8

CAPRI. (2020). CAPRI-Bluedot Insights Instapoll. In *Facebook*. https://www.facebook.com/capricaribbean

Demerouti, E., Nachreiner, F., Bakker, A., & Schaufeli, W. B. (2001). The Job Demands-Resources Model of Burnout. *Journal of Applied Psychology., 86*(3), 499–512.

Dirani, K. M., Abadi, M., Alizadeh, A., Barhate, B., et al. (2020). Leadership Competencies and the Essential Role of Human Resource Development in Times of Crisis: A Response to Covid-19 Pandemic. *Human Resource Development International, 23*(4), 380–394.

Eurofound and the International Labour Office. (2017). *Working Anytime, Anywhere: The Effects on the World of Work*. European Union, Luxembourg, and the ILO. http://eurofound.link/ef1658

Gostin, L., & Wiley, L. (2020). Governmental Public Health Powers During the COVID-19 Pandemic: Stay-at-Home Orders, Business Closures, and Travel Restrictions. *JAMA – Journal of the American Medical Association, 323*(21), 2137–2138.

Greenhaus, J. H., & Beutell, N. J. (1985). Sources of Conflict Between Work and Family Roles. *Academy of Management Review, 10*(1), 76–88.

Hibbert, K. (2020, February 16). Jamaica Lagging Behind on Flexi-Work. *Jamaica Observer*.

ILO. (2020a). *Managing Work-Related Psychosocial Risks During the COVID-19*. ILO. www.ilo.org/publns

ILO. (2020b). Teleworking During the COVID-19 Pandemic and Beyond A Practical Guide. : ILO. https://www.ilo.org/wcmsp5/groups/public/---ed_protect/---protrav/---travail/documents/publication/wcms_751232.pdf

Jamaica Productivity Centre. (2020). *COVID-19 Impact Survey Preliminary Summary*. https://www.mlss.gov.jm/wp-content/uploads/2020/11/JPC-COVID-Survey-Summary-2020.pdf

JMEA. (2020). Since Covid-19, 68 Percent of Manufacturers Report Negative Cashflow JMEA Latest Impact Survey. *Jamaica Observer*. http://www.jamaicaobserver.com/sunday-finance/Since-COVID-19?profile=1442

Kniffin, K. M., Narayanan, J., Anseel, F., Antonakis, J., et al. (2020, June). COVID-19 and the Workplace: Implications, Issues, and Insights for Future Research and Action. *American Psychologist*. https://doi.org/10.1037/amp0000716

MOHW. (2019). *Vision for Health 2030 Ten Year Strategic Plan 2019–2030*. MOHW. https://www.moh.gov.jm/wp-content/uploads/2019/05/MOHW-Vision-for-Health-2030-Final.pdf

PIOJ. (2009). *Vision 2030 Jamaica. National Developemnt Plan*. PIOJ.

PIOJ. (2015). Economic and Social Survey: Jamaica. *Planning Institute of Jamaica*. http://www.pioj.gov.jm/

PIOJ. (2019). *Jamaica Voluntary National Review Report on the Implementation of the 2030 Agenda for Sustainable Development June 2018*.

PSOJ. (2020). *The Workforce During COVID-19 Keeping the Team Motivated and Productive While at Home*. https://psoj.org/wp-content/uploads/simple-file-list/SMEMemoKeepingTeamMotivatedandProductiveatHome-v5-1.pdf

Ricketts, H. (2020). The Government of Jamaica (GoJ) COVID-19 Pandemic Stimulus Response: An Opportunity to Regulate and Formalize? *International Journal of Sociology and Social Policy, 40*(9/10), 979–993.

Sasaki, N., Kuroda, R., Tsuno, K., & Kawakami, N. (2020). Workplace Responses to COVID-19 Associated with Mental Health and Work Performance of Employees in Japan. *Journal of Occupational Health, 62*(1), e12134. https://doi.org/10.1002/1348-9585.12134

Schaufeli, W., & Taris, T. (2014). A Critical Review of the Job Demands-Resources Model: Implications for Improving Work and Health. In G. Bauer & O. Hämmig (Eds.), *Bridging Occupational, Organizational and Public Health: A Transdisciplinary Approach* (pp. 43–68). Springer.

Shockley, K., & Shen, W. (2016). Couple Dynamics: Division of Labor. In L. T. Allen & T. D. Eby (Eds.), *Oxford Library of Psychology. The Oxford Handbook of Work and Family*. Oxford University Press.

Sinclair, R., Allen, T., Barber, L., Bergman, M., et al. (2020). Occupational Health Science in the Time of COVID-19: Now More than Ever. *Occupational Health Science, 4*(1–2), 1–22. https://doi.org/10.1007/s41542-020-00064-3

Sorenson, H. (2016). *Best Practices for Managing Telecommuting Employees*. Capella University.

STATIN. (2020a). *2020 KAP Survey Report Assessing the Knolwedge, Attitudes and Practicies of Jamaicans Regarding the Novel Coronavirus – Covid-19*. https://statinja.gov.jm

STATIN. (2020b). *Labour Force Statistics*. Statistical Institute of Jamaica. https://statinja.gov.jm/LabourForce/NewLFS.aspx

Tuzovic, S., & Kabadayi, S. (2020). The Influence of Social Distancing on Employee Well-Being: A Conceptual Framework and Research Agenda. *Journal of Service Management*. https://doi.org/10.1108/JOSM-05-2020-0140

U.S. Department of Labor. (2020). *Families First Cononavirus Response Act: Employee Paid Leave Rights*.

UNESCO. (2020). *Adverse Consequences of School Closures*. United Nations Educational, Scientific, and Cultural Organisation. https://en.unesco.org/covid19/educationresponse/consequences

UNICEF. (2020). *The Effect of the COVID-19 Pandemic on Jamaican Children Preliminary Results*. https://www.unicef.org/jamaica/reports/effect-covid-19-pandemic-jamaican-children-preliminary-results

WHO. (1995). *Global Strategy on Occupational Health for All*. WHO. https://apps.who.int/iris/bitstream/handle/10665/36845/WHO_OCH_95.1.pdf?ua=1

WHO. (2010). *Healthy Workplace Framework and Model: Background and Supporting Literature and Practices*. WHO. http://www.who.int/occupational_health/healthy_workplace_framework.pdf

CHAPTER 20

COVID-19 and Food Security in Fiji's Urban Areas: The Rise of Urban Farming and Home Gardening as a 'New Way of Life'

Gregoire Randin and Peni Turagabaleti

Introduction

Fiji is a Small Island Developing State (SIDS) in the South Pacific with approximately 110 inhabited islands. Its population of 935,000 mostly live on the country's two major islands, Viti Levu and Vanua Levu. The capital city is Suva, with a metropolitan population of over 300,000. About 60% of Fijian people are Indigenous Fijians (iTaukei) and about 35% are Indo-Fijians (Fijians of Indian origin). Fiji was colonised in 1871 by the British, and gained independence in 1970. The country is usually a major tourism destination in the region, although its growth slowed

G. Randin (✉)
University of Sydney, Sydney, NSW, Australia
e-mail: gran7344@uni.sydney.edu.au

P. Turagabaleti
University of the South Pacific, Suva, Fiji
e-mail: gran7344@uni.sydney.edu.au

Y. Campbell, J. Connell (eds.), *COVID in the Islands: A comparative perspective on the Caribbean and the Pacific*,
https://doi.org/10.1007/978-981-16-5285-1_20

down in the past decade. About half the population are involved in agriculture, which roughly corresponds to the number of people living in rural areas.

Although Fiji had only a few cases of COVID-19 (less than forty as of December 2020) and managed the virus' spread successfully, its impact on the nation's economy was dramatic. Fiji is heavily reliant on tourism, and COVID-19-related global and national air travel restrictions created a massive surge in unemployment rates which became a threat to food security for an important percentage of the population. In urban areas, where people are usually not self-sufficient, the government and communities were quick to respond to that threat by enforcing urban agriculture practices, and using any available piece of land for this purpose. In the last century, ethno-geographer Randy Thaman saw urban farming, or, as he called it 'urban food gardening' or 'homegarden agroforestry' as an important pillar of sustainable development in the South Pacific, and especially in Fiji, that would not only address nutrition-related health problems, food security, poverty alleviation and trade deficits, but also help protect and enrich the cultural traditions of those who were increasingly out-migrating from rural areas and embracing urban living (Thaman, 1975, 1995). In this century that vision has somewhat faded, as even the sites of impoverished squatters came under pressure, and local food 'safety nets' disappeared along with contributions to household incomes and urban food supply (Thornton, 2009). Although the 2010s may have seen a further decline in urban farming in Fiji, Thaman's vision of urban gardening is now more relevant than ever in the case of COVID-19 and food security in Fiji, as we emphasise in this chapter. Thus, we argue that the socio-economic impact of the virus has triggered the emergence of urban farming in greater Suva as an effective way of ensuring food security and has created a mini 'new way of life'.

Methodology

In order to build the foundation of this chapter, we used a qualitative methodological approach; data were gathered via first-hand observation in the city of Suva, analysis of media coverage and policy as well as formal and informal interviews conducted with people were practicing forms of 'COVID-related urban farming'. Eleven formal interviews were conducted across all wards of Suva city between April and November 2020. All participants were working or middle-class Indigenous Fijians (iTaukei), and practiced urban farming besides their regular employment; seven were male and four were female. Indo-Fijians were also active participants in urban food gardening. Interviews were conducted in English (although Fijian words were regularly used) and were semi-structured and open-ended. All participants agreed to participate to the study, and we guaranteed their anonymity.

Central to our perspective was the idea that epidemics do not create 'new' situations but rather exacerbate social realities that were already existing, or, as anthropologists reflecting on COVID-19 have put it: 'When a new virus disease emerges, people rely on pre-existing and competing cultural explanations of infectious diseases' (Ennis-McMillan & Hedges, 2020). In this chapter, we therefore also provide some elements necessary to understand the wider context in which urban farming as a response to COVID-19 has emerged.

COVID-19 and the Threat to Food Security: The Reinforcement of Urban Farming in Suva

Despite the few cases of COVID-19 in Fiji, the pandemic undoubtedly and indirectly uncovered the vulnerability of Fiji's food system. The most vulnerable people were those living within urban and peri-urban communities who were not self-sufficient. Urban farming fast became a key instrument in their response to the pandemic and a survival instrument in the face of global crisis.

In 2017, 2.3 per cent of the population lived in extreme poverty in Fiji (World Bank, 2017), but this number was expected to increase. The unfolding COVID-19 pandemic had a far-reaching impact on Fiji's food system. While Fiji is self-sufficient in many food items, it nevertheless depends on imports, including cereals (especially rice), vegetable oils and vegetables, among others (Farrell et al., 2020). There is also a major difference between rural and urban areas, the latter being much more dependent on processed and purchased food (Sherzad, 2020). Imported food was limited by transport problems during the pandemic and some food prices rose, contributing to worsened food security at at time when it was already declining (Sleet, 2019).

Both the rise in value and appreciation of locally-grown food and the practice of urban agriculture are not a totally novel situation but are embedded in a pre-existing, national context and value system, even if lying dormant and largely ignored prior to the pandemic (not, for example, being a response to climate change). The context in which their new significance occurs is one where an old debate has been overshadowing discourses and policy-making around development issues since Fiji's independence in 1970: the articulation (or lack thereof) of a 'modern' way of life and a 'traditional' way of life (Ravuvu, 1988). Across the country, different categories of actors and social groups view development differently; some rather favour a market-based, neo-liberal and monetary approach of development (a 'modernist' approach'), where others favour a decolonizing, pre-colonial tradition-focused and subsistence agriculture-focused approach (a 'traditionalist' approach) (Ravuvu, 1988). COVID-19, a foreign-born virus brought through a modern form of transport (air travel) that principally hit modern institutions (an economy based on international tourism), seemed to have suddenly weakened the value of the 'modernist' approach of development and to have, at least momentarily, re-valued the notion of agriculture. In some rural areas of Fiji, COVID-19 clearly resulted in increased subsistence farming activities and has, momentarily at least, damaged the trust that people had put in modern institutions, such as the cash income that comes through wages (Randin, 2021).

We believe that this phenomenon has occurred, to a significant extent, in urban areas, and spread to the nation as a whole, as in the following statement from Fiji's Permanent Secretary for Agriculture, Ritesh Dass, where he stated that agriculture should be Fiji's 'new way of life' in a COVID-affected nation:

> Agriculture should be our way of life and not just an activity or a profession. Fiji has proven its resilience in dealing with the crisis quite rapidly and this is what we should be proud of as a nation (quoted by MaiTV, 2020)

Urban farming involves individuals, families or communities, who grow crops and/or keep livestock, using otherwise unused urban and peri-urban spaces, for food production, either for consumption or to sell the products on the roadside or at local markets. It is often seen as a viable safety net for the urban poor (Ngoga & Delbridge, 2020), especially in times of pandemics, and it has the potential to provide a readily accessible food source as well as being a source of income, for those deeply affected by crisis. More generally urban farming and urban gardening are seen as perhaps the most culturally and cost-effective means of simultaneously addressing many of the economic, cultural, nutritional, and environmental problems arising out of urbanisation and globalisation (Thaman et al., 2006).

With increasing job losses, many Fijian families took up urban farming practices, which rapidly changed Suva's landscape, as some of the participants themselves observed:

> People are starting to plant in their garden. It's new. (Suva dweller, 2020)
>
> Before the COVID, not many people were planting. After COVID people started to plant in the backyard. Now you can see many backyards have been used. Before it was just a waste. (Suva dweller, 2020)
>
> It is also a wake-up call for the people during this COVID. And now, people are looking to the right and left, after losing their jobs or having cut working hours, looking to the backyard is the only way (to survive). So they make use of it. If you look around, many backyards are changing compared to before. (Suva dweller, 2020)

Most of the participants saw this practice as a pathway towards ensuring a steady supply of nutritional food, as well as a means of securing an extra source of income when needed (for instance by selling products at the market, or to family and friends).

> Most of the people (do the planting) because they try to put food on the table because they lost their job. So that's one of the ways where they can still put food on the table for the family. Some are also working less hours. (Suva dweller, 2020)
>
> Mostly people plant vegetables. Some use it for food and also in terms of getting some money. (Suva dweller, 2020)

Nearly all the new urban farmers were not doing urban farming as a form of 'replacement' of their previous job, but as a supplementary activity for whenever they were not working:

> We do the planting early morning, in the afternoon, during the weekends and holidays. (Suva dweller, 2020)
>
> When we come back from work, (we) can spend one or two hours in the backyard. (Suva dweller, 2020)

The rise of urban gardening was paralleled by the emerging value of what is called in Fijian ***kakana dina*** (literally meaning 'true/real food'), which mostly comprised root crops such as ***dalo*** (taro), yams, cassava and sweet potatoes and closely resembled what constituted the Fijian traditional pre-colonial diet: "Within the category of food (*kakana*), Fijians distinguish between "true food" (*kakana dina*) and the relish which accompanies it (*na kena i coi*). The subcategory of "true food" includes such root crops as taro, yams, sweet potatoes, and the more recently introduced *tavioka* (cassava), some tree crops such as breadfruit and plantains, and wild yams. Some of these are seasonal (domesticated and wild yams, breadfruit) while others are not (taro and *tavioka*)" (Turner, 1984). Most people told us that they were indeed planting a majority of root crops, and many stated that they were seeing COVID as an opportunity to go back to a *kakana dina* diet. This pattern returns to the 'modernist' versus 'traditional' values previously discussed, where the latter is, once again, reinforced. Some people mentioned the re-connection to an ancestral diet, while others just highlighted the perceived superior nutritional values of the root crops that compose the *kakana dina* diet:

> We plant *kakana dina*. In Fiji it's mostly the dalo, cassava, sweet potatoes; it's like the main thing here, the food that feeds you well, you know the kind of food our ancestors used to eat. So the coronavirus, it teaches us a lesson; like we should learn from it, instead of buying everything. (Suva dweller, 2020)
>
> We plant mainly dalo and cassava, yeah, the *kakana dina*. That's the most useful food for us, like this we need to buy less rice or flour... (Suva dweller, 2020)

Some urban farming practitioners stated that people usually favoured sweet potatoes, as they are faster to mature and ready for consumption in about three months as opposed to a longer period for *dalo* (about eight months) and cassava (about six months). However, our observations across Suva, suggested the two latter were more prevalent than the former (see Fig. 20.1). Intrguigly sweet potato is a relatively recent arrival in Fiji compared with *dalo*.

Although *kakana dina* crops seemed present in nearly all of the participants' practices of urban agriculture, throughout Suva city most people also planted other vegetables. The Fiji government provided starter packs which went beyond what was usually labelled as *kakana dina*; some of these packs contained 12 different crops, that included several short-term crops, *bele* cuttings (Hibiscus manihot), tomato, cabbage, and bean seeds (Movono, 2020). One of the major reasons behind the planting of vegetables was their easy conversion into cash money (being faster to grow), where *kakana dina* is rather preferred for consumption.

Fig. 20.1 Urban farming sites in Suva. From top, left to right: Taro plantation in Samabula ward, cassava plantation along the road in Muanikau ward, cassava plantation in Tamavua ward, yam and cassava plantations in a yard in Muanikau ward. (Credits: *G. Randin, P. Turagabaleti, November 2020*)

> Ok, we plant like, cabbage, tomatoes, some cassava, dalo. So, for some seeking a fast way to get money, they will plant vegetables. That's a very fast way to make money. (Suva dweller, 2020)
>
> We started doing some plants there, like bananas, cabbage, carrots, lettuce. They can feed us and we can sell them too if we are short on money. (Suva dweller, 2020)

It became common to walk through different parts of Suva and find urban farms in every corner; cassava, taro, yam, sweet potato, bread fruit, and plantain were planted in nearly every plot of land, whether they were the size of a football field or of a single car parking space.

People perceived the new urban farming as contributing very positively to Fijian society, corresponding to the Permanent Secretary's statement on agriculture as Fiji's 'new way of life'. Many of the 'new urban farmers' we interviewed stated that the practice 'brought people together' behind the idea of fighting the effects of the virus through locally grown, organic food. When asked if he encountered theft in their new gardens, one participant stated that it would not happen in such a context:

> We never had any problems with people stealing the crops. At this time, people are getting together, help each other. (Suva dweller, 2020)
>
> We see that everyone is working together, they understand the value of not only working together as one, also because the crisis we are in, they understand that keeping healthy and looking out for one another is more important. (Suva dweller quoted by ABC, 2020).

Others also mentioned the creation of a new barter system organized through Facebook, where members could exchange vegetables with other goods or food:

> (…) We have a system that is going on here in Fiji which is the barter system; whenever you have enough vegetables you can share it (the information) through social media so people who want to buy some groceries can exchange it with something they have. (Suva dweller, 2020)

Sharing food and bartering were both part of ***na solesolevaki*** (working together) that emerged more strongly and proved valuable coping mechanisms although conflicts over land and the theft of valuable crops challenged such benefits (Wairiu et al., 2020). Another Suva resident argued that urban farming practices were good for people's mental health in a time where stress and uncertainty were unusually high:

> It brings a lot of positive things for the people. It keeps the people busy instead of getting into depressions, instead of thinking about the impact of COVID. Instead, they are doing something, and later they find out the outcome (of their planting) and they are so excited. The main thing is that it gives food and money too. (Suva dweller, 2020)

Others stated that urban farming contributed positively the looks of the visual environment in Suva and increased people's well-being and satisfaction when walking outside.

> Before backyards are wasted. Now you can see plots of vegetables. It is looking good too for the environment around. (Suva dweller, 2020)

This 're-enchantment' with agricultural practices in Suva also represented a form of re-connection between urban practices and rural practices, where urban dwellers revalued some aspects of a more obviously rural society, including a more traditional '***vaka-vanua***' (way of the land) where ***vanua*** encompasses land, traditions and people, both natural and cultural elements. In villages *vanua* notions of 'communality', sharing one's resources and 'togetherness' were crucial for people's various survival needs (Ravuvu, 1983), and the virus brought back this value (in a novel form) to urban areas: 'Vanua notions of 'togetherness' and self-sufficiency present in rural communities also became central to the discourses of many urban households' (Randin, 2021). Several urban dwellers mentioned this re-connection betwen rural and urban knowledge and lifestyles, and the idea that urban people have had to 're-learn' forgotten traditional practices that may be one or two generations away

> In the village (it) is their daily life to do the planting like this. Only the people in urban areas they start to learn this again. But everyone knows how to do it, it's in the blood. (Suva dweller, 2020)

Yet most urban residents did not abandon urban practices for rural ones; they simply shifted some of their interests and activities towards 'usually typically rural' practices and realities. Such a radical shift was even more evident when people who lost their job due to COVID-19 decided to return to rural areas (Wairiu et al., 2020).

Policy makers were relatively quick to undertake actions to reinforce and foster these practices and support the positive perceptions of people concerning urban farming (Sherzad, 2020; Randin, 2021). The government of Fiji encouraged all Fijians to 'go local' and plant vegetables to ensure access to safe, adequate, and healthy food. In the early stages of the virus in Fiji, the government also gave free seeds to urban families, to encourage urban farming (Government of Fiji, 2020) while the unemployment rate was increasing. Indeed, according to a statement from the Prime Minister of Fiji, Mr. Voreqe Bainimarama, over 115,000 Fijians, or one-third of Fiji's workforce lost their jobs, or have had their hours cut within a few months as the economy collapsed (Krishant, 2020).

As part of this 'home gardening' initiative, the Minister of Agriculture, Dr. Mahendra Reddy, mentioned in an interview that farmers could also choose crops such as rice, taro, sweet potatoes, and cassava. As of May 2020, at least 5000 households had benefited from the free seeds policy (Movono, 2020). Later, Dr. Reddy pleaded to market vendors not to take advantage of consumers by increasing prices in these times of necessity. The Ministry of Agriculture recorded a corresponding increase in entrepreneurial farming amidst COVID-19 (Vacala, 2020). In both urban and rural areas, wherever there was land, people proved willing and enthusiastic to turn it into agriculture.

COVID-19: A Catalyst for a National Nutritional Shift?

Agriculture, of course, cannot be separated from food, as the primary goal of the former is the production of the latter. Health is also part of the equation, as it is intrinsically linked to people's diet. The dietary shift as a response to food insecurity after COVID-19 was embedded in the same socio-environmental context as the emergence of urban agriculture. In the past twenty to thirty years, much research on issues revolving around food security, and the role of imported and processed food versus traditional food, has attributed the rise of non-communicable diseases (NCDs) in Fiji to the adoption of a modern diet to the detriment of a traditional one (Waqanivalu, 2010). Consequently, although food availability in Fiji is generally good 'Fiji suffers from a double burden of over and under-nutrition' (Sleet, 2019). Adult obesity affects about 35 percent of Fijians (a little less than the percentage of the population living in urban areas, although there are also obese people in rural areas) and non-communicable diseases, such as type 2 diabetes, also concern about a third of Fiji's population. Similarly, micronutrient deficiencies are also common, with many nutrients consumed below recommended levels.

Many researchers directly attribute Fiji's poor nutrition to an increasing dependence on cheap imported food and a decreased intake of traditional food (Waqanivalu, 2010; Sleet, 2019). Waqanivalu (2010) associates the increased consumption of processed food to an increase in the value and 'status' of the latter: 'Promotion of traditional foods has fallen by the wayside. They are unable to compete with the glamour and flashiness of imported food'. In an article exploring the potential scenarios that the COVID crisis would provoke in the Pacific region, Farrell et al. (2020: 786) predicted that the unprecedented situation could result in a shift towards a healthier diet if people were to effectively embrace locally-produced, organic food production and consumption: 'Bolstering regional production and intraregional trade in a currently import-dependent region could strengthen the regional economy, and provide the health benefits of consuming locally produced fresh foods – as well as decreasing reliance on global supply chains.' (2020: 786). Based on our observations and interviews, this accurately predicted some of the consequences of COVID-19 for the people in urban Fiji.

People with whom we interacted usually highlighted the potential positive outcome that 'COVID backyard farming' could bring to health. We witnessed on many occasions that 'eating healthier' was a motivation to practice urban farming, as was evident in the following statements:

> Now people care more about eating healthy food, from their own garden. (Suva dweller, 2020)
>
> We can learn from this eh. Eating more healthy and our own food. We don't need to go to the store for everything. So COVID is good too in that sense. (Suva dweller, 2020)
>
> It's good that we grow our own food. Before, we usually (had to) wait for people (our relatives, friends) who come from the village to bring us cassava. Now we can plant it by ourselves, eat healthy food from here. (Suva dweller, 2020)

However since urban households now experienced reduced incomes (with lost employment and reduced market sales) the trend towards locally grown food was also matched by a trend towards the consumption of the cheapest processed foods, especially noodles and snack foods, alongside other cheap goods, alongside a decline in the diversity of diets that conteracted the benefits from planted foods. Indeed the worst food insecurity was in peri-urban areas where urban farming was most common (Wairiu et al., 2020).

The consumption of 'true food' nonetheless regained value during the pandemic and what people most planted in greater Suva, began to blur the pre-existing distinction between 'urban diet' and 'rural diet'. Urban farming as a 'new way of life' had four main pillars: the rise in the value of ***kakana dina***, the rise in the value of a healthy diet, the rise in the value of 'locally produced/organic food' and the rise in the value of having extra means of food and economic security (Fig. 20.2), without any particular hierarchical order.

Conclusion: COVID-19 as an Opportunity for Social Change?

Even though the government was quick to react and enact policies that were well adapted to people's needs and that certainly contributed to the emergence of urban farming at the beginning of the pandemic, policymakers largely failed to continue either the implementation of such policies or to evaluate them over time. No official research was undertaken on the phenomenon, despite its significance. The lack of policy on urban farming in Fiji, given its intensification, may well create environmental and possibly

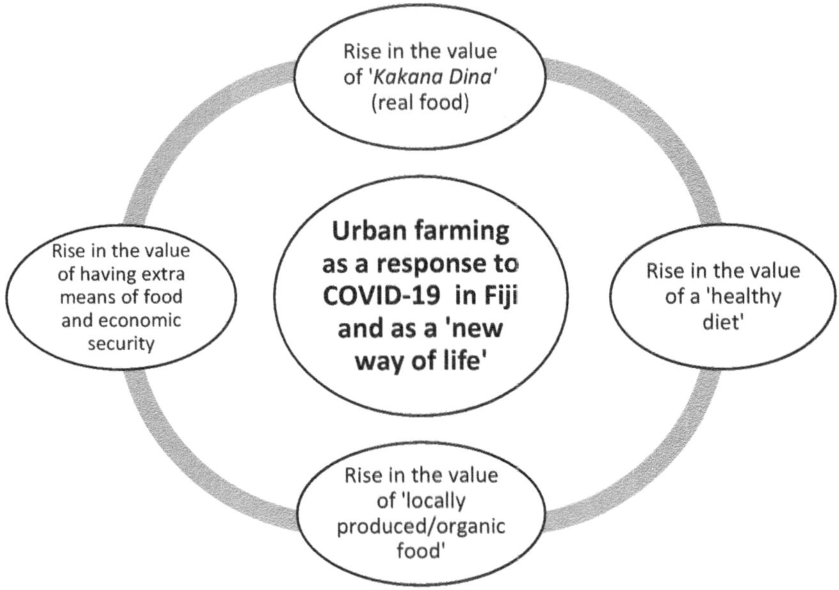

Fig. 20.2 The constituents of urban farming as a COVID-19 response. (Credits: *G. Randin, P. Turagabaleti*)

health problems (since urban plantations must be at least 7.5 metres away from the roadside to be safe for consumption) (Thaman, 1995). Indeed a range of factors have always stood in the way of urban agriculture including poor soils, costs, availability of land and water, theft, lack of planting materials, and lack of government assistance (Thaman et al., 2006; Thaman, 1995) and urban agriculture has been positively discouraged (Thornton, 2009). The very limited involvement of government in providing seeds and support was thus exceptional.

By contrast, as the involvement of so many households indicated, the revitalisation and rise of urban gardening was stimulated at the grass (taro) roots. This emergence of urban farming as a social response to COVID-19, food security and diet offered a possible pathway to a positive social change in Fijian society through a 'neo-traditional approach' (Shah et al., 2018), where the consumption of locally-grown healthy 'true food', and the safe practice of semi-subsistence urban agriculture could further develop in a future free of COVID-19. It is impossible to predict with certainty whether these changes will last in a post-COVID world, especially because urban farming is an 'extra' supplementary way of producing and securing food, and it has not changed or replaced previous forms of employment or food habits, while it is not always feasible to combine this with formal employment or increase the diversity of production. We asked participants what they thought would happen after COVID-19, and their answers varied; some predicted that people would be 'too busy to continue', whereas others stated that people would continue because of the unexpected benefits the practice brought them in sociality, income and nutrition:

> Only the working people, when they get the work back, it will be hard. (Suva dweller, 2020)
>
> It can only stop if people are going back to work and don't have time to do the gardening again; but I think that it will stay long. (Suva dweller, 2020)
>
> I think it will keep on, because most of the people now are seeing the positive outcomes (Suva dweller, 2020)

Even if the of rise of urban farming were not to profoundly change Fiji's urban society beyond the end of the virus, and its implications, it will undoubtedly enhance urban people's resilience and preparedness for future crises, and may even provide solid foundations for future socio-economic and dietary revolutions in the island nation.

References

Ennis-McMillan, M., & Hedges, K. (2020). Pandemic Perspective, Responding to COVID-19. *Open Anthropology, 8*(1). https://www.americananthro.org/StayInformed/OAArticleDetail.aspx?ItemNumber=25631

Farrell, P., Thow, A. M., Wate, J. T., Nonga, N., et al. (2020). COVID-19 and Pacific Food System Resilience: Opportunities to Build a Robust Response. *Food Security, 12*(4), 783–791.

Government of Fiji. (2020). Statement by the Minister of Agriculture Dr Mahendra Reddy on Food Security and Agriculture Growth Expansion Plan. (Online), *The Fijian Government.* Available at: https://www.fiji.gov.fj/Media-Centre/Speeches/Statement-by-the-Minister-for-Agriculture-Dr-Mahen. Accessed 2 Nov 2020.

Krishant, N. (2020). PM Confirms 115,000 Fijians Have Lost Their Jobs or Have Had Their Hours Cut as a Result of COVID-19. (Online), *Fiji Village.* Available at: https://www.fijivillage.com/news/PM-confirms-115000-Fijians-have-lost-their-jobs-or-have-had-their-hours-cut-as-a-result-of-COVID-19-8fxr45/. Accessed 2 Nov 2020.

Movono, L. (2020). Free Seeds from Government for Urban Farming. (Online), *Mai Tv*. Available at: https://maitvfiji.com/free-seeds-from-government-for-urban-farms/. Accessed 2 Nov 2020.

Ngoga, T. H., & Delbridge, V. (2020). *Urban Agriculture: A Viable Safety Net for the Urban Poor During Times of Crisis*. (Blog), International Growth Centre. Available at: https://www.theigc.org/blog/urban-agriculture-a-viable-safety-net-for-the-urban-poor-during-times-of-crisis/. Accessed 3 Nov 2020.

Randin, G. (2021). COVID-19 and Food Security in Fiji: The Reinforcement of Subsistence Farming Practices in Rural and Urban Areas. *Oceania, 90*(S1), 89–95.

Ravuvu, A. (1983). *Vaka i Taukei. The Fijian Way of Life*. USP Institute of Pacific Studies.

Ravuvu, A. (1988). *Development or Dependence*. USP Institute of Pacific Studies.

Shah, S., Moroca, A., & Bhat, J. (2018). Neo-Traditional Approaches for Ensuring Food Security in Fiji Islands. *Environmental Development, 28*, 83–100.

Sherzad S. (2020). *Impacts of COVID-19 on the Food Systems in the Pacific Small Island Developing States (PSIDS) and a Look into the PSIDS Responses*, FAO. http://www.fao.org/in-action/food-for-cities-programme/news/detail/en/c/1278570/

Sleet, P. (2019). Fiji: Poor Nutrition and Agricultural Decline Has Caused Food Security Slump, Future Directions International

Thaman, R. (1975). *Urban Gardeningin Papua New Guinea and Fiji: Present Status and Implications for Urban Land Use Planning*. University of the South Pacific.

Thaman, R. R. (1995). Urban Food Gardening in the Pacific Islands: A Basis for Food Security in Rapidly Urbanising Small-Island States. *Habitat International, 19*(2), 209–224.

Thaman, R., Elevitch, C., & Kennedy, E. (2006). Urban and Home Garden Agroforestry in the Pacific Islands: Current Status and Future Prospects. In B. Kumar & P. Nair (Eds.), *Tropical Homegardens: A Time-Tested Example of Agroforestry* (pp. 25–41). Springer.

Thornton, A. (2009). Garden of Eden? The Impact of Resettlement on Squatters' 'agrihoods' in Fiji. *Development in Practice, 19*(7), 884–894.

Turner, J. (1984). True Food and First Fruits: Rituals of Increase in Fiji. *Ethnology, 23*(2), 133–142.

Vacala, K. (2020). COVID-19 has Affected Nutritional Intake of Fijians. (Online), *Fiji Broadcasting Corporation*. Available at: https://www.fbcnews.com.fj/news/health/covid-19-has-affected-nutritional-intake-of-fijians/

Wairiu, M., Iese, V., Navunicagi, O., Fesaitu, J., et al. (2020). *Assessing Nutrition and Socio-Economic Impact of COVID-19 on Rural and Urban Communities in Fiji*. USP Pacific Centre for Environment and Sustainable Development.

Waqanivalu, T. (2010). Pacific Islanders Pay Heavy Price for Abandoning Traditional Diet. *Bulletin of the World Health Organization, 88*, 484–485.

World Bank. (2017). *Republic of Fiji Systematic Country Diagnostic*. World Bank.

CHAPTER 21

Catching Fish in COVID-19 Currents: Food Security and Governance in Rural Communities in Solomon Islands

Anouk Ride, Hampus Eriksson, Jillian Tutuo, Chelcia Gomese, and Delvene Boso

Introduction

The year 2020 was marked by disruption and change but there were some constants in Solomon Islands, a least developed Melanesian country neighbouring Papua New Guinea and Vanuatu in the western Pacific. Solomon Islands has a population of around 700,000, of whom about

A. Ride (✉)
Australian National University, Canberra, ACT, Australia
e-mail: A.Ride@cgiar.org

H. Eriksson
University of Wollongong, Wollongong, NSW, Australia
e-mail: hampus@uow.edu.au

J. Tutuo • C. Gomese • D. Boso
WorldFish, Honiara, Solomon Islands
e-mail: J.Wate@cgiar.org; Gomese@cgiar.org; D.Boso@cgiar.org

Y. Campbell, J. Connell (eds.), *COVID in the Islands: A comparative perspective on the Caribbean and the Pacific*,
https://doi.org/10.1007/978-981-16-5285-1_21

90,000 live in the capital, Honiara, with most living on around 40 populated archipelagic islands. The economy is based on agriculture and timber, with a limited 'modern' sector. The Solomon Islands remained COVID-19 free until October when a returning student brought the virus from the Philippines, but quarantine prevented any spread. Prior to lockdowns and closures in March, in Honiara, one of the key items people stocked up on was canned fish, while in coastal areas fresh fish remained a staple. As is the case in other countries in Melanesia, most people (80%) are reliant on food they can harvest or grow to sustain themselves and sell at informal markets, while urbanisation and reliance on the formal economy, driven by timber and fish exports, continues to grow.

Fish and other aquatic foods are the most accessible and widely consumed animal source food by coastal people in the Solomon Islands (Albert et al., 2020). This broadly includes fin-fish species, shells and other invertebrate species, as well as processed fish meat, notably the popular and locally produced tin-tuna made by SolTuna. Patterns of consumption and access to the nutrients within aquatic foods vary depending on location, with some islands being more or less reliant on reef fish, oceanic fish, seaweeds, mangrove fruits and bivalves. Societies in Solomon Islands have organised themselves around rural food production. The bartering and trade of produce between people residing in coastal areas and those residing in inland areas shaped the history of interaction between peoples and distribution practices influenced access to nutritious indigenous foods; the exchange of roots and fish is the heartbeat of the Melanesian food system (Ross, 1978).

Almost everyone in Solomon Islands eats fish regularly, but most people are not fishers. Fish are caught, distributed and then acquired by consumers through purchase, gifts or bartering. This system of producing and distributing fish connects remote sources of supply with urban areas of demand and generates indispensable value, both in the form of fish-based livelihoods for the many people involved, and for food and nutrition security in island populations. In the Pacific as a whole, per capita coastal fisheries production is declining (Gillett, 2016). Some estimates state that within a decade an estimated additional 115,000 tonnes of fish per year will be needed across the region for good nutrition (Bell et al., 2009). In addition to this coastal fish supply gap, the region experienced a rapid 30% drop in domestic crop production during the 1980s which has not been recovered. These declining trends in domestic food production are synchronous with the doubling of food imports (Farrell et al., 2020). Shifts from domestic production of nutritious crops and weakening fisheries to imported nutritionally unrewarding foods is symbolic of a contemporary transition to "modern" and convenient foods that has been underway for some time as Solomon Island populations have urbanised (Andersen et al., 2013). These trends have been accompanied by growing recognition of coastal fisheries crucial importance as sources of aquatic foods and livelihoods to the ocean people in the Pacific.

Coastal Fisheries, Food Security and COVID-19 Currents in Solomon Islands

Given the constant presence of fish in livelihoods and food security, it is worth considering its role in crises – such as disasters, conflict and other economic and political disruptions. Pacific SIDS are no strangers to sudden shocks; four of the six nations most at risk from natural hazards and societal vulnerability in the world are Pacific Island nations (World Risk Report, 2019). Solomon Islands is number three on the list.

Living through shocks and disruptions builds resilience (Maru et al., 2014). This resilience can come in the form of social capital (such as cooperation through extended family networks) and traditional or normative practices (such as sharing resources) or in the form of preparedness systems in formalized institutions and processes. In Solomon Islands, environmental conditions are favourable for a range of crops, and the majority of the population (around 92%) is engaged in agriculture in some way on the 1.1 million ha of agricultural land in use (SIG, 2019). Fishing is almost equally widespread and productive through the archipelago and in the 1.3 million km sq Exclusive Economic Zone. In rural areas of Solomon Islands, 68% of households report catching fish or shellfish, and even in urban areas one-third of all households are engaged in this activity (SINSO, 2013). Fish and fisheries are an important source of resilience in many coastal areas during periods of hardship in Melanesia (Eriksson et al., 2017).

It is against this backdrop of shifting domestic production and distribution demands for coastal fish that the COVID-19 pandemic disrupted health, food and economic systems.

The economic disruptions and population movements in Solomon Islands had not been experienced on this scale since the civil conflict from 1998 to 2003, which involved armed militants from two different islands enacting violence and a general breakdown in law and order (SITRC,

2013). Although the pandemic did not create a direct health crisis in most of the Pacific, governments and regional institutions responded by developing mitigation and adaptation measures. In Solomon Islands, the government instituted shutdowns, closed borders and cut salaries of public servants.

Under the State of Emergency (SOE), declared by the Governor General on March 25 as a response to the crisis spurred by COVID-19, the government encouraged urban dwellers without formal employment to return to their provinces. The suspension of schools and education institutions and the limits on movement and large gatherings of people meant many that Solomon Islanders left Honiara. Another key driver behind internal migration was the crackdown on the informal sector. Street side and suburban marketing, a source of income for many households, was banned in Honiara under the SOE measures, then quietly returned without sanctions. The informal sector, including subsistence production and sale of goods at the household level, has a significant impact on the fortunes of provinces and the nation. Wages and salaries (cash and in-kind) made up 24% of total income of households in 2013, followed by income from household-based businesses, at 22% (SINSO, 2013). Both directions by the government and changes in daily lives initiated widespread migration to the provinces and increased populations in the small towns and villages.

The newly developed Solomon Islands government economic stimulus package plan included an emphasis on supporting the primary production sectors of fisheries and agriculture through grants to individuals and organisations. As the spending of these funds fell outside usual ministry projects, there was some concern about transparency of the package in the media (including over 50 articles in a national newspaper "Solomon Star"). At the time of writing this chapter, data about the package's specific spending on fisheries was not available, other than that allocated to the fish processing company SolTuna.

At the end of November 2020 legitimate concerns remained about the state of food and nutrition security. A report published in May 2020 (Eriksson et al., 2020) provided a rapid assessment of how the COVID-19 situation affected food security, particularly the disruption caused by government directives for people to return from Honiara to the provinces, and the broader economic effects, such as job layoffs, prompted by the crisis. WorldFish staff telephoned male and female community leaders to discuss local food supply and consumption in their locations, and examine how the economic and health crisis brought on by COVID-19 was experienced by rural coastal community members.

Data were collected from 20 communities in Malaita and Western Province between April and May 2020. These two provinces are among the most populous and most reliant on fish as a staple food in the nation. A total of 35 people (15 women, 20 men) were interviewed. Each interview was carried out by a WorldFish staff member with previous training and experience in survey data collection and who was known to the interviewee through previous interactions on WorldFish projects. Interviewers carried out the interviews in Pijin, the country's lingua franca, or in local languages. Topics for survey questions included population movement, food production, adequacy of food supply, changes to fish catches and sales, food prices and changes to fishing rules and management. The survey used was later shared and promoted under the coordination of the locally managed marine area (LMMA) network (SPC, 2021). Although the survey was designed to take less than 10 minutes, oral storytelling styles of communication meant that it often took longer. People are identified by gender but not village or other identifiers.

Thirty-one of the respondents (91%) noted that rural populations had increased (Fig. 21.1). Five of these respondents noted that the influx was minimal, and two noted that populations decreased again because some people had returned to Honiara as schools re-opened. Sixteen respondents noted food shortages (47%), while 18 noted that there was enough food in the village (53%). Seven of the respondents that reported enough food had concerns about long-term food security.

Women were slightly more likely to be concerned with the quantity of food, but this was attributed to a range of factors, including the timing of

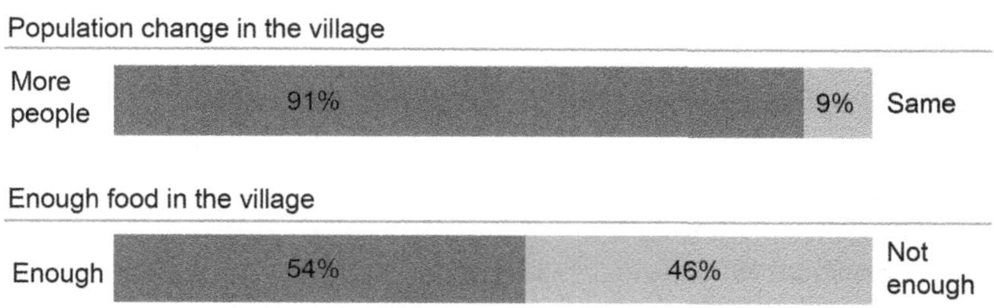

Fig. 21.1 Responses to population changes and pressure on food in villages

the survey and women's larger burden of being responsible for food for households. Surveys conducted later in May tended to be more negative about the food situation than those conducted a few weeks earlier.

> *Most families skipped breakfast and lunch. Only dinner is usually served. But during worst cases, even families go without food for a whole day. This situation is getting worse.* Children *went to school without even having breakfast. Some even survive on dry coconut. This is due to unfavorable weather (rain) experience during the month of April.* (female respondent, Malaita Province)

Some of the strategies to make food last (other than producing more) included minimizing consumption, skipping meals, reducing sales and bartering different types of food (Fig. 21.2). While most strategies were household-based, some were broader, extending into clusters of households or whole communities:

> *Meetings were held to discuss issues of concern, one of which is to advise families to minimize the harvest of food gardens and food* consumption. *Currently, families are harvesting just enough to satisfy their main meals.* (male respondent, Malaita Province)

All but one respondent reported some form of adjustment in their food production, with 34 (97%) specifically mentioning increased production of food through gardening. Increased production in family gardens was the most common practice; however, group gardens, where clusters of families, either related or in a particular place, would work together to plant and harvest, were highlighted as a strategy by three respondents.

> *Everyone focuses on putting more effort into maintaining or extending food gardens. Families plan to increase garden production to ensure food is available*

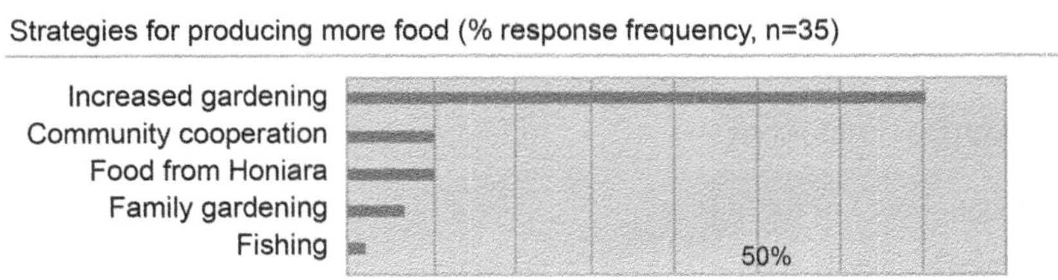

Fig. 21.2 Broad strategies to produce enough food in villages

in the coming months, not only for household consumption, *but also for marketing.* (male respondent, Malaita Province)

Several respondents mentioned that people from town had arrived in the villages without tools and knowledge for gardening. In some locations, those that arrived from Honiara brought supplies with them; however these supplies became depleted and food shortage concerns were raised in this context by three respondents.

> *People did bring food supplies back home, while some unfortunate ones lost most of their food supplies during the MV Taimareho incident [a shipping accident resulting in people overboard and deaths from drowning]. As people already spend about 1 to 2 months at home, they exhausted all supplies, and to make things worse they did not have food gardens, so they rely on relatives for survival.* (male respondent, Malaita Province)

Everyone emphasised staple root crops (sweet potato, giant swamp taro and cassava) as most important for food supply (Fig. 21.3). This was expected given the prominence of root crops in Melanesian diets and local knowledge and the skills for their production. The frequent emphasis on giant swamp taro is important, as it is commonly seen as a "back stop" for slipping into greater food insecurity. The prevalence of these crops in our survey was probably influenced by the information predominantly coming from coastal Malaita, where mangroves and sago swamps are prevalent and

Fig. 21.3 Most important foods for local food and nutrition security in villages, including sweet potato, giant swamp taro and cassava

important in traditional food production. The use of sago starch, a traditional famine food, was not mentioned by respondents.

People pointed to the increased production and/or consumption of locally grown root crops as a result of potential shortages in imported rice. In some instances, rice was viewed as a supplement in order not to deplete food storage in gardens.

> *We are going back to local food, like* bananas, sweet potatoes *and* cassava. *We were already doing that, but more and more people are eating more local food apart from* rice. *This is so our body will be healthy and be able to fight against COVID-19 (information from awareness carried out in the villages). Each family is organized with making their own gardens.* (male respondent, Western Province)

For other food crop production, there were differences between villages and geographies—likely to do with growing conditions and existing village gardening practices. Banana was mentioned by two respondents as part of the priority crops. There was no mention of popular and common cash crop varieties like watermelon, tomato, capsicum, spring onion or papaya, indicating that in periods of uncertainty root vegetables were a primary source of food security.

The coincidence of the COVID-19 pandemic with the cyclone season conflated pressures to create a period of hardship in certain areas, and the influx of people coincided with the period where food production was low because of seasonal changes.

> *[Families] must plant* pana *(small* yams*) and* yams *for the months of March to April during which people always experience food shortages.* (female respondent, Malaita Province)

Across most villages, respondents noted that communities were falling back on, and supported by, traditional practices like bartering and nurturing extended family systems to meet extra demands for fish and garden produce.

> *There is enough food for everyone because we are organised in the community, and we still use cultural practices where families look after each other, especially when someone returns to the village.* (male respondent, Western Province)

Overall rural people have high adaptive capacity and local food system practices acted as an important safety net. Many communities had local fisheries management areas, with rules such as areas closed to fishing (tabu areas), restrictions on time and gear for fishing and some species for which fishing was banned. Twenty of the respondents resided in villages with Community Based Resource Management (CBRM) committees. Out of these, nine reported proactive responses: awareness and information (five respondents), increased enforcement (two), including in some cases by-laws with police backing and organised harvests for community needs (two). One village reported introducing its own rules around social movement in order to minimise risk and improve health awareness as important related activities. Seven respondents reported committees being "inactive" or doing nothing, and in three sites (15% of communities) active conflict situations meant management was not happening.

> *So far, reports of continuous poaching reach the committee. Most reported cases occur during the night. It has been a practice since the management plan was launched. But the good news is so far there are no reports of increased poaching related to the increase in numbers of people coming back to the communities. Most of the people who came back from Honiara are not fishermen.* (male respondent, Malaita Province)

All respondents from villages with CBRM plans stated that people were aware of local rules. However, breaking of fishing rules was widespread and reported in eight of the 12 villages that were practicing CBRM. The infringements related to undersized catches, poaching and disobeying fishing rules. Significantly, such local, rather than national, situations often led to changes in fishing effort and breaking of rules. In some communities, for example, local fisheries management had "broken down" because of tribal conflicts prior to the COVID-19 situation. Fishers' conduct toward local fisheries management was not evidently directly linked to recent changes in villages because of COVID-19, as was indicated:

> *Unfortunately, due to conflicts that arose during the national general election, opposing parties take advantage of the situation and disobeyed the rules that were agreed upon by the community. Since then, the management committee is yet to meet and revisit their plans.* (male respondent, Malaita Province)
>
> *Just two weeks ago, there was a dispute between two of the several tribes that live here. Attempts to resolve the issue by neutral tribes fall on deaf ears, so the managed area was forced to open. Knowing that the area was under*

> *management for years attracted lots of fishermen from distant communities as well as neighboring communities.* (male respondent, Malaita Province)

Although it was anticipated that people would report that fishing efforts had increased to ensure food production for household consumption and cash sales, there were limited and varied increases in fishing efforts, with less than half (46%) of people reporting an increase in fishing (Fig. 21.4). In a few cases, fishing appeared to have declined, notably in Western Province where there was reduced demand for fish from tourism establishments now without visitors and restrictions on movement due to fear of contracting the virus from people coming across the Papua New Guinea border. In villages where fishing had increased, this was explained by the growing need for food and cash.

> *Most who came back from Honiara, the only livelihood that sustains them is* fishing. *If they are to wait for their gardens to be ready, they will starve to death. Therefore, most of them resort to* fishing *as the immediate means to tackle shortage of food. Catches were sold, and the* money *earned is spent on local food or other imported basic food items.* (male respondent, Malaita Province)

Fig. 21.4 Responses on fishing and distribution of fish in villages

> *At the moment, there is still enough food. But if this situation gets worse and the number of people returning to the village continues to increase, food will not be enough. The supply of food from existing gardens now is at maximum. If harvesting increases further, overharvesting will occur, meaning families will go hungry in the weeks to come.* (female respondent, Malaita Province)

Sixteen people (47%) noted that more fish were caught in the villages, while 17 (50%) reported no difference. The catch patterns appeared to vary according to local fishing practices. One village reported that more shells were being collected, while another reported an overall 50% increase in catches. The motivation for increasing fishing efforts was to ensure food supply for own consumption in a period of disrupted cash flow.

> *People tend to fish when they do not earn enough* money *to buy tinned fish due to the cashflow problem ... Increase in fish caught is estimated in the range of 10 to 25 percent.* (male respondent, Malaita Province)

Changes in the marketing of fish varied between villages. Reduced volumes of fish for sale were reported by 26% of people, while 31% reported more fish for sale. Similarly, fish seems to have become cheaper in some places (26% of responses) and more expensive in others (21% of responses). People reported that arrivals from town were either inexperienced fishers or did not have fishing gear, so the influx of people did not necessarily lead to more fishing but to more demand for fish and agricultural crops.

> Fishing *at the moment still remains the same. This is because most people who came back to the village do not own a canoe or even basic* fishing *gear, so they cannot fish. They are yet to fully adapt to village life.* (male respondent, Malaita Province)

Two villages reported more fishing efforts and catches at a nearshore fish aggregating device (FAD). This relatively simple technology is already in use across many villages to attract pelagic fish closer to shore in reach of dugout canoes (Fig. 21.5). The Malaita Provincial Fisheries office had constructed and deployed FADs at several villages in 2020 together with community members.

Fig. 21.5 Fisheries' Aggregation Devices, Malaita Province. (*Photos: Matthew Ishihanua, Malaita Provincial Fisheries Officer, May 5–6, 2020*)

Impacts of COVID-19 Currents on Gender and Fisheries

Women's main source of cash income across the islands has been through informal marketing. In many cases the post-COVID-19 closures of informal markets in urban areas had little effect on women in rural areas as they were more reliant on markets within their village or province. However, at a meeting of rural women in Western Province in October 2020, a key concern was that due to the elevated risk of COVID-19 transmission from people over the border in Papua New Guinea, markets had been restricted and there was also an element of fear over travelling to markets due to possibly coming across traders from Bougainville who might have had the virus.

Restrictions on sales of betel nut had more of an impact on respondents in Malaita, in particular. Young women were less likely than adult married women to engage in marketing fresh produce, but those who did in urban areas were often selling betel nut (World Bank, 2019). The production and sale of this stimulant is a major economic activity and one which the authorities attempted to limit and stop during the shutdowns and state of emergency period, particularly in Auki town and Honiara, partly because the habitual chewing and spitting of betel nut might contribute to the spread of diseases. In 2012–2013, local trade in betel nut was estimated at over $100 million SBD (SINSO, 2013). Disruptions to household economies from changing conditions in betel nut trade during 2020 were noted, particularly in Malaita:

> *Most people depend on the sales of betel nut to pay for imported goods. But due to the* state *of emergency situation, most betel nut vendors lost their means of earning an income and therefore cannot afford to pay for imported food items, which they normally depend on. But with the reopening of the new site for betel nut sales, they should be able to pay for basic food items.* (male respondent, Malaita Province)

Similar to the marketing situation, the national cessation of schools meant that a shift in daily lives and practices among youths was reported. One respondent reported that more youths were engaging with coastal marine resource collection by gleaning the mangroves and reefs. This is a common method for low-input harvesting, making it important for community members, such as youths, who did not have access to fishing gear. Young people rarely participated in decision-making about local fisheries and were also less likely to be involved in decisions around marketing (World Bank, 2019). This means young people are probably taking up roles on the margins of village economies, and, without active inclusion, may struggle to raise individual incomes needed, for example, to support their technical and tertiary studies.

As in many other Melanesian contexts, women were more likely to be caring for children and the elderly, and more likely to be financially dependent on others, such as husbands or parents with salaries. If the economic hardship caused by the COVID-19 situation continues, an important issue will be the effects of job and income losses on dependants, particularly in Malaita, which has the second-highest dependency ratio (96 dependents to every 100 working people) and fertility rate (5.6 children to each woman) in Solomon Islands. The impact of internal circulation of people and of reduced household income on women and children's health and economic well-being will require monitoring in areas of high dependency and vulnerability.

While in most communities, food supply was a family concern, in a few of the communities group gardening across several families and group fish harvests to support community activities and services occurred. One group harvest was organised to buy supplies for the local clinic.

> *Our CBRM committee often meets to discuss matters related to how benefits can be shared or given to other important community groups—for example, clinic, church and school.* (female respondent, Malaita Province)

Despite evident leadership and social capital as sources of resilience, strain was also apparent in some places. For example, unequal access to food and concerns about how long food supplies could last were fuelled by theft and conflicts.

> *People are uncertain about how long their food gardens can last them ... Last week, some students from [village name] mentioned their food from gardens were being stolen.* (female respondent, Malaita Province)

While local governance exists in all communities, its strength in terms of including all people in decision-making, preventing and responding to emerging issues and facilitating sustainable use of common resources varies across the Solomon Islands. Governance to manage the strains of 2020 and its COVID-19 related disruptions was key to adaptation.

Strategies for Fish and Food Security from COVID-19 to the "New Normal"

While many analyses of the effects of COVID-19 focus on employment and job losses, in the Solomon Islands context, a focus on food security is perhaps most apt in assessing impacts on the majority of the population (80% of the population being rural and engaged in subsistence food production practices). Enhancing storage, processing, and distribution of food commodities is vital in mitigating food and nutrition security impacts from changing food production patterns and during the current COVID-19 crisis (Farrell et al., 2020). Most households do not have access to cold storage facilities for food and the drying or local processing of food is largely based on local traditional practices and now found only sporadically across the islands. Imported rice and locally produced canned tuna remain staples, although consumption declined with reduced cash flow. A market survey conducted by WorldFish in 2020 indicated that while local market sellers at formal markets reduced prices to cope with the reduced cash flow of customers, some stores put up prices to offset increased shipping costs and delays. This put both vendors and consumers in a difficult position to acquire and ensure nutritious meals.

At a supply level, availability of fresh fish remained constant while there were indications of lack of other food commodities due to the increased population caused by urban-rural migration. Marketing and markets provide a structure for distribution of fish and nutritious local fresh vegetables and fruits. Upgrading handling practices, infrastructure, storage facilities, and transportation is a recognised need for safer distribution of nutritious aquatic foods that can decrease waste and power rural village economies as they supply urban areas of demand (Kruijssen et al., 2013). That may occur in the new future.

While Melanesian PICs have significant populations engaged in subsistence agriculture, national economic fortunes are tied to international commodity exports and global industries. Key exports from Solomon Islands (fish, logs, copra and cocoa) have all taken a downturn, with the economy contracting by 6% in 2020 (ADB, 2020). The temporary nature of royalties and economic benefits from logging, coupled with the downturn in both log prices and government revenue in 2020, adversely affected some rural communities, particularly those that relied on marketing produce to logging camps in Malaita and Western Province. Rindah Melsen, a leader from Nusatuva Community in Western Province, explained the double impact of reduced markets and reduced cash flow at a WorldFish panel in October:

> *During COVID-19 our experience was we went to Ringi to do our market but then people at this time do not pay very much, the [logging] company has certain rules, only so many people can be there and go and take* money *and come back. It's really affected us, and the company introduced* lockdown *and after a short time our area will be locked down and we will not be able to go there and take our income, the company has rules times to go in and market. Cash flow is really slow inside our community, not many people come to pay, because people keep their cash, keep a bit of income they have. They do not spend a lot, so when they do not spend a lot on our things, women have to go work hard in the garden to produce to survive from their gardens.*

It is evident from this study that most communities are capable of increased food production during shocks like COVID-19, although because of the geographic and cultural diversity of the country there may be communities that have reduced capabilities due to land degradation (such as after bauxite mining) and land scarcity, and local factors such as land insecurity and disputes, and sea level rise. Much of the capability to produce food is due to the resourcefulness of rural women, as pointed out by Edlyn Hauona, a President of Zone 1 WARA women's group at the WorldFish panel:

> *No matter what, women at home prepare for food security. If COVID-19 breaks out, we have food security for our families, every woman does agriculture.*

However, it cannot be assumed that all people are equally capable of meeting household members' needs, and urban residents returning to rural areas after, for example, losing formal employment, may be less likely to have the skills and knowledge to supply their families with food. Those on the fringes of rural economies, such as youth, who typically have reduced access to productive resources and inputs and less options for cash income, face further obstacles of self-sufficiency. Initiatives to kickstart youth into agriculture (such as a UNDP funded kava production project) and provide inputs to communities in order to increase production (such as the Malaita Provincial Government 100 days planting program which included provision of seeds and tools) proved highly valuable interventions to protect agricultural livelihoods and food supply.

In relation to fisheries, while gleaning (collecting species from the shore and reef) is accessible to many, fishing with gear and at sea requires specialised knowledge and skills. Fewer people therefore took up fishing than agriculture in times of increased demand for food and food sales. Where fishing did not require gear (such as collecting shellfish and crabs) it did increase, while demand for fish caught by regular fishers living in rural areas increased as more people needed food. However, since disruption to village economies had limited people's ability to buy fish, so fishers were less inclined to go out to sea.

In times of crisis, regular forms of resilience need to be recognised and used. One such response is through community fisheries management committees, which, with various levels of agency and responsiveness in different locations, form a key local intervention regarding how local marine species are used. In some instances, CBRM committees had enabled whole communities to better respond to the current situation by sustaining harvests to feed larger populations, contributing money to health centers, and continuing fish sales in local villages and the towns of Auki and Gizo. In other cases, management had broken down and was unresponsive due to conflict issues unrelated to the COVID-19 pandemic. Good governance of fisheries has been a key factor in the resilience of coastal communities, and the ability of local markets to maintain their trade, and thus in response to COVID-19. An increasing body of evidence indicates that local resource management inclusive of women is more robust (Westermann et al., 2005; Leisher et al., 2016). In the Solomon Islands context, it is important to balance the different needs of women and men, and different ethnic groups and to give a voice and a productive role to the majority of the population, who are youths. CBRM sites with a proactive response to the

COVID-19 situation, were also places with active representation of women in the committees. However, further research is needed to track the different experiences of men, women and youths in community management and the use of fish and marine resources.

References

Albert, J., Bogard, J., Siota, F., McCarter, J., et al. (2020). Malnutrition in Rural Solomon Islands: An Analysis of the Problem and Its Drivers. *Maternal & Child Nutrition, 16*, e12921.

Andersen, A. B., Thilsted, S. H., & Schwarz, A. (2013). *Food and Nutrition Security in Solomon Islands. CGIAR Research Program on Aquatic Agricultural Systems*. WorldFish.

Asian Development Bank. (2020, July 31, Friday). Solomon Islands Economy to Fall by 6% in 2020. *Solomon Times*.

Bell, J. D., Kronen, M., Vunisea, A., Nash, W. J., et al. (2009). Planning the Use of Fish for Food Security in the Pacific. *Marine Policy, 33*, 64–76.

Eriksson, H., Albert, J., Albert, S., Warren, R., et al. (2017). The Role of Fish and Fisheries in Recovering from Natural Hazards: Lessons Learned from Vanuatu. *Environmental Science and Policy, 76*, 50–58.

Eriksson, H., Ride, A., Boso, D., Sukulu, M., et al. (2020). *Changes and Adaptations in Village Food Systems in Solomon Islands: A Rapid Appraisal During the Early Stages of the COVID-19 Pandemic* (pp. 2020–2022). WorldFish Program Report.

Farrell, P., Thow, A.-M., Wate, J. T., Nonga, N., et al. (2020). COVID-19 and Pacific food System Resilience: Opportunities to Build a Robust Response. *Food Security, 12*, 783–791.

Gillett, R. D. (2016). *Fisheries in the economies of Pacific Island Countries and Territories*. The Pacific Community.

Kruijssen, F., Albert, J. A., Morgan, M., Boso, D., et al. (2013). *Livelihoods, Markets, and Gender Roles in Solomon Islands: Case Studies from Western and Isabel Provinces*. CGIAR Research Program on Aquatic Agricultural Systems. Project Report: AAS-2013-22.

Leisher, C., Temsah, G., Booker, F., Day, M., et al. (2016). Does the Gender Composition of Forest and Fishery Management Groups Affect Resource Governance and Conservation Outcomes? A Systematic Map Protocol. *Environmental Evidence, 5*(6), 1–10. https://doi.org/10.1186/s13750-016-0057-8

Maru, Y. T., Smith, M. S., Sparrow, A., Pinho, P. F., & Dube, O. P. (2014). A Linked Vulnerability and Resilience Framework for Adaptation Pathways in Remote Disadvantaged Communities. *Global Environmental Change, 28*, 337–350.

Ross, H. M. (1978). Baegu Markets, Areal Integration, and Economic Efficiency in Malaita, Solomon Islands. *Ethnology, 17*(2), 119–138.

Secretariat of the Pacific Community/Locally Marine Managed Area Network. (2021). *COVID-19 Pandemic Highlights the Resilience of Local Seafood Economies in the Pacific*. SPC.

Solomon Islands Government. (2019). *Report on National Agricultural Survey (2017)*. Solomon Islands Governmnt.

Solomon Islands National Statistics Office. (2013). *Solomon Islands 2012–2013 Household Income and Expenditure Survey: National Report*. Solomon Islands Government.

Solomon Islands Truth and Reconciliation Commission. (2013). *Confronting the Truth for a Better Solomon Islands, the Final Report of the Solomon Islands Truth and Reconciliation Commission*. Solomon Islands.

Westermann, O., Ashby, J., & Pretty, J. (2005). Gender and Social Capital: The Importance of Gender Differences for the Maturity and Effectiveness of Natural Resource Management Groups. *World Development, 33*(11), 1783–1799. https://doi.org/10.1016/j.worlddev.2005.04.018

World Bank. (2019). *Enhancing the Economic Participation of Vulnerable Young Women in Solomon Islands*. World Bank.

World Risk Report. (2019). https://reliefweb.int/sites/reliefweb.int/files/resources/WorldRiskReport-2019_Onlineenglish.pdf

CHAPTER 22

Of Isolation and Atolls: Coping with Covid-19 in Manus, Papua New Guinea

Jacqueline Lau and Sarah Sutcliffe

Introduction

The spread of COVID-19, and policies across the world to contain it, have continued to have wide-reaching impacts on small-scale fishing communities; from the 'triple economic shock' (Triggs & Kharas, 2020) of changed demand, supply and finances, to extensive environmental, geopolitical, societal and technological consequences (World Economic Forum, 2020). COVID-19 has been 'a harbinger of massive and life altering changes … in small-scale fisheries and coastal fishing communities around the world' (Bennett et al., 2020: 1). In bringing these abrupt changes, COVID-19 has illuminated significant gaps in people's capacity to adapt to shocks, that is their adaptive capacity. Six key domains support adaptive capacity; assets, flexibility, learning, organization, agency, and socio-cognitive constructs (such as previous experiences of risk) (Cinner et al., 2018; Cinner

J. Lau (✉) • S. Sutcliffe
Centre of Excellence for Coral Reef Studies, James Cook University, Townsville, QLD, Australia
e-mail: jacqueline.lau@jcu.edu.au; sarah.sutcliffe@my.jcu.edu.au

Y. Campbell, J. Connell (eds.), *COVID in the Islands: A comparative perspective on the Caribbean and the Pacific*,
https://doi.org/10.1007/978-981-16-5285-1_22

& Barnes, 2019). The COVID-19 pandemic has highlighted whether and how households are able to draw on these domains of adaptive capacity in times of shocks, and the degree to which outside forces (such as imposed isolation) thwart coping mechanisms. Better understanding of the impacts of COVID-19 in the Pacific is highlighting how local adaptive capacities may help or be hindered by this unique and isolating shock.

In the Pacific, most countries—including Papua New Guinea—successfully implemented policies to minimise transmission in the early stages of the pandemic, including hard border closures and limitations on internal travel. Preventing a virus taking hold in the region was critical. Most Pacific Island health systems were and are ill-equipped to cope with an outbreak. Limited testing and contact tracing capacity, insufficient medical supplies, healthcare workers and hospital beds, and community conditions which make household protection difficult, would likely have led to rapid spread and high mortality (OECD/WHO, 2020). While interventionist policies averted a direct health crisis, they continued to have severe secondary consequences for small-island states with a high dependence on international tourism, food imports and remittances (Farrell et al., 2020; Hickey & Unwin, 2020). With a lack of social protection, countries in the Pacific are ill-equipped to support their citizens to navigate the economic fallout from isolation policies (Edwards, 2020). Many people across the Pacific lost incomes, and food became less accessible and less affordable—a critical shock to already food and nutrition insecure communities with pre-existing social-ecological vulnerabilities (Connell et al., 2020).

As of mid-December 2020, Papua New Guinea had 725 confirmed cases of COVID-19 (PNG National Department of Health and World Health Organization, 2020). Between March and June 2020, PNG was in a state of emergency that restricted movement between provinces, closed schools and required non-essential workers to stay home. Subsequently, despite a rise in cases, there was a move to a new normal with relaxed restrictions but continuing emphasis on social distancing and other methods to stop the spread of the pandemic. As this new normal unfolded, the COVID-19 pandemic continued to reverberate across aspects of livelihoods, food and nutrition security, and ultimately human wellbeing.

The early COVID-19 restrictions in PNG had some impact on coastal fishing communities. Rapid assessment surveys on the impact of COVID-19 on fishing and coastal communities in the New Ireland and Central Provinces found that food security was a persistent problem (LMMA Network, PNGCLMA, and WCS-PNG, 2020). Around two-thirds of those surveyed reported that not enough food was available in their villages. In Kavieng, New Ireland, the state of emergency coincided with less than average rainfall, which in turn exacerbated the stress of COVID-19 on food security. In these communities, restrictions did not completely disrupt access to markets. Of those surveyed, only around 25% reported that restrictions on movements during the state of emergency prevented them accessing markets to sell produce and seafood.

This chapter develops this limited understanding by examining and interpreting information derived from in-depth interviews conducted in July and August 2020 with people living on an atoll in Manus Province. It provides an in-depth, and open-ended examination of individual and household's experiences of COVID-19 impacts in a context where, like most atolls, Ahus is highly dependent on fisheries, has little arable land, and most services—including the main market, hospital, ATMs, supermarkets, hardware stores and petrol—are in the town of Lorengau, twenty kilometres away on the Manus mainland. The conclusions of this chapter are pertinent to other small, atoll island coastal communities, who may have fewer food safety nets, and more issues of access (Connell et al., 2020). Specifically, we examine how COVID-19 and rules to combat it impacted food and nutrition security, livelihoods, and wellbeing of individuals and households and whether and how households were able to cope. From there, we outline implications of the continuing pandemic on atoll island communities, and explore what these findings suggest about the capacity for households to adapt to COVID-19.

Ahus Island

Ahus is a small atoll island, north of the mainland of Manus Province, Papua New Guinea. Approximately 780 people live on the island. Although some families receive remittances and support from family members who have migrated to work in PNG cities, most families depend on reef and pelagic fisheries for their livelihoods. Ahus islanders troll for pelagic fish, and line or spear-fish for reef fish in coral reefs surrounding the island, and glean for molluscs and echinoderms. The island itself has little arable land. Aside from coconut *(cocos nucifera)* and galip nut *(canarium indicum)* trees, and some small household gardens (including bananas, capsicums, and leafy greens), almost no food is grown on the island. Instead, Ahus islanders get most of their fresh produce at regular markets both on the island itself and in nearby mainland communities, where they also sell fish.

In addition, many islanders market fish and access shops and services at Lorengau, the closest major town.

Ahus, like many PNG communities, is governed through a combination of customary leadership and democracy, through an elected councillor. Coral reef resources are governed by customary institutions of sea tenure and user rights for different areas, times and gears, based on gender and clan identity; the legitimacy of customary rules, and customary leadership more broadly, has shifted as the island navigates social changes (Lau et al., 2020) (Fig. 22.1).

Ahus islanders are not unfamiliar with shocks to their community. In 2009, a king tide flooded houses and covered much of the lagoon's reef in sand, reducing their productivity. Many perceive climate change to be a growing threat. Households are already taking action to adapt to climate change, for instance, by building seawalls, changing fishing practices, or diversifying livelihoods. A recent study found social connections, socio-ecological ties, and perceived power or agency over change, impacted the sorts of adaptive and transformative actions that households took (Barnes et al., 2020). However, the changes wrought by COVID-19 were unique and overlayed with these environmental shocks.

Methods

We conducted qualitative interviews to understand the impacts of the COVID-19 pandemic on livelihoods, food and nutrition security and wellbeing. Qualitative, short-term insights into its impacts are important in guiding policy and supporting targeted larger scale, and quantitative studies (Steenbergen et al., 2020). The research was only possible because of long-term research partnerships with the community. This chapter is based on seven in-depth interviews. which took place in July and August 2020. We interviewed three women and three men, purposefully chosen to represent different ages, clans, and thus perspectives and experiences of the pandemic. We also interviewed an active local leader, an older man, as a key informant to understand the impacts at a community scale. Interviews were arranged at a convenient time for participants, conducted over mobile phones by a Papua New Guinean research assistant. She undertook comprehensive training in qualitative interviewing as part of the project, and had previously worked as a research assistant in the community. Interviews lasted between 30 minutes to one-hour. The research assistant voice recorded and transcribed each interview, and both the first author

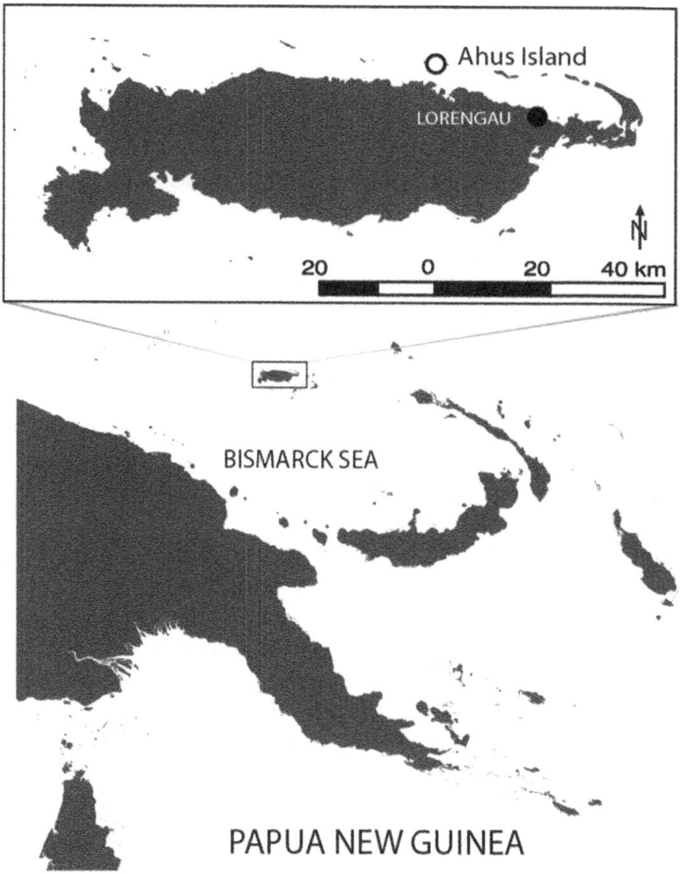

Fig. 22.1 Location of Ahus Island, Manus Province, Papua New Guinea. (From Lau et al., 2020)

and the research assistant translated interviews to English and cross-checked each translation.

Qualitative questions aimed to elicit detailed descriptions of individual and household's experiences across a range of themes including livelihoods, food and nutrition security, and wellbeing. We asked participants to recall their experiences during PNG's state of emergency. The information and conclusions presented in this chapter are based on analysis of the first round of qualitative interviews. These interviews were followed by a second round in October 2020, with a third planned for February 2021.

Markets

COVID-19 impacted both access to and the function of markets, and thus affected the community's predominately fisheries-based livelihoods. Local markets are normally held three times weekly. Mainlanders bring fresh garden produce and sago to sell, or islanders travel to sell fish in two mainland communities. In line with the directives of the state of emergency, local leaders closed the local island markets to avoid large gatherings of people.

> ... they said, there will be no market, it will stop. Because that activity gathers too many people. Too many people go and too many people hang around there. But ... that's the one way for people to get food. So how can I not go to the market? (Man, age 32)

The state of emergency also severely impacted the town market and local mainland markets—the main fish market for many fishing families from the island. Trading hours for stores and market days were reduced. Social distancing and fear of the virus meant that far fewer people attended the markets on the days they were open, resulting in fewer sales and less income for islanders who sell fish there. Decreased sales severely impacted cash flow into the community.

> There weren't a lot of people coming to buy our produce or our fish ... because of the virus, I think everyone was afraid, so not many people came to the market ... Produce and food, and that sort of thing, it all moves with the people. (Man, age 47)

Thus, lack of access to markets, and lack of customers with enough cash to purchase fish meant that, even if people were able to get their fish to the markets, it was not sold at the same volume as prior to the pandemic. The closure of local markets and reduced trading hours in markets and stores in town immediately started to cause food shortages, particularly for those more vulnerable, including the elderly. Quickly, people stopped obeying the rules and the markets began again, but with social distancing rules in place. When the state of emergency lifted, the island's community leaders

Fig. 22.2 Ahus island community members gather at the local island market before Covid-19. People selling vegetables have travelled from a nearby mainland community. (Photo credit: Dean Miller)

worked with community leaders from the mainland to increase the number of local markets, and encouraged the communities to shift back to a barter system while there was little cash (Fig. 22.2).

Livelihoods

Alongside disruptions to markets and a subsequent lack of income, several people emphasised that movement restrictions impacted livelihoods from the sea.

> It's true that if you're in a city, in a town or an urban area, you're more ok than us on islands and in villages. And the mainland is alright as well because they have gardens and that kind of thing. For us on this island, it is hard. Livelihoods are from the sea. (Man, age 44)

Initial movement restrictions, and some confusion about what was safe and allowed during the state of emergency, meant that people fished less than usual, or only went fishing for a short time. For many, fear and uncertainty around the virus led to them fishing less, which also coincided with the low fishing season, and bad weather:

> For fishing, before this thing happened—yes I am aware that the weather was not good—but our people at first were frightened and they weren't fishing enough to sustain their life during this time, when they were scared about this pandemic. But now isn't like that. (Community leader, age 67)

For those who required petrol for outboard motors and trolling, lack of access to petrol became a problem:

> ... when it comes to fishing, the petrol was a problem for us to get, to go out fishing far away. For the canoe, it's ok, we can paddle and go fishing. We got fish and brought them back and they weren't for selling, because how can we paddle over to the mainland and sell them? They were just for the household to eat, and when you eat, you eat fish only. (Man, age 44)

Thus, rather than generating an income, most households fished for food, or, if they were able, for trading with family on the mainland. For instance, alongside the closure of markets in the early stages of the lockdown, movement to the mainland (which is accessible by boat and canoe from the island) was also banned. However, those who were able—predominantly younger fishermen—stopped complying with the bans, and traded with relatives on the mainland. However, only those with family on the mainland could use this tactic, and those with enough strength to paddle across. Elderly people, and people without young men as family members were unable to access garden food in the early stages of the state of emergency, and practices of sharing food and supporting others in the community had also declined (see below).

> But, all the strong young men, the strong men, they paddle canoes. As I told you, they paddled over and exchanged to get these things. And the weak, they couldn't. The weak, like the old men, the old women, and children. When things got hard, they just crossed over to another family and ask. So what can you say, can you just leave them hungry? We have to help them. At least they can stay with the family a bit and eat a little something, or one family doing ok can share a bit of food with them. (Man, age 44)

This suggests that the ban on movement between the mainland and the island had a particularly strong impact. Those who received remittances from family members outside the island were probably buffered from some of these impacts. The same respondent emphasized those that struggled most were "those who didn't have any children living outside the island and sending remittances]", because their only source of support was from the sea.

Informal livelihoods that acted as safety nets were also affected by travel bans. Many women in the community supplement their family's income by marketing small goods (e.g. donuts). Selling betel nut was an informal safety net, particularly for women, for when fishing and fishing income was insufficient:

> When they stopped betel nut in Manus, lots of the women found it hard. When the sea is ok, then it's ok, but if the sea is no good, then we can't even sell betel nut [to make up for it]. If you try and sell betel nut in town the police will get you and lock you up. (Man, age 44)

Bans on movement, and marketing activities had a far-reaching impact on people's livelihoods, and also curtailed existing alternative livelihoods that served as familiar safety nets.

Access to Services

The state of emergency also disrupted access to key services on the mainland. Aside from a health aid-post and numerous small family-run canteens, many key services, such as banks, fuel stations, stores, and the hospital, are on the mainland. When social distancing rules were in place boat transfers to town became very difficult. Rather than 12-15 or so passengers, only 4-7 could travel on each boat, increasing the per-person transport costs. Boats ran less often, and some used smaller motors to save money, meaning that the trip took over 2 hours, compared to around 45 minutes. These changes exacerbated difficulty in accessing the town markets to sell fish, and obtain key commodities, like the petrol needed for trolling.

> Because of the space in the boat, you could only have five people. So, when it was five people to one boat, they couldn't meet the cost of petrol to get from the island to town and to come back. So, the boat owners raised the price to 50 Kina [14 USD]. So, at 50 Kina it will just be five people in the boat. And on the way to town there were marine police checking all the boats. (Woman, age 54)

> Every single boat that goes to town could only take four passengers in one day, in one run. Suppose you want to go on one day, and this day, no, there

are already four people so that's full, so you have to stay and wait for the next run. That's what's happening. (Man, age 32)

Lack of access to services also disrupted people's livelihoods in unforeseen ways. For instance, one respondent actively discouraged her sons from spear-fishing because of the added risk of getting sick and the inability to access the mainland hospital, which had closed to all but urgent patients, in turn leading to less fish in the household.

So I told our family, you can't go to the sea, because if you get sick then how can we go to the hospital? So during that time no one went fishing, and we didn't have money or enough food. (Woman, age 54)

This statement illustrates how people had to make trade-offs between income and food, and the heightened risks brought by reduced access to essential service like health care.

Food and Nutrition Security

The state of emergency immediately led to food shocks in the community. Firstly, when local markets were closed, food became hard to access. The island has little arable land, and the community predominately relies on income from fishing to purchase (or barter) food from the mainland and town. Second, lack of access to markets, led to lack of cash incomes, which decreased people's ability to purchase store foods.

That's the only income we can get … from the sea. We market the fish, we sell the fish … so we can get store food from the income. It helps us to get store food. (Woman, age 25)

These impacted the diversity of people's diets, and people started eating only garden foods, including sago. People also started limiting meal sizes and eating fewer meals.

There was limited food … we had to … look after it well so it could last a long time, for many days more. If we'd gotten food as normal, as if the virus hadn't come, then our food would have run out quickly. So, we reduced food … We'd usually boil four cups of rice, and we reduced it so it went down to two cups of rice. So we'd serve, little, little for each child and each adult. It doesn't matter if you're full up or only just full, that was your share [ration]. (Man, age 44)

Before, we'd all eat rice often. Not now. I've cooked sago over and over, and everyone complains ... but there's nothing else. (Woman, age 32)

Reducing both the quantity and diversity of food consumed to these extremes can have severe ramifications for health.

After local markets re-opened, mainland market and store access and cash flow remained limited, so that people on Ahus were more reliant on directly trading fish for crops with nearby communities. As such, nutritious fruits and vegetables became more accessible than store bought foods with low-nutritional quality. Community leaders explicitly encouraged people to eat more traditional garden food:

During this time, we ate a lot of food from the village. Because we couldn't go to town a lot to find food to eat ... We ate sago, we ate garden food, we ate fish, so we didn't eat tinned fish or that sort of thing. (Woman, age 54)

For me, lots of the time and most times for us now, we have gone back to eating village food. Because rice etc., I've talked about it, getting money during the pandemic was a bit hard. (Man, age 32)

This switch may represent a temporary transition back to traditional foods in contrast with a broad increase in the consumption of processed and imported foods. Increased consumption of fruit and vegetables could have beneficial long-term effects on health and nutrition in the community if this change carries over into life during the 'new normal'.

Social Relationships

New social distancing measures, concern about compliance with rules, and lack of safety nets impacted people's social relationships, trust and wellbeing in the community. Specifically, fears about the spread of the virus and distrust in neighbours' conduct affected relationships across the community:

A few people went to church, and a few people stayed at home. And this freaked us out. If we heard that, oh, people from this family, one or two had gone to town and come back to the village, now we'd be thinking "Aye, that

> guy went to town, and what if they got the virus there and brought it back" ... So the relationships between us, they were spoilt, really screwed. (Man, age 44)

Several respondents additionally emphasised that practices of sharing and supporting others had, because of necessity, started to decline. One respondent was particularly concerned about families not sharing what they had, and interpreted this as greedy or selfish behaviour:

> When the situation happened, plenty of families were, I'd put it like this, they started being selfish and greedy ... food was hard a bit ... if another person wants to come and ask for some sago? There was sago, but they'd say, no we don't have any. There was greedy behaviour when this situation happened. (Woman, 32)

Finally, one woman noted that during the lockdown, bans on movement within the community, and lack of cash meant that incidents related to consumption of alcohol and drunken behaviour had decreased.

> From my viewpoint, when this happened it made lots of men stop drinking as much as before. (Women, 54)

The community leader reemphasised that this change was one good thing resulting from the pandemic.

Isolation and Adaptive Capacity

Having assets to draw on in times of need is an important part of people's adaptive capacity. However, on islands the ability to convert assets or use assets in times of shock depends on the type of asset itself. For instance, owning a boat and outboard motor does not greatly enhance adaptive capacity if petrol is inaccessible. Likewise, having cash does provide an important safety net, but converting cash to food is challenging in circumstances where it's difficult to access markets. Indeed, in analogous circumstances in Vanuatu, dependence on a cash economy and inter-island trade was a key vulnerability to COVID-19 restrictions, resulting in similar food insecurity issues in small, remote islands (Steenbergen et al., 2020). The food shortages and dietary changes reported in Ahus and other communities across the Pacific during COVID-19 compounded existing food and nutrition security challenges arising from economic and environmental changes in the region (Campbell, 2015). Communities on larger islands in PNG, Solomon Islands and Vanuatu were able to turn to household gardens and agriculture as a food and income safety net (Eriksson et al., 2020; Steenbergen et al., 2020). However, lack of assets beyond fishing grounds meant that, unlike other coastal communities in the Pacific, Ahus islanders could not rely on 'fall back' sources of food like home gardens (Eriksson et al., 2020; Farrell et al., 2020; Steenbergen et al., 2020). Fishing livelihoods cannot of themselves provide a healthy diet without adding vegetables through trade. Ahus was particularly vulnerable because of its limited livelihood diversity, isolation and small size. Other atolls with limited opportunities for agriculture faced similar challenges.

Lack of flexibility, in terms of dependence on cash and trade, and lack of on-island food options, make small island communities highly vulnerable when shocks, such as the state of emergency, create isolation. Nonetheless, when markets did reopen in Ahus, the community leaders and the community more broadly were quick to return to traditional bartering systems, and to change the frequency of local markets to make up for lack of access in town. Thus, while dependence on the cash economy can limit flexibility, flexibility can be quickly reinstated by drawing on traditional modes of exchange. The capacity for atoll communities to adapt may also be constrained by a lack of livelihood flexibility. In Ahus, few viable livelihoods exist aside from fishing. In addition, in the case of the COVID-19 state of emergency, alternative livelihoods based on marketing of small quantities of other goods, including handicrafts made on the island, were subject to the same restrictions as marketing fish.

In the context of these vulnerabilities and limits to adaptive capacity, many households had to limit the quality and quantity of their foods to cope with COVID-19 restrictions. This shift has troubling implications for food and nutrition insecurity. Insufficient energy and micronutrient intake can have severe health impacts, particularly for lactating women and young children with high micronutrient needs for growth and development (Black et al., 2008). If the restrictions during state of emergency had been more prolonged, food insecurity would have become an even more critical problem. Indeed, the COVID-19 pandemic will have dire impacts on food security globally (Laborde et al., 2020). Social safety nets are critical to addressing disrupted food access, but are not always well developed in the Pacific (Edwards, 2020). Indeed, although Ahus island had previously received food support during environmental disasters, there was no support during the state of emergency. Nonetheless, one positive outcome during and after the state of emergency was a shift to more nutritious garden foods, when store-bought foods became less accessible and cash harder to come by. The over-consumption of processed store foods has been a growing public health concern across the Pacific in recent years (Popkin et al., 2012).

The socio-cognitive dimensions of adaptive capacity, for example, household and individual perceptions of risk, shaped people's responses to the state of emergency. For instance, early fears about the virus led people to fish less, and some families calculated that the inaccessibility of the hospital made certain types of more physically taxing fishing (notably spearfishing) too risky. Conversely, for others—such as the young men who traded with the mainland when it was banned—the choice not to comply with rules outweighed the risks. The state of emergency rules therefore interacted with family's decision-making in different ways, for different people.

Changing social relations and distrust are particularly concerning and perhaps unique to the sort of shock caused by COVID-19. Social connections and networks are crucial in shaping how and whether people cope and adapt to climate change (Barnes et al., 2020). The nature of the pandemic and rules around movement and distancing meant a rise in distrust and a decrease in face-to-face connections within the community, potentially with negative consequences for the social networks crucial to supporting families through future shocks and through the continuing COVID-19 pandemic. In addition, on Ahus, food sharing is an important aspect of social connection. Stress on food and nutrition security has flow-on effects on social relationships in the community by curtailing people's ability to share food and support one-another. When assets are low, as was the case in Ahus, then social support through networks may also decrease, and this dynamic may create a negative feedback loop until the shock has passed.

New Normal: Ways Forward

The food shocks and isolation experienced during the state of emergency eventually subsided. Indeed, subsequent interviews suggest that access to markets, food and services has returned to a 'new normal' in the community, even as PNG experienced an increase in cases. Nonetheless, the continuing pandemic means that future shocks and stresses are likely across the world, in Papua New Guinea and in Ahus. Household's experiences during the state of emergency emphasise the need to recognise distinct challenges that atoll islands face if they become more isolated. In particular, atoll communities will be highly impacted by rules affecting the ease and accessibility of boat transport, especially if they have a high dependence on fisheries and fish markets for income, outside markets for food, and with little food grown on-island. Rules should be coupled with measures (e.g. food support) to support communities during times when extreme measures are necessary. One option is designing ways and practices for mainland garden produce to reach islands and for fish to be bartered or sold, without overt human contact: precisely the situation of many coastal 'markets' in early contact times. Our findings about the food and nutrition security impacts of COVID-19 highlight the need to ensure that support reaches islands in a timely manner, especially during extreme shocks like the state of emergency. Several interviewees mentioned their confusion over whether there would be government support, and talked about previous food and other relief they'd received in the aftermath of natural disasters. But it is all too likely that small and remote islands be easily ignored and neglected. Finally, there is a need to ensure clear communication about future rules, and to acknowledge trade-offs between social distancing with livelihoods, food and nutrition security and wellbeing. In PNG, many government directives will be enacted through ward development councils and local forms of governance. Fear and anxiety about COVID-19 may have resulted in rules that did not account for impacts on food and nutrition security, and non-compliance with the rules may impact people's trust in the legitimacy of future directives. As PNG's new normal continues to unfold, the COVID-19 pandemic will reverberate across aspects of livelihoods, food and nutrition security, social relationships and ultimately human wellbeing, but almost certainly in unforeseen ways.

References

Barnes, M. L., Wang, P., Cinner, J., Graham, N., et al. (2020). Social Determinants of Adaptive and Transformative Responses to Climate Change. *Nature Climate Change, 10*(9), 823–828.

Bennett, N., Finkbeiner, E., Ban, N., Belhabib, D., et al. (2020). The COVID-19 Pandemic, Small-Scale Fisheries and Coastal Fishing Communities. *Coastal Management, 48*(4), 336–347.

Black, R., Allen, L., Bhutta, Z., Caulfield, L., et al. (2008). Maternal and Child Undernutrition: Global and Regional Exposures and Health Consequences. *The Lancet, 371*(9608), 243–260.

Campbell, J. (2015). Development, Global Change and Traditional Food Security in Pacific Island Countries. *Regional Environmental Change, 15*(7), 1313–1324.

Cinner, J., Adger, N., Allison, A., Barnes, M., et al. (2018). Building Adaptive Capacity to Climate Change in Tropical Coastal Communities. *Nature Climate Change, 8*(2), 117–123.

Cinner, J., & Barnes, M. (2019). Social Dimensions of Resilience in Social-Ecological Systems. *One Earth, 1*(1), 51–56.

Connell, J., Lowitt, K., Saint Ville, A., & Hickey, G. (2020). Food Security and Sovereignty in Small Island Developing States: Contemporary Crises and Challenges. In J. Connell & K. Lowitt (Eds.), *Food Security in Small Island States* (pp. 1–23). Springer.

Edwards, R. (2020). Bubble in, Bubble Out: Lessons for the COVID-19 Recovery and Future Crises from the Pacific. *World Development, 135*, 105072.

Eriksson, H., Ride, A., Boso, D., Sukulu, M., et al. (2020). *Changes and Adaptations in Village Food Systems in Solomon Islands: A Rapid Appraisal During the Early Stages of the COVID-19 Pandemic.* World Fish Program Report: 2020–22.

Farrell, P., Thow, A.-M., Wate, J., Nonga, N., et al. (2020). COVID-19 and Pacific Food System Resilience: Opportunities to Build a Robust Response. *Food Security, 12*, 783–791.

Hickey, G., & Unwin, N. (2020). Addressing the Triple Burden of Malnutrition in the Time of COVID-19 and Climate Change in Small Island Developing States: What Role for Improved Local Food Production? *Food Security, 12*(4), 831–835.

Laborde, D., Martin, W., Swinnen, J., & Vos, R. (2020). COVID-19 Risks to Global Food Security. *Science, 369*(6503), 500–502.

Lau, J., Cinner, J., Fabinyi, M., Gurney, G., & Hicks, C. (2020). Access to Marine Ecosystem Services: Examining Entanglement and Legitimacy in Customary Institutions. *World Development, 126*, 104730.

LMMA Network, PNGCLMA, and WCS-PNG. (2020). *COVID-19 Update #4: PNG.* LMMA.

OECD/WHO. (2020). *Health at a Glance: Asia/Pacific 2020.* OECD.

PNG National Department of Health and World Health Organization. (2020). *Papua New Guinea Coronavirus Disease 2019 (COVID-19) Health Situation Report #51.*

Popkin, B., Adair, L., & Ng, S. (2012). Global Nutrition Transition and the Pandemic of Obesity in Developing Countries. *Nutrition Reviews, 70*(1), 3–21.

Steenbergen, D., Neihapi, P., Koran, D., Sami, A., et al. (2020). COVID-19 Restrictions Amidst Cyclones and Volcanoes: A Rapid Assessment of Early Impacts on Livelihoods and Food Security in Coastal Communities in Vanuatu. *Marine Policy, 121*, 104199.

Triggs, A., & Kharas, H. (2020). *The Triple Economic Shock of COVID-19 and Priorities for an Emergency G-20 Leaders Meeting*. Brookings Institute. Retrieved May 25, 2020. www.brookings.edu/blog/future-development/2020/03/17/the-triple-economic-shock-of-covid-19-and-priorities-for-an-emergency-g-20-leaders-meeting/

World Economic Forum. (2020). *Insight Report: COVID-19 Risks Outlook A Preliminary Mapping and Its Implications*. World Economic Forum.

CHAPTER 23

COVID-19: The Impact of a Complex Disaster on Household Food Security in Caribbean SIDS

Tracy Marshall, Arlette Saint Ville, Lystra Fletcher-Paul, and Wendy-Ann Isaac

Introduction

When the Caribbean region reported its first confirmed case of coronavirus 2019 (COVID-19) in Jamaica on 10 March 2010, the region joined in the fight to reduce loss of life and spread of the novel virus. At the time of the arrival of this complex disaster, disaster agencies were ready to roll out their annual disaster management plans (from June to November) for the 2020 Atlantic hurricane season. Governments' across the region responded by implementing COVID-19 emergency management and public health regulations to control or prevent the virus spread. At the regional level, a 'COVID-19 Response Agri-Food Plan' was developed to

T. Marshall (✉) • A. Saint Ville • L. Fletcher-Paul • W.-A. Isaac
The University of the West Indies, St. Augustine, Trinidad and Tobago
e-mail: tracy.marshall@my.uwi.edu; Arlette.SaintVille@sta.uwi.edu; lystra.fletcher-paul@sta.uwi.edu; wendy-ann.isaac@sta.uwi.edu

Y. Campbell, J. Connell (eds.), *COVID in the Islands: A comparative perspective on the Caribbean and the Pacific*,
https://doi.org/10.1007/978-981-16-5285-1_23

minimise the impact on food security, concentrating on adequate food access and production within the region. National disaster and public health agencies grappled with managing these intersecting hazards using generic policies shared by global public health agencies. Understandably, this meant that the initial response to reduce disaster risk had minimal place-based food security considerations. To explore the elements involved in the shaping of human behaviour in complex and uncertain environments related to disaster risk, this chapter reviews the first wave of the COVID-19 pandemic in the Caribbean region, government strategies and household food security responses to determine how food security was incorporated into national policy interventions and household strategies.

Human Behaviour and Reducing Risk in a Complex and Uncertain Environment

Human behaviour is key to understanding and managing disaster risk, minimising negative impacts and saving lives. This is because coping is less about the magnitude of the hazard but about responding based on local-level vulnerabilities to disaster risk (Tuladhar et al., 2015). Human behaviour refers "to changes of a living being's position or attitude, which describes movements or actions pertaining to a particular environment" (Maturana & Varela, 1992, 137). Behaviour to reduce risk in this context can be influenced by acquired and genetic factors. With the COVID-19 pandemic, acquired factors can be viewed as the stimuli, which influenced human action to reduce risk; genetic factors comprise physical and psychical elements (Ristea, 2013). These characteristics influence risk and safety decisions by altering attitudes and perceptions, suggesting that understanding how national policy responses to the pandemic influenced household food security should centre on acquired and psychical elements of human behaviour and risk reduction.

Of particular interest here is the role played by emotions, such as fear and anxiety and by rationality. Emotions play a critical role in shaping risk perception and contribute to decision-making processes, especially in the case of COVID-19. 'Risk as feelings', a term coined by Loewenstein et al. (2001, 270), is one of two ways humans respond to hazards, and these responses are 'fast, instinctive, and intuitive' (Slovic et al., 2005: S35). A second human response involves rational risk analysis, involving a scientific and logical approach to decision-making. Understanding the influence of rationality and/or emotions in decision-making is particularly important in unpacking how households respond to reduce risk, and access food, especially with the protocols implemented to reduce movement and minimise spread of the virus. It is expected that 'risk as feelings' would trigger recognition of potential threats to food accessibility and availability since food is a basic need. Such a perception would likely give rise to instinctive emotional responses in a complex and uncertain environment. We thus seek to examine what factors accounted for responses to food security during the earliest phase of COVID-19 in the Caribbean, and what roles were played by emotion and rationality in household decision-making. The COVID-19 pandemic's impact was sudden and unexpected, and knowledge of the virus was limited. It might therefore be anticipated that its threat would trigger gut feelings and highly emotional responses rather that involve logical, scientific responses to food security, especially at the household level.

Methods and Islands

Data collection involved primary and secondary data from five countries, with primary data from web searches of each country's government official websites and social media platforms. Secondary data came from three sources for the five countries: the World Food Programme (WFP) Caribbean COVID-19 Food Security and Livelihoods Impact Survey; Google web searches, and Google COVID-19 Community Mobility data on movement. A survey was administered between 1 to 12 April 2020 (Chong et al., 2020). The study began three weeks before 10 March 2020, in advance of official reports of COVID-19 in the Caribbean Region, and went to the end of April 2020, the period after initial protocols were implemented.

The WFP survey provided a rapid assessment of the impacts of the COVID-19 pandemic on livelihoods, food security, and access to markets in the Caribbean for 19 countries (Chong et al., 2020). Data used included variables on sociodemographics, reasons for lack of access to markets, changes to shopping behaviour, food consumption patterns, and livelihood disruptions (Chong et al., 2020). Google COVID-19 Community Mobility data (17 February to 30 April 2020) showed movement of households in relation to government implemented protocols and the first confirmed cases in the countries. Mobility trends were selected for workplaces, groceries and pharmacies, and residential places (Google, 2020). Google Trends was used to investigate web searches by households on the virus within their country and how to prepare for and mitigate the risk to food security, and the extent of interest in preparing for the virus and food supply. This was supported by analysis of national documents from the five countries which had been disseminated to households covering COVID-19 emergency management and public health regulations. These provided insights into how governments perceived the risk to food security. Collectively these captured the four dimensions of food security at the national and household level; availability, access, utilisation, and stability (FAO, 2008). Our conclusions are limited in being representative of the populations in these countries who had access to the various online sources and, in the case of the mobility data, only captured persons who used their location history on smartphones. That may have skewed results to a more affluent population.

The five selected countries represent different perceived livelihood impacts from COVID-19, ranging from severe (Belize), low to moderate (Trinidad and Tobago, Barbados, and The Bahamas), and low to no impact (Jamaica) (Chong et al., 2020). These diverse perceptions reflect similarities and dissimilarities in areas such as food and economic activities. However all are characterised by some degree of food insecurity, and by substantial and growing food imports (mainly of grains but including various kinds of processed food). The shift in food consumption away from local staples and traditional foods in favour of highly processed imported foods which are cheaper, easier to store and have a longer shelf life, has continued (FAO, 2020), and has been matched by a parallel increase in non-communicable diseases.

Belize is a relatively poor coastal country in Central America, bordered by Mexico and Guatemala. Although not an island, as a former British colony it retains institutional links to the Anglophone Caribbean, and is classified as a SIDS. It has a population of approximately 360,000 people of diverse ethnic and cultural backgrounds, with local food choices influenced by Creole, Mestizo, Maya, and East Indian cultures. Some of their dishes include rice and beans, relleno soup (using spice made from charred chillies), tortillas, and corn porridge (GOB, 2020). Its primary economic activity is agriculture. Belize had a higher reliance on informal or casual labour than the other countries in this category. Of the 2515 WFP survey respondents, 643 households were from Belize. The twin islands of Trinidad and Tobago constitute the most southerly island state in the Caribbean archipelago. A combined population of over 1.3 million, are more urbanised than in most Caribbean states, economic activity is focused on oil production and food security is at risk (Shah et al., 2020). As in Belize, multiple ethnicities have influenced the cuisine. However locally unique are doubles (a dish made from flour, channa, and dhal), and pelau. Trinidad and Tobago households represented 535 of the survey data. Barbados, the most easterly island in the Caribbean, has a population of approximately 285,000, an economy oriented to tourism and a traditional cuisine that blends African, Indian and British cultures with popular dishes including cou-cou (cornmeal based dish) and fried flying fish, fishcakes, pudding and souse (pickled pork or other meats served with steamed sweet potato). The Bahamas is an archipelago of group of 700 islands, 30 of which are inhabited by a population of approximately 380,000. Like Barbados, it has a tourism dependent economy, and agriculture is difficult on the low sandy islands (Thomas et al., 2018). Indigenous dishes are based on seafood. Respondents from The Bahamas were 586 and Barbados 537. Jamaica is a large, mountainous island with a population of approximately 2.8 million. Its economy has been heavily reliant on agriculture, but mining and tourism are of greater economic value. Here too, the cuisine reflects various cultures, with popular and distinctive dishes such as ackee and saltfish, curried goat and rice, and jerk chicken. Jamaican households represent 214 of the survey data.

While these supposedly typical local dishes are largely based on local foods an increasing proportion of food in all these countries is imported, and fast foods are ubiquitous. Overall about 50% of Caribbean SIDS import more than 80% of their food and this is reasonably typical of these five countries, so that purchasing power and food trade are crucial to maintaining food security in the Caribbean. Only Belize produces more than 50% of its food, whereas The Bahamas imports a very high proportion of its food although more than half was absorbed in the hotel and restaurant sector. Of these five countries, Barbados has met global hunger targets and, in the other four, less than 10% of the population are underfed (FAO, 2015). Food and nutrition security depends as much on ability to purchase as on ability to produce.

How Was Food Security Incorporated into National Interventions, Other Policies, and Households' Response Measures?

All five countries had confirmed at least one COVID-19 case by mid-March and had responded with varied protocols within an average of nine days. Most countries took proactive measures, beginning with border controls and public information campaigns (Fig. 23.1). Both Belize and

Trinidad and Tobago had taken some actions before the first confirmed case in the country and others took actions preceding the WHO declaration of the Public Health Emergency of International Concern (PHEIC) on 30 January 2020. Further actions included screening arrivals, quarantining arrivals from some or all regions, banning arrivals from some or all regions or total border closure.

The various specific protocols implemented during March and April 2020 (Table 23.1) included mobility restrictions, curfews, and lockdown/ stay at home orders and control of gatherings (with limits to public gatherings, and regulated numbers), closure of businesses and public services, and closure of schools. These protocols directly impacted householders' accessibility to food. Which may have negatively impacted household food security.

A stringency index can be derived to measure the number and strictness of these government policies (Hale & Webster, 2020), with data collected from publicly available information provided by the Oxford COVID-19 Government Response Tracker. In each country, there was an observed increase in strictness over time; Jamaica (25 to 87.04), Trinidad and

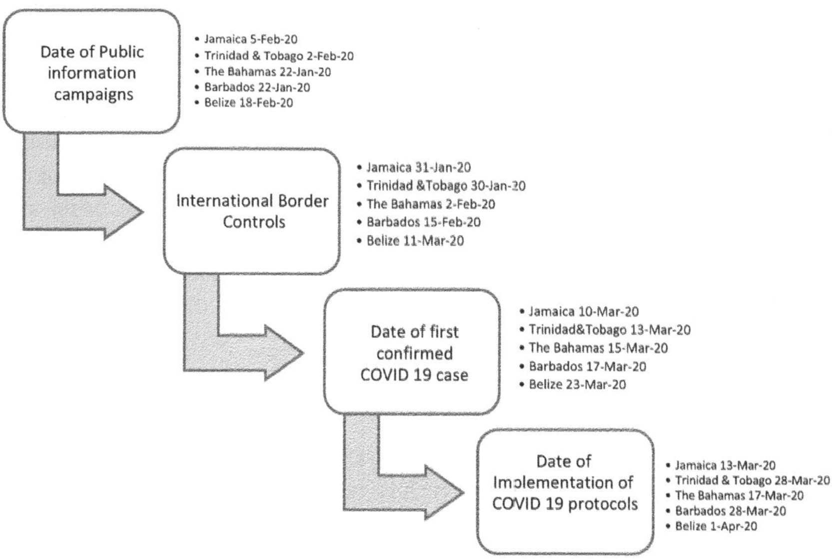

Fig. 23.1 Dates of government policy responses to the threat of COVID-19 case and date of implementation of COVID19 protocols

Table 23.1 Category of protocols implemented by each country from March to April 2020

Country	*Protocols implemented for March-April*	*Control movement into country*			*Control movement in country*				*Control gatherings*		
		Border controls	*Border closure: full*	*Border closure: partial*	*Mobility restrictions*	*Curfews*	*Lockdown/ stay at home order: full*	*Lockdown/ stay at home order: partial*	*Limit to public gatherings*	*Closed businesses or public services*	*Closed schools*
Jamaica	13 March 2020-Declared a disaster area	x	x		x		x		x	x	x
Trinidad and Tobago	28 March 2020	x	x		x		x		x		x
The Bahamas	17 March 2020-State of public emergency proclaimed	x	x		x	x	x		x	x	
Barbados	28 March 2020-Declaration of a public health emergency	x			x	x	x	x	x	x	x
Belize	20 March 2020	x		x		x	x	x	x	x	x

Tobago (20.37 to 90.74), Barbados (38.89 to 88.89), and Belize (54.63 to 76.85), with Trinidad and Tobago having the highest COVID-19 government response on the stringency Index. (No stringency index was calculated for The Bahamas). Protocols became tougher.

Meanwhile, over the early months of the COVID-19 protocols being in place household mobility patterns changed (Figure 23.2) as people responded to the new regulations, with immediately reduced mobility, that were similar for all the countries. That was particularly evident for travel to workplaces in very country, whereas travel to grocery shops, supermarkets, and pharmacies showed minimal change, for Trinidad and Tobago, The Bahamas, and Belize. Barbados and Jamaica did have a slight decrease in visits to grocery stores and pharmacies. Only a sight decrease was expected since these businesses had exemptions to remain open. Conversely, and unsurprisingly, with the implementation of stay-at-home orders and work-from-home policies, it was expected that time spent at home increased, with a spike on and around 13 April 2020 marking the Easter weekend public holidays (Fig. 23.3).

The various specific protocols (Table 23.2) were anticipated to hinder household food security by either limiting access to various sources of food (groceries and markets) or by leading to a reduction in incomes and thus in purchasing power. Households' reported impacts varies somewhat. Without alternative provisions, these protocols to control the spread of COVID-19 posed threats to household food security. Barbados and Belize households claimed the most significant impact from the lack of access to markets. Public transportation limitations had the least impact on households in Barbados. The majority of households across all countries reported livelihood disruptions. The WFP survey data showed that the main household income sources across all countries came from salaried employment and ownership of businesses. Dependence on other income sources such as support from family and friends, remittances, petty trade, and government assistance was less than 10% in each country. Most households reported movement restrictions as a threat to household food security in Barbados and The Bahamas. Households mostly reported loss of jobs or reduced salary in Belize, and impacts from markets or groceries being closed were reported by a majority in Barbados and Jamaica, in comparison to the other countries.

COVID-19 protocols directly supporting household food security were adopted at national and household level in each country (Table 23.3). Policies granted exemptions to farmers, processors, distributors, retailers,

Select [1]Countries as of 30 April 2020	WORKPLACES Mobility trends for places of work	GROCERY &PHARMACY Places like grocery markets, food warehouses, farmer's markets, specialty food shops, food shops, drug stores, and pharmacies	RESIDENTIAL Mobility trends for places of residence
Barbados	-69.1%	-5%	+28.1%
	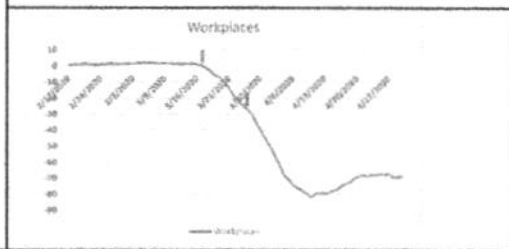	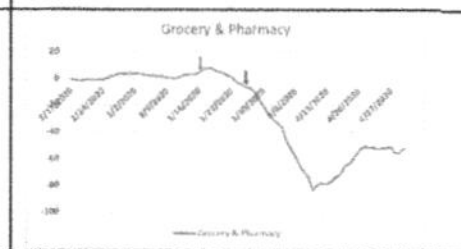	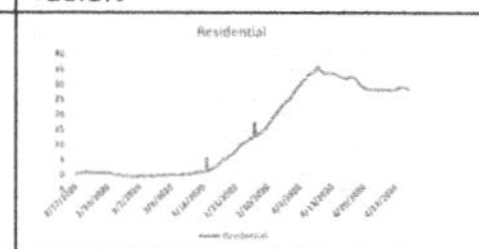
Belize	-58.4%	+1.3%	+23.3%
		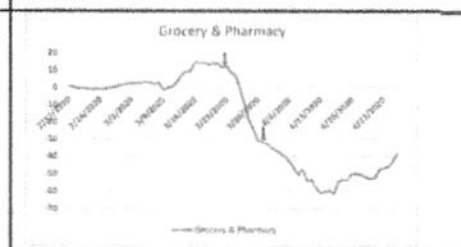	
Jamaica	-48.5%	-0.4%	+ 22.3%
	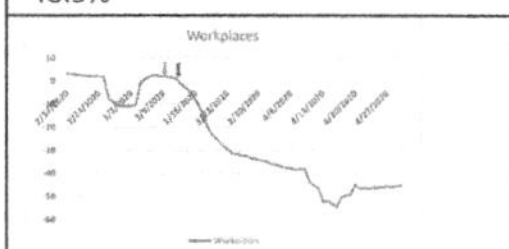	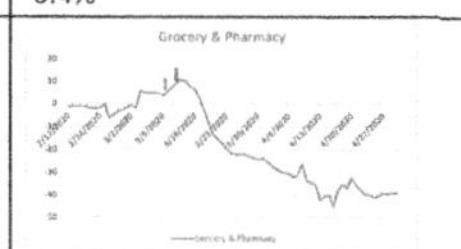	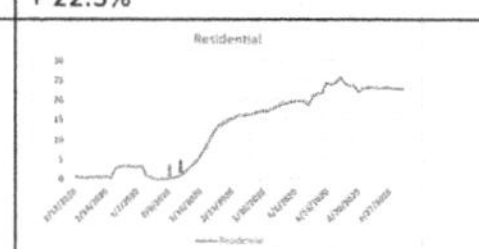
The Bahamas	-63%	+1.4%	+28.2%
		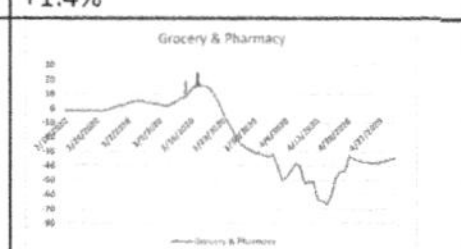	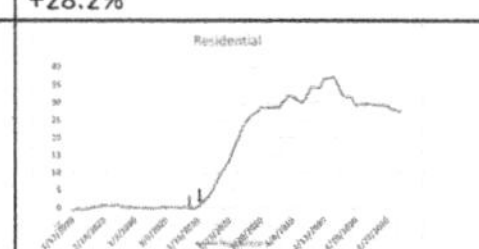
Trinidad and Tobago	-55%	+2.1%	+23.5%
	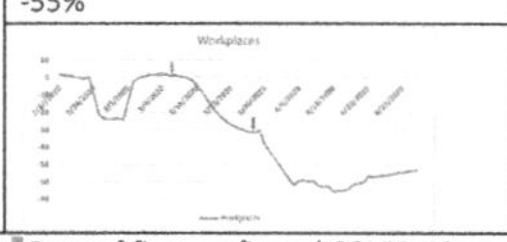	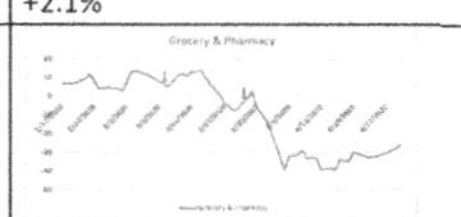	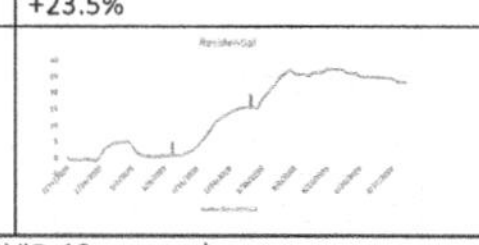
Legend	Date of first confirmed COVID-19 case	Date of implementation of COVID-19 protocols	

Fig. 23.2 The absolute percentage change in mobility trends for the period from February to April 2020. (Our World In Data. https://ourworldindata.org/covid-mobility-trends)

and consumers to continue operations or provide access to goods and services. Access was given to consumers to access these businesses at specified times. Continued access to food imports was maintained by all governments. Trinidad and Tobago and Barbados were notable for their diverse approaches. Trinidad and Tobago explicitly granted wide-ranging exemptions to all food related businesses, that took into consideration the value chain from farm to fork. In Barbados, protocols considered the control and regulation of imports, exports, and prices of staple food items.

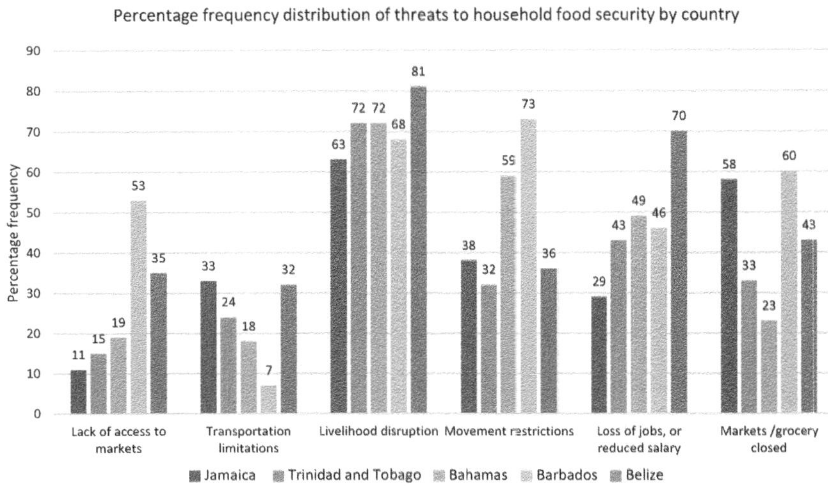

Fig. 23.3 Threats to household food security

Governments also implemented home-gardening initiatives that sought to increase food production at home by households. At a different scale, regional initiatives included the Healthy Caribbean Coalition, and the Organization of Eastern Caribbean States (OECS) Commission, which coordinated activities through Civil Society Organisations. Agri-food risk management plans were also developed in Barbados and The Bahamas (FAO, 2020). Other food security efforts included the distribution of seeds by the Ministry of Agriculture in Trinidad and Tobago and in Barbados the distribution of over 3000 fruit trees through the "The National Tree Planting Project" as a longer-term strategy.

In gaining knowledge of COVID-19 there was an inevitable spike in household web searches from the 10 March 2020 with specific interest in information about the local context (Fig. 23.4). These were broadly similar by country and by specific topic, although the web searches were highest in Trinidad and Tobago and lowest in Barbados, which may indicate differences in the level of perceived risk to the COVID-19 pandemic, which in turn may have been a function of differences in its impact on their national economies.

Web searches by 'country name and coronavirus' reached the highest value on a relative scale on 10 March 2020 the date of the first confirmed

Table 23.2 Protocols that hinder household food security by limiting access or reducing or stopping food production and related services

		Measures implemented from March to April 2020					*Threats to household food security*
		Jamaica	*Trinidad and Tobago*	*The Bahamas*	*Barbados*	*Belize*	
Control movement in-country	**Mobility restrictions**	Public transport by bus limited to seated passengers only, by car one person less than the maximum amount	Public transport by motor vehicle limited to half the passenger carrying capacity of the vehicle	All travel restricted by public bus transportation, mail boat, sailing inter-island			Lack of access to markets, especially households using public transportation, and livelihood disruptions
	Curfew			20 March -31 March 2020 -9:00 pm to 5:00 am	15 April 2020 to midnight on 3 May 2020 -6:00 pm to 6:00 am	8:00 pm-4:59 am for the period of public emergency declared by Proclamation	Movement restrictions, livelihood disruption
	Lockdown/ Stay at home order: full	For 7 days beginning from the 18 March 2020, all persons who can work from home should do so		Effective 3 April 2020, (2) exemptions to leave residence only for urgent medical care at a hospital	Exemptions to leave residence set out by use of a alphabetical system		Movement restrictions, Disruption to livelihoods, loss of jobs or reduced salary

(*continued*)

Table 23.2 (continued)

		Measures implemented from March to April 2020					*Threats to household food security*
		Jamaica	*Trinidad and Tobago*	*The Bahamas*	*Barbados*	*Belize*	
Control gatherings	**Limited public gatherings**	For a period of 7days from 18 March 2020, gatherings in any public place not to exceed 20 persons at a time 25 March to 7 April 2020, gatherings in any public place not to exceed 10 persons and a distance of at least 3 feet from other persons maintained	7 April to 30 April 2020, persons are not to gather in any public place where the number exceeds 10 Later amended to 5 persons	Owners or operators of establishments should restrict customers inside the business to enable 3 to 6 feet between persons		Social gatherings are limited to 25 people Protocols for social distancing should ensure that all customers and staff maintain physical distancing of no less than 3 feet in or outside of the business	Delays, disruptions, and lack of access to markets, transportation limitations, livelihood disruptions
	Closed businesses or public services	Markets shall operate only during the hours of 6:00 am to 2:00 pm	Retail services open until 6:00 pm daily	From 20 March 2020, all businesses, offices, stores, and organisations suspended operations to the general public Effective 3 April 2020, every business, establishment or undertaking remained closed until 5:00 am Monday 6 April 2020	From 15 April 2020 to midnight on 3 May 2020, every non-essential service remained closed	Businesses and offices operated only during the specified period 5:00 am to 7:00 pm	Delays, disruptions, and lack of access to markets, transportation limitations, livelihood disruptions, loss of jobs or reduced salary

Table 23.3 Government policies which supported household food security

	Government protocols that supported household food security	*Food security dimension*	*Food supply chain*
Jamaica	**Businesses exempted from limits to public gatherings**-markets, supermarkets, corner shops which provide food; banking businesses, services connected with the loading and unloading of ships and with the storage and delivery of goods **Lockdowns/stay at home orders:** exemptions given to leave residence once per day for food sourcing (targeted at persons 75>)	Physical availability of food; economic and physical access to food; diversity of diet	Consumers Distributors
Trinidad and Tobago	**Border closures:** remained open to air and sea cargo **Lockdowns/stay at home orders:** exemption granted for workers associated with services including retail services-markets supermarkets, fruit stalls, and vegetable shops or stalls, bakeries, parlours for the provision of food; wholesale stores providing food; manufacturer, transportation, and logistics services related to food; services related to food beverage, agriculture and fisheries, take away or delivery, feed processing **Closed businesses or public services:** exemptions granted to operate the following businesses, 6:00 pm- until – retail services-markets supermarkets, fruit stalls and vegetable shops or stalls, bakeries, parlours for the provision of food	Physical availability of food; economic and physical access to food; diversity of diet	Farmer Processors Distributors Retailer Consumers
The Bahamas	**Border closures:** exemptions granted for transport freight **Lockdowns/stay at home orders:** exemptions granted to operate the following businesses, 6:00 am to 5:00 pm- wholesale or retail grocery stores and farmers' markets, commercial ports and related businesses 9:00 am to 5:00 pm banks 6:00 am to 7:30 pm-drive-thru or take away food vendors	Physical availability of food; economic and physical access to food, diversity of diet	Farmer Consumers Distributors Retailer

(*continued*)

Table 23.3 (continued)

	Government protocols that supported household food security	*Food security dimension*	*Food supply chain*
Barbados	**Closed businesses or public services:** regulations to maintain, control or regulate the supply of goods or services including imports and exports, prices and production of vegetables, and keeping of livestock. **Limit to public gatherings:** movement by alphabetical systems to access markets, supermarkets, and minimarts **Lockdowns/stay at home orders:** exemptions granted to operate the following businesses, 9:00 am to 1:00 pm, or 9:00 am to 4:00 pm- banks and credit unions 9:00 am to 4:00 pm -bakeries and bread depots for the sale of bread 6:00 am and 4:00 pm- bakeries for the baking of bread 8:00 am and 4:00 pm- butchers' shops, delivery services, manufacturers, food processors and distributors (5:00 pm) public markets, supermarkets and minimarts, village shops, vendors, wholesalers of food items and related services 5:00 am and 5:00 pm -farms sugar factories for the entire day	Physical availability of food; economic and physical access to food, diversity of diet	Farmer Processors Distributors Retailer Consumers
Belize	**Border closures:** exemptions granted for cargo flights and cargo shipments **Closed businesses or public services:** exemptions granted to operate the following businesses, 5:00 am to 7:00 pm- bakeries and tortilla factories; agro-processors, food, and carbonated and other beverage manufacturers; agricultural and aquaculture farms; fishers; retailers and wholesaler services, take out services only from restaurants, saloons diners and other similar establishments **Lockdowns/stay at home orders:** exemption granted to purchase essential needs, for an essential worker or employee of an approved business to go to or from work	Physical availability of food; economic and physical access to food, diversity of diet	Farmer Processors Distributors Retailer Consumers

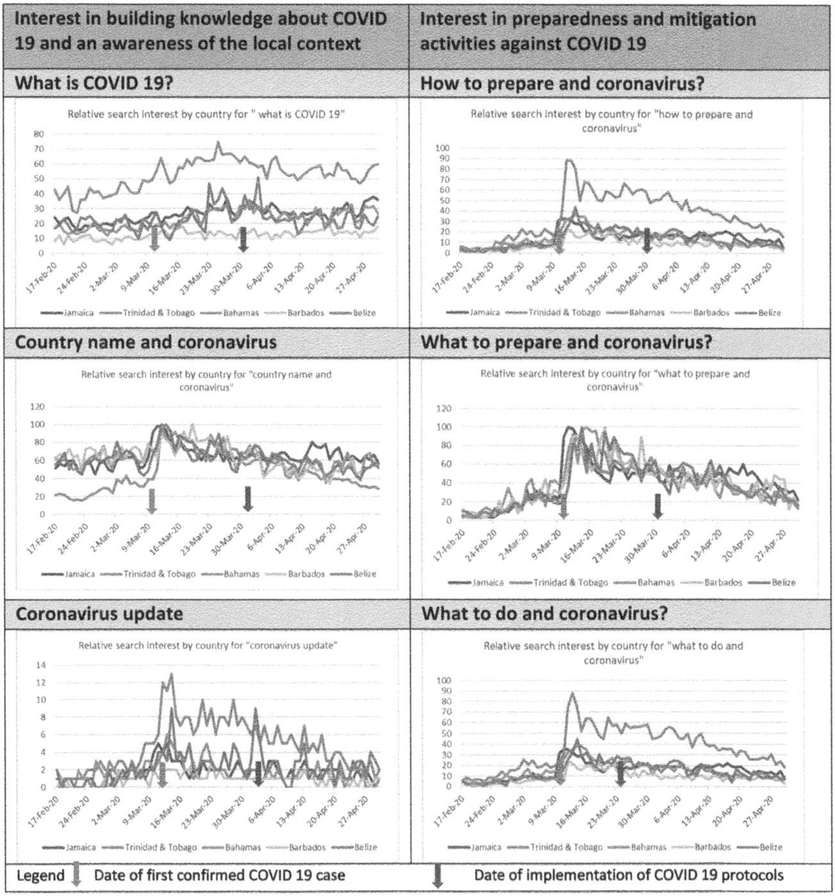

Fig. 23.4 Interest in building knowledge about COVID-19 and preparedness and mitigation activities

case in the Caribbean. This reaction occurred across all countries despite the virus being confirmed only in Jamaica at that time. Unsurprisingly web-users in all countries then showed an interest in preparedness and mitigation activities against COVID-19. Web-users also searched for food related information, extended into 'how to make' masks and hand sanitisers (Fig. 23.5). Likewise, web-users also searched 'what is emergency food supply,' and 'how to get food'.

For more than half of households, food consumption patterns remained normal over this two-month period except in Belize (Fig. 23.5) although

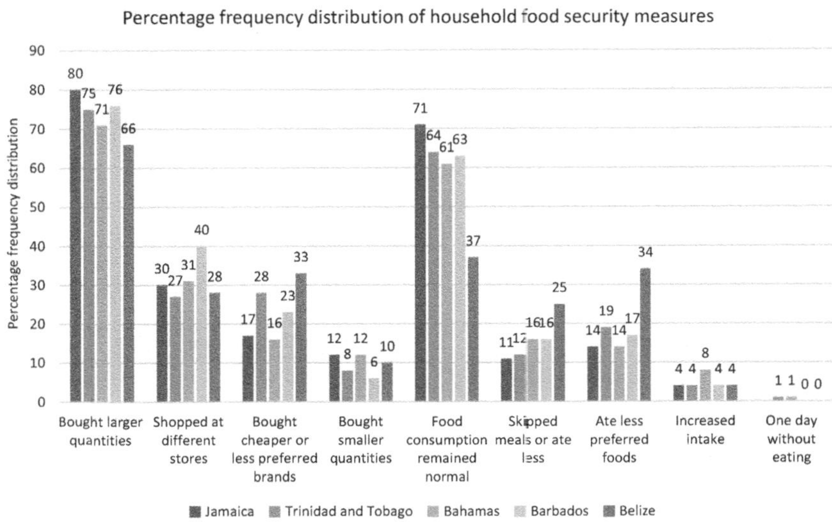

Fig. 23.5 Household food security measures

the majority of the households in all the countries reported purchasing larger food quantities than usual. At the same time there was a tendency to skip meals, eat less preferred foods and buy less-preferred brands as, presumably, income factors began to become significant (Fig. 23.6).

After the implementation of the COVID-19 lockdown protocols, a gradual increase in interest developed on how to make traditional or 'preferred' foods evident in both Trinidad and Tobago and Jamaica (Figs. 23.7 and 23.8). That trend too is likely to be a response to declining incomes and the need for greater self-reliance. By contrast in Barbados and Belize there was no greater interest in shifting towards local foods (Fig. 23.9).

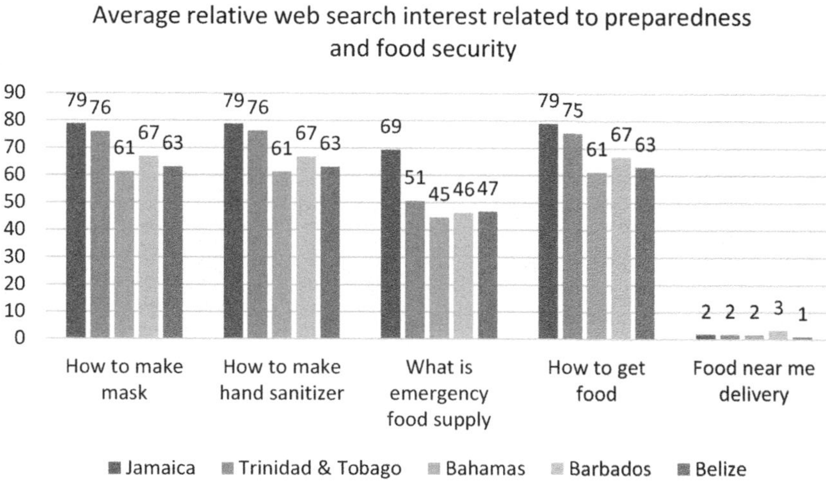

Fig. 23.6 Average relative web search interest related to preparedness and food security

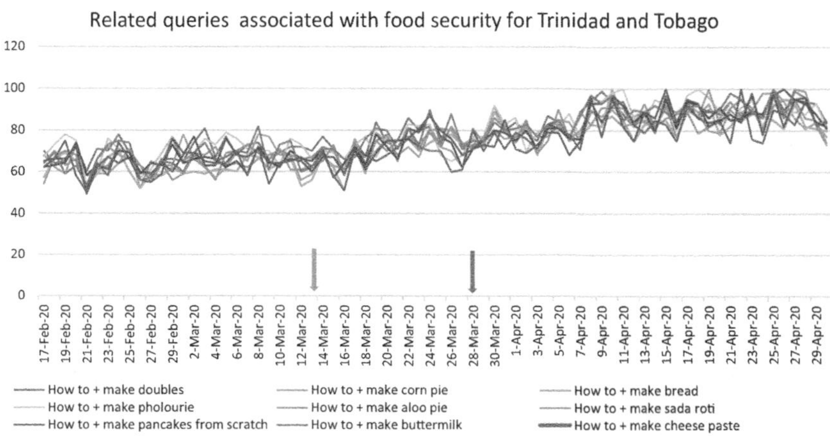

Fig. 23.7 Related queries associated with food security for Trinidad and Tobago

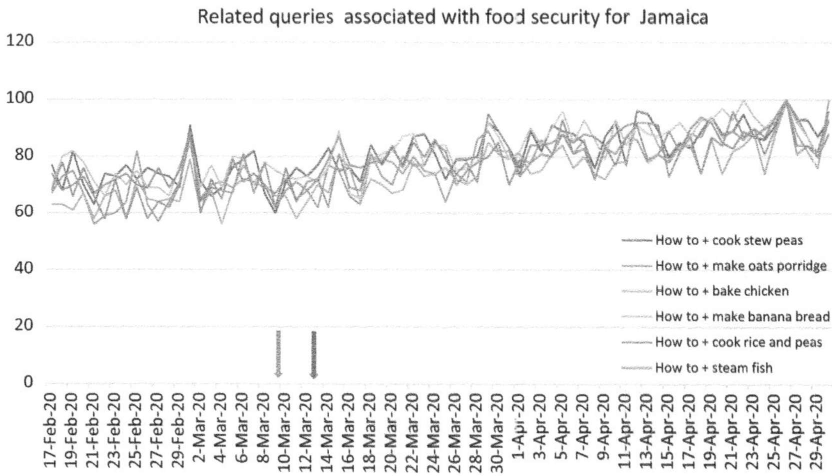

Fig. 23.8 Related queries associated with food security for Jamaica

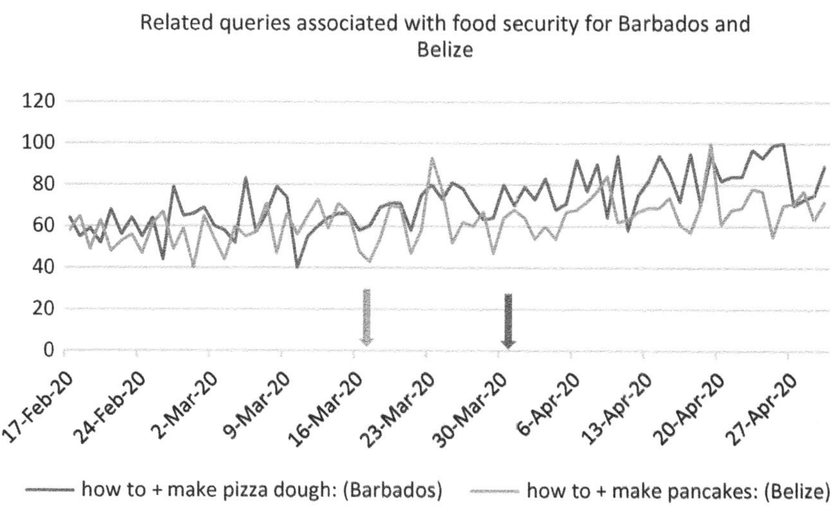

Fig. 23.9 Related queries associated with food security for Barbados and Belize

Policy Considerations to Safeguard Food Security in a Pandemic

This illustration of how COVID-19 as a complex disaster in five Caribbean SIDS increased food insecurity as a secondary impact implies three policy themes. Firstly, interdisciplinary, hazard-specific considerations are essential in planning responses to complex disasters to addressing the vulnerabilities of SIDS; secondly, national government practices are critical to influencing household risk perception and thus to devising responses through approaches to risk governance, that promote safety and compliance, and enable accessibility to crucial service and sources of food; thirdly, though general trends are evident countries vary because of the different extent of local food production and access to incomes.

The lack of interdisciplinary responses to the COVID-19 pandemic meant that the initial strict, public-health focus led to some degree of food insecurity in all five countries, circumstances that were also occurring elsewhere and contributing to insecurity among the more vulnerable (Heck et al., 2020: 825). The conflict between containing the spread of COVID-19 and maintaining access to food at the household level was evident in the initial negative impact of protocols such as transportation limitations and lack of access to markets. Thus the need for weekly sourcing of fresh produce (fruits and vegetables), because of their short shelf life, became more difficult. Disaster management agencies in the Caribbean have advocated the purchase of non-perishable food items (i.e. canned food) as part of conventional household disaster supplies, where such typical Caribbean disasters as cyclones result in electrical power outages and water outages (CDEMA, 2020). The legacy of this pre-COVID perspective may have contributed to households purchasing larger quantities of processed foods, as seen in Fig. 23.5. However, in the context of the triple burden of diseases noted by Hickey and Unwin (2020) in the Caribbean, a more nuanced information campaign around food was needed, so that interventions might implicitly link healthy food policies, nutrition education, and support of local fresh food production and consumption.

In the earliest phase pf COVID-19 response national governments in the Caribbean as elsewhere instigated and strengthened new policy responses and protocols for households (Brousselle et al., 2020; Charoenwong et al., 2020). During March and April 2020, rising stringency indices marked greater strictness of policy responses inducing a heightened risk perception to likely threats from COVID-19. The more dramatic early measures, notably border closures and curfews, were atypical as first responses to potential hazards, as in the case of other recent infectious diseases in the region (e.g. Zika, Chikungunya) or tropical cyclones. These initial actions confirmed the gravity of the pandemic threat. Household compliance with protocols brought reduced mobility, a quick example of isomorphic learning at household level: learning from unwanted events (Toft & Reynolds, 2005) but also from scientific evidence and advice from PAHO, WHO and the international community (Murphy et al., 2020). This external reliance in an unprecedented context may have however limited the opportunity for national governments to use this new knowledge in responding to and managing their localised national vulnerabilities.

Although heavy reliance on food imports by all five countries required air and seaports to remain open, geographic-specific trends in securing preferred foods suggesting the regional variations in the role of place in achieving food insecurity. While impacts were similar across countries, their severity varied; respondents in Barbados reported lack of access to markets, and movement restrictions more so than other countries. In Belize, threats came from livelihood disruptions and loss of jobs or reduced salaries. Understandably, COVID-19 threatened basic human necessities triggering human emotions (Aven & Bouder, 2020, 851), yet cognitive skills and knowledge were part of household risk assessment processes. Most households sought to access knowledge that would inform their responses, enabling better decision-making on food security and strengthening existing or devising new coping strategies. Knowledge is location specific and diversity in local contexts and its implications for risk perception were evident in perceived future livelihood impacts across countries. For example, in Belize respondents were more dependent on informal economic employment and were more affected by losing jobs or reduced salaries, and consequently adopted coping strategies such as skipping meals or eating less. In Belize particularly therefore, fear and anxiety may have played a role in household decision making, resulting in reduced types and diversity of foods with declines in quality and quantity, food skipping meals and hunger.

In each of the countries responses to COVID-19 may have been partly emotional – unsurprising given the mortality rates predicted early on – but they were also logical, where they responded to income deficits, by seeking out cheaper food. They also involved web searches to develop more effective coping strategies (such as local food preparation searches). And with the aid of IT they were increasingly 'fast, instinctive, and intuitive'. Building self-sufficiency in food production and preparation could help manage household food consumption. The range of coping strategies revealed that, even during the exigencies of an unexpected and unpredictable COVID-19, households comprehended the issues that were involved, responded to mobility restrictions, reduced access to store-bought cooked and processed foods and fresh produce, resulting in a myriad of responses such as shopping at different stores, skipping meals, and home gardening. That might have proved challenging and occasionally unpleasant but it demonstrated a widespread resilience to an unprecedented hazard.

References

Aven, T., & Bouder, F. (2020). The COVID-19 Pandemic: How Can Risk Science Help? *Journal of Risk Research, 23*(7-8), 849–854. https://doi.org/10.1080/13669877.2020.1756383

Brousselle, A., Brunet-Jailly, E., Kennedy, C., Phillips, S. D., Quigley, K., & Roberts, A. (2020). Beyond COVID-19: Five Commentaries on Reimagining Governance for Future Crises and Resilience. *Canadian Public Administration, 63*(3), 369–408.

CDEMA. (2020). *Hurricane Preparedness Tips for Homes*. Retrieved from https://www.cdema.org/hurricane-preparedness-tip-for-homes

Charoenwong, B., Kwan, A., & Pursiainen, V. (2020). Social Connections with COVID-19–Affected Areas Increase Compliance with Mobility Restrictions. *Science Advances, 6*(47), eabc3054. https://doi.org/10.1126/sciadv.abc3054

Chong, A., Bailey, S., Ciardi, F., Gouretskaia, E., & Grainger, N. (2020). *Caribbean COVID-19 Food Security & Livelihoods Impact Survey-Regional Summary Report*. Retrieved from Barbados: https://www.wfp.org/countries/caribbean

FAO. (2008). An Introduction to the Basic Concepts of Food Security. Food Security Information for Action: Practical Guides., : FAO.

FAO. (2015). *State of Food Insecurity in the CARICOM Caribbean*. FAO.

FAO. (2020). *Small Island Developing States Response to COVID-19: Highlighting Food Security, Nutrition and Sustainable Food Systems*. FAO.

GOB (Government of Belize). (2020). *Official Government of Belize (GOB) Website*. Retrieved from https://www.belize.gov.bz/

Google. (2020). *COVID 19 Community Mobility Reports*. Retrieved from https://www.google.com/covid19/mobility/

Hale, T., & Webster, S. (2020). *Oxford COVID-19 Government Response Tracker*. Retrieved from https://www.bsg.ox.ac.uk/research/research-projects/oxford-covid-19-government-response-tracker. From Blavatnik School of Government https://www.bsg.ox.ac.uk/research/research-projects/oxford-covid-19-government-response-tracker

Heck, S., Campos, H., Barker, I., Okello, J. J., Boy, E., Brown, L., & Birol, E. (2020). Resilient Agri-Food Systems for Nutrition Amidst COVID-19:

Evidence and Lessons from Food-Based Approaches to Overcome Micronutrient Deficiency and Rebuild Livelihoods After Crises. *Food Security, 12*, 823–830. https://doi.org/10.1007/s12571-020-01067-2

Hickey, G. M., & Unwin, N. (2020). Addressing the Triple Burden of Malnutrition in the Time of COVID-19 and Climate Change in Small Island Developing States: What Role for Improved Local Food Production? *Food Security, 12*(4), 831–835.

Loewenstein, G. F., Weber, E. U., Hsee, C. K., & Welch, N. (2001). Risk as Feelings. *Psychological Bulletin, 127*(2), 267–286. https://doi.org/10.1037/0033-2909.127.2.267

Maturana, H. R., & Varela, F. J. (1992). *The Tree of Knowledge: The Biological Roots of Human Understanding* (Revised ed.). Shambhala Publications.

Murphy, M. M., Jeyaseelan, S. M., Howitt, C., Greaves, N., et al. (2020). COVID-19 Containment in the Caribbean: The Experience of Small Island Developing States. *Research in Globalization, 2*, 1–9. https://doi.org/10.1016/j.resglo.2020.100019

Ristea, I. (2013). Reflections on Mechanisms Influencing Human Behavior. *Procedia – Social and Behavioral Sciences, 92*, 799–805. https://doi.org/10.1016/j.sbspro.2013.08.757

Shah, K., Dulal, H., & Awojobi, M. (2020). Food Security amd Livelihood Vulnerability to Climate Change in Trinidad and Tobago. In J. Connell & K. Lowitt (Eds.), *Food Security in Small Island States* (pp. 219–237). Springer.

Slovic, P., Peters, E., Finucane, M., & MacGregor, D. (2005). Affect, Risk and Decision Making. *Health Psychology, 24*(4), S35–S40. https://doi.org/10.1037/0278-6133.24.4.S35

Thomas, A., Moore, A., & Edwards, M. (2018). Feeding Island Dreams: Exploring the Relationship Between Food Security and Agritourism in the Caribbean. *Island Studies Journal, 13*(2), 145–162.

Toft, B., & Reynolds, S. (2005). *Learning from Disasters: A Management Approach* (3rd ed.). Palgrave Macmillan.

Tuladhar, G., Yatabe, R., Dahal, R., & Bhandary, N. (2015). Disaster Risk Reduction Knowledge of Local People in Nepal. *Geoenvironmental Disasters, 2*(5), 1–12. https://doi.org/10.1186/s40677-014-0011-4

CHAPTER 24

God and the Virus in Papua New Guinea: Outsourcing Risk, Living with Uncertainty and (Re)creating a *Niupela Pasin*

Monica Minnegal and Peter D. Dwyer

Introduction

Early reports on the emergence and global spread of severe acute respiratory syndrome coronavirus 2 recognised that mitigation practices implemented in developed countries might be inappropriate in less-developed countries where access to medical services is problematic or customary social practices are likely to enhance contagion (Bermant & Ssorin-Chaikov, 2020; Dalglish, 2020; Higgins et al., 2020). Drawing on the work of anthropologist Mary Douglas, Brown (2020: 11) argued that studies concerned with COVID-19 should combine attention to

The original version of this chapter was revised. The correction to this chapter can be found at https://doi.org/10.1057/978-981-16-5285-1_30

M. Minnegal (✉) • P. D. Dwyer
University of Melbourne, Parkville, VIC, Australia
e-mail: mmam@unimelb.edu.au; pddwyer@unimelb.edu.au

Y. Campbell, J. Connell (eds.), *COVID in the Islands: A comparative perspective on the Caribbean and the Pacific*,
https://doi.org/10.1007/978-981-16-5285-1_24

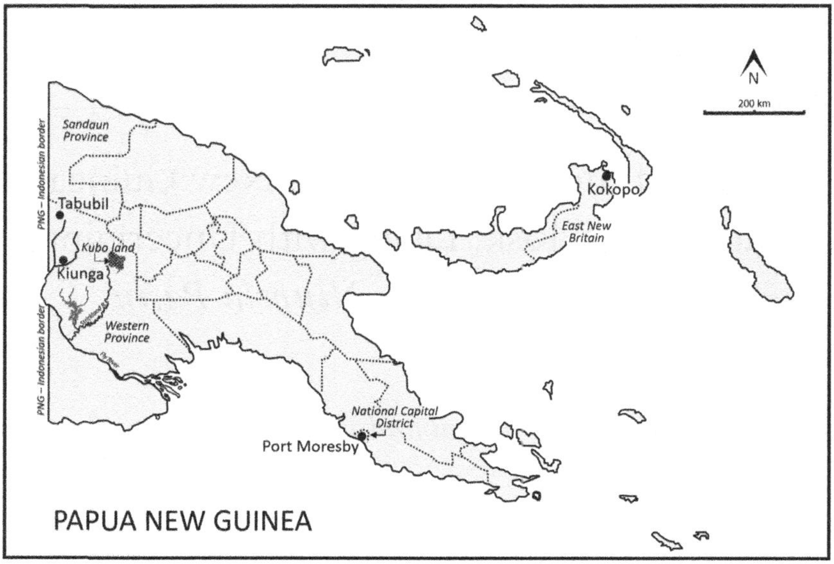

Fig. 24.1 Map of Papua New Guinea showing locations mentioned in the text

associated risks and uncertainties with recognition that the 'rational and the magical are combined in ... everyday practices'.[1]

In this chapter we explore three ways in which COVID-19 has been experienced and responded to in Papua New Guinea (PNG). To date, most reports from that country emphasize potential problems arising from an under-resourced health system and existing, deeply embedded, social practices (Allen, 2020; Kabuni, 2020; Minnegal & Dwyer, 2020a; Rooney, 2020; Troolin, 2020; Yakam, 2020).[2] Here, we seek a more explicit, ethnographic approach. Within the frame of a chronology of virus-related events in PNG, we present stories from three locations (Fig. 24.1). We first explore shifts in Prime Minister James Marape's responses; though

[1] Risk may be quantified, uncertainty cannot. Risk applies to contexts in which actors can assess the likelihood that certain events will occur, where the past is a reliable guide to the future. Uncertainty applies to contexts in which no such assessment is possible, contexts where causality is unclear and the future appears unpredictable (Knight, 1921).

[2] Between April and November 2020, the PNG National Research Institute published 12 Covid-related research articles (Spotlight Volume 13, Numbers 2 to 10 and 14, 15, 17; available at https://pngnri.org/index.php/our-research/home)

based in the capital, Port Moresby, this is a story that spans the nation. Second, we discuss the experiences of people from a remote area of Western Province who found themselves cut-off from access to resources that they had come to accept as essential to well-being. Third, we draw on a report about people in East New Britain who responded to a lockdown by reaffirming past practices of exchange and sociality (Fainu, 2020). We contextualize each of these cases with reference to risk and uncertainty. In addition, we draw out connections between observed responses and Christian belief, as the threat of COVID-19 led people to grapple with issues of authority and authoritative knowledge that are central to both managing risk and addressing uncertainty.

Outsourcing Risk

James Marape became the eighth Prime Minister of Papua New Guinea in May 2019. He was 48 years old, a university-educated Huli highlands man, son of a Seventh Day Adventist pastor and himself a committed Christian. In his maiden speech as prime minister, Marape promised to review ‘outdated’ laws governing resource extraction, to not be pushed around by foreign corporations, and to ensure that his people received a fair share of revenues (Gluyas, 2019). His stated vision was for Papua New Guinea to become ‘the richest, black, Christian nation on planet earth’ (Anon, 2019a). Through the next eight months Marape initiated revisions of the national Mining and the Oil & Gas Acts, and sought to renegotiate various agreements to increase the share of resource rents accruing to the nation. He revisited the Papua Liquefied Natural Gas agreement, postponed settlement of a P’nyang LNG agreement and refused to renew Barrick’s licence to mine gold at Porgera (Anon, 2019b, 2019c). The rhetoric was fine. The sense of concern for his people was fine. But, on the ground, real changes were hard to discern. Outbreaks of murderous violence in his home province were no less frequent (Lyons, 2019). Royalties still were not being received by landowners associated with key gas wells feeding the PNG LNG pipeline (Anon, 2019d). There was no evidence that the long-standing bugbear of corruption in high places was being redressed (Walton & Husnia, 2020).

It might be asked, therefore, what, if anything, had really changed as a result of Marape taking over as Prime Minister. But eight months is not a lot of time to implement a challenging agenda and a rather bewildering vision, so a pessimistic view might be unfair to Marape. And then, in the early months of 2020 the disease that became known as COVID-19 emerged and was declared a global pandemic (World Health Organization, 2020). In PNG, the first case was reported in mid-March (Dwyer & Minnegal, 2020). The infected person was an expatriate mine-worker who was diagnosed soon after arrival in the country and returned to Australia. Given what was happening elsewhere in the world, the level of concern in both government and community was understandably high. A nationwide state of emergency was declared on 24 March. Schools were closed, air travel banned, towns placed in lockdown and 25 March designated by Marape as a day of prayer and fasting (PNG Today, 2020). Ten weeks later, however, there had been only eight reported cases and no deaths. PNG, it seemed, had escaped the pandemic.

But Marape's government remained cautious, on 12 June passing a 'National Pandemic Act' that conferred ongoing powers to restrict freedom of movement and association and only then revoking the State of Emergency (GoPNG, 2020a). It continued monitoring the situation across the country, maintaining an informative web site,[3] and promoting adoption of a ***niupela pasin***: a 'new normal' that prioritized behavioural changes to reduce the likelihood of exposure and infection. Through all this time Marape kept in touch with the people of PNG, through press releases and social media. His own Facebook posts were seen by thousands of people and widely shared. In general, these provided clear and detailed information about current cases and government actions. He would end with words such as 'God has protect[ed] us thus far and he will continue to do so. Thank you and God bless PNG' (Department of Prime Minister & NEC, 2020). But science, and the state, also had an acknowledged role to play.

There was, however, one exception to this clear-sighted approach. In mid-April an infected customs officer visited the place where Marape and his police minister worked (Whiting, 2020). Both men were now at risk and went into self-isolation. On the evening of his first day in isolation, Marape posted to Facebook. In part, he wrote:

> Good evening all ... Future always remain uncertain, but we can aim to be better on earth and into eternity too. Coronavirus has taught humanity that we are mere mortals (that we can die) .... Don't forget Jesus Christ said He will come to redeem humanity .... When you die, your status does not mat-

[3] The web site for the PNG Joint Agency Task Force National Control Centre for Covid-19 is https://covid19.info.gov.pg

> ter, your money will be insignificant, your family and tribe will not be buried with you, it's you and you alone. If you value your life, then spend some time alone in this Corona isolation time to place proper valuation on your life. (James Marape 2020; note that Marape maintains several Facebook pages with slightly different names.)

James Marape, it seems, was understandably anxious. But he bounced back; and so too did the country. Travel restrictions were lifted and schools reopened. By mid-July there had been several weeks with no new cases reported. Then, abruptly, the virus not only reappeared in PNG, it seemed to take off (Fig. 24.2).

By mid-July only 15 cases of COVID-19 had been reported from PNG. Two weeks later, at the close of that month, the count had reached 72 cases and two deaths. With the number of cases rapidly climbing in Port Moresby and the town placed in a second lockdown, in late July Marape posted to Facebook:

> In the midst of Covid 19, or in your moments of down time, where ever you are I know all of you going through hardships, I am still committed to giving my best for you all, but better than me, Jesus gave His all for you. (PM Hon. James Marape, 2020a)

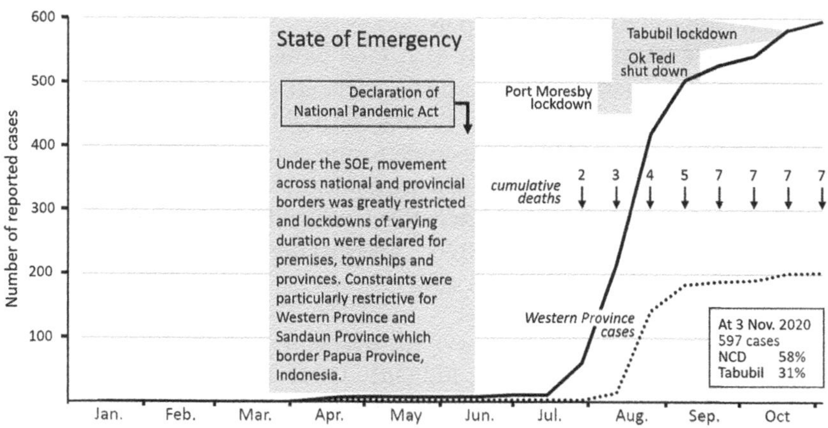

Fig. 24.2 Reported Covid-19 cases and deaths. (Records were compiled from the web site of the PNG National Control Centre for Covid-19 and collated in two-week periods commencing 1 January 2020)

He continued, '... as the threat of Corona faces PNG and the world', by directing his readers to the 'beautifully constructed song' *There was Jesus*:

> When the life I built came crashing to the ground/When the friends I had were nowhere to be found/I couldn't see it then but I can see it now/There was Jesus.

With the chorus:

> In the waiting, in the searching/In the healing, in the hurting/Like a blessing buried in the broken pieces/Every minute, every moment/Where I've been or where I'm going/Even when I didn't know it/Or couldn't see it/ There was Jesus.

By this time, the virus had spread to other provinces. Fly-in workers carried it to the Ok Tedi mining town of Tabubil, in Western Province. Mining operations were suspended for six weeks and the town was placed in lockdown.[4] Through August, the number of COVID-19 cases in PNG continued to increase. The lockdown in Port Moresby had no obvious positive effect. It was lifted after two weeks and, on 27 August, in parliament, Marape 'assured the nation that there will be no more Covid-19 lockdowns' (Kuku, 2020). He said that 'people should learn to live with the pandemic', and asked everyone 'to bear with us as we adjust ... let us help adjust to living with these basic principles and God will continue to help us'.

But Marape wanted to ensure that God would help and, to this end, argued that PNG should be formally declared a Christian country. He observed that the importance of PNG's cultural heritage and Christianity were both mentioned in the preamble to the constitution. He wanted to go further, however, and enshrine Christianity in the constitution itself. By 19 August the PNG National Executive Council had approved a proposal to do this (Anon, 2020; Oge, 2020). A week later, in his statement on the National Day of Prayer, Marape said:

> My brothers and sisters, this day ... is a solemn day of our people coming together in unison, to BOW BEFORE GOD IN HUMILITY AND PRAYER, seeking His forgiveness, love and direction in a turbulent time of confusion, fear and panic. (Marape, 2020a, 2020b)

[4] Though the Ok Tedi mine reopened in mid-September, restrictions on people's movements continued in the mine-controlled town of Tabubil. Children from neighbouring villages were not allowed to attend Tabubil schools to at least 7 October (Ok Tedi Mining, 2020).

Marape was asserting, it seems, that attempts to control spread of the virus by government action – by lockdowns – were useless; people must live with the virus. And, since September, to the time of writing (4 November), this is how Marape has acted. There have been no further substantive references to COVID-19. At a school graduation ceremony, in September, he did not mention the on-going pandemic but rather encouraged students to 'put God first' and to 'walk with Jesus' for the rest of their lives (GoPNG, 2020b). In mid-October, when formally thanking 'United Nations, China, Australia, America and other friends' who 'helped the country cope with the pandemic', he commented that the donation of '30 ventilators will go a long way in helping patients and persons ***suspected of COVID-19***' and said that 'the Government was ***looking beyond COVID-19***' to ensure that 'other essential health care services would not be compromised' (GoPNG, 2020c, ***our emphases***).

This, then, captures the essence of Marape's personal COVID-19 strategy. He distanced himself from personal engagement. He outsourced PNG's response to COVID-19 by outsourcing risk to God, handing over responsibility for managing the threat this disease posed to his people. Marape's perspective is summarised well in a statement published on the government's official COVID-19 website in mid-June: 'the hand of God' is credited with restricting spread of the virus in PNG, while 'natural immunity ... a natural defence mechanism built by God himself' protects those who do become infected (GoPNG, 2020d). Marape acknowledged 'probabilities': he counted the cases and compared the numbers in PNG with those in other places. But to him the future is known, it is secure, for it is in the hands of a caring, loving God.

Living with Uncertainty

Elsewhere in PNG, that trust in God was itself being eroded as the state's response to the threat of COVID-19 exacerbated a growing sense of disconnection from wider social institutions and a loss of faith in the capacity of these institutions to shape events. Kubo people live in a remote corner of Western Province. It was the 1970s before colonial and national governments extended administrative control to this area and the early 1980s before a Christian mission was established there (Minnegal & Dwyer, 2017). Since then population size has more than doubled. People who were formerly scattered have come together in larger villages. Increasingly, men, and sometimes women and children, move between village and town, with a few relocating semi-permanently to Kiunga and even Port Moresby. The boom-bust nature of mineral and gas exploration provides an intermittent cash income for some. But subsistence farming and foraging remain the primary way of life for most and contact with people and places beyond the local, through radio or mobile phone, remains intermittent at best. At Suabi village, with a population approaching 1000, everyone asserts belief in the Christian God and follows one of three Christian denominations. With equal certainty, everyone is sure that most deaths are the result of sorcery attack.

News of COVID-19 came slowly to Suabi and, when it came, it generated ambiguity (Minnegal & Dwyer, 2020b). People heard the government instructions to stay in place, to not move, to reduce interactions and maintain distance from others. Many accepted that the recommendations were sensible. But, in their face-to-face world, it was neither possible nor socially desirable to put them into practice. More worryingly, with limited and unreliable access to phone networks, parents lost contact with town-based children. With schools closed, many who had been boarding in school dormitories were left without secure access to accommodation and food. With planes no longer flying, they had little choice but to walk home; five days walk from the town of Kiunga, passing through – and potentially spreading virus to – several other communities and language groups on the way. Getting home was their priority.

With the return of these youths, however, news of what was happening elsewhere was further reduced. As the national state of emergency continued, and as people at Suabi felt increasingly isolated, we became a conduit for news. Though three and a half thousand kilometres separate Melbourne from Suabi, we received calls when the always erratic network was available, seeking updates on COVID-related news for PNG, Australia, Indonesia and the world. One theme was central in these exchanges. Our friend Martin (as we will call him here) knew that people in the towns were having a hard time, that the lockdown made access to food difficult there. He acknowledged that village people, in contrast, were eating well. But Martin wanted town-folk to understand that, with no planes coming and all travel officially in abeyance, 'it is very hard' now for village people too.

For several decades the lives of Kubo people – even those who remain in villages – has been re-ordered. They already live in, and identify with, a 'new normal'. The relative safety of travel by foot, the weekly touch-down of a Mission Aviation Fellowship plane, the regular appearance of outsiders to service schools or health centres or to explore for gas and minerals, has altered both the catalogue of everyday, and now essential, resources and the paths to their acquisition. Metal tools for gardening, building, hunting and fishing; cooking and eating paraphernalia; western clothing; torches: these are no longer seen as luxuries. But access to them requires that modern kinds of communication and modern kinds of movement remain open.

The experience Martin described during these phone calls was not one in which his people were 'locked down'; it was one in which they were 'locked out'. Just as they were, at last, successfully reaching out to, and encompassing, the modern, their access to that world had suddenly been withdrawn. And an inability to access two particular items became the central representation of this enforced predicament. Time and time again, Martin told us that people at Suabi had run out of salt and soap. 'White people taught us to wash,' he said. 'Now we have no soap and our bodies stink. ... We are not the people we used to be'.

In addition, in these conversations – there was often an audience gathered on the hill where reception was possible – one question kept recurring. 'Has a medicine been found yet, for this disease?' We could feel the disappointment that our answers generated. The disease had not yet reached Kubo people: that was good. But it had defeated white people and their science: that was very bad. White people and their science, it seemed, were less powerful than had been imagined. The certainty, the control over life and death, that these had promised was being eroded. That loss of confidence, in turn, left space to consider other possibilities and revisit past understandings and practices with respect to illness and death.

A concern with sorcery has remained ever-present among Kubo people and their neighbours. In recent decades, however, as people increasingly turned to external authorities – church and state – to resolve tensions, practices intended to ward off the effects of sorcery or to identify practitioners have become less evident. Some, it seemed, had been put aside. But that is again changing. In 2014, several curing dances were held at Suabi on behalf of very sick individuals (Minnegal & Dwyer, 2017: 123–24). Each time, one or two costumed men danced through the night to the beat of a kundu drum, brushing their sago-frond skirts across the body of the sitting invalid, drawing the attention of spirits who might come to their aid. The local pastor did not approve. Within the community some tension was emerging over the efficacy of customary and Christian ideologies. It was evident when people were sick. It surfaced more dramatically when people died. Few doubted – not even the pastor – that sorcery was involved. But, under the influence of Christianity, since the mid-1990s punishment of the sorcerer had been left to God and public blame attributed only after a suspected sorcerer died.

More recently, however, and with increasing frequency, suspect sorcerers have again been subjected to customary tests: the proper cooking of sago, the effective use of fish poison. Further, new forms of sorcery are appearing (Minnegal & Dwyer, 2017: 203–04) and some people are experimenting with new ways to address the threat of sorcery. On the Strickland River, at villages where we first worked in the 1980s and 1990s, a group of seven young men assert that they can identify sorcerers by their 'smell', confessions have been elicited by inflicting pain, and threats of corporal punishment are increasingly common. Women are more likely to be named as suspects than was the case a generation earlier. And people report that sorcerers now may collaborate in perpetrating their evil acts, forming gangs rather than working alone as they did before.

Kubo people are challenged by the seeming inability of science to resolve a global medical problem. They are challenged too by the re-emergence of customary practices that cast doubt on the universality of Christian belief and the effectiveness of Christian practice. Should COVID-19 appear in their midst, and cause sickness or death, they are primed to wonder who it is, among themselves, that has harnessed the power of the virus and turned it upon enemies. The suggestion has been already aired among a neighbouring people and, indeed, more prominently, on a Western Province Facebook group.[5]

In the context of a disease that threatens but is not yet manifest in their midst, Kubo people live with uncertainty. They can only wait and hope, perhaps pray and cast identified sorcerers aside. At this time, they have no obvious path out of a crisis that was not of their making: a crisis that neither church nor state seem able to resolve, a crisis that resists resolution in terms of the precepts of Christianity or science and which, therefore,

[5] Late in August a post to the Facebook group Fly River Forum reported the most recent count of PNG Covid-19 cases and directed attention to the fact that 143 of the 419 were from Western Province. One person responded that 'this Covid 19 lockdown made opportunity to sanguma people [sorcerers] are killing the people'. Another commented: 'Yes … sanguma takim advantage lon this c19 situation lon kilim ol man stap'. A third wrote that he was afraid of sanguma while a fourth cautioned earlier correspondents with the words: 'honest there is a lot of people in North Fly dying for all sorts of reasons'.

tempts consideration of earlier ways of being in the world. Whereas Marape put his trust in God to manage the threat of COVID-19, the people of Suabi – without access to the material comforts that Marape enjoys – must trust their own ability to identify and deal with threats as they arise.

(Re)creating a *Niupela Pasin*

Elsewhere again, in PNG, people have been renewing a sense of autonomy in domains of secular action, and not just spiritual action, as the threat of COVID-19 challenges access to broader economic engagements and thus local livelihoods. Tolai people of East New Britain are famed in both popular and anthropological imaginings for their use of 'shell money', *tabu*. At an earlier time, exchanges of *tabu* were essential to mortuary and marriage rituals and played a role in some seemingly more prosaic utilitarian concerns. By 1991, however, 'modern money' had displaced the role of *tabu* in many secular transactions and the 'importance of *tabu* for the Tolai' now lay in its ritual roles (Simet, 1991: ix). Epstein (1998: 23), too, considered that 'conceptually, morally, and functionally Tolai saw a sharp contrast between their own shell currency' – locally produced, embodying local labour, and circulating within a local sphere of exchange – 'and the recently imported money of the Europeans. From their perspective, money was much inferior. ... Money pertained to mundane things, like the purchase of food' but *tabu* pertained to 'spiritual matters'. At the present time, *tabu* retains significance in marriage and mortuary exchanges and may be used, as well, in birthday celebrations, initiations and the resolution of disputes. It remains integral to a major ritual, the *kinavai*, that, in the past, was one component of a sequence of events performed to celebrate the arrival of ancestors on Tolai land. In contemporary performances 'clans from different villages re-enact the arrival [in the late 1800s] of the missionaries and the bible' (Baker & Sieber, 2019; Sieber, 2019). Clansmen come ashore accompanied by masked figures – spiritual beings named *tubuan* – and distribute *tabu* to the crowd that waits near a large Christian cross.

In August 2020, Kalolaine Fainu (2020) – herself from East New Britain – published an article in *The Guardian*. She wrote of people from villages to the northwest of the capital Kokopo whose access to services, store goods and hard currency had been seriously diminished by the lockdown measures imposed by government. They were left with no way of accessing these material expressions of modernity. Like the Kubo people discussed earlier, they were, in effect, 'locked out'. They responded, Fainu wrote, by reviving earlier patterns of informal exchange. They used 'their tabu collections to buy everyday items because they simply had no money'. As one woman said: 'We couldn't travel to town to sell our produce at the markets or buy things from the shop, so we started to use our tabu more'. Another told her that a few people 'who were able to circumvent the cancellation of public buses brought back store goods and exchanged them with neighbours for tabu'. Fainu reported also that elders 'revived old ways of preserving food' and 'taught younger generations how to weave baskets and fish traps'.

It is likely that the reported COVID-related shift in the use of *tabu* was more quantitative than qualitative. The place of *tabu* in Tolai society is complex and ambiguous. It fulfils ritual functions such as those depicted above but is, simultaneously, implicated in non-customary practices intended to enhance the status and wealth of particular men. '*Tabu* and ritual have returned', Martin (2010: 6) wrote, 'but it is the ways in which *tabu* is acquired today that make its accumulation in large amounts morally dubious in the eyes of many'. Though a separation of the secular and the sacred is unlikely to have been as rigid as some have implied it seems that, irrespective of past or present materialist roles, *tabu* 'moonlight as a portal into a pre-capitalist spirit world we will never see' (Scott, 2020). Thus, the deeper significance of the change Fainu reported lies in a psychological and relational sphere rather than an economic and categorical sphere. For the Tolai people she discussed, it seems that the 'new normal' emerging in the time of Covid-19 has entailed the sense of a partial return to an 'old normal', of a return to an imagined time and place where boundaries between the secular and the sacred were harder to discern and where the source of wealth and well-being lay in one's own labour. As Scott (2020) argued, a 'return to localism' has been 'a resurging theme in the age of Covid'. It is in this sense that Tolai people found their own path to reducing the risk and uncertainty associated with corona virus.

Conclusion: Science, Religion and Tradition

Lorea (2020) directed attention to multiple instances, from across the globe, of the interface between religion and coronavirus and argued that 'it is of fundamental importance in such an extraordinary time … to look at the bigger picture of a dramatically changing world – a world of fragile certainties and desperate calls for comforting predictions'. We have written of three ways in which Papua New Guineans have experienced and responded to COVID-19. They will not be the only ways. In each case, however, the key actors are confronted by collapsing certainties and a desire for a knowable future. In each case, too, in striving to glimpse that future, they grapple with issues of authority. Marape turns to the ultimate authority of God. The Kubo people with whom we have worked are thrown into confusion when the authority of Western systems of knowledge – science or religion – appears to fail. And on East New Britain, perhaps more hopefully, the demonstrable weakening of an imposed economic system opens space for recapturing the authority of earlier, and deeper relational values.

The Kubo and Tolai cases play out on a less public stage than do the thoughts and actions of Marape and, for that reason at least, are less subject to public scrutiny. In both cases, however, COVID-19 has generated not just practical challenges but also growing uncertainty about access to, and the authority of, overarching structures – of justice for Kubo, of economy for Tolai. A renewed emphasis in these places on local autonomy, in domains of spiritual action and secular action respectively, can be traced to a loss of trust in 'modern' epistemologies among Kubo people, a loss of trust in the currency of modern livelihoods among Tolai. COVID-19 has not, as yet, played itself out in Papua New Guinea. What then is the likely aftermath of a pandemic that has had a greater impact on people's understandings than on their physical well-being? For Papua New Guineans, as for so many people the world over, like the virus itself, the transformative potential of COVID-19 is hidden. In which directions will people turn for sources of authority, in which directions will trust be restored?

Coda

In late October news outlets reported on claims by a team of PNG university scientists to have discovered a cure for COVID-19 and the likelihood that government would fund their research to the tune of K10.2 million (Elapa, 2020). The scientists asserted that their research 'was conducted in light of [the] Prime Minister's vision to Make PNG become the richest black Christian nation in the world' and gave 'thanks to our God Almighty for giving us the grace to come up with the drugs for Covid19' (Kerenga, 2020). Prime Minister Marape recommended through the National Executive Council that the money be awarded to Niugini Biomed Limited, a company that had been registered under that name on 17 August 2020. But the news proved controversial, eliciting much negative comment on social media. Marape's public response to these criticisms was more cautious; he spoke of the potential to draw on intellectual and natural resources of the nation rather than of a specific treatment for COVID-19. 'I am not a Prime Minister to kill ingenuity, research, science and study', he wrote (PM Hon. James Marape, 2020b). 'We are a nation of huge biodiversity, in fact, with five to seven per cent of the world's biodiversity. Our ancestors lived with malaria, snake bites and all manner of tropical diseases. I am certain that we can find new medicines, amidst all this biodiversity, in our Blessed land.' Final approval of the grant came on 5 November. The 'bulk of the K10.2 million would be channelled towards rebuilding of science faculty laboratories' at the University of Papua New Guinea to facilitate planned and future research (GoPNG, 2020e).

Acknowledgements Thanks to Dylan Gaffney and Tom Powell Davies for their invitation to present an earlier version of this paper to the University of Cambridge 'Risk and Renewal in the Pacific' seminar series.

References

Allen, B. (2020). *Coronavirus Covid-19 in Papua New Guinea*. Australian National University, Department of Pacific Affairs. In Brief 2020/6.

Anon. (2019a). Marape New PNG Prime Minister. *The Australian*. Available at https://www.theaustralian.com.au/world/james-marape-new-png-prime-minister/news-story/6d80ca592b4d2d9e926cc9ee656e4210. Accessed 22 Oct 2020.

Anon. (2019b, October 17). Gas Project to Be Standalone. *The National*. Available at https://www.thenational.com.pg/gas-project-to-be-standalone/. Accessed 22 Oct 2020.

Anon. (2019c, August 31). Porgera's New Lease. *The National*. Available at https://www.thenational.com.pg/porgeras-new-lease/. Accessed 22 Oct 2020.

Anon. (2019d, November 18). Court Halts PDL7 Account Opening Exercise. *Post Courier*. Available at https://postcourier.com.pg/court-halts-pdl7-account-opening-exercise/. Accessed 22 Oct 2020.

Anon. (2020, August 19). PM: PNG Is All Christian. *The National*. Available at https://www.thenational.com.pg/pm-png-is-all-christian/. Accessed 22 Oct 2020.

Baker, S., & Sieber, C. (2019, March 19). The Spiritual Shell Money of Papua New Guinea. *Hakai Magazine*. Available at https://www.hakaimagazine.com/videos-visuals/cash-with-spiritual-currency/. Accessed 21 Oct 2020.

Bermant, L. S., & Ssorin-Chaikov, N. (2020). Introduction: Urgent Anthropological COVID-19 Forum. *Social Anthropology, 28*, 218–219.

Brown, P. (2020). Studying COVID-19 in Light of Critical Approaches to Risk and Uncertainty: Research Pathways, Conceptual Tools, and Some Magic from Mary Douglas. *Health, Risk & Society, 22*, 1–14.

Dalglish, S. L. (2020). COVID-19 Gives the Lie to Global Health Expertise. *The Lancet, 395*, 1189.

Department of Prime Minister & NEC. (2020, April 16). *Facebook Post.* Available at https://www.facebook.com/SirManasupeHaus/photos/a.836303540053943/1132942113723416/?type=3. Accessed 22 Oct 2020.

Dwyer, P. D., & Minnegal, M. (2020). COVID-19 and Facebook in Papua New Guinea: Fly River Forum. *Asia and the Pacific Policy Studies, 7*(3), 233–246. https://doi.org/10.1002/app5.312

Elapa, J. (2020, October 29). K10.2 Million for Covid-19 Procurement, Treatment Approved. *Post Courier.* Available at https://postcourier.com.pg/k10-2-million-for-covid-19-procurement-treatment-approved/. Accessed 30 Oct 2020.

Epstein, A. L. (1998). Tubuan: The Survival of the Male Cult Among the Tolai. *Journal of Ritual Studies, 12*(2), 15–28.

Fainu, K. (2020, August 22). The Return of Shell Money: PNG revives old ways after Covid's Blow to Economy. *The Guardian.* Available at https://www.theguardian.com/world/2020/aug/22/the-return-of-shell-money-png-revives-old-ways-after-covids-blow-to-economy. Accessed 21 Oct 2020.

Gluyas. A. (2019, May 31). New PNG PM Strives for Economic Independence Through Mining Reform. *Australian Mining.* Available at https://www.australianmining.com.au/news/new-png-pm-strives-for-economic-independence-through-mining-reform. Accessed 22 Oct 2020.

GoPNG. (2020a). *National Pandemic Act.* Available at http://www.parliament.gov.pg/uploads/acts/20A_08.pdf. Accessed 8 Nov 2020.

GoPNG. (2020b, September 28). *Marape Motivates Students to Remain Focused Spiritually.* Available at https://covid19.info.gov.pg/index.php/2020/09/28/marape-motivates-students-to-remain-focused-spiritually/. Accessed 22 Oct 2020.

GoPNG. (2020c, October 13). *Marape Expresses Gratitude to United Nations and China for Donating 30 Ventilators.* Available at https://covid19.info.gov.pg/index.php/2020/10/13/marape-expresses-gratitude-to-united-nations-and-china-for-donating-30-ventilators/. Accessed 22 Oct 2020.

GoPNG. (2020d, June 17). *PM Acknowledges God's Hand in Covid-19 Response.* Available at https://covid19.info.gov.pg/index.php/2020/06/18/pm-acknowledges-gods-hand-in-covid-19-response/. Accessed 7 Nov 2020.

GoPNG. (2020e). *Cabinet Approves K10. 2 Million for Covid-19 Research.* Available at https://covid19.info.gov.pg/index.php/2020/11/06/cabinet-approves-k10-2-million-for-covid-19-research/. Accessed 9 Nov 2020.

Higgins, R., Martin, E., & Vesperi, M. D. (2020). An Anthropology of the COVID-19 Pandemic. *Anthropology Now, 12*(1), 2–6.

Kabuni, M. (2020). *Covid-19 in Papua New Guinea: The State of Emergency Cannot Fix Years of Negligence*. Policy Forum, Asia & The Pacific Policy Society. Available at https://www.policyforum.net/covid-19-in-papua-new-guinea/. Accessed 27 Oct 2020.

Kerenga, B. (2020). *Media Statement*, Niugini BioMed Limited. Available at https://www.facebook.com/permalink.php?story_fbid=882098682324465&id=349747055559633&__cft. Accessed 31 Oct 2020.

Knight, F. H. (1921). *Risk, Uncertainty and Profit*. Houghton Mifflin Company.

Kuku, R. (2020, August 27). No More Lockdown: PM. *The National*. Available at https://www.thenational.com.pg/no-more-lockdown-pm/. Accessed 22 October 2020.

Lorea, C. E. (2020). Religious Returns, Ritual Changes and Divinations on COVID-19. *Social Anthropology, 28*, 307–308.

Lyons, K. (2019, July 10). At Least 15 Women and Children Killed in Tribal Massacre in Papua New Guinea. *The Guardian*. Available at https://www.theguardian.com/world/2019/jul/10/at-least-15-women-and-children-killed-in-tribal-massacre-in-papua-new-guinea. Accessed 22 Oct 2020.

Marape, J (2020a, April 17). *Facebook Post*. Available at https://www.facebook.com/james.marape.106/posts/245632890145610. Accessed 22 Oct 2020.

Marape, J. (2020b, August 26). *Prime Minister James Marape's Unedited Grand Statement Delivered Today During the 'national day of prayer'*. Available at http://www.chr.org.au/places/docs/PNG-STATEMENT.pdf. Accessed 22 Oct 2020.

Martin, K. (2010). The Death of the Big Men: Depreciation of Elites in New Guinea. *Ethnos, 75*, 1–22.

Minnegal, M., & Dwyer, P. D. (2017). *Navigating the Future: An Ethnography of Change in Papua New Guinea*. ANU Press.

Minnegal, M., & Dwyer, P. D. (2020a, April 16). God, Health and COVID-19 in Remote Papua New Guinea. *Devpolicy Blog*. Australian National University. Available at https://devpolicy.org/god-health-and-covid-19-in-remote-papua-new-guinea-20200416/. Accessed 27 Oct 2020.

Minnegal, M., & Dwyer, P. D. (2020b). Communication, Privilege and the Ironies of Isolation: From Melbourne to Papua New Guinea. Available at https://medanthucl.com/2020/04/17/communication-privilege-and-the-ironies-of-isolation-from-melbourne-to-papua-new-guinea/. Accessed 22 Oct 2020.

Oge, R. (2020, August 26). Against Amending the Constitution to Make PNG a Christian Country. *Devpolicy Blog*. Available at https://devpolicy.org/against-amending-the-constitution-to-make-png-a-christian-country-20200826/. Accessed 21 Oct 2020.

Ok Tedi Mining. (2020, October 7). *Media Release*. Available at https://oktedi.com/ok-tedi-addresses-community-concerns-regarding-covid-19-restrictions/. Accessed 26 Oct 2020.

Papua New Guinea Today. (2020, March 23). *Papua New Guinea Locks Down Effective Today Amid COVID-19 Fears*. Available at https://news.pngfacts.com/2020/03/papua-new-guinea-locks-down-effective.html. Accessed 24 Oct 2020.

PM Hon. James Marape. (2020a, July 27). *Facebook Post*. Available at https://www.facebook.com/MarapeJames/photos/a.802481266817831/1148142602251694/. Accessed 22 Oct 2020.

PM Hon. James Marape. (2020b). *PM Marape: Government Supports Medical/Scientific Research*. Available at https://www.facebook.com/MarapeJames/posts/1227552980977322. Accessed 9 Nov 2020.

Rooney, M. L. (2020, August 3). What's More Important: Defence Funding on Manus or Helping PNG Fight COVID-19? *Devpolicy Blog*, Australian National University. Available at https://devpolicy.org/whats-more-important-defence-funding-on-manus-or-helping-png-fight-covid-19-20200803/. Accessed 14 Nov 2020.

Scott, B. (2020). DFMM Episode 2: Cere-Money vs. Money in Papua New Guinea. Available at https://brettscott.substack.com/p/dfmm-episode-2-cere-money-vs-money. Accessed 22 Oct 2020.

Sieber, C. (2019). *In Papua New Guinea, a Tribe Still Uses Shells as Money*. Available at https://www.vice.com/en/article/9kp5ye/papua-new-guinea-tolai-tribe-shell-money. Accessed 23 Oct 2020.

Simet, J. L. (1991). *Tabu: Analysis of a Tolai Ritual Object*. Unpublished PhD thesis, The Australian National University.

Troolin, D. E. (2020). Distantly United: Papua New Guinean Relationality in the Face of COVID-19. *Anthropology Now, 12*(1), 84–90.

Walton, G. W., & Husnia, H. (2020). Boom and Bust? Political Will and Anticorruption in Papua New Guinea. *Asia & the Pacific Policy Studies, 7*(2), 187–203.

Whiting, N. (2020, April 17). Papua New Guinea's PM James Marape Tested for Coronavirus After Worker at COVID-19 Operations Centre Tests Positive. *Australian Broadcasting Corporation News*. Available at https://www.abc.net.au/news/2020-04-17/papua-new-guinea-coronavirus-centre-lockdown-confirmed-covid-19/12156768. Accessed 22 Oct 2020.

World Health Organization. (2020, March 11). *WHO Director-General's Opening Remarks at the Media Briefing on COVID-19*. Available at https://www.who.int/dg/speeches/detail/who-director-general-s-opening-remarks-at-the-media-briefing-on-covid-19%2D%2D-11-march-2020. Accessed 22 Oct 2020.

Yakam, L. C. (2020). *Approaches to the COVID-19 Pandemic in Papua New Guinea: Government Measures and Public Responses*. Department of Pacific Affairs, Australian National University, In Brief 2020/28.

CHAPTER 25

Igat wei bilong lukautim mipela yet istap (We Can Look After Ourselves): Community-Based Organisations Responding to COVID-19 in Papua New Guinea

Verena Thomas, Jackie Kauli, Eriko Fufurefa, Angela Apa, and Umba Peter

Introduction

Papua New Guinea (PNG) is the largest nation in the Pacific Islands. Because of its diverse population and geography, and the limited capacity of its health system, it experienced a range of impacts from the COVID-19 pandemic. The PNG government implemented a two months long State

V. Thomas (✉) • J. Kauli
Queensland University of Technology, Brisbane, QLD, Australia
e-mail: verena.thomas@qut.edu.au; j.kauli@qut.edu.au

E. Fufurefa • U. Peter
Kafe Urban Settlers' Women's Association, Goroka, Papua New Guinea

A. Apa
Kup Women for Peace, Kundiawa, Papua New Guinea

Y. Campbell, J. Connell (eds.), *COVID in the Islands: A comparative perspective on the Caribbean and the Pacific*,
https://doi.org/10.1007/978-981-16-5285-1_25

of Emergency and imposed a lockdown for the population relatively early in the crisis. At the time, many other countries adopted similar policy responses to the unprecedented global pandemic. However, in PNG, it quickly became clear that the lockdown was not sustainable for the many people relying predominantly on the informal sector to maintain their livelihoods, especially when the government did not have social protection mechanisms for most of its population. Increasingly, the lack of access to income and sufficient food presented a significant risk that exacerbated existing inequalities and social issues such as gender-based violence and sorcery accusation related violence.

This chapter emerges from a collaboration between human rights defenders (Fufurefa, Apa, Peter), and their organisations Kafe Urban Settlers' Women's Association (KUSWA) and Kup Women for Peace (KWP), and communication researchers (Thomas, Kauli) who have been collaborating on an action research project that has developed communication strategies to address sorcery accusation related violence and gender-based violence in Papua New Guinea. Through this research project, the Yumi Sanap Strong (Let's Stand Strong Together) initiative was established in 2017 (https://www.yumisanapstrong.org). Some of its key outputs have included documenting the experiences of human rights defenders through digital storytelling, and the collaborative development of a creative toolkit to communicate current PNG laws to community groups (Kauli & Thomas, 2018). The project harnesses storytelling and relational approaches based on indigenous knowledge systems to support community action and resilience.

Prior to COVID-19, our teams had planned to implement training activities with key community influencers using the *Communicating the Law Toolkit* (Kauli & Thomas, 2018). We initially thought that COVID-19 would cause some minor delays in our program and were waiting for the crisis to be resolved. But when borders closed and lockdowns were imposed, our approach and program goals had to shift significantly. Because our teams operated across PNG and Australia, one of our first challenges was communication and getting information to better understand the impacts of COVID-19 on the PNG population in rural and urban areas.

For the two community-based organisations (CBOs) KUSWA and KWP, the COVID-19 pandemic meant that funding mechanisms were initially put on hold. There was limited access to finance and other resources to support the families of their staff. Through sporadic text messages and phone calls, which were often hampered by muffled noise due to bad reception, we shared information with staff of these CBOs. They were struggling to have enough food in their houses, to move around and make things work. CBO directors Fufurefa and Apa had to look after their own families, but also felt an immense responsibility for people in their surrounding community. While we all worked together to address immediate needs, we also started to develop programs that could work in the COVID-19 environment. We invested in more smartphones, internet data and solar chargers. Collectively, we set up a WhatsApp group to ensure ongoing communication.

As information about COVID-19 rippled throughout the country, KUSWA and KWP received support from donors, organisations and government to lead and implement COVID-19 awareness programs. KUSWA's and KWP's documentation of and reflections on their COVID-19 awareness activities provided a picture of the challenges that communities were facing that were not necessarily visible in media representations. From this, we discussed valuable lessons and the important role the organisations were playing in their communities during that time.

In this chapter, we collectively present some of the impacts of the COVID-19 lockdown in PNG, and the role that KUSWA and KWP played within their communities. We discuss the role of human rights–based CBOs in responding to the COVID-19 pandemic, emphasising their ability to harness community relations, contextualise information and integrate gendered approaches to violence prevention and emergency responses. Through establishing cross-sector collaborations, KUSWA and KWP increased community resilience and mitigated potential impacts on communities and vulnerable groups. We discuss recommendations for how public health communication and public administration can respond to a crisis by integrating CBO experience and expertise.

COVID-19 and Challenges for Vulnerable Groups in PNG

As of 22 November 2020, PNG had recorded a total of 612 COVID-19 cases and seven deaths (Department of Health & World Health Organisation, 2020: 1). Despite this relatively low number of recorded cases, in a country with a population of 9 million, the pandemic, and its associated lockdowns between March and August 2020, significantly affected the country's population, particularly interrupting the lives – and livelihoods – of those in rural and urban settlement communities. More than 80% of the PNG population live in rural areas and informal settlements with limited to no access to basic services and facilities. Communities are heavily reliant on the ability to generate their own income on a daily basis through trading fresh produce and self-made goods at markets. The lockdown disrupted the supply chains for food and other essential products and left many families without sufficient food (Kopel, 2020). This put a further risk on people's health and increased the potential to cause social disharmony in communities and families.

PNG's lockdown restrictions "disproportionally affected the poor and vulnerable" and those who were already disadvantaged, such as residents of urban settlements (Kopel, 2020: 1). With gender-based violence being a key development challenge in PNG, Hukula (2020: 1) argues that "access to services for vulnerable groups such as victims of gender-based violence and persons living with disability should be a key consideration for the government". Described as a catalyst for exacerbating domestic violence (Kopel, 2020), the lockdown further prevented women and children living in abusive circumstances from leaving their homes and from seeking sanctuary, which means they are even more at risk of harm. Additionally, elderly folk and people living with disabilities, two of the most marginalised groups in PNG, were already living with challenging medical circumstances and are at a higher risk of being affected by the pandemic.

The risks among vulnerable populations are further exacerbated when communication and prevention measures do not provide equitable access (see Hukula, 2020; Kaut-Nasengom, 2020). Kuman (2020), reporting on people who were turned away by immediate family members due to misconceptions about the virus, argues that public health messages about COVID-19 must "take into account people's sociocultural and religious beliefs and backgrounds". Otherwise, inappropriate messaging might increase people's vulnerability. It is therefore important to recognise the various information needs of different groups. Additionally, there have been reports that police and authorities have been heavy-handed in their enforcement of the State of Emergency, presenting further challenges to people already under stress and dealing with uncertainty (Kopel, 2020).

The COVID-19 pandemic challenged the human rights discourse as governments struggled to achieve the appropriate balance between individual rights, such as freedom of movement or the right to education, and ensuring that populations were kept safe and protected. Human rights and gendered approaches by civil society and particularly women's rights organisations provided an important foundation to develop strategies to support vulnerable groups in PNG.

CBOs Advocating for Human Rights in PNG

Civil society groups have played a key role in responding to human rights issues in PNG. Much of the advocacy of civil society groups has focused on addressing gender inequalities and the rights of children. In particular, high rates of gender-based violence and the visibility of sorcery accusation related violence in the media led to increased advocacy from civil society over the past decade, resulting in legislative changes in family protection and related violence between 2013 and 2016 (Biersack et al., 2016). Because there are challenges with enforcing the laws, most work at the community level is supported by human rights defenders and civil society organisations, including CBOs (UNDP, The Equality Institute, & Department for Community Development and Religion, 2016).

A CBO can be defined as a "public or private non-profit organisation that represents a community or a specific part of a larger community, and targets meeting a specific need in that community" (Adebayo et al., 2018: 482). With regards to gender-based violence, for example, referral pathways to access support services are provided to survivors of violence, and CBOs work closely with village courts, the police, and health centres. Human rights defenders work with the CBOs to translate and interpret human rights concepts for each local community, ensuring that each community is provided with services and prevention strategies that address that community's specific challenges.

CBOs fulfil an important function in times of disaster because they can tap into their local networks and relationships. However, because CBOs are located within their communities, they are most likely themselves affected (Drennan & Morrissey, 2019), as was the case with both KUSWA and KWP during the COVID-19 lockdown. A challenge is also provided through the often-hierarchical government response during a pandemic, and the need to activate existing networks on the ground. This is particularly important when there is a risk of exacerbating existing inequalities, and the largest group affected by such inequalities are arguably women and girls (Cousins, 2020). As such, a gendered and human rights approach is essential to responding to the COVID-19 pandemic in the Pacific. CBOs hold a strategic place in this response, given the low capacity in government health systems and social welfare, and the need to take care of vulnerable groups. Reflecting on the experiences of CBOs and their interaction with stakeholders and local communities provides important understanding for designing locally embedded emergency response and public health strategies.

Methodology

This project explored the response to the COVID-19 State of Emergency by two CBOs in the PNG Highlands region: KWP in the Simbu Province and KUSWA in the Eastern Highlands Province (EHP). Both organisations are women-led human rights organisations that focus on providing prevention and support services to support women and families affected by violence. To better understand how CBOs could be supported in the future, the project sought to understand the role that CBOs played during this pandemic, including the challenges they faced and the impacts they had.

Both KWP and KUSWA are part of the Highlands Women Human Rights Defenders Movement. KUSWA was started in 2001 in the Banana Block urban settlement in Goroka town by eight women who were survivors of domestic violence and wanted to alleviate their own suffering and the suffering of others in similar situations. KUSWA's mission is to empower and develop women's full potential to live productive lives in EHP through active promotion of health, gender equality, full participation, and life skills activities. Over the past 20 years, KUSWA expanded its services, which include counselling, violence prevention and supporting referral pathways for survivors of violence, while lobbying for social and legal change at provincial and national levels. KWP was formed in 2000 with the intention of "***daunim hevi bilong ol mama***" (reducing the sorrow of mothers) following 30 years of chronic tribal violence within the Kup region, a sub-district in the Simbu Province in the PNG Highlands. Women leaders from warring clans put aside tribal allegiances and joined forces to reverse trends of tribal fighting and associated gender violence. The Kup Women have had remarkable success in stopping tribal fighting in the Kup region and in promoting peaceful community development. Both KUSWA and KWP network with multiple partners across the Highlands and provide vital support to families and communities.

Data collection was undertaken in collaboration with the CBOs, using a reflection-in-action and on-action approach (Schön, 1982), in addition to conducting qualitative interviews with key stakeholders and community members. Staff from KUSWA and KWP documented their activities and interactions with communities and reflected on their activities in subsequent reports and interviews. Through the existing Yumi Sanap Strong program, which seeks to address sorcery accusation related violence, KUSWA and KWP implemented COVID-19 prevention strategies in their programs and training with key influencers. Stakeholders from the district administration, police and other participants in COVID-19 awareness programs were interviewed and provided reflections on the way they interacted with the CBOs. Community members, both in EHP and Simbu, were also interviewed about the impact the State of Emergency had on them and the support they were able to receive through their interactions with the CBOs. Interviews were conducted in the PNG lingua franca, Tok Pisin, and in English.

The project sought to identify vital aspects that emerged from the responses developed by KUSWA and KWP. While data was collected in the PNG Highlands, the themes are relevant to other Pacific Islands or other remote and urban settlement communities in countries with limited social government support systems. In the following section, we present the findings from the interviews and reflections, before discussing some of the key components that emerged from the CBOs' work in the COVID-19 response.

Impacts of COVID-19 State of Emergency as Reported by Participants

The impact most referred to by participants was the loss of daily income due to the shutdown of market activities. This was particularly so for those from rural areas, who relied on selling their food at urban markets, as well as for those living in informal settlements, who relied on daily informal market activities:

> When COVID-19 happened, we faced many struggles in our community. We sustain our lives through our daily markets. When the government put the ban in place, we faced many hardships, to find enough food and look after our families for their daily needs. We do not have access to any income. (Resident, Banana Block, EHP, Male)

This difficulty affected families and children, who were facing food shortages and had little opportunity to find ways to sustain themselves. One of the residents and volunteers for KUSWA described the agitation among youth who were on the verge of moving towards criminal activities to ensure food supplies for families. In particular, Fufurefa observed that children were affected:

> I've seen that many families are affected – they don't have enough food in their house. Some live without food in their house for some nights, and they go out looking for families, asking for food and asking for some support for their family and children to look after themselves during that time.

In rural areas, it was women who carried the burden because they usually worked in the food gardens, took care of the family and had to ensure ongoing food supply for their families. In addition to trying to maintain their livelihoods, people in rural areas suffered from the lack of transport and the closure of markets and stores, preventing access to essential items. These issues were compounded by the fear of contracting the virus and also the fear of accessing health services, due to misinformation, not receiving any direct information or relying on information from others in the community. When deaths were reported in the newspaper, people related it to their previous experiences:

> It was similar to when HIV came to our country – people talked about how it kills people. This scared me, and we feel it is happening in our community, so it is creating fear. (Community Youth Leader, Simbu, Male)

KUSWA reported that many women did not go to the hospital when they needed care, due to their fear of contracting the virus:

> Even if our children were sick, we were not taking them to the hospital, in case someone who has been infected would be at the hospital, and so we could contract the virus. We were really scared, so we just stayed at the house. (Resident, Banana Block EHP, Female)

When KUSWA staff identified women in need, they supported them with transport and access to health services that many of the women required earlier. Fufurefa spoke of one time when KUSWA staff tried to transport a woman to local medical services: "We were too late. The woman passed away in our car on the way to hospital."

In the Simbu province, fears were heightened when deaths occurred. Subsequent community discussions about *sanguma* (sorcery) and related accusations were mentioned as an issue that needed to be carefully managed by the KWP team. Here, the need for providing direct information to communities and engaging in conversations was important in mitigating potentially larger negative impacts.

Vulnerable community members, including women and girls, people living with disabilities and those relocated due to sorcery accusations or tribal fights, were most at risk of being neglected and experiencing further violence during the lockdown. Before exploring how KUSWA and KWP addressed some of the challenges of protecting vulnerable members of the community, we briefly discuss the positive aspects mentioned by interview participants because these also informed mitigation strategies.

Due to limited cash availability, there was less noise in urban communities, like the Banana Block, being caused by intoxicated people late at night. Little money was available to spend on additional leisure activities and this made people think about how to manage their funds:

> The little money we had, we had to think about how to budget, how we would have enough food, with the markets and some of the stores being closed. (Resident, Banana Block, EHP, Male)

In particular, participants commented on the positive change that came from men spending more time with their wives and children:

> When this disease was not here, fathers or men would usually go to town to drink beer, waste money, and they don't go back to the village. Now, during lockdown, men stayed at home; women, too, stayed at home. So, the men had to help their wives and children to do work in the garden and manage the work at home together. (Public servant, EHP, Male)

This observation was confirmed by KUSWA and KWP staff who saw the advantage of men and women working closer together in looking after their homes and families:

> Some of the men, they don't really go and settle with their families within their homes. Just at certain times they would pop up to see their families. But the COVID lockdown brought their families together. (Umba Peter, KUSWA Male Advocator)

> During COVID-19, the men spend their time with their wives going to the garden, making gardens, giving support to their wives, bringing the children to school and caring for the little ones, doing their laundry and all these things. That was the advantage side. (Angela Apa, KWP Director)

One of the participants said women and men would now walk together, when usually they would be doing separate activities. He interpreted this new behaviour as due to the fear of COVID-19.

Those who had received information about hygiene also considered that this was an advantage. They reflected on previous behaviour and said that adapting regular hand washing routines and being aware of their health was beneficial in their lives:

> We learnt good hygiene. We do not wash our hands as regularly usually, before we eat or cook. We learnt better hygiene, to wash our hands regularly and look after our health, so we stay well. (Resident, Banana Block, EHP, Female)

Other identified benefits were in the way that organisations and government departments worked together, which we will explore in the context of the work that KUSWA and KWP implemented.

Programs Implemented by KUSWA and KWP

KUSWA and KWP both received funding support to work in partnership with government, health departments, non-government organisations (NGOs) and other community-based partners to implement programs for communities. KUSWA and KWP staff also had to take care of their own immediate communities and families. Their main activities consisted of raising awareness of COVID-19 at community market spaces and schools in urban and rural areas, supporting emergency services for women and providing training to stakeholders and communities. The programs were generally designed to make use of the existing resources within communities, such as the construction of tippy-taps (a hands free way to wash hands especially appropriate where running water is absent, usually operated by a foot lever):

> We received training from KWP in how to wash our hands regularly, and after going to the toilet, to hang up a container and use soap, and how to

cough into our elbows. We really received great training. (Community Leader, Simbu, Male)

As mentioned earlier, KUSWA and KWP provided emergency services, particularly for women with limited access or at risk of violence. Due to their remit of addressing human rights and gender inequalities, both organisations integrated gender awareness and education into their COVID-19 programs:

> ... we integrate [these topics] with their gender roles and responsibilities in these times [during the COVID-19 pandemic]. How men can look after women, how women can look after men and then look after their families and within their households – there is always old women and men within their households. And, also, how they can look after their communities. (Umba Peter, KUSWA Male Advocator)

The CBOs were resourceful, adapting and designing their own communication strategies to ensure their communities were safe. They drew on different information they received from the World Health Organization and other government departments, via radio and newspapers, and they translated this information into appropriate messages for their communities. They then took time to ensure that health information was received and understood (see Fig. 25.1). Collectively, and in partnership with other

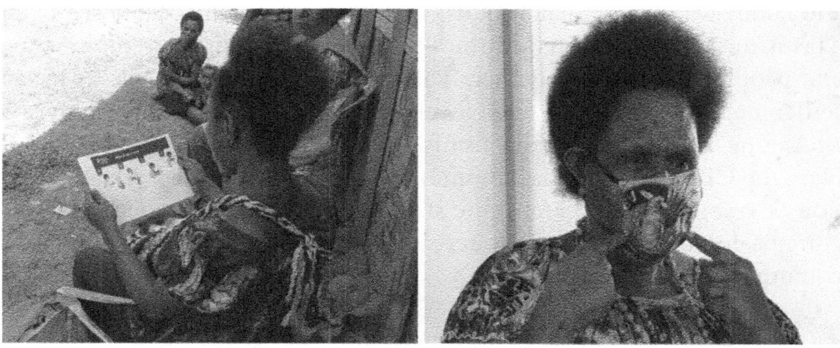

Fig. 25.1 Eriko Fufurefa providing information to women in Konamempi, Henganofi District, and demonstrating the wearing of facemasks at a Police Training in Goroka, Eastern Highlands Province. (Photos taken by Umba Peter and Elias Alex)

organisations, KUSWA and KWP reflected on the most impactful ways of providing the information. By speaking the language of the communities, for example, they gained respect and attention from people:

> Some were talking in Tok Pisin, some in English, some of us, we speak in *tok ples* (local language). So, when people in the community hear us and see that people from other areas have also come, they have much respect and gather to listen to what is being said. (Public servant, EHP, Male)

A key message from the CBOs was for people to care for each other, based on Melanesian values of family and community, including encouraging the sharing of food and providing support for each other in looking after children. This message was possible once initial feelings of fear had been addressed and information about virus protection had been provided. The need to work as a community and maintain awareness of vulnerable groups was crucial since no other support was provided. Public servants who partnered with KUSWA and KWP commented that they worked with these two organisations because of the trust-based relationships these organisations had developed with communities.

Challenges and Opportunities Experienced by the CBOs

As many of us around the world experienced, public messages about care and family were somewhat at odds with messages about social distancing. This is further complicated in the Pacific, where families are often large, and people live closely together. The distancing measurements in urban settlements are simply not feasible and this presented further challenges.

Among the challenges of misinformation, fear and lack of food supplies, the CBOs realised that communities simply did not have the foundation or resources to adhere to the standard guidelines provided. Fufurefa visited a community with no water supply and reflected on the words of a community member there:

> … you people are coming to the community to talk about the health measures and how we will prevent COVID. But you have to come and give us water supply into the community, so that we can have access to water to manage our health system. Otherwise, you people should go away. Don't come and talk about how we will prevent COVID in the community.

The CBO staff used their negotiation skills developed over many decades experiencing peace mediation in communities. They listened to community members to discuss solutions that worked for them. There was also limited personal protective equipment available, including for the health workers, so KUSWA and KWP tapped into their women's networks to sew face masks. While community members expressed frustration with the government in terms of the COVID-19 response, the CBOs encouraged communities to work together and harness their existing resources to implement preventative processes. As KWP staff member Agnes Sil emphasised: "*Igat wei bilong lukautim mipela yet istap*" (We can look after ourselves).

The kind of police brutality documented in news reports was not reported in our interviews. Instead, communities welcomed the police presence to ensure law and order issues were addressed; furthermore, police worked closely with the CBOs to implement COVID-19 awareness programs. However, the CBOs emphasised the need for police, and even health workers, to be trained to alleviate the challenges and fears of the community in appropriate ways. Police appreciated working with the CBOs because key information was incorporated into their programs:

> From what I can see, KUSWA plays an important role in addressing any kind of situation. They can fit in any situation. I work closely with them and I have seen how they can address issues regarding violence against women, domestic violence, gender-based violence, tribal fighting – anything, you name it. (Police Sergeant, EHP, Female)

CBOs travelled with teams made up of police officers, health officers and district staff to communities to provide health information (see Fig. 25.2). Limitations in reaching remote communities and finding appropriate local community solutions led KUSWA and KWP to train local facilitators and leaders in communicating messages clearly and supporting people in their own and neighbouring communities. Kup leaders also increased their own communities' COVID-19 awareness and ensured that accusations related to sorcery talk were diffused:

> The KWP women taught us to put a stop to any accusations or talk about sorcery, so we work towards reducing any such rumours, and educate others in our community. When we leaders provide this guidance, community members all listen to us and stay peaceful. (Community leader, Simbu, Male)

Fig. 25.2 Umba Peter speaking at Kesawaka, Henganofi District, Eastern Highlands Province. (Photo taken by Kenneth Manman)

KUSWA and KWP staff observed that each location faced different challenges. This situation highlights the need for contextualised responses to ensure that communities were open to receiving information and removing misunderstandings, that vulnerable groups were identified and looked after in the community, and that strategies were adapted to what was possible and feasible. Fufurefa observed that the partnerships established in response to COVID-19 led to new and better working relationships:

> Through this coronavirus [pandemic], I've seen the health department really come out and they worked really closely with the local NGOs to address this issue. This was the first time I've seen this. It was a good process, so we have to adopt this mechanism and make it workable.

Fig. 25.3 Paul Bayuwe, District Administrator, Henganofi District, speaking at a joint health awareness program at Fore Market, Fayantina, Eastern Highlands Province. (Photo taken by Kenneth Manman)

This cross-sector partnership was also important in demonstrating the seriousness of the COVID-19 pandemic to the community, and it ensured that teams could harness their different strengths to work with communities, and diffuse information at prominent meeting points such as markets (see Fig. 25.3).

Discussion

Pandemics are most dangerous for those already at risk due to existing health issues or marginalisation within society (Michener et al., 2020). The response to COVID-19 in PNG showed that vulnerable people were not always considered in standard government approaches to public health. As CBOs, both KWP and KUSWA have demonstrated that they fill an important gap in connecting and working with community to implement emergency responses. These CBOs draw on established community knowledge and social capital. This approach builds resilience and can further present "community members as active agents rather than passive victims" (Murphy, 2007). CBO staff highlighted the importance of tapping into community knowledge and relations in the absence of government support.

In the context of PNG, KUSWA and KWP drew on Melanesian values of harnessing and strengthening reciprocal relationships and kinship networks for support (Narokobi, 1980). In the context of urban settlements, for example, *haus makets* (small stalls located near people's homes) have become important spaces where sharing of resources is observed and negotiated to provide a safety net for community members (Rooney, 2019). However the limitations of using spaces such as *haus makets* and informal markets during the lockdown posed significant challenges to the community support systems. As our research demonstrated, to re-activate the positive support mechanisms within communities, misinformation or lack of information needed to be addressed.

Henderson (2020: 3) argues that contextualisation, such as "understanding social practices, as well as myths and misconceptions that are circulating" is "vital to successful public health communications". CBOs are in a position, not only to respond to people's fears, scepticism and interpretations, taking into account their worldviews, but also to provide access to support systems and address issues of power within their communities. KUSWA and KWP diffused narratives of sorcery and related accusations, and worked with communities who had limited access to water. Here, interpersonal communication within a social context – extended by the CBOs through training of community leaders and facilitators – proved to be a key aspect (see Ackerson & Viswanath, 2009), supported by the trust CBOs received due to their long-term community work. KUSWA's and KWP's ability to understand context across diverse communities means that these organisations (and others like them) formed a crucial component of public health initiatives in a crisis.

An awareness of existing social inequalities and how these might be exacerbated through the pandemic is necessary. Women and girls were disadvantaged due to misinformation, limited access to health services and transport, and the fact that they carried most of the burden of providing food to their families. However, there were also positive examples where men stepped in to assist their wives and spent more time with their families. As women's human rights organisations, both KWP and KUSWA increased awareness of crucial information about the role of men and women, and people living with disabilities. Human rights defenders working in their communities are the frontline people to address issues around gender-based violence (Biersack et al., 2016). These workers are aware of the importance of addressing gender issues through relational and holistic community models (Crook et al., 2016). Through their existing partnerships, KUSWA and KWP staff ensured that referral pathways to access different support services were accessible as much as possible (see Hukula, 2020). As such, gendered approaches to COVID-19 awareness and service provision during the pandemic further supported resilience and protective mechanisms in the community.

The experiences of the CBOs during the pandemic also generated new partnerships and new opportunities of working together, supporting cross-sector collaboration. The crisis forced people to think creatively about how they could best provide support. For example, KUSWA staff highlighted how valuable new partnerships were formed with the Provincial Health Department. The CBOs' social and cultural understandings also played an important role in training provided to public servants such as police and health officers.

Lastly, our joint project on addressing sorcery accusation related violence adapted to the new context. The documentation that KUSWA and KWP provided through digital media technologies such as photographs and mobile discussion groups emerged as important contributions in understanding the work of human rights defenders in this new context (Yumi Sanap Strong, 2020). This allowed the sharing of experiences at global scale and provided a perspective not reflected in media reports (see Thomas & Kauli, 2020). In the context of an increased localisation of aid (Chen & Cook, 2020), the experiences of CBOs provide important results for further exploration, but, more importantly, they provide potential lessons for public administration in responding to emergency and crisis situations.

Recognising the Role of CBOs in Emergency Response

In this chapter, we have highlighted the work of two CBOs, KUSWA and KWP, in the Highlands of PNG, that through their work and engagement made a valuable contribution in responding to COVID-19 in several parts of the PNG highlands. In a country with limited social protection mechanisms and diverse communities, KUSWA and KWP responded to the needs and contexts of each of their local communities, worked with each community to harness existing strengths and encouraged collaboration to protect vulnerable groups. This has highlighted in particular the need for human rights organisations to address issues that are exacerbated by COVID-19, such as gender-based violence and, in the context of PNG, violence related to sorcery accusations. COVID-19 measures have been integrated into the CBOs programs and, vice versa, COVID-19 awareness must also address and be aware of existing community issues that become more pronounced during a health crisis.

Governments such as the PNG government often develop 'standard' responses to crises in line with other countries and global guidelines. Our research and that of others demonstrates that the approach taken to public health messaging in PNG and the lockdown implementation caused initial fear and put people in high-risk situations. An appropriate national response was required. CBOs should have been part of the government's plan and response from the start, so that existing community structures and knowledge could be harnessed, and local needs addressed. When the local government of the Eastern Highlands Province recognised the importance of CBOs such as KUSWA during the pandemic, the CBO provided an important link and bridge to community. Governments can also benefit from CBOs' existing communication and mediation processes, which can be used to form strong partnerships between government departments and NGO networks.

Acknowledgements The authors would like to acknowledge the contribution of partner organisations in the Eastern Highlands and Simbu provinces, in particular the COVID Response Team Henganofi, the District Development Authority Henganofi, the Eastern Highlands Provincial Health Authority, Oxfam PNG, Community Development Agency (Simbu), Pacific Women Shaping Pacific Development (supported by the Australian Government through the Papua New Guinea-Australia Partnership) and the Highlands Women's Human Rights Defenders' Network. We would like to thank participating communities, as well as KUSWA and KWP staff and volunteers for their contribution.

References

Ackerson, L. K., & Viswanath, K. (2009). The Social Context of Interpersonal Communication and Health. *Journal of Health Communication, 14*(S1), 5–17.

Adebayo, O. W., Salerno, J. P., Francillon, V., & Williams, J. R. (2018). A Systematic Review of Components of Community-Based Organisation Engagement. *Health & Social Care in the Community, 26*(4), e474–e484. https://doi.org/10.1111/hsc.12533

Biersack, A., Jolly, M., & Macintyre, M. (2016). *Gender Violence & Human Rights: Seeking Justice in Fiji, Papua New Guinea and Vanuatu*. ANU Press.

Chen, C., & Cook, A. D. (2020). Humanitarian Assistance in the Asia-Pacific During COVID-19. *NTS Insight, IN20*(06). Retrieved from https://reliefweb.int/sites/reliefweb.int/files/resources/NTS-Insight-Humanitarian-Assistance-in-the-Asia-Pacific-during-COVID-19-Aug2020.pdf

Cousins, S. (2020). COVID-19 Has "devastating" Effect on Women and Girls. *The Lancet, 396*(10247), 301–302.

Crook, T., Farran, S., & Roëll, E. (2016). *Understanding Gender Inequality Actions in the Pacific: Ethnographic Case-Studies & Policy Options*. Publications Office of the European Union.

Department of Health, and World Health Organisation. (2020). *Papua New Guinea Coronavirus Disease 2019 (COVID-19) Health Situation Report #48*. Retrieved from Port Moresby. https://www.health.gov.pg/covid19/PNGSR48PCOVID-19(2020-11-22).pdf

Drennan, L., & Morrissey, L. (2019). Resilience Policy in Practice – Surveying the Role of Community Based Organisations in Local Disaster Management. *Local Government Studies, 45*(3), 328–349.

Henderson, L. (2020). Covid-19 – We Are Yet to Witness a Cohesive Government Communications Strategy. *BMJ*. Retrieved from http://bura.brunel.ac.uk/handle/2438/20777

Hukula, F. (2020). The Potential Impact of COVID-19 on Vulnerable Groups. *National Research Institute of Papua New Guinea: Spotlight, 13*(7), 1–4. Retrieved from https://pngnri.org/images/Publications/Spotlight_Vol_13_Issue_7_The_potential_impact_of_COVID-19_pandemic_on_vulnerable_groups_.pdf

Kauli, J., & Thomas, V. (2018). *Communicating the Law: A Participatory Communication Toolkit for Human Rights Defenders in Papua New Guinea*. Oxfam PNG/Australia, Highlands Women Human Rights Defenders Movement, Queensland University of Technology.

Kaut-Nasengom, S. (2020). Potential Impact of COVID-19 on Persons with Disabilities in Papua New Guinea. *National Research Institute of Papua New Guinea: Spotlight, 13*(14), 1–3. Retrieved from https://pngnri.org/images/Publications/Spotlight_Vol_13_Issue_14_Potential_impact_of_COVID-19_on_persons_with_disabilities_in_Papua_New_Guinea.pdf

Kopel, E. (2020, October 25). *The Impact of COVID-19 on Livelihoods in Papua New Guinea*. Retrieved from https://www.policyforum.net/the-impact-of-covid-19-on-livelihoods-in-papua-new-guinea/

Kuman, G. (2020). *COVID-19 Messages and PNG Sociocultural Beliefs*. Retrieved from https://devpolicy.org/png-messages-covid-19-beliefs-20200622/

Michener, L., Aguilar-Gaxiola, S., Alberti, P. M., Castaneda, M. J., et al. (2020). Peer Reviewed: Engaging with Communities – Lessons (Re) Learned from COVID-19. *Preventing Chronic Disease, 17*(E16), 1–8.

Murphy, B. L. (2007). Locating Social Capital in Resilient Community-Level Emergency Management. *Natural Hazards, 41*(2), 297–315.

Narokobi, B. (1980). *The Melanesian Way.* Institute of Papua New Guinea Studies.

Rooney, M. N. (2019). Sharing What Can Be Sold: Women Haus Maket Vendors in Port Moresby's Settlements. *Oceania, 89*(2), 154–167.

Schön, D. A. (1982). *The Reflective Practitioner: How Professionals Think in Action.* Basic Books.

Thomas, V., & Kauli, J. (2020). Strengthening the Voices of Human Rights Defenders in the Media: A Case Study on Addressing Sorcery Accusation Related Violence in Papua New Guinea. *Pacific Journalism Review, 26*(1), 86–104.

UNDP, The Equality Institute, and Department for Community Development and Religion. (2016). *Painim Aut Na Luksave – Understanding Gender-Based Violence to Secure Sustainable Development in Papua New Guinea.* UNDP.

Yumi Sanap Strong. (2020). *Responding to COVID-19: Kafe Women's Association Participates in WOW Global24.* Retrieved from https://yumisanapstrong.org/responding-to-covid-19-kafe-womens-association-participates-in-wow-global-24/

CHAPTER 26

Safeguarding Human Rights in a Global Pandemic: Reflections on the Caribbean's Response to COVID-19

Alecia Johns

Introduction

The COVID-19 pandemic remains the single most globally devastating health crisis in decades. In an effort to stem the tide of this deadly virus, governments across the Caribbean (and the world) have resorted to strict lockdown measures which necessarily place restrictions on citizens' rights to, *inter alia*, freedom of movement, assembly and religion. In addition, the pandemic has served to exacerbate existing inequalities with the result that critical social and economic rights – to health, education, work, and an adequate standard of living – are also at risk for millions in the region. This chapter examines the extent to which select nations in the Caribbean have managed to safeguard individuals' rights in the midst of this novel and rapidly evolving pandemic. It examines the main facets of a holistic rights-based response, which seeks not only

A. Johns (✉)
Independent Researcher, Tortola, British Virgin Islands

Y. Campbell, J. Connell (eds.), *COVID in the Islands: A comparative perspective on the Caribbean and the Pacific*,
https://doi.org/10.1007/978-981-16-5285-1_26

to limit infringement on civil and political rights, but to also take positive measures in order to secure social and economic rights and the welfare of those most vulnerable.

The Need for a Holistic Rights-Based Response to COVID-19

When considering the intersection between the pandemic and human rights, one's first thought may be of the intractable tension in which public health measures, designed to restrict movement and interaction, are at odds with individuals' personal freedoms to move, associate, and assemble. However, this tension is only part of the problem. Also at stake are key social and economic rights, such as the rights to health, education, work, and an adequate standard of living. These socio-economic rights are recognised in a number of international human rights instruments including, most notably, the International Covenant on Economic, Social and Cultural Rights (ICESCR), to which a majority of Commonwealth Caribbean states are party. All Commonwealth Caribbean states are party with the exception of Antigua & Barbuda, St Kitts & Nevis, and St Lucia. A number of these rights are also guaranteed in the constitutions of some Caribbean states such as the right to health and/or a healthy environment, the right to work, and the right to education.

These *"second generation"* rights are often viewed as less enforceable or less justiciable, that is capable of being decided by legal principles or by a court of justice, than their civil and political counterparts (Fredman, 2008). However, as the UN Committee on Economic, Social and Cultural Rights (1998) (the "CESCR") points out, the two sets of rights are in fact indivisible and interdependent. Few events which have highlighted this interconnectedness quite as much as the COVID-19 pandemic. First, the pandemic has underlined the importance of civil rights such as freedom of movement for the exercise of socio-economic rights to work, education and access to food. The right to work is in turn integral to an adequate standard of living and, as the Supreme Court of India recognised in *Tellis v Bombay Municipal Council*, it is an essential facet of the right to life itself since "no person can live without the means of living". Secondly, the pandemic has made plain that several underlying socio-economic determinants are central to effective compliance with public health measures and consequently in securing the rights to health and life (Sekalala et al.,

2020). Access to clean water is necessary for frequent sanitisation. Social distancing and stay at home orders presume access not only to housing, but also to accommodation which is adequately spacious to meet this need. However, quite divorced from these directives is the reality that an estimated 35% of the population of Latin America and the Caribbean does not have access to safely managed water services (PAHO, 2020) while countless others face homelessness or overcrowded housing. As the United Nations (2020a) points out, poverty itself is an enormous risk factor.

A holistic rights-based approach is therefore necessary in order to ensure a commitment to providing these underlying determinants for compliance, as well as mitigating the social and economic hardships caused by lockdown measures (which have often served to exacerbate existing inequalities). This gives rise to a positive duty on states to adopt targeted measures in order to assist those most vulnerable, and those worst affected by the pandemic (Human Rights Watch, 2020a).

This chapter will briefly examine aspects of the governmental responses in the Caribbean, primarily in Jamaica and Trinidad and Tobago, and the extent to which these island nations have managed to safeguard civil and political and socio-economic rights in the midst of an unprecedented pandemic. Both countries, like most in the English-speaking Caribbean, are constitutional democracies based on the Westminster parliamentary model and 'defined by an over-emphasis on European retention' (Wheatle & Campbell, 2020: 19). They possess a constitutionally entrenched bill of rights and a robust independent judiciary charged with safeguarding this.

Civil and Political Rights: Personal Freedoms Versus Public Health

Following the declaration of COVID-19 as a global pandemic by the WHO on 11 March 2020, and the emergence of imported cases across the region, Caribbean governments began instituting lockdown measures in an effort to prevent community spread. In Trinidad and Tobago, the first such law was a regulation made pursuant to the Public Health Ordinance dated 19 March 2020 by which bars and restaurants were ordered closed. This was followed by a series of additional regulations, also issued in March, which provided for, among other things, the closure of all theatres, clubs and betting houses, a limit of ten persons at religious services, the closure of all air and sea ports, work from home orders for those not providing essential services, and a five-person limit on public gatherings. Similar lockdown measures were imposed by nations across the region. These were inclusive, in some instances, of strict curfews in order to restrict residents' hours of movement. The British Virgin Islands represented one extreme case in which a 24-hour curfew was imposed in April during which residents were not allowed to leave their homes at all for 15 consecutive days.

These restrictive measures were gradually relaxed over time with several successive regulations and orders being introduced as new data emerged. For example, in Jamaica, approximately 15 Orders were made by the Prime Minister under the Disaster Risk Management Act between March and November 2020. Similarly, in Trinidad and Tobago, a total of 32 Regulations have been made by the Minister of Health in the same period. The restriction on religious services in Trinidad went from an absolute cap of ten persons in March, to the allowance of 50% capacity pursuant to the November regulations.

Notwithstanding the time-limited nature of these restrictions, they undoubtedly caused an interruption to the full exercise and enjoyment of individuals' freedoms of movement, association, assembly and religion. However, these rights are not absolute and may be limited by the state provided that such limitation is "reasonably justifiable" in a democratic society: this is the language employed in the rights' limitation clauses of a number of Commonwealth Caribbean constitutions including Trinidad and Tobago, Jamaica, and St Lucia. Other constitutions, such as that of Barbados and Antigua and Barbuda, provide expressly for limitations which are "reasonably required" in the interests of, among other things, "public health".

In assessing whether a limitation is justifiable, courts across the region have resorted to widely accepted tests of necessity and proportionality in order to balance legislative objectives against individual harms. This proportionality analysis generally examines the following questions in respect of the impugned law: (i) whether the legislative objective was a legitimate one and sufficiently important to justify limiting the right; (ii) whether the measures designed to meet the objective were rationally connected to that aim; (iii) whether the means employed were no more than necessary and (iv) whether there is overall proportionality between the deleterious and salutary effects of the law in question (Barak, 2012). This four-part test was advanced by the Supreme Court of Canada in *R v Oakes* 26 DLR (4th) 200 and has been widely referenced by courts in the region: see, for example, *Julian Robinson v Attorney General* and *Jason Jones v Attorney General of Trinidad and Tobago*.

It remains to be seen how most courts in the region will approach this balancing act in the context of COVID-19 restrictions. The High Court of Trinidad and Tobago has however had occasion to rule on two constitutional challenges brought in response to the government's COVID-19 regulations. Both cases are instructive. In *Suraj v Attorney General of Trinidad and Tobago*, the claimants were arrested in April for an alleged violation of the regulation which limited public gatherings to five people. Each was kept in custody for several days in what they described as tightly cramped, inhumane conditions. The claimants' main contention was that the Public Health Regulations were unlawful on the basis that they were made by the Minister of Health, were not approved or passed by a special majority in Parliament, and were therefore outside the scope of proper "legislation" which could impinge on individuals' rights pursuant to section 13 of the constitution. The claimants argued in the alternative that the regulations were a disproportionate and unjustified interference with their rights to life, liberty and security of the person, protection of the law, and freedoms of movement, association, and assembly.

On the question of the legality of the Public Health Regulations, the court ruled that the Public Health Ordinance (the parent legislation pursuant to which the regulations were made) was saved law pursuant to the constitution's savings law clause, which provides that pre-independence legislation shall not be invalidated on account of an inconsistency with the fundamental rights provisions of the constitution. The continued existence and applicability of such savings law clauses in Caribbean constitutions have been the subject of both academic debate (see, for example, Wheatle (2020) and O'Brien (2020)) and intense judicial scrutiny. The extent to which they fetter the court's ability to fully safeguard constitutional rights has been rightly decried. The court in *Suraj* openly noted its "discomfort" with the continued existence of saved law "58 years after independence" but acknowledged that "judicial will, creativity and ingenuity" can only go thus far. *Suraj* therefore represented a lost opportunity for a full and robust proportionality analysis in this sphere. However, the court went further to assess whether the Regulations were properly within the scope of the legislative power delegated to the Minister of Health pursuant to section 105 of the Public Health Ordinance. The court ruled that the Regulations were lawful and that there was no breach of the separation of powers doctrine on account of the Minister of Health making the

Regulations. This was a power which was appropriately delegated to him given that the virus was a "moving target" such that rapid legislative changes would be required, which was not conducive to a process of Parliamentary approval at each instance. This finding bears relevance across the region given that COVID-19 regulations have often been made by members of the executive pursuant to powers delegated under the relevant public health or disaster management statute. For example, in Jamaica, successive orders were made by the Prime Minister pursuant to legislative powers conferred upon him by section 26(2) of the Disaster Risk Management Act.

The court went on to briefly outline, in observations that did not form a necessary part of its decision, what its ruling would have been had a proportionality analysis been required. The court first outlined that a significant measure of deference had to be accorded to the executive and legislature given the need for quick responses which bear upon various economic, social, and political factors. Secondly, the court noted that the uncontradicted evidence of the Chief Medical Officer was that restricting gatherings and enforcing social distancing were critical components in controlling the spread of the disease. The court also took judicial notice of the fact that many of the measures imposed were similar to those taken in other democratic states such as "some parts of the United States, European countries, Commonwealth, and Caribbean States". Further, the fact that the regulations had been time-limited in each instance and regularly amended, suggested that there was ongoing analysis in order to take into account prevailing circumstances. The court consequently noted that it would have concluded that the regulations were a proportionate response to the management of the pandemic.

The case of *Suraj v AG* was heard together with that of *Maharaj v AG*, in which the claimant alleged that the Public Health Regulations were unlawful to the extent that they sought to criminalise non-compliance with published "Guidelines for Places of Worship" on the basis that these guidelines fell short of the legal certainty required for the imposition of criminal sanctions and were therefore in breach of his right to freedom of conscience and religion. The High Court ruled in favour of Mr. Maharaj on this point, noting that criminal statutes must be clearly expressed as mandatory commands; the form and substance of the "guidelines" fell afoul of this settled legal principle. However, in so ruling, the court was careful to point out that it saw no issue with the substance of the guidelines, which promoted hygienic public safety measures and which gave religious bodies an option to continue their observances albeit in a limited way. The court observed that there was no fundamental threat to the claimant's core right to practice his religion, which he was still able to do albeit in the physical company of fewer persons. However the court was not engaged in a full-scale proportionality analysis given the presence of the savings law clause.

It remains to be seen how the intractable tension between right to freedom of religion and COVID-19 restrictions will be assessed by other courts in the region. Judges in South Africa and the United States have faced this question. In *Mohamed v The President of the Republic of South Africa*, the South African High Court ruled that lockdown regulations which restricted religious gatherings were reasonably justifiable given that they were time-limited and necessary in order to protect the rights to health and life. The claimants were practicing Muslims who argued that the regulations criminalised what they deemed to be a compulsory act of worship: performing five daily prayers in congregation and at a mosque. The court outlined that the requested exemption for places of worship would have opened the floodgates and would have made the requisite social distancing impossible to enforce. The court similarly expressed deference for regulations which had been crafted with the "advice of experts" and, after extensive consultation, noted further that every citizen was called upon to make sacrifices "for the greater good". The United States Supreme Court similarly echoed the need for deference to the legislature in this area in its refusal to grant injunctive relief in order to prevent enforcement of California's regulation that places of worship operate at 25% capacity. In *South Bay United Pentecostal Church v Governor of California*, Chief Justice Roberts outlined, "The precise question of when restrictions on particular social activities should be lifted during the pandemic is a dynamic and fact-intensive matter subject to reasonable disagreement ... When [elected] officials undertake to act in areas fraught with medical and scientific uncertainties, their latitude must be especially broad. Where those broad limits are not exceeded, they should not be subject to second-guessing by an unelected federal judiciary". This does however contrast with the position taken by the same court in the decision of *Roman Catholic Diocese of Brooklyn v Governor of New York*, where injunctive relief was granted to prevent enforcement of a strict 10 – 25 person capacity limit for religious services, pending the determination of an appeal on the question of whether these restrictions infringed the First Amendment.

These decisions will undoubtedly prove instructive to courts in the region, which have in the past engaged in comprehensive comparative analysis when ruling on constitutional rights. In this respect, the Canadian decision of *Taylor v Newfoundland* provides the fullest example of the proportionality analysis which courts in the region may be called upon to engage given the judicially acknowledged similarity between the limitation provision in the Canadian Charter of Rights and that of constitutions in the region. In *Taylor*, the applicant challenged the legality of an order made by the Chief Medical Officer of Newfoundland and Labrador which temporarily restricted non-residents' entry into the province. As a result of the restriction, the claimant was denied entry to the province in order to attend her mother's funeral. The court ruled that the claimant's right to freedom of movement had been infringed. However, this infringement was held to be demonstrably justified in response to the pandemic. Unsurprisingly, the court held that the prevention of the spread of COVID-19 was a pressing and substantial legislative objective. Further, with reference to empirical evidence, the court was satisfied that there was a rational connection between travel restrictions and containment of the disease. As to whether the travel restriction was "necessary" or the least restrictive means to achieve that legislative aim, the court noted that less stringent measures of self-isolation and social distancing were not viable substitutes given the difficulties associated with enforcing those measures. The court also expressed the need for deference in this sphere given that "courts do not have the specialised expertise to second guess the decisions of public health officials". However, the court noted that notwithstanding this deference, the court must not abdicate its responsibility as guardian of the Constitution and rule of law. At the final stage of the proportionality analysis, the court opined that the harm done by the travel restrictions (mental anguish to some) did not outweigh the collective benefit to the population of reducing the spread of COVID-19. The court held that the claimant's Charter right to freedom of movement "must give way to the common good".

These decisions reveal that courts were generally unlikely to strike down temporary, time-limited restrictions which were based on medical and scientific evidence, given the countervailing collective interest of preventing the spread of an unpredictable and deadly disease. However, as time passes, courts will need to be astute to the danger of some of these restrictions creeping into a "new normal" of restricted rights on a more permanent basis (Pugh, 2020: 2). In this regard, there has been controversy surrounding more permanent legislative measures in the region such as St Lucia's COVID-19 (Prevention and Control) Act. The Act was passed in September and will be in place for a period of two years. The Act mandates the wearing of masks and physical distancing in public places as well as compliance with all COVID-19 protocols issued by the Chief Medical Officer. Section 58 provides that where a police officer has reasonable cause to believe that a person contravenes the Act, he may stop, question, detain and arrest the person if he considers it necessary to do so. Further, any person who contravenes the Act commits an offence and is liable to a fine not exceeding one thousand dollars or a maximum of six months' imprisonment. These measures raise questions regarding the suitability of criminal sanctions in the context of imprecise directives to "physically distance" in public, as well as the propriety of the broad police powers conferred by the Act. While the Opposition in St Lucia has threatened a constitutional challenge to this Act (George, 2020), it remains to be seen whether this will be issued and, if so, how the court will view these more permanent restrictions.

Social and Economic Rights and the Protection of Those Most Vulnerable

As outlined above, the majority of states in the region possess international obligations to guarantee their citizens' social and economic rights pursuant to the ICESCR. The precise scope of the domestic recognition of these rights in the region, as well as the fraught questions of their enforceability and justiciability on the domestic plane, are beyond the scope of this chapter. As James (2020) points out, after an in-depth examination of these very questions, there remains a great deal to be done in order to fully secure the progressive realisation of these rights in the region. In this chapter, focus will be placed on states' international obligations and, where relevant, their constitutionally enshrined socio-economic rights.

Article 11 of the ICESCR provides that every person has a right to an adequate standard of living, including adequate food, clothing, housing and the continuous improvement of living conditions. Article 6 recognises the right to work and the opportunity to gain a living as one freely chooses. Lockdown measures severely infringed these two rights for many people in the region. Tourism remains a large sector in many Caribbean economies, yet thousands of service workers in the hospitality and tourism industries were out of work for many months. The measures disproportionately affected those with low paying jobs who were unable to work from home. In order to mitigate the effects of this, a number of governments provided economic relief to those in need. In April 2020, the Government of Jamaica launched a stimulus response titled "COVID-19 Allocation of Resources for Employees (CARE)". The CARE program consisted of a wide range of grants in order to cover a broad cross-section of those affected. These included five-month cash transfers for employees in the tourism industry and other sectors who had lost work as a result of the pandemic, grants for small businesses including those operating in the tourism sector, and deferral of student loan payments (Ricketts, 2020). In an effort to maintain transparency, and to encourage formalisation of those in the informal sector, there were strict eligibility requirements including the provision of Tax Registration Numbers (TRNs) and where applicable, business registration licenses. However, as Ricketts (2020) points out, the CARE program did not operate as sufficient incentive for formalisation with the effect that many workers in the informal economy did not benefit. Similarly, in Trinidad and Tobago, the government's public assistance grants for those in need have been noted to have 'left behind' many self-employed individuals, including migrants and refugees working in small businesses (United Nations, 2020b: 25).

The right to education, as recognized in article 13 of the ICESCR, has also been impacted for millions of students across the region. In Jamaica, this right is constitutionally enshrined; the Charter of Rights guarantees the right of every child to "publicly funded tuition in a public education institution at the pre-primary and primary levels". The Constitution of Trinidad and Tobago guarantees the right of a parent or guardian to provide a school of his choice for the education of his child or ward. However, as Jaramillo (2020) points out, millions of students in the region missed nearly five months of school, and distance learning strategies intended to maintain a degree of continuity were unevenly applied and may serve to further exacerbate existing educational gaps and inequalities. Electronic learning programs require (at the very least) access to suitable devices, stable internet, and a quiet environment conducive to learning. Many from low income homes struggled to provide these basic resources for their children. Measures adopted by the government of Jamaica included a laudable partnership with the private sector called "One Laptop or Tablet per Child" aimed at securing devices for 100,000 students in need.

As Jaramillo (2020) notes, comprehensive government responses are required which ought to include plans for the safe, limited re-opening of schools, ensuring access to the Internet and suitable devices for communication with teachers, as well as measures to ensure the emotional well-being of students, parents, and teachers.

The rights to equality and non-discrimination also give rise to additional positive obligations to protect those most vulnerable (Campbell, 2020), which in the context of the pandemic includes prisoners and other detainees of the state, as well as migrants and refugees (United Nations, 2020a: 11-12). Trinidad and Tobago's response in respect of both groups thus came into sharp focus. In *Dexter Simon v Commissioner of Prisons*, five inmates of the Maximum Security Prison (MSP), who were remanded for years awaiting trial, applied for judicial review of the government's decision to exempt prisons from the Public Health Regulations on the basis that this violated their rights to equality of treatment, equality before the law, and the right to life. The applicants contended that the MSP was unsanitary. They outlined that there was no consistent supply of running water and an inadequate supply of soap for necessary sanitization. Further, it was argued that they were not provided with face masks and the over-crowded cells did not allow for social distancing. The government's evidence was that protocols had been put in place in order to secure the safety of prisons, such as the implementation of virtual visitation and virtual court attendance for detainees, new prisoners were quarantined for 14 days, and screening and sanitisation was conducted in respect of all personnel entering the prisons on a daily basis. With respect to masks, the government's position was that this presented a security risk in that it could obscure a large portion of prisoners' faces and may therefore facilitate misconduct. Ultimately, the court was satisfied with the extent of these measures and declined to grant the applicants leave to apply for judicial review, noting that the government's measures were sufficient as evidenced by the fact that there had been no COVID-19 outbreak at MSP to date, or even a single positive case among prisoners or staff. The High Court's ruling was delivered on 8 October 2020. Six weeks later, over 80 prisoners of the MSP tested positive for COVID-19 (Khan, 2020). It remains to be seen how the court will deal with any subsequent challenges relating to this outbreak, and whether more stringent protective measures would be required of the state.

In Jamaica, a COVID-19 outbreak at the Tower Street Correctional Facility led to human rights activists renewing calls for the early release of low-risk prisoners in order to minimize overcrowding in state facilities (*Jamaica Gleaner*, 2020). As Amon (2020) notes, respecting the right to health of those detained may require states, in part, to achieve physical distancing through the early release of many arbitrarily detained, such as those in pre-trial detention. It is noteworthy that the applicants in *Dexter Simon* were all remanded in pre-trial detention for periods ranging between 4 and 17 years: an unacceptable state of affairs by any measure. A failure to control the spread of COVID-19 in prisons may also constitute a violation of the prohibition against torture, inhuman treatment and degrading punishment, which is recognised internationally as a non-derogable right for which there can be no derogation (where the law may not be obeyed), not even in periods of national emergency (Lebret, 2020).

Finally, with respect to migrants and refugees, it remains vital that the tightening of borders and travel restrictions do not prevent people fleeing war or persecution from obtaining the protection guaranteed to them under international human rights law (United Nations Human Rights Commission, 2020). The Government of Trinidad and Tobago came under severe scrutiny for the deportation of 258 Venezuelans who were said to have illegally entered the country during the pandemic and whose presence was alleged to be compromising efforts to contain the disease (Amnesty International, 2020a). Moreover 93 of these individuals were deported just two days after a UN-appointed fact finding mission on Venezuela found reasonable grounds to believe that authorities under Nicolas Maduro have committed grave human rights violations which could amount to crimes against humanity, including unlawful executions, enforced disappearances, and arbitrary detentions (Human Rights Watch, 2020b). The government was urged to comply with its non-refoulement obligations by protecting asylum seekers and others from deportation. Following this international outcry, in November 2020, 16 children and 12 adults were returned to Trinidad and Tobago following their deportation to Venezuela (Amnesty International, 2020b).

Conclusion

The COVID-19 pandemic has left virtually no aspect of societal operation unscathed. Its pervasive impact necessarily gave rise to a myriad of human rights' implications which touch and concern both civil and political and socio-economic rights. In respect of restrictions on personal freedoms, courts conferred a high degree of deference on authorities given the complex and rapid determinations necessary to combat a moving target. Nonetheless, courts across the region ought to remain astute to safeguard against disproportionate measures, and the gradual introduction of more permanent measures which seek to normalise an erosion of fundamental rights in the long term. Thus far, Caribbean governments have generally instituted moderate, time-limited measures which have been subject to continuous revision and progressive relaxation. In respect of social and economic rights, reasonable efforts have been made with the limited economic resources available. However, more needs to be done in certain states to secure the right to health of those most vulnerable, including prisoners and refugees. These goals remain challenging for small island developing states in the region, whose heavy reliance on tourism (in already fledging and largely undiversified economies) serve to stymie rapid adaptation during crises. Ultimately though, a holistic rights-based response remains necessary given the interdependence and indivisibility between socio-economic and civil and political rights. The pandemic, like any threat, also creates a concomitant opportunity: for renewed focus on these crucial social and economic rights which are essential for future risk minimisation in the face of further public health crises.

References

Amnesty International. (2020a, August 6). *Trinidad and Tobago: Deportation of 165 Venezuelans Violates International Law*. https://www.amnesty.org/en/latest/news/2020/08/trinidad-tobago-deportation-165-venezuelans-violates-international-law/

Amnesty International. (2020b, November 25). *Open Letter to Dr Keith Rowley, Prime Minister of Trinidad and Tobago: Return of Deported Children Gives Your Government a Second Chance*. https://www.amnesty.org/en/latest/news/2020/11/trinidad-tobago-return-deported-children-second-chance/

Amon, J. (2020). COVID-19 and Detention: Respecting Human Rights. *Health and Human Rights Journal, 22*(1), 367–370.

Barak, A. (2012). *Proportionality: Constitutional Rights and Their Limitations*. Cambridge University Press.

Campbell, Y. (2020). *Citizenship on the Margins: State Power, Security and Precariousness in 21st Century Jamaica*. Palgrave Macmillan.

Committee on Economic, Social and Cultural Rights. (1998). *General Comment No. 9: The Domestic Application of the Covenant*. E/C.12/1998/24.

Fredman, S. (2008). *Human Rights Transformed: Positive Rights and Positive Duties*. Oxford University Press.

George, M. (2020, October 7). SLP Contemplates Court Challenge to COVID-19 (Prevention and Control) Bill. *The Voice*.
Human Rights Watch. (2020a). *Human Rights Dimensions of COVID-19 Response*. https://www.hrw.org/node/339654/printable/print
Human Rights Watch. (2020b, October 8). *Open Letter to Dr Keith Rowley, Prime Minister of Trinidad and Tobago*. https://www.hrw.org/news/2020/10/08/open-letter-dr-keith-rowley-prime-minister-trinidad-and-tobago
Jamaica Gleaner. (2020, October 4). 'A Death Sentence' – Advocate Renews Call for Release of Low-Risk Prisoners Amid COVID-19 Outbreak at Tower Street.
James, W. (2020). Social and Economic Rights in the Caribbean. In R. Albert, D. O'Brien, & S. Wheatle (Eds.), *The Oxford Handbook of Caribbean Constitutions* (pp. 455–481). Oxford University Press.
Jaramillo, S. (2020). *COVID-19 and Primary and Secondary Education: The Impact of the Crisis and Public Policy Implications for Latin America and the Caribbean*. United Nations Development Program.
Khan, R. (2020, November 14). 64 MSP Prisoners Positive for COVID-19. *Trinidad and Tobago Guardian*.
Lebret, A. (2020, May 4). COVID-19 Pandemic and Derogation to Human Rights. *Journal of Law and the Biosciences*, Advance Online Publication. https://doi.org/10.1093/jlb/lssa015.
O'Brien, D. (2020). The Interpretation of Commonwealth Caribbean Constitutions: Does Text Matter? In R. Albert, D. O'Brien, & S. Wheatle (Eds.), *The Oxford Handbook of Caribbean Constitutions* (pp. 555–581). Oxford University Press.
PAHO. (2020). *2030 Agenda for Drinking Water, Sanitation and Hygiene in Latin America and the Caribbean: A Look from the Human Rights Perspective*. PAHO.
Pugh, J. (2020). The United Kingdom's Coronavirus Act, Deprivations of Liberty, and the Right to Liberty and Security of the Person. *Journal of Law and the Biosciences, 7*(1), 1–14.
Ricketts, H. (2020). The Government of Jamaica (GoJ) COVID-19 Pandemic Stimulus Response: An Opportunity to Regulate and Formalize? *International Journal of Sociology and Social Policy, 40*(9/10). 979–993.
Sekalala, S., Forman, L., Habibi, R., & Meier, B. (2020). Health and Human Rights Inextricably Linked in the COVID-19 Response. *BMJ Global Health, 2020*, 5. https://doi.org/10.1136/bmjgh-2020-003359
United Nations. (2020a). *COVID-19 and Human Rights: We Are All in This Together*. United Nations. https://www.un.org/victimsofterrorism/sites/www.un.org.victimsofterrorism/files/un_-_human_rights_and_covid_april_2020.pdf
United Nations. (2020b). *Socio Economic Response to COVID-19: Assessment and Framework (Trinidad and Tobago)*. https://reliefweb.int/sites/reliefweb.int/files/resources/TTO_Socioeconomic-Response-Plan_2020.pdf

United Nations Human Rights Commission. (2020, May 13). *COVID-19 Guidance: Human Rights at the Heart of the Response*. https://www.ohchr.org/Documents/Events/COVID-19_Guidance.pdf

Wheatle, S. (2020). Constitutional Principles: Forging Caribbean Constitutionalism. In R. Albert, D. O'Brien, & S. Wheatle (Eds.), *The Oxford Handbook of Caribbean Constitutions* (pp. 79–104). Oxford University Press.

Wheatle, S., & Campbell, Y. (2020). Constitutional Faith and Identity in the Caribbean: Tradition, Politics and the Creolisation of Caribbean Constitutional Law. *Commonwealth & Comparative Politics, 58*(3), 344–365.

CHAPTER 27

To Comply or Not to Comply: State Resistance and Exceptions to COVID-19 Rules & Regulations in Jamaica

Yonique Campbell and Anthony Harriott

Jamaica applied strict regulatory measures at the beginning of the COVID-19 pandemic and has relied extensively on the Disaster Risk Management Act (DRMA), to manage the spread of the virus. This was the case in several Caribbean states. Across the globe, countries have had to adopt similar measures but have varied the degree of restrictions on the freedom of movement of their citizens in an effort to balance lives and livelihoods as well as regard for the political culture and the political strength of libertarian currents within their populations. A reality of life in Jamaica and much of the Global South is that a significant proportion of

Y. Campbell (✉)
University of the West Indies, Kingston, Jamaica
e-mail: yonique.campbell02@uwimona.edu.jm

A. Harriott
Institute of Criminal Justice and Security, University of the West Indies, Kingston, Jamaica

Y. Campbell, J. Connell (eds.), *COVID in the Islands: A comparative perspective on the Caribbean and the Pacific*,
https://doi.org/10.1007/978-981-16-5285-1_27

the labour force make their living on the streets and in crowded market places. Balancing lives and livelihoods is thus a delicate political act.

Jamaica, therefore, took a measured approach and resorted to lock-down as a last resort and only to prevent community spread. Total lock-downs were thus restricted to specific locales. The largest administrative and only unit to be locked-down was the parish of St. Catherine. Jamaica and other Caribbean countries such as Barbados, Trinidad and those in the Eastern Caribbean, because of their financial and technical capacity to respond to the pandemic, and limited numbers of hospital beds and ICU facilities, treated these measures not as a last resort but as a priority. The Jamaican state sought to encourage adherence to the standard-prescribed COVID prevention measures but moral suasion was not enough. Considerable state coercive action was required. This effort to balance lives and livelihoods, self-regulation and state-regulation defined Jamaica's distinctive response to the pandemic.

Despite appeals from various state actors for citizen self-regulation, it was obvious that some citizens were more capable and willing than others to comply with DRMA. Those among the urban poor, for example, argued that the seemingly measured restrictions such as curfews would nevertheless affect their livelihoods in very negative ways. Working from home was not applicable to people making their living from street vending. A cushion was needed. An initial ten billion dollars Allocation of Resources for Employees (CARE) programme, described by the Minister of Finance and Public Service as a temporary cash transfer programme to individuals and businesses to cushion the economic impact of the Covid-19 pandemic, which included a direct cash payment, may have acted as an important but not sufficient tool. Moreover, many of those in the informal sector were effectively excluded because they lacked the required documentation to claim this cash payout (Ricketts, 2020). This incapacity of the government to fully meet the needs of the people strengthened the case for those groups claiming the right to an exception to COVID-19 protocols on economic grounds. Given the limited capacities of Jamaica, even the most finely calibrated balancing of lives and livelihoods would have created stress on the population and on the state administrative systems.

Whether on the grounds of economic necessity, lifestyle or religion, the arguments given by different groups about why they believed they should be an exception to various protocols related to social distancing, mask wearing, curfews and quarantine orders make Jamaica an interesting case study. It permits a probe not just of the general problem of non-compliance but more specifically how non-compliance is justified when it carries grave consequences including death.

Many citizens believe that they are exceptional and ought to be free to disregard the COVID-19 rules. We argue that making the case for exceptionalisms may be seen as a method of negotiating non-compliance to COVID-19 rules without directly confronting the authority of the state. It is a method of undermining the authority of the state and negotiating freedom of action without being directly confrontational. This method erodes the state system but it allows the state to co-opt any serious movement against it. This process of negotiation and co-option is one of the keys to understanding the absence of any serious national protest contesting COVID-19 protocols in Jamaica. Despite disagreements and non-compliance with state regulations, serious protests or direct confrontation between state and citizens has been largely avoided during the pandemic and those which have occurred have been few, localised and limited to spontaneous occurrences at entertainment events.

Consistent with the aim of examining the reasons and meanings associated with non-compliance, we selected sites of high intensity non-compliance that were presumed to be correspondingly rich in the justifications for non-compliance. Late in 2020, the main sites of non-compliance were particular denominations of the Christian church, a general elections campaign and selected communities of the urban poor. A purposive high intensity (extreme case) snowball sample with the following string structure was therefore adopted: church congregationalists who exhibited a specified level of religiosity (attended church at least twice per month in 2019, prior to COVID-19); party activists in the 2020 election campaign involved in face-to-face party proselytising attempts to win votes (campaign periods have long been taken as periods of exception); and residents in selected communities of the urban poor. Each of these strings were then subdivided into male and female strings, the rationale being a presumed gender variation in risk-taking, alongside possible variations in within-group or sub-group behaviour and justifications by gender that may illuminate meanings (via comparison). At the starting position we identified eight people, two in each of the four categories and of these two – one female and male. In snowballing, we filtered out the compliant since we were interested in non-compliant individuals. All respondents were over 18 years of age at the time of the interviews. Pseudonyms were used for all participants in keeping with ethical considerations. Translations from Patois to English were done only where necessary to facilitate a wide range of readership. We were supported in our fieldwork by Shar-Lee Amori, our Research Assistant.

COVID-19 Regulations in Jamaica

The constitution and the law, more generally, set the context in which the government sought to enforce and ensure compliance to COVID-19 rules. In Jamaica, the main law governing regulation of COVID-19 is the DRMA. Since the first case of COVID-19, recorded on the 10th of March 2020, and late in 2020, the government had promulgated 17 Orders under the Act, with the aim of controlling the spread of the disease. Empowered by section 26(2) of the DRMA, the Prime Minister was responsible for making these orders. Curfews, lock-down measures, quarantine of individuals and entire communities, mask wearing, social distancing, and work-from-home orders have all been imposed under this law. Jamaicans have been required to adhere to protocols related to public gatherings, funerals, and hosting entertainment events. Jamaica adopted the standard WHO protocols. For example, the latest regulations gazetted by the Government of Jamaica in 2020 mandated:

> gatherings in any public place during the period from November 1, 2020, to November 30, 2020, shall not exceed 15 persons at a time; (b) each person at a gathering in a public place shall maintain a distance of at least 182.88 centimetres (or 6 feet) from other persons ("social distancing"); and (c) when in a public place (which, for the avoidance of doubt, includes a workplace and a licensed public passenger vehicle), each person shall wear a mask fitted to that person's face so as to cover that person's nose and mouth. (Ministry of Justice, 2020)

In the face of numerous breaches, the police force, which was authorised to detect breaches and enforce regulations under the Disaster Risk Management Act and take appropriate actions, tried to increase their enforcement efforts, arresting over 2000 persons for breaches of the DRMA between March and November 2020. More than 60 of these arrests were related to entertainment events (Jamaica Information Service (JIS), 2020). Entertainers and their patrons continued to breach curfew orders, social distancing protocols, limits on the number of persons that may attend events and other rules related to hosting events.

Despite these arrests and, although empowered under the DRMA, which states that 'where any member of the public fails to comply with the direction of the police, that individual commits a breach of section 27(1)(b) of the (DRMA) and is liable to prosecution under section 52(b) of the Act', non-compliance remained pervasive. The Act itself hobbled its enforcement by requiring the police to warn first offenders. This requirement ensures reasonableness on the part of law enforcement but in so doing necessitated an added effort to differentiate the incorrigibles from those who were amenable to compliance if a small measure of coercion was applied to them.

Jamaicans who reside abroad have flouted the regulations that require persons entering the country to subject themselves to home quarantine for 14 days. In March 2020, 30 cases were linked to a church convention held in the small but main-road town of Sandy Bay in the parish of Clarendon. This event resulted in the entire Sandy Bay community being placed under quarantine (Jamaica Observer, 2020). Curfews were routinely breached. Over the National Heroes weekend holiday, for example, the security forces warned over 300 persons for breaches of curfew orders in St. Catherine (Jamaica Gleaner, 2020).

In light of these breaches, and consistent with a spirit of negotiation various statements were made by the leadership of the police as well as the Prime Minister urging voluntary compliance with DRMA and COVID-19 regulations and, in the case of the police, expressing the will to enforce the law. The Commissioner of Police, Antony Anderson (a former Chief of Defence Staff), against the grain of police culture, stated boldly that 'we will be relentless in our fight to rid the country of lawlessness, especially those who defy orders of the Disaster Risk Management Act and we will be out there in our numbers' (JIS, 2020). Anticipating the negative effects that non-compliance would have on the community-spread of the virus, the Prime Minister stated that.

> I'm actually disappointed that, as a country, we have to turn to arrests. I'm very disappointed. But I would not be fulfilling my job as leader of the country if I were to allow any form of indiscipline to be rewarded by [culpable persons] not facing the consequences of the law, especially in this time of emergency.

Breaches of COVID-19 protocols and these statements by state actors must be assessed against a broader, perennial problem of non-compliance and challenges to law enforcement in Jamaica (Campbell, 2020; Harriott, 2000), especially to public order type legislation and regulations, evident across several areas: road traffic violations, illegal vending, littering, informal settlement and violation of noise abatement laws. Our aim in this paper is not to explain all non-compliance but rather the justifications for COVID-specific non-compliance. Recent studies on COVID-19 have noted that perceived levels of risks (Clark et al., 2020; Webster et al., 2020), socio-demographic factors, occupation or economic concerns (Bodas & Peleg, 2002; Kuiper et al., 2020; Webster et al., 2020), compensation, or lack thereof, for lost wages can all affect willingness and ability to comply.

Making the Case for Exceptions

Using data collected in the field between September and October 2020, we will now explore how the case for an exception was made by the three sets of respondents. In seeking to negotiate non-compliance, actors argue for: sites of exception, events exception and lifestyle-occupational exceptionalism. As a site of exception, the church is regarded as a sacred space, where an expression of fear of COVID-19 may be taken as an indicator of weak faith. The church makes the case for an exception on the basis that it 'abides by a higher order' than that of the Jamaican state. Most respondents were adamant that state regulations had no place in the church, as a place of worship:

> The house of the Lord is a sacred and holy place. No unholy sinful thing can enter there. (John)

> I go to church to praise God, nothing will hinder me from raising my voice. Sometimes you can't even breathe in the masks; they are uncomfortable. And the blood of Jesus will protect me. COVID is not bigger and 'badder' than my God. (Sister Mary)

> When the Holy Ghost takes control, all these weak laws and rules have no place in God's house. And the truth is Andrew [the Prime Minister] and Tufton [the Minister of Health] say every day that most of us will eventually catch the virus anyway, so why should I stress, especially in God's house. If I am to get it, I will get it and the grace of God will see me through. (Sister Mary)

The protection from God and the superiority of the laws of God are central to narratives of non-compliance and the intolerance of especially mask wearing and social distancing in the church

> We know the law, but the house of the Lord abides by a higher order. I would say that most feel as comfortable as I do that they are protected when in church.
>
> We feel protected. If a member has the disease, we feel protected against it. Can you imagine me preaching that the blood of God saves through a mask? The blood of God conquers all and walk around sanitizing people. The Lord will protect us, but stand 6 feet apart. That would be hypocrisy at its finest ... If you cannot trust that God will protect you in his holy place, stay home and worship. We do stream our services on Facebook and other platforms. (John)

Like other groups, members of non-compliant churches are very skilful in their claim-making. They limit their non-compliance to sacred space-place and are compliant beyond it – especially in public spaces that are patrolled by the police. The state is expected to recognise and respect the sites that are ruled by their God's "higher authority." And, in turn, the congregants are encouraged to respect the sites that are ruled by the lower authority of the state. The understanding is that the state will not directly confront rule breakers within the walls of the church (there was at least one instance of this, which drew broad condemnation). The church, too, understands that in order to successfully make the case for an exception, its members must limit their rule-breaking. Non-compliance in the church should not to be taken as a sign of generalised non-compliance. In fact, most research participants expressed their willingness to comply with COVID-19 rules outside the church. These churches were not therefore sites of a generalised politicized confrontation with state authority. This attitude was expressed as follows:

> Church people are not stupid. Most if not all of us are cautious elsewhere, but just with God in that intimate space, we don't really enforce it. Yes, my daughter had cancer last year, so we do the little extra to protect her. The rest of us are healthy as far as we know. Again, we are not stupid. Just in church. And we are not trying to break laws outside of church either (Sister Ann). Yes, we wear our mask on the church bus. We have to drive along Holland Bamboo and police are always out there so in the bus we practice

our thing [social distancing]. We are not unruly. We just trust God especially at church.

There is coexistence but it is not tension free. Tensions between the noncompliant churches and the state grew when the sacred spaces became the sites of "superspreader" events. Their activities exceed their spaces when its consequences exceed the understood boundaries – as infections. These instances were never escalated as political conflict between church and state.

Similarly, the urban poor make a case for exceptionalism, arguing that they are an exception because of their oppression, economic precariousness, disenfranchisement and the cost of compliance, in the form of sacrificing their livelihoods. There was a general concern with the cost of compliance and strong arguments that the state must recognise that the conditions of the urban poor were not conducive to compliance. The trade-off for marginalisation of the poor, high-density house occupancy, poor quality education and inadequate provision of public goods is non-compliance or selective compliance with state regulations. Respondents expressed an unwillingness to bear the costs of masks and sanitizers and bemoaned the fact that adhering to COVID-19 rules meant choosing between public health and economic survival:

A woman who lives down the road sews and sells [masks]. I bought three for my children, that is money for one week's worth of food. And, hand sanitizer is also too expensive. The money I would use to buy that I would need to feed my children. I used to work with a small company, but they have had to close down since the country lockdown [affected business]. I cannot seem to find any work since then. So when the government says to wear a mask, sanitize and social distance, I pay no attention to that, because they do not know how hard it has been. I have had to borrow a lot of money [from friends & family] just to feed my children. I will go to jail when I cannot pay them back (Sandy).

The amount of bills I have and can barely pay. It is the small shop down the road that helps. Soon I will not even be able to afford chicken parts and I will not be able to take anything from the shop. To be honest, maybe if I was not worried about how I would survive day to day I would pay more attention to what the government is saying, but it is hard. It is really hard. I have not worked from March (2020), so I really cannot spend my time worried about COVID. It is not that I do not care, but the money I would take to buy sanitizer and mask, I use that for food. If I were to get some regular help, maybe things would change. Look, I bought masks for my children so it is not that I do not understand the seriousness of the matter. It is just hard. COVID is not on my mind, and I am not the only one (Ann).

We just ensure we wash our hands when they are dirty. The sanitizer is expensive, I feel like my Member of Parliament should provide those things. Our money is for our survival. (Paul)

Spartacus made similar claims:

These little parties on a Wednesday and weekend are how a lot of young people make some money. All the bars and cook shops. If you start enforcing social distancing and mask wearing and the other regulations, a lot of people will lose the small income from these events. I think too that is why the police do not really bother us. If they start doing that, we will have no choice but to obey or go to jail, but until then those laws do not apply in the ghetto.

Like the church's approach, the case is made that spatial and social norms among the urban poor are incompatible with state COVID-19 regulations. Some respondents even expressed the more generalised view that the law was in tension with the lived realities of the urban poor. It is a critique of the law as not being sufficiently "common." Social norms, therefore, play a crucial role in negotiations and decisions about compliance (Brown, 2009; Gray, 2004). Paul does not think the rules are in alignment with 'ghetto livity':

This is the ghetto. We do not really comply with those things unless Babylon [the police] are in our faces and forcing it, and even then we are not really afraid of that. Even the same police come to a bar here and do not wear masks, so why must we? My friends are not into the mask wearing either.

As I said, this is the ghetto, so people here do not pay much attention to those things. Nobody here is really afraid of that. We do not get up everyday thinking about that. Money is all that is on our minds. We just have to do what we have to do. And if we get [COVID] and feel sick we simply go to the University Hospital. If people are worried about that, just like you wear your mask and have your little bottle of sanitizer on your waist, that is what they should do and leave us alone.

Those things do not happen in the ghetto. Have you ever been to one of our parties yet? Social what? Remember there is a set of laws for Jamaica and another for people living in the ghetto.

This idea of 'one law for Jamaica and one for ghetto people' (which was repeatedly expressed by respondents among the urban poor) reflects an impulse to not just negotiate within the established boundaries but to push beyond them and contest the law itself. This is, however, subordinate to the more dominant tendency to negotiate the exception and live with the existing laws. The dominant method for the church, in matters of COVID compliance, was to contest space (and outside of COVID public morality laws) while for the urban poor it was an aspiration to subordinate state law. The notion of 'two laws' lays bare the incomplete decolonisation project which aimed to use the imported legal system, which mimics the UK Westminster model, as the basis for citizenship and obedience (Wheatle & Campbell, 2020). Among other forces, significant class inequalities have undermined this.

> This is the ghetto ... Do you think this is uptown? Social distance for what reason? Do you see anybody else around here social distancing? You said your friend recommended that you talk to me; was she social distancing? The government must be more realistic and stop with the stupidity. Jamaican people will only social distance when they know that if they do not they will get in trouble, or they cannot get a service like at the bank or the tax office. Down here, that does not really apply to us. (Spartacus)

The narratives of exception among the urban poor are also embedded in the realities of low rates of conviction for breaches of various COVID-19 regulations, which suggest that people can reasonably predict that they will not be punished for non-compliance. In spaces outside the community and under the watchful gaze of their critics, the police were more keen to be seen to be enforcing the rules. This attracts a greater willingness to comply:

> First of all, the taxi drivers here will still carry the maximum number of people possible in their car. So if you want to social distance it is better for you to walk because they will tell you to get out if you complain. As to the mask. I carry it with me when I go [into] the town and places like that. The police in these places do not really care about us, but in places like Half-Way-Tree you have to be careful because they do not hesitate to lock you up, especially if they know you are from the ghetto. So to avoid problems you wear it. But, I see a lot of people not wearing their masks in those places, so I want to know what all the fuss is about. They should lock us all up.

> That will not happen in these areas. You see those police officers social distancing? Look where their masks are, around their necks and they are in people's faces laughing. Andrew Holness [the prime minister] does not realise those rules do not apply in the ghetto. Even curfew, almost every night a party is kept here and everybody come her to drink and dance with women. They do not enforce anything so why should we stress?
>
> Taxi drivers only stress passengers (to wear masks) when they are driving on the main road because if the police stop then they are in problems. Otherwise, the lowered numbers of passengers and other regulations they are to follow, like no air conditioning etc., they do not really adhere to.

It is clear that while many people did not comply with COVID-19 protocols in their communities and in spaces where the chances of being arrested and charged were low, they were more willing to comply in spaces where the police had a record of actually enforcing the laws.

Political campaigns were also regarded by participants as sites of exception and a space where COVID-19 rules were, therefore, not applicable. Having lifted some COVID-19 restrictions the government, under pressure from some segments of the public about the risks to public health, was forced to revise rules relating to mass gatherings, community meetings and motorcades. For the government, making the usual exception for itself during the election period proved difficult in a pandemic. Exciting the general public with political campaigns that draw on Jamaican music and language, motorcades, rallies and political stunts proved unusually difficult. Many people blamed the government for the spike in COVID-19 cases which followed the election campaign period.

Despite the election being held during the COVID-19 pandemic, adherence to the rules was a challenge and was seen as contrary to the traditional ways in which campaigns are conducted. While people were forced to comply in the polling stations, there was very little compliance outside of this:

> When the election fun hits you, nobody remembers about social distancing and mask wearing. Remember that election campaigning and road marches are filled with music and alcohol, and the police pay us no attention. COVID measures would have just ruined election festivities, so of course we were not paying that much attention ... I did not go anywhere and see people actually wearing their masks and social distancing the way they should. Everybody had it on their foreheads, neck, hands. So they were not thinking about me. And like I said, when the excitement of election hits with music,

> debates, conversation, people really do forget about what the Prime Minister said on TV. (Joe Brown)

Elections can be contentious in Jamaica, and it is expected, based on the political culture, that political supporters who are seeking to unseat the party in power will strongly resist government rules and regulations or see them as oppressive and 'a trick'. Elections by their very nature produce a general environment of mistrust. One respondent saw the protocols as a 'con':

> No, the supporters of the Jamaica Labour Party (JLP) cannot con us. We are not fools … Most of the people I was around did not wear a mask, so they felt no responsibility for me. Or they do not trust the JLP either. (Neil Maw)

> Alright. So the JLP supporters they say are capitalist. That would mean they only care about money, so even the sale of masks and expensive hand sanitizers benefit their wealthy friends, because you have to go into their pharmacies to buy them and they are not cheap. That's why I do not trust them. COVID they say is like the flu, so if you take some medication and rest when you catch it, you will be alright. What is all this talking and lockdown for? And see, Andrew [the Prime Minister] called an election when he knew people were going to congregate, so COVID cannot be that serious. (Anthony Clarke)

In negotiating non-compliance and making the case for an exception, the particularities of not just the class system, political behavior during the pandemic, institutional norms and the costs of compliance, but also the risks to individuals, were made. The lower the perceived risks to groups or individuals, the higher the probability of non-compliance and the more likely it is that groups will also use this to explain why they should be regarded as an exception to state regulation. Many among the urban poor did not regard COVID-19 as a serious threat. It was not something they are afraid of:

> This is just like the flu and then you are fine again. Maybe one or two will die, but I heard they had other sicknesses. I do not have asthma or any such thing. And [marijuana] is good for all of those things. Young people in other countries are dying a lot, yes, but I think Jamaica is handling the virus better than most other countries, so I am okay as far as I am concerned. (Paul)

> I would not have missed that [a party], and before you ask, nobody wore a mask or social distanced at that party. We came to party, and when we are under the influence of alcohol we are not thinking about anything the government said. Come to the party, have fun, drink rum and go home. Even the police stopped by and had a drink, a dance and left. They did not even mention the mask wearing. Only the government is concerned about that. We are fine. (Paul)

The lack of COVID-19 cases in some of the churches, where research participants were drawn from, was also used as a way of justifying and defending the church as an exception:

> And I feel like if you come to church then you are to understand that we are under the protection of the Holy Ghost and COVID cannot touch us there. And up until now, no church member has caught it. God is good. (Sister Mary)

> And I don't know of any member who has caught the virus. (John)

People participating in the recent elections also assessed their risk based on their age and physical proximity to the spaces where the disease is most prominent:

> I am not really afraid of those things. We are young and healthy so we are fine. Catch it, build our immunity and then we are out on the road again. Simple ... Just like the flu, COVID can get here, but we are far [from major towns] and we really do not go into the town unless necessary. I never thought about that part anyway, but maybe. (Neil Maw)

> I live with my little son and he is young and healthy. He is alright. He also doesn't really wear his mask. (Richard Brown)

> ... Do you see how far from the town we are ... Here we do not really believe in COVID. Not in the way Tufton [Minister of Health] is on TV every day talking about it. So we are not too concerned (Joy).

Conclusion

Unlike some of the more developed countries and more mature democracies, notably the USA and the Netherlands (but evident in most western democratic states), which have experienced violent libertarian resistance to the standard WHO recommended COVID prevention measures, Jamaica has been largely free of such manifestations. Jamaica's stability under the stress of the COVID crisis has been sustained in part because non-compliance is negotiated as limited exceptions that may be accommodated by the state. Challenging important minorities – whether religious groups or 'ghetto' residents – would be likely to damage state stability (and disadvantage governments in elections despite, in this context, the likelihood of higher COVID-19 mortality rates.) One outcome was that the police force only acted in some limited areas.

A further consequence of non-compliance and exception-making is likely to be the low uptake of vaccination in Jamaica, contrary to its past vaccination record. Amongst all the countries included in vaccination forecasting Jamaica has the lowest uptake (35%) in the Americas. Estimates done by the Ministry of Health suggest only 60% compliance among healthcare personnel, police, the military and the correctional services, which are priority groups. Accommodating for exceptionalism may come at a further cost.

We suggest that these COVID-specific exceptions are but expressions of a larger tried and tested repertoire of resistance to state control that has been exhibited across a larger field of socio-political action by the Jamaican people. The government's policy of balancing lives and livelihoods (which predated the position of the WHO) was grounded in an accurate estimate of the livelihood strategies of the majority of Jamaicans and the realities of socio-political power of the poor in Jamaican democracy. In a state where many are poor and marginalised, and employed in the informal sector, a strategy that avoided excessive lockdowns made political sense. Whether it saved lives, as people could eke out an informal existence, or increased the mortality rate is a moot point.

Acknowledgments We wish to thank Ms. Shar-Lee Amori, a graduate student in the Department of Government at the University of the West Indies, for her assistance in the field.

References

Bodas, M., & Peleg, K. (2002). Self-Isolation Compliance in the COVID-19 Era Influenced by Compensation: Findings from a Recent Survey in Israel. *Health Affairs, 39*(6), 936–941. https://doi.org/10.1377/hlthaff.2020.00382

Brown, E. M. (2009). Outsourcing Immigration Compliance. *Harvard Law School Program on Risk Regulation Research Paper, 8–7*(February), 1–60. https://doi.org/10.2139/ssrn.1131022

Campbell, Y. (2020). *Citizenship on the Margins: State Power, Security and Precariousness in 21st Century Jamaica*. Palgrave Macmillan.

Clark, C., Davila, A., Regis, M., & Kraus, S. (2020). Predictors of COVID-19 Voluntary Compliance Behaviors: An International Investigation. *Global Transitions, 2*(June), 76–82. https://doi.org/10.1016/j.glt.2020.06.003

Gray, O. (2004). *Demeaned but Empowered: The Social Power of the Urban Poor in Jamaica*. University of the West Indies Press.

Harriott, A. (2000). *Police and Crime Control in Jamaica: Problems of Reforming Ex-Colonial Constabularies*. University of the West Indies Press.

Jamaica Constabulary Force. (2020, December 4). *JCF Commits to Intensified Enforcement of the Disaster Risk Management Act*. Jamaica Information Service. https://jis.gov.jm/jcf-commits-to-intensified-enforcement-of-the-disaster-risk-management-act/

Jamaica Gleaner. (2020, October 21). Partygoers Caught Breaching COVID Rules Booked for Court November 24. Jamaica-Gleaner.com. https://jamaica-gleaner.com/article/news/20201020/partygoers-caught-breaching-covid-rules-booked-court-november-24

Jamaica Observer. (2020, August 6). *Sandy Bay, Clarendon Under Community Quarantine*. http://www.jamaicaobserver.com/latestnews/Sandy_Bay,_Clarendon_under_community_quarantine?profile=1228

Kuiper, M. E., de Bruijn, A., Folmer, C. R., Olthuis, E., et al. (2020). The Intelligent Lockdown: Compliance with COVID-19 Mitigation Measures in the Netherlands. *Amsterdam Law School Research Paper No* 2020-20, (May), 1–38. https://doi.org/10.2139/ssrn.3598215

Ministry of Justice. (2020). *MOJ Response: Disaster Risk Management Orders*. https://moj.gov.jm/content/covid-19

Ricketts, H. (2020). The Government of Jamaica (GoJ) COVID-19 Pandemic Stimulus Response: An Opportunity to Regulate and Formalize? *International Journal of Sociology and Social Policy, 40*(9/10), 979–993.

Webster, R. K., Brooks, S. L., Smith, L. E., Woodland, L., et al. (2020). How to Improve Adherence with Quarantine: Rapid Review of the Evidence. *Public Health, 182*(May), 163–169. https://doi.org/10.1016/j.puhe.2020.03.007

Wheatle, S., & Campbell, Y. (2020). Constitutional Faith and Identity in the Caribbean: Tradition, Politics and the Creolisation of Caribbean Constitutional Law. *Commonwealth & Comparative Politics, 58*(3), 344–365.

CHAPTER 28

Towards a Resilient, Inclusive and Green Recovery in the Caribbean

Ainka Granderson and Nicole Leotaud

Introduction

The COVID-19 pandemic provided a stark reminder of the vulnerability of small island developing states (SIDS) in the Caribbean Community (CARICOM) and their highly open economies where the decline in global economic activity compounded domestic impacts (IMF, 2020a). The 15 Member States of CARICOM are geographically very diverse. They consist of the arc of smaller islands and archipelagic states in the eastern Caribbean including Antigua and Barbuda, Bahamas, Barbados, Dominica, Grenada, Montserrat, St. Kitts and Nevis, Saint Lucia, St. Vincent and the Grenadines and Trinidad and Tobago; the larger northern states of Haiti and Jamaica; and the continental countries of Belize in Central America and Guyana and Suriname in South America. All CARICOM Member States are independent, except for Montserrat as an overseas territory of

A. Granderson (✉) • N. Leotaud
Caribbean Natural Resources Institute (CANARI),
Port of Spain, Trinidad and Tobago
e-mail: ainka@canari.org; nicole@canari.org

Y. Campbell, J. Connell (eds.), *COVID in the Islands: A comparative perspective on the Caribbean and the Pacific*,
https://doi.org/10.1007/978-981-16-5285-1_28

the United Kingdom, which funds over 60% of the island's budget via official grant transfers (Caribbean Development Bank, 2019). However, all these states have relatively small undiversified economies dependent on tourism, service industries and the productive sectors, including agriculture, fisheries, mining and petroleum, but with limited domestic markets. All are highly reliant on international services and trade and thus the economic strength of trading partners.

This chapter examines the impact of COVID-19 and responses to it across the broad region and examines the possibilities of an economic recovery and long-term development that may be more pro-poor, inclusive, environmentally sustainable and resilient, and how this would necessitate rethinking current governance models. It also considers the possibility of more effective and appropriate involvement of civil society in development. We seek to examine the potential for the adoption of local ecosystem-based solutions by civil society organisations, local communities and small and micro-enterprises (SMEs) that would deliver ecological and socio-economic resilience.

COVID-19 Impacts on CARICOM SIDS and the Triple Crises

While the region's morbidity and mortality levels have been relatively low (CARICOM, 2020a), policy responses to reduce COVID-19 spread and save lives have had significant socio-economic consequences, especially for those already living in poverty, informal and unskilled workers and other vulnerable groups (UN ECLAC, 2020a). Restrictions on travel and non-essential activities, quarantine and social distancing measures devastated tourism and related service industries. Working remotely and reduced operating hours and closures of non-essential businesses, especially of small and micro enterprises (SMEs), have had widespread and ongoing impacts on the workforce.

COVID-19 brought on an economic crisis across the region due to a reduction in the output of key economic sectors, tax revenues and foreign exchange earnings, and an increase in health and social expenditures. Severe reductions in growth in Gross Domestic Product (GDP) were expected for 2020, with some recovery in 2021 (Table 28.1). In particular, COVID-19 travel reductions and restrictions brought the contraction of tourism-dependent economies in Bahamas, Belize, Jamaica and the

Table 28.1 Real GDP growth in the independent CARICOM countries (%, year-over-year) (IMF, 2020a)

				Projections	
	2017	*2018*	*2019*	*2020*	*2021*
Caribbean – tourism-dependent					
Antigua and Barbuda	3.1	7	3.4	-17.3	4.7
Bahamas	3.1	3	1.2	-14.8	4.6
Barbados	0.5	-0.6	-0.1	-11.6	7.4
Belize	1.9	2.1	-2	-16	8
Dominica	-9.5	0.5	8.4	-8.8	3.3
Grenada	4.4	4.1	3	-11.8	3
Jamaica	0.7	1.9	0.9	-8.6	3.6
St. Kitts and Nevis	-2	2.9	2.8	-18.7	8
Saint Lucia	3.5	2.6	1.7	-16.9	7.2
St. Vincent and the Grenadines	1	2.2	0.4	-7	3.7
Caribbean – commodity exporters					
Guyana	3.7	4.4	5.4	26.2	8.1
Haiti	1.2	1.5	-1.2	-4	1.2
Suriname	1.8	2.6	0.3	-13.1	1.5
Trinidad and Tobago	-2.3	-0.2	0	-5.6	2.6

islands of the eastern Caribbean. The economies of Guyana, Haiti, Suriname and Trinidad and Tobago, all based on exports of raw materials such as bauxite, gold, petroleum and timber and agricultural commodities, were somewhat buffered from the impacts of COVID-19 on tourism. However, depressed economic growth and the reduction in global oil demand was expected to have a significant impact on Trinidad and Tobago's economy and the growing petroleum sector in Guyana and Suriname. Remittances are also important in several CARICOM Member States (for example representing 30% of the economy of Haiti) and anticipated reductions were expected to have a significant impact due to depressed growth in developed countries and the effects on the Caribbean diaspora from COVID-19 (IMF, 2020b). Reductions may not necessarily eventuate.

According to the United Nations Economic Commission for Latin America and the Caribbean (UN ECLAC), heavy job losses and an overall reduction in income occurred for CARICOM and the wider region, with 11.6 million more unemployed in 2020 than 2019 in Latin America and the Caribbean (UN ECLAC, 2020a). COVID-19 will probably result in the worst recession in the region in a century, with a 9% contraction in regional GDP in 2020 and a 4.4% increase in poverty (UN ECLAC, 2020a), with more severe recessions and job losses projected in the tourism-dependent CARICOM countries.

Losses of jobs and income also led to increased food insecurity, compounded where food imports were reduced or disrupted, particularly as the primary source, the United States, was also hit hard by COVID-19. Although certain countries have significant agricultural sectors, such as Belize, Dominica, Grenada, Guyana and St. Vincent and the Grenadines, approximately 80% to 90% of all food consumed in the region is imported (Ewing-Chow, 2020).

With CARICOM states facing multiple impacts, their ability to mount effective responses to and recovery from COVID-19 was varied and based on their levels of economic and social development. Haiti is the only country classified as a least developed country and the poorest country in the Western Hemisphere, and quickly received aid to bolster its health services. Other states are classified as upper middle income, with Bahamas and Trinidad and Tobago being classified as high income, and were therefore limited in their access to overseas development assistance despite poverty and income inequality persisting across the region. Many CARICOM states are also highly indebted, with average debt estimated at 75% of GDP at the end of 2019 (Inter-American Development Bank, 2019). Countries therefore had limited fiscal space and ability to invest in the health and social protection programmes required to respond to a pandemic and thus implemented strong policy responses in an attempt to minimise the number of cases faced by weak public health systems, including border closures, mandatory quarantines, social distancing and mask wearing in public, curfews, closures and work at home orders for non-essential businesses, with ensuing socio-economic consequences.

COVID-19 has compounded and is interlinked with pre-existing crises due to climate change and biodiversity loss (OECD, 2020a, 2020b). SIDS are highly vulnerable to climate change and related disasters. Rising sea levels, increased air and sea surface temperatures, rainfall variability, ocean acidification and more intense hurricanes and storms are expected to trigger a series of interconnected biophysical and socio-economic impacts affecting their economies, food and water security, health, infrastructure and natural ecosystems (Nurse et al., 2014; Climate Studies Group Mona, 2020). If greenhouse gas emissions continue unabated, the cost to CARICOM SIDS in terms of increased hurricane damage, infrastructure damage and loss of tourism revenue is projected to reach US$22 billion by 2050 or 10% of the current regional economy (Caribbean Community Climate Change Centre, 2012). Further, biodiversity and ecosystems providing critical goods and services supporting the economies, livelihoods and well-being of people in CARICOM SIDS are being lost and degraded due to ad hoc development, pollution, invasive species and disease, resource overexploitation and natural disasters like earthquakes and volcanoes. This has resulted in declining ecosystem cover and health and species diversity (UNEP – Caribbean Environment Programme, 2020).

Much of the burden of these impacts from COVID-19, climate change and biodiversity loss has been and will be borne by the poor and the most vulnerable and at-risk groups. COVID-19 will disproportionately impact women, young and low-skilled workers and those working in the informal sector (IMF, 2020a). In the CARICOM SIDS context, those most vulnerable to the triple threats from COVID-19, climate change and biodiversity loss are poor, marginalised rural communities and resource users like small-scale farmers, fisherfolk and natural resource-based community micro-enterprises that depend on ecosystem goods and services for their livelihoods. Yet, the most vulnerable have limited voice and engagement in decisions about environmental governance and climate actions and limited access to finance to support their own local actions (SEI, 2017; CANARI, 2020a, 2020b, 2020c).

COVID-19 has presented an unusual opportunity to address these multiple and intertwined risks from the climate and biodiversity crises, and entrenched socio-economic inequities, as governments focus on responding to and recovering from COVID-19 impacts. Based on a review of COVID-19 response policy trackers (IMF, 2020b; KPMG, 2020; UN ECLAC, 2020b), and the World Bank (2020) database of projects, initial responses from CARICOM governments to COVID-19 largely focused on emergency responses, including support to the health sector and to businesses and households, with limited evidence of any inclusion of green and climate-focused approaches. CARICOM governments have largely shaped their COVID-19 recovery with limited stakeholder participation which could have helped ensure that recovery responded to the needs of the most vulnerable. This chapter calls for a pro-poor, inclusive, environmentally sustainable and climate resilient approach to COVID-19 recovery, a rethinking of the current governance model and recognition of civil society's role in supporting economic recovery and long-term development in CARICOM SIDS. These solutions not only address climate change and natural disasters, but provide co-benefits for biodiversity conservation, food and water security and sustainable livelihoods.

Enabling Pro-Poor, Inclusive, Green and Resilient Recovery

Setting the Framework

Universal calls have been made by global agencies to 'build forward better' in response to COVID-19 and strong leadership and social dialogue around these ideas are needed in CARICOM SIDS. Putting biodiversity conservation, climate resilience and economic inclusiveness at the centre of COVID-19 recovery is critical to address the multiple and intertwined risks from biodiversity loss, the climate crisis and systemic social inequities (Clegg et al., 2021; Engel et al., 2020; IMF, 2020a; UN, 2020). In a region already facing critical sustainable development challenges, the pandemic has brought to the fore socio-economic and structural inequities and underlying governance issues related to accountability, transparency and shared decision-making, highlighting the need to rethink economic development and governance models. A focus on a pro-poor, inclusive, green and climate resilient COVID-19 recovery can put CARICOM SIDS on a path towards long-term socio-economic transformation and sustainability and ensure fair, equitable and just outcomes.

Recovery can draw on previous dialogues and initiatives aligned to calls for economic reform to achieve socially just and environmentally sustainable development outcomes, which have been growing since the Rio+20 conference. In the CARICOM SIDS, early interest in green economy approaches has shifted to involve blue economy approaches where harnessing ocean resources represents a significant and emerging opportunity for economic development (Caribbean Development Bank, 2014; Clegg et al., 2021; Patil et al., 2016), although the concept and its application are still being explored (Clegg et al., 2021). Regardless of nomenclature, the core principles underlying the green and blue economy are what should drive the economic transformation (OECS, 2020a; Leotaud et al., 2021). The five principles distilled by the Partners for Inclusive Green Economy (2019) provide a useful framework; *The Wellbeing Principle* envisages an economy that enables all people to create and enjoy genuine, shared prosperity; *The Justice Principle* promotes equity within and between generations, equitable distribution of opportunity and outcome and is inclusive and non-discriminatory; *The Planetary Boundaries Principle* safeguards, restores and invests in nature and Earth's systems and recognises the diverse values of nature in sustaining society; *The Efficiency and Sufficiency Principle* highlights that meeting basic needs and services is essential to human dignity, and *The Good Governance Principle* outlines the need for strong collaborative institutions and public participation to complement enlightened leadership with societal demand.

These ideas are starting to be reflected in policy, for example, in the Regional Green-Blue Economy Strategy and Action Plan of July 2020, adopted by the Organisation of Eastern Caribbean States (OECS).[1] Its goal is transformation to "an economically viable, climate resilient, economy that is based largely on sustainable domestic production and inward investment and pursues long-term prosperity for its entire population, in particular its most vulnerable, through socio-economic inclusiveness, responsible management of its coastal, marine and terrestrial ecological resources, and transparent, participatory governance" (OECS, 2020a: 5). That strategy emphasises that transformation would require a very broad agenda of engagement and collaboration, resource mobilisation, building human capital, adopting a data-driven approach and accessing and developing new technology, institutional and regulatory reform, monitoring and adapting based on learning (OECS, 2020a). However, as CARICOM SIDS shift their focus away from emergency responses to recovery, fundamental economic transformation will be needed and can draw from global and Caribbean thinking and practice to build on the 'green shoots' (OECS, 2020a) of initiatives grounded in core principles for inclusiveness, sustainability, climate resilience and good governance.

The OECS report assessing COVID-19 impacts proposed short and medium term recommendations for Eastern Caribbean governments, and emphasised that "the post-COVID-19 future for the OECS involves taking full advantage of the opportunity of its disruptions and the putting in place of nothing less than a different development paradigm" (OECS, 2020b: 12), with elements that speak directly to self-sustainability, economic justice with entrepreneurship and wider economic participation, a more inclusive social safety net, decent work in the tourism sector,

[1] The OECS is an eleven-member grouping of islands in the Eastern Caribbean comprising Anguilla, Antigua and Barbuda, British Virgin Islands, Dominica, Grenada, Guadeloupe, Martinique, Montserrat, Saint Lucia, St. Kitts and Nevis and St. Vincent and the Grenadines.

building of human capacity through educational reform, financial systems providing equitable access to capital at all levels and regional integration with, "an inclusive dynamic that ensures shared benefits to all" (OECS, 2020b: 12). The report emphasises the importance of a focus on long-term sustainability and resilience and proposes the establishment of a well-resourced OECS Resilience Fund that would support national and regional measures.

Barbados was an early leader in conducting national scoping studies on the green and blue economy, and then adopting and mainstreaming green and blue economy pathways in its development policies and plans, including in its National Strategic Plan 2006–2025 and via establishing a Ministry of Maritime Affairs and the Blue Economy in 2018. That placed it in a strong position to direct COVID-19 recovery investments and, indeed, a number of responses reflect green approaches. Its fiscal stimulus Barbados Employment and Sustainable Transformation (BEST) Plan includes transformation of the tourism sector by committing to greening through water conservation and water harvesting measures where applicable, and the installation of renewable energy capacity to reduce reliance on fossil fuels (CARICOM, 2020b). Barbados also made global news for its promotion of the 'Barbados Welcome Stamp' encouraging international visitors to stay and work visa-free for up to one year as a stimulus plan to support its vulnerable tourism sector (Taylor, 2020), which had provided 41% of GDP prior to COVID-19 (Mooney & Zegarra, 2020).

Similar plans exist in other countries. Grenada has leveraged its collaboration with the World Bank in developing its Blue Growth Coastal Master Plan (Patil & Diez, 2016) with the World Bank US$25 million credit for Grenada's COVID-19 response and recovery, which refers to enhancing climate resilience and enhancing long-term sustainability. Guyana will need to manage both its expected oil boom and COVID-19 recovery as it implements its 20-year Green State Development Strategy Vision 2040 (Government of Guyana, 2019).

Jamaica is in the process of developing a Green Economy Investment Strategy, which builds on existing policy frameworks such as the National Investment Policy of Jamaica and Vision 2030 National Development Plan and can inform priorities for COVID-19 recovery when completed. The May 2021 draft proposes a broad definition of "green investment" beyond the traditional focus on renewable energy to encompass actions "reducing the carbon footprint of the economy, enhancing the sustainability of the use of natural resources, facilitating adaptation to climate change, and promoting greater economic inclusiveness" (Witter, 2021: 34). This broader thinking can be usefully applied to design of investment in COVID-19 recovery.

It remains uncertain if those countries yet to commit to economic transformation would be as ready to direct COVID-19 recovery investments to measures focused on environmental conservation, low-carbon pathways and climate resilience. For example, Trinidad and Tobago's Roadmap guiding the response to COVID-19 (Government of the Republic of Trinidad and Tobago, 2020) mentions the need to increase electricity generation via renewable energy and rehabilitating groundwater resources but misses the opportunity to frame the recovery priorities as rebuilding depleted natural capital and reorienting to low-carbon strategies. In a country very dependent on the fossil fuel sector, and with high levels of energy consumption and a predominant focus on industry and manufacturing, the shift to low-carbon development will require major restructuring and investment.

Considerations for Inclusive Responses

COVID-19 recovery roadmaps and packages need first and foremost to target those most in need and provide decent work and economic opportunities for all. The recovery must address the deepening and emerging inequalities in terms of economic security of at-risk groups, including households living near or below the poverty line, marginalised rural and indigenous communities, persons with disabilities, women, youth, socially displaced and illegal immigrants, unemployed and non-essential workers. SMEs, especially micro-enterprises and those in the informal sector, are more vulnerable (Montmasson-Clair et al., 2019). Attention should be paid to natural resource-based enterprises and livelihoods given that these face additional threats from climate change and biodiversity loss.

CARICOM government's capital expenditure projects to stimulate the economy should deliver on key development priorities such as improved food, energy and water security and create employment with a focus on marginalised and at-risk groups and local economic development. Employment opportunities should not draw people away from existing businesses and sustainable livelihoods but create new opportunities to address high unemployment and poverty. COVID-19 has brought the issue of food security to the forefront of national and regional attention (OECS, 2020b) and support to agriculture and the small-scale fisheries sector is being provided in some responses, for example through direct support in Haiti (IMF, 2020b) and St. Vincent and the Grenadines (KPMG, 2020). Enhancing these sectors, while ensuring that their operations are environmentally sustainable, is a universal challenge.

A key focus should be to support and strengthen the institutional framework to stimulate local SMEs (Leotaud et al., 2021), especially to support local agriculture and fisheries for food security and natural resource-based micro-enterprises which are the foundation of livelihoods and local economies in rural communities but are also the most vulnerable to the impacts of climate change and biodiversity loss. Support should include appropriate tax incentives, and provision of enabling legal and policy frameworks, appropriate funding and technical assistance for access to markets, business development and innovation (Edwards et al., 2019). SMEs often operate on an informal and sub-optimal level and are excluded from accessing critical support services that would help them develop, be sustainable and deliver socio-economic benefits in poor communities (Montmasson-Clair et al., 2019). Despite being a significant component of the economy, Caribbean governments have generally paid inadequate attention to the micro-enterprise sector, and particularly have not addressed the challenges of informality. A Caribbean Development Bank (2016) study estimated that micro, small and medium enterprises contributed 50% of GDP and created 45% of jobs in the CARICOM region. It further recognised the importance of the informal sector noting that there were nearly three times as many informal enterprises operating in the countries studied. Informal micro-enterprises, including those owned by women, are particularly important for economic empowerment of poor households and need greater attention.

Stimulus packages have been widely offered to SMEs within CARICOM. Barbados created a Green and Digital Investment Fund accessible to tourism-related firms, small businesses and manufacturing and agricultural enterprises to aid in their transformation to greener and more digitised business processes (KPMG, 2020). Jamaica provided grants to the marginally self-employed and informally employed and to registered SMEs (KPMG, 2020). Trinidad and Tobago has supported enhanced access to finance for individual and small businesses via credit unions, with a reduced lending rate (KPMG, 2020). Bahamas has provided business continuity loans to SMEs and expanded its youth apprenticeship programme, albeit focused narrowly on the construction sector (KPMG, 2020). Haiti has provided relief grants to SMEs experiencing challenges to sustain operations and retain employees and for training and development (IMF, 2020b). The outcomes of such strategies are not yet apparent, while they should be accessible to micro-enterprises and the informal sector if they are to embrace economic participation of the most vulnerable. Support programmes may be difficult to access by entrepreneurs, with limited capacity to fill out complex forms online, so are not usually available to informal enterprises. Many formal and informal micro-enterprises are already excluded from traditional finance and credit mechanisms and lack the financial cushion to weather the COVID-19 storm. Longer term recovery from COVID-19 will require greater attention to the SME and informal sectors, including the rural nature-based niche sector which is marginalised from opportunities under traditional business support programmes, never an easy task.

Removal of fiscal subsidies that result in harm to local businesses, citizens and the environment and re-investing these funds in sustainable COVID-19 recovery are also important (OECD, 2020b). This could include removing subsidies for imported fuel and non-essential foods and redirecting funds towards projects/programmes for local SME development that delivers triple bottom-line economic, social and environmental co-benefits.

Considerations for Green and Resilient Responses

Another key area is investing in low-carbon and climate resilient solutions to stimulate the economy. For example, Barbados, which is already a regional leader in solar energy, announced a US$360 million Green Energy Park supplying 30 megawatts of energy as part of its COVID-19 economic stimulus package (KPMG, 2020). Scaling up initiatives to adopt renewable energy and energy efficient measures, such as solar hot water and photovoltaics (PV), energy efficient lighting and appliances, and electric and hybrid vehicles for residential and commercial use, can provide low-carbon pathways to COVID-19 recovery. But making green technology accessible to poor households and micro-enterprises will also be needed for an equitable recovery.

Investing in green and blue infrastructure by protecting and restoring natural ecosystems and critical ecosystem services, including carbon capture, is also key (Browder et al., 2019). For example, Trinidad and Tobago's COVID-19 recovery plan has identified the importance of rehabilitating groundwater assets as critical supporting infrastructure for economic development (Government of the Republic of Trinidad and Tobago, 2020). This plan could address longstanding issues by including attention to watershed rehabilitation that would provide co-benefits including flood mitigation and supporting agroforestry for food security and sustainable rural livelihoods. Such nature-based solutions, which involve the conservation, sustainable management and restoration of ecosystems, also provide mitigation co-benefits as forests and mangroves act as important carbon sinks that help with carbon capture and mitigate climate change and also support livelihoods of vulnerable rural communities (Kapos et al., 2019). Nature-based solutions are also often cheaper than conventional engineered solutions (Browder et al., 2019; Kapos et al., 2019). This reduces the need for CARICOM governments to source loans from multilateral banks and other financiers and take on significant debt. Climate-proofing critical infrastructure, such as roads, bridges and air and sea ports, as part of government's capital expenditure projects, can also create employment opportunities and build resilience to climate change and disasters.

Debt for nature/climate swaps are being proposed as a possible avenue for financing green, resilient and just COVID-19 recoveries (De Wit et al., 2020; Steele & Patel, 2020), and are valuable given the limited fiscal space in CARICOM SIDS that are already constrained by high debt burdens and facing reductions in remittances, overseas development assistance and tax revenues. To help these vulnerable states face the COVID-19 crisis, CARICOM leaders have called for concessional financing, debt relief, and the use of COVID-linked financial instruments where the payback obligations followed the country's medical and economic recovery (UN ECLAC, 2020c). The World Bank's US$25 million credit for Grenada's COVID-19 response and recovery is interest-free with a maturity of 40 years, including a grace period of 10 years, and in a country listed by the International Monetary Fund (IMF) as being in debt distress, with a projected increase in debt to 69% of GDP in 2020 (IMF, 2020a). Debt for nature/climate swaps are an innovative financial instrument that could be considered in the COVID-19 recovery toolbox of CARICOM governments.

Considerations for Participatory Governance

Ensuring there is an open and transparent process for decision-making and participatory governance is also key to a COVID-19 recovery that meets the needs of the most vulnerable and wider citizenry in CARICOM

SIDS. A 'whole of society' approach will better enable a full and effective recovery and the transition to more inclusive, green and resilient development models. For example, to develop its national COVID-19 recovery plans, multi-stakeholder forums are being used in Grenada and the Trinidad and Tobago government set up a national committee of experts and invited public submissions. But further efforts are needed to facilitate deeper social dialogue at local, sub-national and national levels that will include perspectives from CSOs, local communities and SMEs to ensure that plans address real needs on the ground.

However, despite widespread recognition that COVID-19 will have broad and deep socio-economic impacts, many public sector responses have been formulated with limited engagement of those who will be most impacted and in the absence of broad-based public dialogue on appropriate responses to this crisis and directions for national development. CARICOM governments need to do more to engage civil society and the private sector in meaningful collaborations and partnerships to implement COVID-19 recovery solutions. These can be inspired by models such as the longstanding Barbados Social Partnership, a tripartite relationship between the public and private sectors and labour to facilitate dialogue on economic development, although this has been recently critiqued as excluding civil society involvement in governance, emphasising the challenges to participation (Hinds, 2019).

Rethinking Civil Society's Role in COVID-19 Recovery and Development

To promote pro-poor, inclusive, green and resilient development, requires a shift from the conventional top-down approach to a participatory approach to governance that engages actors across all levels of society and sectors as part of the COVID-19 recovery and future development. There needs to be recognition of the significant role that civil society can play as strategic partners in designing and implementing recovery measures (UN ECLAC, 2020d) as well as holding governments and their development partners accountable for delivery of COVID-19 recovery and other commitments. Civil Society Organisations (CSOs) and local communities and resource users can play a role as innovators, trusted local partners for implementation of projects and programmes, and advocates and mobilisers of resources that enable effective action. In this way, they can add value to and complement the work of governments and the private sector for a 'whole of society' approach.

As innovators and solution providers, Caribbean CSOs and local communities have piloted new and innovative approaches for biodiversity conservation, climate change adaptation and mitigation and sustainable livelihoods that can be scaled up as part of COVID-19 recovery and the transition to a blue-green economy (CANARI, 2020a, 2020b, 2020c; OECS, 2020a). For example, Fragments of Hope, a community-based organisation in Palencia, Belize, has pioneered the creation of coral nurseries with temperature tolerant species and transplanting of fragments for coral reef restoration and climate resilience. Since 2006, they have worked with fishers and divers that they have trained from nearby communities, including women and youth. They have transplanted over 50,000 coral fragments from nurseries to various sites, including Laughing Bird Caye National Park, that now have thriving corals and a thriving dive tourism industry. Fragments of Hope was awarded the UN Climate Solutions Award in 2017 for their innovative work (UNFCCC, 2017). This is currently being scaled up and replicated throughout the CARICOM and wider Caribbean region via various initiatives.

The Caribbean Natural Resources Institute (CANARI) has supported development of local community micro-enterprises based on the use of ecosystem goods and services, for example in small-scale fisheries, eco-tourism, crafts, natural pharmaceuticals, food and cosmetics. Entrepreneurs have used an innovative tool called the Local Green-Blue Enterprise Radar that CANARI developed to help them identify areas where they could improve their business practices to enhance delivery of economic, social and environmental co-benefits alongside good governance (CANARI, 2020b). CANARI also developed a climate proofing tool for SMEs (Sandy & Dardaine-Edwards, 2017), which supports them to assess the risks they face along their value chain due to climate change and to identify ways they can build resilience while delivering additional value to their business.

CSOs can play a role as trusted partners for implementation, building on their close work with local communities, enterprises and resource users over many years. They have an in-depth understanding of local context, how to build local capacity and effectively engage local communities and resources, including the most vulnerable and marginalised, to ensure projects and programmes are delivered to meet local needs and have concrete impacts on the ground. Some CSOs also have direct co-management responsibilities over significant areas of national territories and can deliver solutions in these areas. For example, national trusts and other CSOs have enabled the protection and sustainable use of biodiversity and natural resources in protected areas in partnership with national governments and local communities and resource users for over 20 years. The Bahamas National Trust (BNT) manages the country's 32 national parks covering over 2 million acres (BNT, 2017), and the Caribbean Coastal Area Management (C-CAM) Foundation manages the Portland Bight Protected Area in Jamaica, covering 4.7% of the island and 47.6% of the island shelf (C-CAM, 2013).

As advocates and mobilisers of resources, CSOs can also play a key role in promoting pro-poor, inclusive, green and resilient development through changes in policy, practice and financing. A well-known and successful example is the 1.5tostayalive campaign spearheaded by regional CSO, Panos Caribbean, that helped CARICOM negotiators get global commitment to the 1.5°C target in the Paris Agreement and leverage additional technical assistance and funding for low-carbon and climate resilient development in SIDS (CANARI, 2017). The campaign has further advocated for climate justice and on-the-ground actions that meet the needs of poor and marginalised communities that have contributed least to climate change but are most affected by it, which is relevant to COVID-19 responses focused on building climate resilience.

CSOs are therefore well placed to deliver solutions tailored to local contexts and with economic, environmental and social co-benefits if they are properly involved as partners in development and delivery of COVID-19 responses. They can also hold governments and their development partners accountable for delivery of COVID-19 recovery and other key commitments on biodiversity, climate change and the 2030 Agenda and its Sustainable Development Goals (SDGs) (CANARI, 2020c). However, civil society leadership and capacity needs to be strengthened for them to effectively fulfil these three key roles as innovators, trusted local partners for implementation and advocates and mobilisers of resources. CSO-led initiatives in the Caribbean region are often small-scale, short-term, piecemeal and lack strategic impact. While they seek to address gender and other social inequities in their work, most CSOs do not have specific training and procedures in place to effectively enable a gender-responsive and fully inclusive approach (Dulal et al., 2009; Dunn, 2013). This is generally due to a lack of funding and resources with few funding streams available for learning and capacity building of CSOs to facilitate scaling up and replication based on best practices (Stockholm

Environment Institute, 2017). Additional mechanisms need to be created to facilitate meaningful engagement and development of partnerships for CSOs to play a role in implementing COVID-19 responses. For example, Trinidad and Tobago's roadmap for COVID-19 recovery highlights the role of CSOs as providers of social protection services (Government of the Republic of Trinidad and Tobago, 2020) but does not venture further into recognising potential roles in other areas of socio-economic development.

Conclusions

The COVID-19 pandemic and the significant investment of national governments and their development partners earmarked for recovery efforts provided a crucial window of opportunity towards realising CARICOM's climate ambitions, delivering the SDGs and safeguarding the biodiversity and natural resources that are the basis for key economic sectors like agriculture, fisheries and tourism and are essential for the well-being and livelihoods of people in these SIDS. A pro-poor, inclusive, green and resilient approach to COVID-19 recovery can support the transition to a long-term economic development model that ensures a more just, equitable and sustainable future. This can be achieved when guided by a strategic national plan to drive economic transformation to 'build forward better' with recognition of ecological limits and economic justice and good governance priorities. Focusing on supporting SMEs and informal businesses, investing in renewable energy and green and blue infrastructure, and leveraging the power of civil society to deliver solutions on the ground are some concrete strategies that can be used by all CARICOM SIDS. This needs to be done, recognising the diverse natural resource base and socio-economic and cultural contexts for each country, with strategies and priorities determined and driven by local people through inclusive processes.

Although the specific challenges and the opportunities differ for each CARICOM member state in responding to the COVID-19 crisis, the Caribbean SIDS have an opportunity to undertake a long-term social and economic recovery programme that is green and asserts that the economy should serve people within ecological limits, broadens economic opportunity, focuses on ensuring that no one if left behind, adopts a 'whole of society' approach and commits to participatory governance. They need to move beyond recovery to building resilience not only to pandemics, but also to the biodiversity and climate crises.

Bold steps taken now to transform development pathways in CARICOM SIDS hold the promise of strengthening food, energy and water security, diversifying economies, providing decent work and economic opportunities, and protecting and restoring the invaluable natural capital upon which the region's economies, livelihoods, well-being and culture depends. That will not always be easy but it is demonstrably necessary.

References

Bahamas National Trust (BNT). (2017). *Bahamas National Trust Strategic Plan 2018–2022.* https://bnt.bs/wp-content/uploads/2019/02/BNT-Strategic-Plan.pdf

CANARI. (2017). *Effective Organisational Capacity Building of Civil Society Organisations: Lessons from the Climate ACTT Project* (CANARI Policy Brief 24). https://canari.org/pb24/effective-organisational-capacity-building-of-civil-society-organisations/

CANARI. (2020a). *Rising to the Climate Challenge: Coastal and Marine Resilience in the Caribbean* (CANARI Issue Paper 2). https://canari.org/wp-content/uploads/2020/08/CANARI-Coastal-Marine-Resilience-Issue-Paper.pdf

CANARI. (2020b). *The Local Green-Blue Enterprise Radar: A Tool to Support Community Enterprises.* CANARI. https://canari.org/wp-content/uploads/2017/08/CANARI-LGE-Radar-Toolkit1.pdf

CANARI. (2020c). *A Trinidad and Tobago Civil Society Report: Spotlight on Sustainable Development Goal 13 Climate Action.* CANARI. https://canari.org/wp-content/uploads/2017/09/A-Trinidad-and-Tobago-Civil-Society-Report-Spotlight-on-SDG-13-Climate-Action-1.pdf

Caribbean Coastal Area Management Foundation (C-CAM). (2013). *Climate Change Risk Assessment Report: Portland Bight Protected Area.* Prepared by the Climate Studies Group Mona. https://savegoatislands.org/wp-content/uploads/2014/04/PBPA_Climate_Change_Risk_Analysis_CCAM.pdf

Caribbean Community (CARICOM). (2020a, December 29). CARICOM: COVID-19 Dashboard. *CARICOM Today.* https://today.caricom.org/covid19/regional/

Caribbean Community Climate Change Centre (CCCCC). (2012). *Delivering Transformational Change 2011–2021. Implementing the CARICOM Regional Framework for Achieving Development Resilient to Climate Change.* CCCCC. https://www.caribbeanclimate.bz/the-cost-of-inaction-on-climate-change-in-the-caribbean/

Caribbean Development Bank (CDB). (2014). *A New Paradigm for Caribbean Development: Transitioning to a Green Economy.* CDB. https://www.caribank.org/publications-and-resources/resource-library/thematic-papers/study-new-paradigm-caribbean-development-transitioning-green-economy

CARICOM. (2020b, September 18). BEST' Plan – A Stimulus Package for the Barbados Tourism Sector. *CARICOM Today*. https://today.caricom.org/2020/09/18/best-plan-a-stimulus-package-for-the-barbados-tourism-sector/

CDB. (2016). *Micro, Small and Medium Enterprise Development in the Caribbean: Towards a New Frontier*. CDB. https://issuu.com/caribank/docs/micro__small_and_medium_enterprise_/82

CDB. (2019). *Montserrat Economic Brief 2019*. CDB. https://issuu.com/caribank/docs/cer_2019_montserrat

Clegg, P., Mahon, R., McConney, P., & Oxenford, H. A. (Eds.). (2021). *The Caribbean Blue Economy*. Routledge.

Climate Studies Group, Mona - University of the West Indies. (2020). *State of the Caribbean Climate (SOCC) Report*. Prepared for the Caribbean Development Bank. Available at http://socc.mona.uwi.edu/report

De Wit, W., Freschi, A., & Dobrin, G.-C. (2020, September 29). Debt-for-Nature Swaps: A Possible Solution to a "green and just recovery" from Covid-19. *Dalberg Blog*. https://dalberg.com/our-ideas/debt-for-nature-swaps-a-possible-solution-to-a-green-and-just-recovery-from-covid-19/#:~:text=As%20the%20Covid%2D19%20crisis,for%20investment%20in%20the%20environment

Dulal, H., Shah, U., & Ahmad, N. (2009). Social Equity Considerations in the Implementation of Caribbean Climate Change Adaptation Policies. *Sustainability, 1*(3), 363–383.

Dunn, L. (Ed). (2013). *Gender, Climate Change and Disaster Risk Management* (Working Paper Series No. 7). Institute of Gender and Development Studies, Mona Unit, University of the West Indies and Friedrich Ebert Stiftung. http://library.fes.de/pdf-files/bueros/fescaribe/10711.pdf

Edwards, A. D., Sandy, K., & Leotaud, N. (2019). *Opportunities for Local Green Enterprises in Trinidad and Tobago*. CANARI. https://canari.org/wp-content/uploads/2017/08/Diagnostic-study-TT-green-enterprises-March-2019.pdf

Engel, H., Hamilton, A., Hieronimus, S., Nauclér, T., et al. (2020). *How a Post-Pandemic Stimulus Can Both Create Jobs and Help the Climate*. https://www.mckinsey.com/business-functions/sustainability/our-insights/how-a-post-pandemic-stimulus-can-both-create-jobs-and-help-the-climate

Ewing-Chow, D. (2020, March 31). Five Ways that COVID-19 Has Changed What Food Insecurity Looks Like in the Caribbean. *Forbes*. https://www.forbes.com/sites/daphneewingchow/2020/03/31/five-ways-that-covid-19-has-changed-what-food-insecurity-looks-like-in-the-caribbean/?subId3=xid%3Afr1585745902158dgi&sh=2362ed64143e

Government of Guyana. (2019). *Green State Development Strategy Vision 2040*. https://doe.gov.gy/gsds

Government of the Republic of Trinidad and Tobago. (2020). *Roadmap for Trinidad and Tobago Post COVID-19 Pandemic*. https://planning.gov.tt/sites/default/files/Report%20of%20the%20Roadmap%20to%20Recovery%20Committee_1st_.pdf

Greg, B., Suzanne, O., Irene, R. B., Todd, G., & Glenn-Marie, L. (2019). *Integrating Green and Gray: Creating Next Generation Infrastructure*. World Resources Institute.

Hinds, K. (2019). Barbados: Civil Society Inclusion with Social Partnership. In *Civil Society Organisations, Governance and the Caribbean Community* (pp. 101–131). https://link.springer.com/chapter/10.1007%2F978-3-030-04396-4_5

Inter-American Development Bank (IDB). (2019). Regional Overview. *Caribbean Region Quarterly Bulletin*, *8*(4): 2–4. IDB. https://publications.iadb.org/publications/english/document/Caribbean_Region_Quarterly_Bulletin_Volume_8_Issue_4_December_2019_en.pdf

International Monetary Fund (IMF). (2020a, October). *Regional Economic Outlook. Western Hemisphere: Pandemic Persistence Clouds the Recovery*. IMF. https://www.imf.org/en/Publications/REO/Issues/2020/10/30/Regional-Economic-Outlook-October-2020-Western-Hemisphere-Pandemic-Persistence-Clouds-the-49797

IMF. (2020b, December 18). Policy Responses to COVID-19. *Policy Tracker*. https://www.imf.org/en/Topics/imf-and-covid19/Policy-Responses-to-COVID-19

Kapos, V., Wicander, S., Salvaterra, T., Dawkins, K., & Hicks, C. (2019). The Role of the Natural Environment in Adaptation. *Background Paper for the Global Commission on Adaptation*. Rotterdam and Washington, DC: Global Commission on Adaptation. https://gca.org/reports/the-role-of-the-natural-environment-in-adaptation/

KPMG. (2020, December 22). Government Response – Global Landscape: An Overview of Government and Institution Measures Around the World in Response to COVID_19. *Government Stimulus Tracker*. https://home.kpmg/xx/en/home/insights/2020/04/government-response-global-landscape.html

Leotaud, N., Girvan, A., & Jattansingh, S. (2021). The Blue Economy Winners and Losers in the Wider Caribbean. In P. Clegg, R. Mahon, P. McConney, & H. A. Oxenford (Eds.), *The Caribbean Blue Economy*. Routledge.

Montmasson-Clair, G., Patel, M., Mudombi, S., Jattansingh, S., Granderson, A., & Leotaud, N. (2019). *All Is Not Green: Climate Change Adaptation and Small Business Resilience in Low- and Middle-Income Countries*. Paper Commissioned by the Global Commission on Adaptation. https://www.tips.org.za/research-archive/sustainable-growth/green-economy-2/item/3716-all-is-not-green-climate-change-adaptation-and-small-business-resilience-in-low-and-middle-income-countries

Mooney, H., & Zegarra, M. A. (2020, March 16). COVID-19: Tourism-based shock scenarios for Caribbean countries. *Caribbean DEVTrends Blog*. https://blogs.iadb.org/caribbean-dev-trends/en/covid-19-tourism-based-shock-scenarios-for-caribbean-countries/

Nurse, L., McLean, R. F., Agard, J., Briguglio, L. P., et al. (2014). Small Islands. In V. R. Barros et al. (Eds.), *Climate Change 2014: Impacts, Adaptation, and Vulnerability* (pp. 1613–1654). Cambridge University Press.

Organisation for Economic Co-operation and Development (OECD). (2020a). *Biodiversity and the Economic Response to COVID-19: Ensuring a Green and Resilient Recovery*. https://read.oecd-ilibrary.org/view/?ref=136_136726-x5msnju6xg&title=Biodiversity-and-the-economic-response-to-COVID-19-Ensuring-a-green-and-resilient-recovery

OECD. (2020b). *Building Back Better: A Sustainable, Resilient Recovery After COVID-19*. https://read.oecd-ilibrary.org/view/?ref=133_133639-s08q2ridhf&title=Building-back-better-_A-sustainable-resilient-recovery-after-Covid-19

Organisation of Eastern Caribbean States (OECS). (2020a). OECS Regional Green-Blue Economy Strategy and Action Plan. OECS Commission: .

OECS. (2020b). *COVID-19 and Beyond: Impact Assessments and Responses*. Organisation of Eastern Caribbean States (OECS) Commission. https://www.oecs.org/our-work/knowledge/library/business/covid-19-and-beyond-impact-accessment-and-responses

Partners for Inclusive Green Economy. (2019). *Principles, Priorities and Pathways for Inclusive Green Economies: Economic Transformation to Deliver the SDGs*. https://www.greeneconomycoalition.org/assets/reports/GEC-Reports/Principles-priorities-pathways-inclusive-green-economies-web.pdf

Patil, P. G., & Diez, S. M. (2016). *Grenada – Blue Growth Coastal Master Plan*. The World Bank Group. http://documents.worldbank.org/curated/en/358651480931239134/Grenada-Blue-growth-coastal-master-plan

Patil, P. G., Virdin, J., Diez, S. M., Roberts, J., & Singh, A. (2016). *Toward A Blue Economy: A Promise for Sustainable Growth in the Caribbean: An Overview*. The World Bank.

Sandy. K., & Dardaine-Edwards, A. (2017). *Building Resilience and Adding Value to Local Green Enterprises: Developing a 'climate-proofing' Methodology* (CANARI Technical Report No. 403). CANARI.

Steele, P., & Patel, S. (2020). *Tackling the Triple Crisis. Using Debt Swaps to Address Debt, Climate and Nature Loss Post-COVID-19*. International Institute for Environment and Development (IIED) Issue Paper. https://pubs.iied.org/16674IIED/

Stockholm Environment Institute (SEI). (2017). *Climate Finance in the Caribbean Region's Small Island Developing States* (SEI Working Paper No. 2017–08). https://mediamanager.sei.org/documents/Publications/SEI-WP2017-climate-finance-caribbean.pdf

Taylor, A. (2020, July 16). Barbados Wants You to Work From Its Beaches During the Pandemic. *Washington Post*. https://www.washingtonpost.com/world/2020/07/16/barbados-work-remote-coronavirus/

United Nations (UN). (2020). *Policy brief: The impact of COVID-19 on Latin America and the Caribbean*. July 2020. https://www.un.org/sites/un2.un.org/files/sg_policy_brief_covid_lac.pdf

United Nations Economic Commission for Latin America and the Caribbean (UN ECLAC). (2020a). The Social Challenge in Times of COVID-19. *Special Report COVID-19*, 3. https://repositorio.cepal.org/bitstream/handle/11362/45544/1/S2000324_en.pdf

UN ECLAC. (2020b, December 29). *COVID-19 Observatory in Latin America and the Caribbean: Economic and Social Impact*. https://www.cepal.org/en/topics/covid-19

UN ECLAC. (2020c, April 29). Borrowing Is Not an Option for Caribbean Countries, Access to Concessional Funding and Debt Relief Is Urgently Needed to Face the COVID-19 Crisis. *UN ECLAC Press Release*. https://www.cepal.org/en/pressreleases/borrowing-not-option-caribbean-countries-access-concessional-funding-and-debt-relief

UN ECLAC. (2020d). Political and Social Compacts for Equality and Sustainable Development in Latin America and the Caribbean in the Post-COVID-19 Recovery. *Special Report COVID-19*, 8. https://repositorio.cepal.org/bitstream/handle/11362/46146/1/S2000672_en.pdf

United Nations Environment Programme – Caribbean Environment Programme (UNEP CEP). (2020). *State of Marine Habitats in the Wider Caribbean*. UNEP CEP.

United Nations Framework Convention on Climate Change (UNFCCC). (2017). *More Corals = More Fish, Belize*. https://unfccc.int/climate-action/momentum-for-change/women-for-results/more-corals-more-fish

Witter, M. (2021, May 2). *A Green Economy Investment Strategy for Jamaica (Draft)*. Prepared for the Ministry of Economic Growth and Job Creation and the National Environment and Planning Agency.

World Bank. (2020). *Database of Projects and Operations*. World Bank. https://projects.worldbank.org/en/projects-operations/project-country

CHAPTER 29

Aftermath: Towards a 'New Normal'?

John Connell and Yonique Campbell

As these chapters have shown, SIDS exhibit diversity and variety, successes and failures, frustrations and resilience, in the face of COVID-19. No islands are the same and a world of difference, whether in geography, economics or politics, separates the Bahamas from Vava'u and Trinidad from Tuvalu. Meanwhile change continues. New virus spikes occur and new solutions are offered. While viable vaccines are now accessible, islands and SIDS still live with uncertainty, however much island governments and households are thinking ahead and, indeed, because they are thinking ahead. Vaccine uptake might be a more widespread challenge should 'vaccine nationalism' diminish the chances of SIDS acquiring vaccines quickly. Even optimism over imminent access to a functional vaccine raises questions over its delivery to remote corners of archipelagic SIDS. In the meantime, the virus is mutating, contributing one more degree of uncertainty.

J. Connell (✉)
School of Geosciences, University of Sydney, Sydney, NSW, Australia
e-mail: john.connell@sydney.edu.au

Y. Campbell
University of the West Indies, Kingston, Jamaica
e-mail: yonique.campbell02@uwimona.edu.jm

Y. Campbell, J. Connell (eds.), *COVID in the Islands: A comparative perspective on the Caribbean and the Pacific*,
https://doi.org/10.1007/978-981-16-5285-1_29

COVID-19 drew initial attention to widespread problems of health care, revealing not merely that most SIDS had no familiarity with such terms as PPE or social distancing (as in most other countries) but that SIDS had a particularly high incidence of health problems, especially NCDs, with ensuing problems of co-morbidity, and that only small proportions of not particularly large national budgets were devoted to health care. Little of that was spent on public health care, and especially on primary health care, whereas too much had to be spent on NCDs and on referrals: a distortion of resource use favouring a few. In the SIDS as elsewhere little had been learned from recent more localised epidemics, such as Ebola, SARS and MERS, but just a little from Zika, that were seemingly obscure tropical diseases from Asia and Africa. That had resonance; the Spanish Flu of 1918 that was truly global had largely been consigned to history as an unusual incident from another era.

The pandemic intensified numerous socio-economic crises in SIDS such as unemployment, exhaustion of remittances and savings by family members, domestic violence, food insecurity, inadequate infrastructure (including housing) and access to services (such as clean water), political tensions, digital and rural-urban divides, all circumstances often replicated on a grander scale elsewhere, and widely recognised as failures of neo-liberal growth models (Karim et al., 2020; Büscher et al., 2020; Cohen & Mata-Sánchez, 2021). Widespread food insecurity reflected and reinforced inequity, as the outcome of lost livelihoods and difficulties in accessing housing, sanitation and other basic needs (Klassen & Murphy, 2020). COVID-19 highlighted and exacerbated existing vulnerability, precarity and disadvantage as differences and divisions were reinforced, pointing to 'structural violence', where disease disproportionately affected the poor and marginalised (Farmer, 2001). Effective, equitable post-pandemic restructuring confronted some basic and wide-ranging structural conditions.

One consequence of COVID-19 was some withdrawal from, if not rejection of, globalisation through broken transport and commodity chains, lost tourism markets and rigid borders. Conversely, more traditional ways – bartering and the resuscitation of home gardens and subsistence fisheries – suggested local solutions. That was scarcely by choice and proved hard after urban lives where, for example, fishing techniques and social relationships had been forgotten. Moreover self-reliance brought a degree of parochialism: strangers brought germs.

In lockdown especially, with economies that had crashed, it might have seemed like an opportunity to build on these changes and re-examine social and economic development strategies. In practice, under pressure, that was far from easy when immediate health problems, rising unemployment and other concerns necessarily diverted attention, alongside inherent constraints to change and flexibility.

Could a 'return' towards subsistence and more local economies be viable and supported by national and regional policies? How could health systems be strengthened? Is that possible in the straightened circumstances that SIDS face? What risk mitigation strategies would yield the best results? Will small island states forge a common path to recovery, strengthening regional relations? Or will they imagine a much different future? Diverse possible futures reflect the politics and geography of the islands (as independent states or overseas territories, single islands or archipelagoes), social and economic status, scale and structure of development. What role will the private sector play in revival? Could it be innovative? Might there then be a 'new normal' that would be clean and green, self-reliant and equitable, or will there be a desperate clutching at any straws to get underway again. What sort of light, if any, is at the end of the tunnel?

Build Back Better

Underpinning government responses to the pandemic was a degree of optimism that some long-term benefits might be possible and that island states might 'build back better': the phrase and intended outcome after other hazards. A 'new normal' might even be superior, with a higher regard, and possibly even greater financial support for public health care systems and for health workers, at every scale (Bourgeault et al., 2020). By default SIDS have entered into a new era where health care costs are more substantial, already evident in the cost of maintaining testing and tracing regimes alongside vaccinations. Yet there has been no obvious concern for greater and more focused health expenditure – despite the constant balancing act between lives and livelihoods – and no models have come from more developed countries. Nonetheless it seems possible that health systems will be better supported – perhaps under pressure from aid donors – and that they will focus more strongly on primary health care (and seek some belated decline in NCDs) since vulnerability to COVID-19 and other possible epidemics is greater amongst those with NCDs (Jackson-Morris et al., 2020), some reduction in costly referrals, and the retention of health staff for under-resourced hospitals. That is less likely to spill over into better workplace health and safety.

The prospects for economic restructuring are less evident. Very clearly, almost everywhere, many were biding time – workers waiting to return to their market stalls or to their positions in hotels, taxis drivers anxiously waiting for custom, shopkeepers simply wanting to open their doors again ... not looking for something new but just impatient for the revival of a familiar economy that worked to provide livelihoods Governments, with responsibility for national budgets (and perhaps even concerned over aid fatigue, as donor states built their own fractured economies) were even more anxious to revive economies (which, in several cases, provoked new COVID-19 spikes and second waves). Conceivably the SIDS might have entered a new era where governments are more strategic and interventionist, as they have had to be for health, but that may not last and its direction is far from clear.

Back to the Land?

For some recourse came in turning around. Disrupted communications had usually resulted in broken commodity chains, sometimes lost markets and scarce imported foods (although that was largely the result of declining incomes), so that agriculture revived. This too was partly by default, out of necessity as much as choice, but it was unusual for so many Caribbean and Pacific governments to genuinely give priority rather than lip service to agricultural development (and even for agricultural development officers to leave their offices). Urban-rural migrants, reversing a long directional trend, returned to agriculture and fisheries where they could. That contributed to food security, may have benefited nutrition and offered hope. But not everywhere – especially in the Caribbean – was there productive land to return to, and hard-won agricultural livelihoods in difficult environments had long been forsaken. Farmers became the least likely to experience worsened food security, unlike urban populations. Exceptionally, as in Vava'u, where government support was forthcoming and fortuitous niches opened up, agriculture even became more profitable.

With disrupted transport, government advocacy in at least some Pacific SIDS for people to return 'home' to safer and more productive lives, and tourism dead, island life took on a smaller, more intimate and introspective scale. With closed borders there was little choice. Physical return marked some return to 'tradition', from belief systems to agricultural and fishing practices. Barter resumed, local social groups came to the fore to renew social cohesion and engage in collective action (sometimes as work groups). Metaphorical safety nets were provided. Island communities are vulnerable to shocks, and dependence on the cash economy limits flexibility, but that could be reinstated by drawing on traditional modes of exchange. The revival of bartering may strengthen local social networks and result in a greater respect for tradition and a determination to retain a strong subsistence base, that would support the quest for food security and more adequate nutrition (Farrell et al., 2020). Up to a point 'embedded, inclusive, often informal and unruly, economies, rooted in mutualism and solidarity, have flourished' (Leach et al., 2021). While much of that was positive, it was not always so. Alongside forgotten fishing techniques and old norms of reciprocity and sharing, land tenure posed problems for some, and return brought occasional tensions as social relations were renegotiated. Greater pressure on scarce resources contributed to land disputes, theft of food, tensions over intrusion into marine protected areas, all of which easily eroded trust.

Whatever the resultant satisfactions of being at 'home', this was scarcely a rejection of, or disdain for, globalisation, since that was how tourism – the basis of most island economies – had boomed. Youth were frustrated over lost urban, 'modern' opportunities and reluctant to 'regress' even when that might be temporary. Indeed even the workers who had contributed to agricultural expansion in Vava'u chafed at the closed border and their inability to migrate to better paid employment overseas.

Return to rural areas often placed more burdens on women as, usually, the primary food producers. More widely, in the SIDS as elsewhere, COVID-19 emphasised both the manner in which women usually bore the brunt – sometimes very directly – of lost employment and livelihood opportunities, or indirectly through dropping out of education, matched by a rise in inequity and inequality. That emphasized the gender gap, circumstances not unique to the SIDS (e.g. Dang & Nguyen, 2021) and that had also occurred in countries experiencing previous pandemics (Wenham et al., 2020). Restructuring that would enable greater opportunities for gender and youth posed obvious difficulties in what were often patriarchal and hierarchical societies. Moreover the uneven consequences of COVID-19 brought questioning of the extent to which people 'were all in this together' or had their own agendas. Even at local level, new forms of cooperation and collaboration could sit uneasily with sometimes rigid structures of rural social relations.

Just as the pandemic marginalised women and youth, and occasionally raised tensions even in villages, it drew attention to other forms of uneven development, evident particularly in the virtual disappearance of the informal sector and the consequential rise of poverty. Class realities and rigidities sometimes became evident, alongside the rural-urban digital divide. Disadvantaged citizens in Jamaica, and elsewhere, expressed the view that 'social distance is for the middle class and rich' and 'we're (not) all in this together'. The pandemic burden was unevenly shared. Government policies gave little attention to the living conditions, usually high-density housing, of people living in marginalised 'informal' urban spaces, such as those in Kingston, Jamaica, in the informal settlements of Port Moresby or the social housing of Auckland. For those at the lower end of the stratum, compliance sometimes entailed a trade-off between public health and economic survival, and scarcely offered a different future. Inequality was most obvious in urban areas as the informal sector disappeared and unemployment soared, public health provision was inequitable, social welfare provision thin and weak social networks provided limited support. Three hundred years ago, Daniel Defoe's novel *Journal of a Plague Year* (1722) – more popular than ever in 2020 – had intimated that plague was no social leveller but that it struck hardest those who lacked money and resources. In a different era, and in reality, the conclusion remained the same.

The Trouble with Tourism

At the core of most SIDS economies is tourism, a sector emblematic of globalisation, as frustrated attempts to generate domestic tourism showed, and perhaps also the epitome of unsustainability. Pacific SIDS especially are littered with the failures of ecotourism projects, where local people sought to be involved in sustainable development. Marketing and access were impossible and resorts outnumbered them. Might there then be possibilities for SIDS to value and triumph local social relationships and the need to restructure the tourism sector, to avoid intense competitiveness at every scale?. A quickly produced Vanuatu Tourism Recovery Plan pointed out:

> Our tourism industry has been developed in a way that has undermined our subsistence capacities which has made some communities and households dependent and vulnerable. With the absence of formal safety nets in Vanuatu, informal community and culturally based social protection systems (brought about through the traditional economy) are more vital than ever before. (Vanuatu Department of Tourism, 2020: 5)

This idealistic plan sought to move away from 'business as usual' to promote local linkages (from agriculture to soap manufacture), smaller-scale tourism and local values. It gained little traction. Elsewhere public-private partnerships were promoted – somewhat innovative for tourism. Yet in SIDS most tourism, and related activities from hire cars to airlines, is foreign-owned, so that it is usually true that 'international tourism constructs as it commodifies, alienates as it appropriates, and dominates as it penetrates. Local authority is undermined, local empowerment is difficult to sustain, and local environments are changed for ever' (Conway, 2002: 120). That was never more evident than in the Bahamas as it sought to open up, while cruise ship companies are far beyond local regulation. Here it is a truism that people – and countries – have sought to make their own livelihoods but not necessarily under conditions of their own choosing. Tourists too are subject to the whims of economics and international fashion. Tourism development has rarely been about local participation, empowerment and equity.

Almost inevitably it is in the tourist economy particularly that the urgent push for development has been centred: perhaps an addiction to an old normal or a realisation that this was still the core of economic growth (and locked down people the world over would want to spread their wings as soon as the pandemic was over, if not sooner). Desperation for development meant the too-rapid resurrection of the industry in several SIDS and a spike in COVID-19 cases. Former tourism workers may have their own distinctive approach to future employment in the industry. Where they had returned to rural activities, many were enjoying a break, while simply waiting for the old order to resume and the restoration of valuable and often congenial employment. For all that a new mantra of 'diversification' began to characterise island tourist economies, it was not a time for experimentation or change.

Green Dreams

The dominance and structure of tourism offered real challenges to restructuring and rethinking development. That suggested both that similar challenges might face other activities – in the race to rebirth – but also that seemingly old-fashioned core activities might offer better prospects for a new normal. Envisaging a new and different future was neither easy nor immediately practical; visions of greater self-reliance were scarce as few wished to escape modernity however that might be interpreted.

Nonetheless some Caribbean economies had been moving towards greater sustainability (Chap. 28), despite islands being particularly difficult places in which to achieve sustainability (Connell, 2018). For some a cautious optimism existed. As Silliniu Lina Chang, in Samoa, suggested: 'COVID-19 has presented us and our blue Pacific family with a rare opportunity – a time for all of us to reset. Think outside of our comfort zone and look beyond to the neighbour that is in need' (quoted in UN Women, 2020). That will require a vision that is starkly different from current reality, to focus on equity, sustainability, and global justice (Sasale, 2020) and a focus on health and other basic needs at a variety of scales.

By contrast, new economic strategies, while distinctive and innovative, could be ultimately conventional. The Northern Marianas was banking on its newly established cannabis industry (from growing to marijuana retail shops and lounges) to counter the collapse of tourism (Radio New Zealand, 28 October 2020). Vanuatu and St Lucia sought to increase their semi-legal passport sales to honorary citizens. Barbados and Montserrat created a 'Welcome Stamp' for remote digital workers. Other strategies were more typical, banking on past experience, and focusing on agriculture and eventual tourism.

Digital futures were considered but while the crisis established remote working and a new acronym – WFH: working from home – as a 'new normal' in many parts of the world, it offered much less to SIDS with quite different economic and employment structures. However, the rise of, and familiarity with, social media, and the value of telephony in transferring remittances and information, suggested a more socially valuable digital future, perhaps with much touted telehealth finally becoming of significance. But digital divides and the absence of supporting resources need to be overcome.

Various constraints to change typify SIDS, from international markets and supply chains (whether for tourists or commodity sales) to some degree of dependence on aid. As the rush to resume international tourism demonstrated, this was not the time to consider new strategies, and as several chapters have indicated, enforced periods of reflection did not generate radical difference. However, in contexts as different as southern Auckland and the Eastern Highlands of PNG, they did indicate that people were perfectly capable of taking control of their own lives and contexts where opportunities existed.

External ties were not necessarily to be rejected. In the Pacific SIDS, where remittances proved invaluable, unique arrangements were made to ensure that 'normal' guestworker migration was resumed as quickly as possible (with special charter flights facilitating mobility). Islanders from virus-free SIDS overcame a fear of COVID-19 to participate. Similar enthusiasms were replicated elsewhere (Griffith, 2020). Metropolitan donor countries focused aid on health system strengthening, and a redirection towards social protection; aid was not to be spurned despite accompanying strategic objectives.

The Old Was Normal

Ultimately the old order was normal and familiar. It might have worked far from perfectly, but change was not always easy to contemplate, especially by vested interests. The world did not change after the Spanish Flu. Capitalism was not in crisis. Local people proved creative and resilient in the face of COVID-19 but resilience was not necessarily directed to creating something new. It is nonetheless here, on a small scale, in the restoration of local social relations and the revitalisation of local exchange and trust, that hope remains, anchored in indigenous knowledge. Enforced (re)thinking, even through a year of tribulations, could not necessarily produce new and better outcomes or new certainties.

Might it therefore be simply the old normal that returns? Arundhati Roy's questioning that began this book suggests optimism but implies uncertainty. There seems no obvious reason why islands and island states should now choose new development paths. Nor can they easily make unilateral decisions. Indeed in straightened times political and other conflicts may worsen. To reiterate Neelesh Gounder's (2020) point, government – in this case in Fiji – displayed 'no strategic repositioning'; 'new' was almost entirely rhetorical. By and large the pandemic had not changed the way governments conceptualised development, and how that might involve changing thinking on the part of the private sector (much based overseas), or where that entailed new opportunities and motivations, or about the socio-ecological circumstances from which epidemics emerged or about equity, resilience, devolution and sustainability and how that might shape new notions and practices of trust between citizens and governments. There is scant evidence that beleaguered health workers, invaluable on the frontline against the pandemic, will be treated with greater respect let alone higher wages. Or that the public service will be valued more highly. Or that social protection policies will be enhanced. Or that the informal sector will be better supported. Agriculture at least may be better funded to boost domestic production and assist food security, and so contribute to meeting the crucial SDGs 1 and 2 (no poverty and zero hunger). Greater regional collaboration and cooperation between SIDS will need to restored, after the partial and temporary resort to nationalism (but, as that was being stressed, the important Pacific Islands Forum Secretariat has split apart, through a collapse of regional trust).

If this were not implicitly challenging enough, sunk physical costs mark an old order where, despite rising underemployment, many – often the more powerful – were able to work though the pandemic in unchanged workplaces. An Asian Development Bank article, late in 2020, examining the economic challenges that COVID-19 had posed for the Micronesian SIDS, included the sub-heading – 'Keeping the lights on while awaiting the "new normal"'- but focused on conventional macro-economic policies to revive the economy without any further reference to a 'new normal' (Tinio & Rabanal, 2020). The possibility of change had been reduced to a buzzword while the recommended short-term stimulus packages would slowly enable the SIDS to return to an old reality. A release into consumption seemed likely to follow, after the stringencies of lockdown. Green dreams could founder amidst nostalgia for, and against the exigencies and collapsed certainties of, daily life.

The COVID-19 crisis has been seen as a forerunner of a potentially greater crisis – the impact of climate change. Indeed 2020 may also be remembered as the warmest year in human history. The philosopher, Bruno Latour (2020), has queried whether the pandemic was simply a dress rehearsal for impending climate change catastrophe, and others have pointed to similarities in terms of the need for collective action (Oldekop et al., 2020, Cole & Dodds, 2021). The Secretary General of the Pacific Islands Forum, Meg Taylor, stated: 'The COVID-19 public health emergency and its ensuing humanitarian and economic fallout offers us a glimpse of what the global climate emergency can become if it is left unchecked and if we do not act now' (quoted in McAdam & Pryke, 2020). At much the same time the then President of Palau, Tommy Remengesau, argued that Pacific leaders should not lose sight of another priority 'losing our relationship with nature', by tackling ocean pollution, overfishing, climate change and loss of biodiversity (PacNews, 25 September 2020) and the President of the Marshall Islands, David Kabua, stressed that 'the pandemic is a reminder of how connected we are .. we have an unprecedented chance to transform the global economy' (Radio New Zealand, 21

September 2020). Few argued with such statements but remedies were rarely a policy priority at any scale.

Ironically the global response to the COVID-19 pandemic led to a sudden initial reduction of both GHG emissions and air pollutants, partly because air transport ground to a halt and, although this cooling trend was soon offset, an economic recovery oriented towards 'green' environmentalism and reductions in fossil fuel uses (Forster et al., 2020), would be beneficial to the SIDS. New relationships will undoubtedly emerge in these strategic arenas, as social relations evolve in other arenas, within a restructuring – not always intentional – of political and economic systems. A 'new normal' without conviction will emerge. The COVID-19 pandemic will continue to reverberate across lives and livelihoods, social, economic, political and ecological relationships, and ultimately human wellbeing, in unforeseen and unplanned ways. That implies uneasy times and frustrations for those seeking to build back better. The much anticipated 'new normal' may look in the end remarkably like the 'old normal': a mirror rather than a portal.

References

Bourgeault, I., Maier, C., Dieleman, M., Ball, J., et al. (2020). The COVID-19 Pandemic Presents an Opportunity to Develop More Sustainable Health Workforces. *Human Resources for Health, 18*(83).

Büscher, B., Feola, G., Fisher, A., et al. (2020). Planning for a World Beyond COVID-19: Five Pillars for Post-Neoliberal Development. *World Development, 140*, 105357.

Cohen, J., & Mata-Sánchez, N. (2021). Challenges, Inequalities and COVID-19: Examples from Indigenous Oaxaca, Mexico. *Global Public Health, 16*(4), 639–649.

Cole, J., & Dodds, K. (2021). Unhealthy Geopolitics: Can the Response to COVID-19 Reform Climate Change Policy? *Bulletin of the World Health Organisation, 99*, 148–154.

Connell, J. (2018). Islands. Balancing Development and Sustainability? *Environmental Conservation, 45*(2), 111–124.

Conway, D. (2002). Tourism, Agriculture and the Sustainability of Terrestrial Ecosystems in Small Islands. In Y. Apostolopoulos & D. Gayle (Eds.), *Island Tourism and Sustainable Development* (pp. 113–129). Praeger.

Dang, H., & Nguyen, C. (2021). Gender Inequality During the COVID-19 Pandemic: Income, Expenditure, Savings and Job Loss. *World Development, 40*, 105296.

Farmer, P. (2001). *Infections and Inequalities: The Modern Plagues*. University of California Press.

Farrell, P., Thow, A.-M., Wate, J., Nonga, N., et al. (2020). COVID-19 and Pacific Food System Resilience: Opportunities to Build a Robust Response. *Food Security, 12*, 783–791.

Forster, P., Forster, H., Evans, M., et al. (2020). Current and Future Global Climate Impacts Resulting from COVID-19. *Nature Climate Change, 10*, 913–919.

Gounder, N. (2020, April 17). Fiji's Economic Freefall, and the Government's Response. *ANU DevPolicy Blog*.

Griffith, D. (2020). Working Through the Pandemic: What Crises Tell Us About Economics. *Anthropology Now, 12*(2), 18–25.

Jackson-Morris, A., Nugent, R., Ralston, J., Cavalcanti, O., & Wilding, J. (2020). Strengthening Resistance to the COVID-19 Pandemic and Fostering Future Resilience Requires Concerted Action on Obesity. *Global Health Action, 13*, 1804700.

Karim, R., Islam, M., & Talukder, M. (2020). COVID-19's Impacts on Migrant Workers from Bangladesh: In Search of Policy Intervention. *World Development, 136*, 105–123.

Klassen, S., & Murphy, S. (2020). Equity as Both a Means and an End: Lessons for Resilient Food Systems from COVID-19. *World Development, 136*, 105104.

Latour, B. (2020, March 26). Is This a Dress Rehearsal? *Critical Inquiry*.

Leach, M., MacGregor, H., Scoones, I., & Wilkinson, A. (2021). Post-Pandemic Transformations: How and Why COVID-19 Requires Us to Rethink Development. *World Development, 138*, 105233.

McAdam, J., & Pryke, J. (2020, October 27). Preparing for When Climate Change Drives People from Their Homes. *The Interpreter*.

Oldekop, J., Horner, R., Hulme, D., et al. (2020). COVID-19 and the Case for Global Development. *World Development, 134*, 105044.

Sasale, M. (2020). COVID-19: Can This Crisis Be Transformative for Global Health? *Global Public Health, 15*(11), 1740–1752.

Tinio, C., & Rabanal, R. (2020). Addressing the Economic Changes of COVID-19 in the Federated States of Micronesia and the Marshall Island s. *Pacific Economic Monitor*, 11–13.

Vanuatu Department of Tourism. (2020). *The VSTP Tourism Crises Response and Recovery Plan (2020–2023)*. Vanuatu Department of Tourism.

Wenham, C., Smith, J., Davies, S., Feng, H., et al. (2020). Women Are Most Affected by Pandemics – Lessons from Past Outbreaks. *Nature, 583*, 194–198.

UN Women. (2020). *Across the Pacific Crisis Centres Respond to COVID-19 Amid Natural Disasters*. United Nations.

Correction To: God and the Virus in Papua New Guinea: Outsourcing Risk, Living with Uncertainty and (Re)creating a *Niupela Pasin*

Monica Minnegal and Peter D. Dwyer

Correction to:

Chapter 24 in: Y. Campbell, J. Connell (eds.), *COVID in the Islands: A comparative perspective on the Caribbean and the Pacific*, https://doi.org/10.1007/978-981-16-5285-1_24

The original version of this chapter was revised as the word "draft" was inadvertently present in the figure 24.1 and 24.2 that has been updated.

The updated version of this chapter can be found at
https://doi.org/10.1007/978-981-16-5285-1_24

Y. Campbell, J. Connell (eds.), *COVID in the Islands: A comparative perspective on the Caribbean and the Pacific*,
https://doi.org/10.1007/978-981-16-5285-1_30

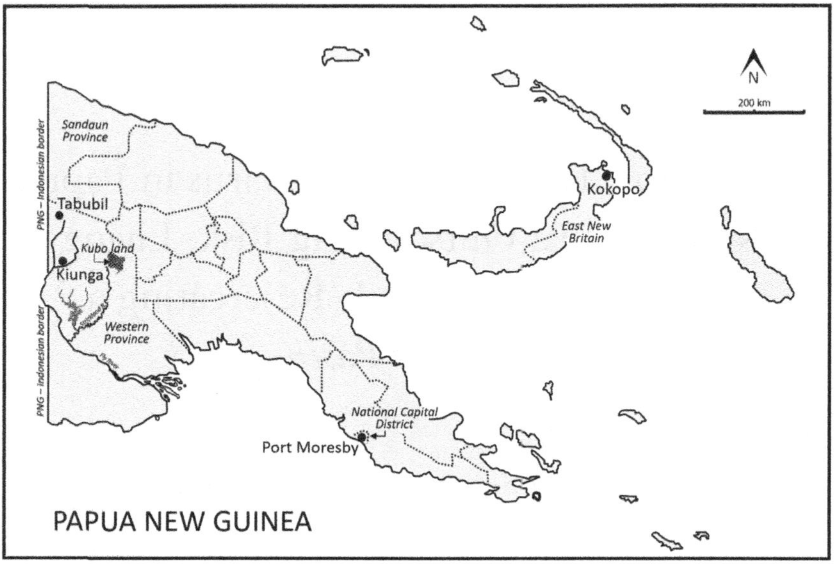

Fig. 24.1 Map of Papua New Guinea showing locations mentioned in the text

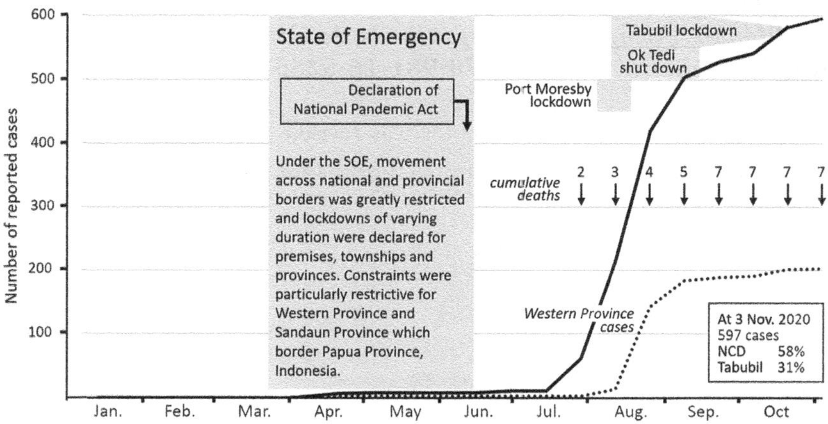

Fig. 24.2 Reported Covid-19 cases and deaths. (Records were compiled from the web site of the PNG National Control Centre for Covid-19 and collated in two-week periods commencing 1 January 2020)

Index[1]

[1] Note: Page numbers followed by 'n' refer to notes.

Y. Campbell, J. Connell (eds.), *COVID in the Islands: A comparative perspective on the Caribbean and the Pacific*, https://doi.org/10.1007/978-981-16-5285-1

N

O

P

The manufacturer's authorised representative in the EU is Springer Nature Customer Service Centre GmbH, Europaplatz 3, 69115 Heidelberg, Germany. If you have any concerns regarding our products, please contact ProductSafety@springernature.com

Printed and bound by CPI Group (UK) Ltd, Croydon, CR0 4YY

30/09/2026

02225327-0018